Introduction to
Distribution Logistics

THE WILEY BICENTENNIAL–KNOWLEDGE FOR GENERATIONS

*E*ach generation has its unique needs and aspirations. When Charles Wiley first opened his small printing shop in lower Manhattan in 1807, it was a generation of boundless potential searching for an identity. And we were there, helping to define a new American literary tradition. Over half a century later, in the midst of the Second Industrial Revolution, it was a generation focused on building the future. Once again, we were there, supplying the critical scientific, technical, and engineering knowledge that helped frame the world. Throughout the 20th Century, and into the new millennium, nations began to reach out beyond their own borders and a new international community was born. Wiley was there, expanding its operations around the world to enable a global exchange of ideas, opinions, and know-how.

For 200 years, Wiley has been an integral part of each generation's journey, enabling the flow of information and understanding necessary to meet their needs and fulfill their aspirations. Today, bold new technologies are changing the way we live and learn. Wiley will be there, providing you the must-have knowledge you need to imagine new worlds, new possibilities, and new opportunities.

Generations come and go, but you can always count on Wiley to provide you the knowledge you need, when and where you need it!

WILLIAM J. PESCE
PRESIDENT AND CHIEF EXECUTIVE OFFICER

PETER BOOTH WILEY
CHAIRMAN OF THE BOARD

Introduction to Distribution Logistics

Paolo Brandimarte

Politecnico di Torino
Torino, Italy

Giulio Zotteri

Politecnico di Torino
Torino, Italy

WILEY-INTERSCIENCE
A John Wiley & Sons, Inc., Publication

For general information on our other products and services or for technical support, please contact our
Customer Care Department within the United States at (800) 762-2974, outside the United States at
(317) 572-3993 or fax (317) 572-4002.

Wiley also publishes its books in a variety of electronic formats. Some content that appears in print may
not be available in electronic format. For information about Wiley products, visit our web site at
www.wiley.com.

Wiley Bicentennial Logo: Richard J. Pacifico

Library of Congress Cataloging-in-Publication Data:

Brandimarte, Paolo.
  Introduction to distribution logisttics / Paolo Brandimarte, Giulio Zotteri.
      p. cm. — (Statistics in practice)
  Includes bibliographical references and index.
  ISBN 978-0-471-75044-4 (cloth)
  1. Network analysis (Planning)—Mathematics.   2. Production
scheduling—Statistical methods.   3. Business logistics—Statistical methods.
4. Traffic flow—Mathematical models.   5. Physical distribution of
goods—Mathematics.   6. Distribution (Probability theory)   I. Zotteri, Giulio,
1970– II. Title
  T57.85.B726 2007
  658.4'032—dc22   2007006008

Printed in the United States of America.

10 9 8 7 6 5 4 3 2 1

Contents

Preface

We teach a course on *Distribution Logistics* at the Faculty of Engineering and Management at Politecnico di Torino. After an initial teaching experience based on assembling diverse material from various origins, we reluctantly accepted the painful idea of writing our own textbook.

Many books have been published on Supply Chain Management and Distribution Logistics, and the list includes some truly excellent ones. Still, we felt that there was some place for this book. We perceived a sort of dichotomy between very advanced books aimed at mathematically gifted (possibly Ph.D.) students, and all-encompassing manuals, which did cover a lot of topics in an excellent manner, but did not emphasize the quantitative approach in a way that we consider suitable for Engineering students. Our hope was to write a book that (i) focuses on a rather narrow set of themes related to Supply Chain Management; (ii) is quantitatively oriented, while still not neglecting issues that are difficult to quantify; (iii) shows how to build models to make logistic decisions, but still discusses practical issues and uses real-life examples to hopefully guide the reader through the hazards of Mathematics, Statistics, and Optimization.

In what follows, there is extensive use of tools from Probability, Statistics, and Mathematical Programming. In order to make the book as self-contained as possible, and to enlarge its potential audience, we included extensive appendices on these topics. Thus, while the book requires some level of mathematical maturity, it can be used by students (both at graduate and undergraduate level) in such diverse areas as Engineering, Business Administration, Economics, Mathematics and Statistics, and (last but not least) by the potential users of the proposed methodologies.

We do not want to encourage an uncritical use of algorithms and sophisticated models in place of intuition and common sense. A classical example, from the Just in Time folklore, is the Japanese attitude towards setup times: There is little point in building complicated mathematical models to manage production with setup times, if they can be eliminated by proper improvement of manufacturing. However, there is a steady increase of commercially available software packages including quantitative-based procedures, and we think that a good working knowledge of quantitative models and methods is needed, first of all, to use these tools with care and to be fully aware of their up- and down-sides. The limitations of mathematical modeling lead many students to identify *theoretical* with *not practical*. On the contrary, as James Clerk Maxwell (apparently) put it,

> *There is nothing more practical than a good theory.*

There is no contradiction between *good* theory and *good* applications, ruling out those bad "applications," in which it is hard to see *what* was applied exactly. A suitably simplified, but formal representation is valuable in understanding the nature of problems, in assessing tradeoffs, and in developing solution approaches or alternative strategies. This is not to say that there is no danger in relying only on quantitative modeling, or that they guarantee the success of our endeavors: One can tackle the wrong problem, or solve it on the basis of unreliable data, or finding a theoretically optimal solution which cannot be applied, due to some neglected organizational constraint. Still, only an expert and competent decision maker can find the right balance between conflicting requirements, possibly adapting the proposed solution to fully account for unmodeled features of a real-life problem. A strong background in quantitative modeling allows a practitioner to make the most out of them or, when the context so dictates, to knowingly avoid the use of an inappropriate tool.

When tackling any management problem, a practitioner needs a clear view of the environment in which a firm operates and of its positioning in terms of strategies and competition levers. All of this conjures up the idea of something inherently "creative" and definitely in contrast with the "mechanistic" flavor of quantitative approaches. Again, this is a false myth. It is often said that in mathematics there can be no opinions, and this is certainly true for low level algebra. But the way mathematics is *applied* to tackle a relevant problem does require a fair share of creativity and ingenuity: We must spot the subset of relevant variables, the objective to pursue, and the options at hand. All of this is far from boring routine, and we hope that in writing this book we can share our enthusiasm for tackling and modeling distribution problems with the reader.

The book consists of eight chapters:

Chapter 1 is an overview of Supply Chain Management, with the aim of providing the appropriate context and to draw the line between what is included in the book and what is not.

Chapter 2 deals with distribution network design problems. With respect to the following chapters, this one deals with issues at a strategic level, and it relies more heavily on mathematical programming models.

Chapter 3 is dedicated to several forecasting models. We include classical topics in time series forecasting, with some additional emphasis on initialization and testing issues. We also deal with regression based modeling, and forecasting demand for new products.

Chapter 4 offers basic background in classical inventory control models assuming deterministic demand, such as the Economic Order Quantity model and some of its variant. With respect to standard literature, we also emphasize parameter uncertainty issues, multi-item problems, and mathematical programming models for multi-period problems.

Chapter 5 covers several models for inventory management subject to uncertain demand, assuming a single facility.

Chapter 6 outlines issues in multiechelon inventory systems. This is a very difficult topic, requiring considerable background; hence, we have limited the treatment to some simple cases in order to let the reader appreciate the issues involved.

Chapter 7 covers incentive issues in a supply chain where multiple actors interact with conflicting views and objectives. This is a relatively unusual topic in distribution logistics books, bordering with Industrial Economics. Unlike other chapters, the models we present here are not really operational, but aim at shedding some light on basic problems and concepts that can be used to tackle them.

Chapter 8 is relatively independent from previous ones, as it provides the reader with the essential background on the operational problem of Vehicle Routing. This problem lends itself to quite sophisticated combinatorial optimization strategies, but just provide the reader with the basic knowledge to understand the basic strategies that are used within some commercially available software packages.

Finally, we have included two relatively extensive appendices on Probability and Statistics and Mathematical Programming.

As you may see, we do not cover *physical* distribution logistics and materials handling. There are other important topics which are omitted, such as supplier management, and discrete event simulation. Indeed, we did not aim at writing a comprehensive manual dealing with all of the topics which are related to Supply Chain Management. There are voluminous handbooks which have been written with this aim, whereas we wanted to provide students and practitioners with a solid background on quantitative approaches, in order to pave the way for their extension and adaptation to real life problems, with all of their nuances and peculiarities.

We should also mention that, in our teaching, we complement our lecture notes with the discussion of business cases, mostly from the rich library of the Harvard Business School, and software development laboratories. The interactive nature of these fundamental additional activities does not lend them to textbook coverage.

Despite the best of our efforts, typos and mistakes are a fact of life when writing lengthy books. We will be grateful to readers who will be kind enough to share their opinions or criticism, and will point out our mistakes. A list of errata will be posted and maintained on the following Web page:

> `http://staff.polito.it/paolo.brandimarte/`

Courtesy of Murphy's law, our Web manager will decide that all of our URLs have to change a few days after publication of the book. An up-to-date link will be maintained on the following Wiley Web page:

> `http://www.wiley.com/mathematics`

We also plan to post some supplements on topics that we have omitted in order to keep the book to a manageable size. Some web sections are already integrated within the book, and they are characterized by a section numbering starting with 'W' (e.g., web section W.2.5). Overly technical sections are left as supplements at the end of chapters; they are characterized by a section numbering starting with 'S' (e.g., supplement S.5.8)

As a final remark, although the book is the result of a joint effort, chapters 1, 2, 8, and appendices A and B can be attributed to the first author (PB); chapters 3, 4, 5, 6, and 7 can be attributed to the second author (GZ).

PAOLO BRANDIMARTE
`paolo.brandimarte@polito.it`

GIULIO ZOTTERI
`giulio.zotteri@polito.it`

Politecnico di Torino
May 2007

1

Supply Chain Management

1.1 WHAT DO WE MEAN BY LOGISTICS?

Logistics has quite a long history, whose origins predate by far the initial attempts to make it "scientific." Many engineering schools were born because of the need for building better military fortifications and weapons. Logistics followed a pattern common to that of many fields in engineering: Military applications gave an important impulse to its development. While relatively small armies in the past could sustain themselves also by robbing local populations, proper management of supplies was required at later times to support larger armies in need for ammunition and a significant amount of food. Napoleon, who is acknowledged with the motto "An army marches on its stomach," is considered an innovator in this respect, because (what we now call) supply chain management afforded his armies a far greater degree of mobility than his rivals. Logistics has played an increasing role in later conflicts, like the American Civil War (ACW), where transporting supplies and troops was accomplished by an array of transportation means including supply wagons, rail, ships, and (in the Western Theater) rivers. The role of logistics can be appreciated by considering how the availability of supplies is of no use if the supplies cannot be routed to destination, whereas clever organization may make good enough use of scarce resources. A paradox in Confederate logistics during ACW was that an economy strong in agriculture and weak in industrial power, compared to its Union counterpart, succeeded in maintaining a flow of weapons and ammunitions, whereas troops often starved because of lack of food.[1] Indeed, some military academics are reported to say that "amateurs study tactics, professionals study logistics."

Military applications continued to play a prominent role in the development of scientific logistics in the 20th century.[2] The quantitative approach to management problems is typically associated with Operations Research, whose origin can be attributed in part to the need of managing the supply chain across the Atlantic Ocean during World War II.[3] However, we should not think that the scientific approach to logistics is that recent. For instance, the well-known Economic Order Quantity (EOQ) formula for inventory management dates back to the early 20th century, since it was published in 1913[4]; furthermore, the manifesto of Taylorism[5] was published in 1911, but its roots can be traced back to a rationalization process in manufacturing, which had been quite active during the 19th century.

Given this long history, we should not be surprised that the term "Logistics" has now a rather wide and often ambiguous meaning. Indeed, several professional and academic organizations have attempted to draw the line, pointing out what we should mean by this term. The U.S. Council of Logistics Management proposed the following definition:

> *Business logistics is the term describing the integration of two or more activities for the purpose of planning, implementing and controlling the efficient flow of raw materials, in-process inventory and finished goods from the point of origin to point of consumption. These activities may include, but are not limited to customer service, demand forecasting, distribution communications, inventory control, material handling, order processing, parts and service support, plant and warehouse site selection, procurement, packaging, return goods handling, salvage and scrap disposal, traffic and transportation and warehousing and storage.*

The term *business logistics* emphasizes a separation from other fields, such as urban transportation, which could be included in a more general notion of logistics. The definition we have reported is not very recent, as it dates back to 1979, but it includes both management issues and material handling issues, which are more physical in nature. This book is only concerned with management issues, not with physical activities which might be labeled as

---

[1]See: R.K. Krick, The Power of the Land, in: A. Sheehan-Dean (editor), *Struggle for a Vast Future: the American Civil War*, Osprey Publishing, Oxford, 2006.

[2]Those of us who are sane enough not to appreciate the grim arts of war too much, may find some consolation in thinking that the same approaches can be used to route huge amounts of essential supplies, in a short time span, to areas struck by natural disasters.

[3]Another element in the birth of Operations Research was queuing theory, initially developed to model telephone traffic. It is worth remembering that the celebrated simplex method to solve linear programming problems was developed in 1947 by George Dantzig, who worked for U.S. Air Force.

[4]See: F.W. Harris. How many parts to make at once. *Factory: the Magazine of Management.* Vol. **10**, 1913, pp. 135–136. Reprinted in *Operations Research*, 1990, Vol. **38**, pp. 947–950.

[5]F.W. Taylor. *The Principles of Scientific Management.* Harper & Row, New York, 1911.

"industrial" logistics. This is certainly *not* to say that industrial logistics has a lesser role, or that there is no interconnection between hardware and managerial issues. Some management activities have no sense if the underlying physical process is not properly designed and if certain technologies are not exploited. Our aim is to define a consistent and relatively limited scope, in order to offer a pedagogical treatment of selected material at a suitably deep level, rather than offering a superficial handbook covering all possible topics. As we stress below, solid foundations are essential to any practitioner, as general principles have to be twisted and adapted to many diverse and peculiar settings, and a superficial listing of cookbook recipes is actually of little use, if not counterproductive in case these recipes are applied improperly.

Apparently, the definition above includes too many things. However, modern integration trends have given rise to Supply Chain Management (SCM) as an almost all-encompassing discipline. On the supply side of the chain, increasing emphasis is given to supplier relationships management, purchasing, and contract design. On the other end of the spectrum, customer relationships management (CRM) is another example of an issue which is gaining relevance. Information Technology (IT) had a dramatic impact too, thanks to the rise of Internet, which made electronic commerce, online auctions for products and services, and the sharing of large databases possible. As far as information systems are concerned, the introduction of Enterprise Resource Planning (ERP) systems has made the case for the interconnection with other functional areas, such as manufacturing,[6] accounting, etc. And if this does not look confusing enough, the list of complications could go on and include other factors:

- The reduced lifespan of products and the need for customization imply that the supply chain has to be continuously redesigned. Even product design may interact with logistics. For instance, design for supply chain management has been successfully applied by Hewlett-Packard.[7]

- Globalization has introduced a new array of risk factors which impact SCM, such has exchange rate risk and, at a higher level, political risk.

- The availability of several transportation modes and the concentration of production into large sites have a deep impact on transportation management.

---

[6]Indeed, in many practical settings, we cannot deal with distribution logistics without paying due attention to production. From a methodological point of view, many models and modeling techniques we illustrate in the book are often included in books on manufacturing management.

[7]See: H.L. Lee, C. Billington, and B. Carter, 1993, Hewlett-Packard Gains Control of Inventory and Service through Design for Localization. **Interfaces**, Vol. 23, pp. 1–11.

- Revenue and yield management[8] have a prominent role in the air transportation and in the service industry, but they are likely to see an increased role in distribution too (think of price cuts at the end of the selling season in many retail chains).

- Environmental issues dictate that we also pay due attention to *reverse logistics*.

All of the above, and more, has something to do with Supply Chain Management. Trying to cover such a wide spectrum or topics and issues in one book is a hopeless endeavor, unless one is willing to just compile a list of buzzwords. We believe that students (and practitioners) should have a firm grasp of basic principles of distribution logistics. Armed with a solid background, they can tackle new developments with confidence. Quantitative models and methods play a fundamental role in developing basic skills, and indeed this book is more quantitative oriented than others in this area. However, we did not aim at writing a high-level research survey for Ph.D. students. We only outline problems and solutions, using both toy examples to build intuition and real cases when appropriate. Moreover, we should never forget that quantitative models may be implemented in a computer program, but they are ultimately applied by people. People have incentives, possibly unwritten ones; this applies both to single individuals and to organizations. Indeed, distribution logistics typically crosses borders between organizations, and understanding incentives and organizational barriers is a prerequisite to successfully apply any "scientific" solution.

1.1.1   Plan of the chapter

After insisting on what we do *not* include in the book, we would better explain what we *do* include. This chapter lays down the foundations for the next ones, according to the following plan.

- A distribution network is characterized by a physical arrangement of facilities, such as warehouses and transit points, on a possibly wide geographical area. In section 1.2 we illustrate typical structures of distribution networks. The physical arrangement of facilities does not tell the whole story, as goods flow in the network by some transportation means (e.g., trucks or rail). Inventory and transportation management strategies contribute to the definition of a distribution network. Furthermore, information flows must be described too.

- When designing a distribution network, we should make our decisions in a way that supports a specific strategy. There is no single "one-

---

[8]Revenue and yield management are essentially dynamic pricing policies. They have a prominent role in the case of goods which cannot be stored, such as seats on an aircraft; transportation services can also be priced dynamically, as well as perishable items.

best-way" strategy that works in all possible settings. A strategy is a compromise between the need of achieving a good competitive position, according to a selected profile, and the need of keeping costs low. Competitive factors, cost drivers, and possible strategies are outlined in section 1.3.

- A distribution network typically includes locations in which goods are stocked. Common wisdom maintains that inventories are the source of a long array of evils and should be kept as low as possible. In fact, inventories are a source of many relevant costs, but they play specific roles in achieving a certain competitive position. Hence, they must be properly managed and we should have their functions very clear in mind. Section 1.4 illustrates the roles of inventories.

- A recurring theme in this book is uncertainty. Demand uncertainty is the single most relevant complicating factor in distribution logistics. Good forecasting procedures may be used to predict future demand, but they can only reduce rather than eliminate uncertainty. Even if uncertainty cannot be eliminated, it can be managed. In section 1.5 we start outlining a few ways to deal with uncertainty.

- Goods move on a distribution network, from factories in which they are produced, through warehouses and transit points, to retail stores. Managing transportation is another relevant piece in the overall puzzle. Section 1.6 illustrates some basic ways to define a transportation strategy.

- The flow of goods is what is typically associated to logistics, but the flow of information is just as important. Any decision procedure is based on some piece of information, but without information sharing, certain procedures are simply not feasible. Information sharing may be difficult in a large firm consisting of several branches, let alone a supply chain involving different firms. Furthermore, assigning decision rights in a supply chain involving several actors is not a trivial task. Section 1.7 outlines a few issues related to information, incentives, and decisions.

- The structure of a network is something that should not change too quickly, since the decision to build a facility may be made considering a relatively long time horizon, say years. A recent tendency is to lease warehouses, which contributes to shorten the time span of these decisions.[9] Nevertheless, moving all the goods from an old warehouse to a new one is not something we want to do on a monthly basis. On the contrary, a change in the inventory management strategy can be

---

[9]Another factor which calls for frequent changes in the supply chain is the reduced life-cycle of products.

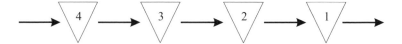

Fig. 1.1   Linear logistic structure.

achieved on a shorter time span, and transportation must be managed daily. Hence, different decisions may have different time horizons and pertain to various hierarchical levels. In section 1.8 we introduce strategic, tactical, and operational decisions. These should be regarded as loose guidelines, since sometimes it is hard to draw the line between the levels, due to tight interactions between different them.

• There are some recurring expressions in Distribution Logistics, and more generally in Operations Management, such as make-to-stock, make-to-order, push, and pull. They have raised quite a bit of controversy, as sometimes they are used ambiguously. Indeed, they do not really define specific strategies, but they do define *attributes* of possibly hybrid decision strategies. In section 1.9 we illustrate the meaning of these terms as features of decision strategies.

• Last but not least, to tackle all of the above problems we may take advantage of models and methods. Quantitative approaches play a prominent role in the book, which is not to say that they should be applied with a blind faith in their power. Section 1.10 helps in classifying quantitative models, including those which are quite useful but are not dealt with here; the most notable example is discrete event simulation.

1.2   STRUCTURE OF PRODUCTION/DISTRIBUTION NETWORKS

From a physical point of view, a supply chain consists of possibly several stages where items are produced, transformed, assembled, packaged, and distributed to consumers. The simplest structure is illustrated in figure 1.1, where we see a linear arrangement of nodes. Each node in this chain can be more or less complex. The first node is likely to be a factory, where items are produced; we deal with this node as a black box, but a manufacturing system would consist in turn of several machines, laid out according to a certain pattern. From our distribution point of view, these details are not quite relevant per se. However, the arrangement of the manufacturing system has a definite impact on performance measures such as flow time, i.e., the time that an order takes to go through all of the stages required by its technological cycle. The manufacturing flow time is clearly relevant from the supply chain point of view. Thus, we do not investigate the internal structure of the nodes and treat them as black boxes. However, the performance (cost, lead time, etc.) of each

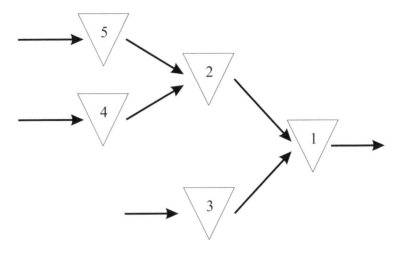

Fig. 1.2 Supply chain structure with assemblies.

black box is very relevant to us. The network could be extended to the left, and include the production of raw materials, but any analysis has to focus on a portion of the overall chain. Proceeding to the right in the figure, we may find other stages at which material is transformed; we should pay attention to the increase in value of the product, which affects the overall economic performance of the network. After the whole chain of transformations, the products may flow through other stages, at which material is simply stocked in a warehouse, until the retail store is reached. Factories may have inventories too, both inbound and outbound.

Along a linear chain, we may have transformations and transportations of items. However, assembly of components into end items is a common occurrence. When items from different sources are assembled, we get a converging structure like that illustrated in figure 1.2. Readers with a manufacturing background could be tempted to interpret the convergent network in the figure like a bill of materials, i.e., a technological representation of how an end item is obtained by assembling components and possibly complex subassemblies. Actually, what we are representing here is the geographical structure of the network, where components can be produced in a continent and assembled in another one. In a convergent network, we clearly see the need for proper synchronization in the material flow: If we miss even one, possibly low-cost component, we cannot assemble the product we need.

Finally, figure 1.3 illustrates an arborescent (or divergent) network which is typical of pure distribution. Here node 1 could be a large warehouse located near a factory producing an item, nodes 2 and 3 might be regional warehouses, and the remaining nodes could be retail stores (in a real network, there would be much more retail stores than depicted). In a pure distribution

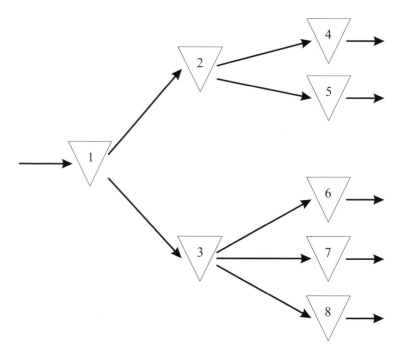

Fig. 1.3 Pure distribution (arborescent) network.

network, the product is always the same.[10] However, whenever material is transferred downstream, we commit it to a certain section of the network. Such an allocation decision is absent in the previous cases, and it must be made with care when material availability is scarce. One could wonder why intermediate stages are needed; after all, they are a cost. We will consider the roles of intermediate stages in depth in chapter 2. Intermediate nodes might help the company in exploiting economies of scale in transportation and/or to reduce the impact of demand uncertainty. We should note that intermediate nodes can be distribution warehouses, but they can be also simple transit points with no facility to store inventory; alternative terms for the last case are "transshipment nodes" or "cross-docking platforms."

The three structures we have illustrated are just basic prototypes. A real-life supply chain is a hybrid of all of them, with many variations. For instance, in the distribution network of figure 1.3, material flows downstream according to a regular pattern, stage by stage. In practice, some retail stores could be served directly from node 1. We will see that this depends on the demand

---

[10]The lack of physical transformations does not imply that the cost of items does not change; as an example, consider customs duties we may have to pay when crossing certain borders.

volume; when this is large enough, we do not need intermediate nodes to take advantage of economies of scale in transportation. For example, we might be in a position to fill a full truck leaving the warehouse to visit a given store. Another variation, with respect to scholastic cases, is the reverse flow of materials. In the previous figures, we see material flowing downstream, but recycling and the need to collect waste call for proper management of *reverse logistics*. The increasing concerns for the environment make such issues more and more relevant. Finally, we may have flows of materials between peer nodes, i.e., stages which are located at the same level in the network. These *lateral* shipments can be used to reallocate material among stores of large retail chains in case one is experiencing a stockout and another is overstocked.

A network design problem calls for structuring a possibly large supply chain, locating facilities, deciding their capacity, and optimizing the transportation of material among them. This is a very difficult task, as we shall see in chapter 2. Fortunately, we are often interested in the *partial* redesign of a network, which makes the task considerably easier. However, the shorter and shorter life cycle of products calls for the continuous redesign of supply chains.[11]

1.3 COMPETITION FACTORS, COST DRIVERS, AND STRATEGY

When managing a supply chain, the natural aim is providing the customer with a suitably good service, and doing so at a suitably low cost. By "good service," we mean that the customer should get *what* she wants, *when* she wants it, and *how* she wants it. Other factors could be relevant, such as after-sales service, but even if we focus on the minimal set of attributes that make a good service, we see that there is no single dominant strategy: There is no possibility of being first in class along all possible dimensions, at a reasonable cost. What we need is a clear view of the dimensions on which we compete, in order to get priorities straight. In the following sections we illustrate a few examples of attributes which define competition factors; then we list a few sources of cost that we must keep under control; finally, we illustrate how all of the relevant dimensions can be traded off, by prioritizing competition factors to define a strategy.

1.3.1 Competition factors

Say that you are a customer wishing to buy a certain good. What are the attributes that are important to you? Probably, most people would point out **quality** requirements. Of course, quality of the goods is key factor influencing

---

[11]Hewlett Packard provides an excellent example of using software tools based on quantitative methods to design supply chains in a dynamic environment; see [2].

consumers' choice, but it is itself a complex concept encompassing multiple dimensions. The quality of a car can be measured through the number of safety features, the top speed, gas consumption, acceleration, etc. Moreover, quality can be measured through the *target quality* (i.e., the quality the product should have, according to its design) and *conformance quality* (i.e., the ability of the single item to meet the target quality over time). Also, the quality of the good could be traded off against price, depending on which market segment we want to address. Moreover, quality is relevant not only in terms of goods, but also in terms of *service*. Indeed, there are complementary services which may contribute to establish a reputation. Consumers can return the merchandise they bought to many mail retailers (as well as to brick and mortar retailers in countries such as the U.S.A.). Other services are more and more relevant in times of increasing environmental concerns; we have already mentioned the role of reverse logistics and the possibility of returning packaging materials, used products, etc., which contribute to the positive image of an environmentally responsible supplier. After-sales services are specifically important for durable goods whereas installation support is very important for complex systems such as high-end audio and video systems.

If we think of distribution services *per se*, fast delivery may be important, but *dependability* may be even more. So, waiting for a long **Delivery Lead Time** (DLT) may be unpleasing, but a very uncertain and unreliable DLT may be even more annoying. In fact the possibility of tracking shipments or to check order status, possibly via Web, is typically offered by couriers, such as DHL and Fed Ex, by Internet-based sellers, and service centers of non-durable goods. From the consumer's point of view, DLT must be zero for some products: No one would like to wait a few days for a bottle of milk. However, the DLT for milk is not zero from the point of view of the retail store or of other actors along the supply chain. Yet, we will see that a zero DLT may make the management of inventories much easier. On the one hand a non-zero DLT provides us with some advance information that can help us improve performance (e.g., reduce inventories or increase service level). On the other hand, exploiting this information is all but trivial and complicates modeling substantially.

At the other end of the spectrum, engineered-to-order items have a long DLT: No one would expect to find a radar system on the shelves. In between these extreme cases, there is an array of intermediate possibilities. DLT is linked to the structure of the network, the transportation means adopted, and the inventory levels and their deployment in the network. If large amounts of goods are held near the customers (say at the stores), DLT is short; it can also be reduced if quick but costly transportation services are used. So, we see that there is a tradeoff between DLT and different types of cost.

Example 1.1 CHL is an Italian retail chain of information technology products. An important feature of its strategy is that it does not maintain inventories at retail stores, which are just used to collect orders and to deliver items to customers. This results in a significant reduction in inventory levels, which is particularly relevant for items characterized by very fast obsolescence and thus high cost of inventories.   □

Another relevant competitive weapon is **assortment**, i.e., the variety of products offered. For a manufacturer, this means offering a large catalogue and the possibility of customizing an end item according to customers' wishes. For a retail store, this means offering a large set of alternative items on its shelves. In both cases, we see that variety comes at a cost. Also, we can trade off assortment with DLT. If products are customized to order, we need some time for this operation and customers shall be willing to wait. If you offer a large assortment with zero DLT, you have to keep a lot of items in inventory, each one with a possibly low and hardly predictable demand. However, variety may be an important and valuable asset to attract customers. Indeed, there may be a positive feedback, when variety increases demand, thereby easing some of the difficulties associated with low levels of demand.

Another relevant feature of the supply chain is the **flexibility**, that is the ability to adapt to changes and exceptional conditions. For example, a flexible supply chain can fulfill an extremely important order in an exceptionally short time. We can have different kinds of flexibility according to the variable that raises the need for a change. We call *product flexibility* the ability to adapt the product to customers' needs. For example, the ability to configure the product to customer specifications might be crucial for complex products such as furniture or cars. A company that carries inventories of components and assembles them to order usually can achieve a great deal of flexibility with limited resources (provided customers are willing to wait while components are being assembled). Think of the large number of different sandwiches one can prepare with just a few basic components! We call *flexibility to product innovations* the ability to manage the introduction of a new product. To achieve this kind of flexibility the company might need to buy flexible production systems and might want to carry components over, that is use components and subsystems from previous generations of the product. Such kind of flexibility is more and more important nowadays given the growing importance of new products and product novelty. We call *delivery flexibility* the ability to adapt deliveries to customers' needs. For example, the ability to deliver rush orders or manage luggage of VIP clients with a tight connection in a hub-airport might be crucial. We call *volume flexibility* the ability to increase/decrease production and distribution quantities on a short notice. This flexibility is expecially valued in markets with a sharply seasonal pattern, such as Christmas gifts, etc. This flexibility can be gained through both spare resources (e.g., spare capacity), flexible resources (e.g., temporary workers), or appropriate planning (e.g., we might produce/distribute all products with a predictable demand before the peak of the season so that during the peak we can use the limited production/distribution capacity to manage just the uncertain part of demand).

1.3.2   Cost drivers

Keeping costs under control is a fundamental factor in supply chain competition. We should state quite clearly that cost minimization per se need not be a winning strategy; a strategy is a good tradeoff between the objective of minimizing costs and the objective of maximizing other competitive performance metrics such as quality, delivery, service, etc. Keeping this in mind, we should list the typical cost drivers in supply chain management, in order to set the stage for decision-making approaches. Before doing so, we should classify costs according to a couple of dimensions.

- *Costs can be linear or nonlinear.* Consider an arbitrary activity (e.g., how many parts we make or buy), and denote its level by a decision variable by x. A linear cost function is something like $f(x) = cx$, where c is a unit cost. More generally, if we have N activities indexed by i, a linear cost function has the form $f(\mathbf{x}) = \sum_{i=1}^{N} c_i x_i$; note how linearity implies that costs pertaining to different activities are simply added. Otherwise, we have to deal with a nonlinear (possibly discontinuous) cost function. Examples of nonlinear cost functions are $f(x) = x^{0.6}$ or $f(x_1, x_2) = x_1 x_2$. Consider, for instance, purchasing large amounts of some component; a discount might be offered if the purchased quantity is above a given threshold. In such a situation, we have an economy of scale; diseconomy of scales occur when scaling an activity level up increases the related cost more than proportionally. Interactions among activities may also result in a nonlinear total cost function.

  In practice, costs are always nonlinear, but sometimes they can be suitably approximated by linear functions, at least for small variations of the level of activity (say number of units purchased or produced). When formulating an optimization model (see appendix B), keeping everything linear is an important concern in order to limit the computational effort required for solving the model. Even when assuming a linear cost function is too far from reality, nonlinear costs can be approximated by piecewise linear functions (see section 2.3) whereas in the general case they can be fairly different.

  We may also recall two important concepts. Consider a generic cost function $c(x)$. The value of the first-order derivative $c'(x)$ is called **marginal** cost. The marginal cost is constant for a linear cost function, but not in general. The **average** cost is $c(x)/x$; we may see that average and marginal cost are the same for a linear cost function.

- *Costs can be fixed or variable.* In accounting, a cost is **fixed** if there is nothing we can do about it in the short term.[12] For instance, the

---

[12]Strictly speaking, accounting professionals use "period" and "product" costs.

cost of a plant is fixed from the point of view of short-term operations (consider, e.g., rent, depreciation, or cost of fixed personnel). The direct production cost is **variable**, since we can change it through production decisions on a much shorter time scale. Of course, in the long run all costs are variable, so the distinction is a matter of time scale. Nevertheless, such fixed costs do not (or at least should not) influence current decisions; they may contribute a constant term to an objective function in an optimization problem, but this does not change the optimal solution. In the short run, these fixed costs are constant, no matter what the short term decisions are. So in a way they are simply irrelevant for decisions making processes. Sometimes, the term *sunk* cost is used to refer to a cost which has been paid and no future decision has any influence on it.

In this book, we will use fixed/variable costs with a slightly different meaning. If a cost function can be expressed by

$$c(x) = \begin{cases} F + cx & \text{if } x > 0, \\ 0 & \text{otherwise,} \end{cases}$$

we refer to F is the fixed cost. Hence, what we mean by "fixed cost" is a cost that does not depend on the value of a decision variable, provided it is strictly positive.[13] The typical example of fixed cost in this vein is a fixed ordering cost, i.e., a cost that we pay whenever we order, whatever amount we order. Clearly, such costs might encourage ordering larger quantities, resulting in economies of scale.[14] Hence, fixed cost in this sense do influence decisions, unlike fixed costs in the accounting sense (for the sake of clarity in the remainder of this book we will call these sunk costs).

Fixed costs may result in piecewise constant cost functions. Consider the cost of transporting an amount x of some good, and assume that there is a fixed cost component, that we pay for each truck we use. Depending on x, we may have to use one truck or two. This induces a discontinuity

---

[13]Sometimes, the term fixed *charge* is used to avoid ambiguity.

[14]Notice that in Economics the term "economies of scale" has a slightly different meaning, since they are regarded as a long term phenomenon. When we face economies of scale, the long term average cost decreases as the production volume (per unit of time, say per year) increases. When economists say "it is a long term effect," they really mean that we can observe such a reduction in the average cost when we compare different plants (or, more generally, infrastructures) with different capacities. On the contrary, the effect of the production volume on the costs of a given plant is a short term effect. As such, economists do not consider this to be related to economies of scale. In this book we use the term economies of scale in a broader sense. Therefore, in this book the economies of scale lead to the reduction of the average unit price and might be due to the dilution of some fixed costs when the level of activity (say the production volume, the purchase quantity, or any other relevant level of activity) increases. We disregard the distinction between short term decisions and long term ones.

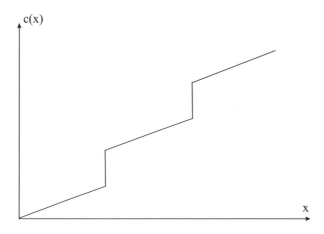

Fig. 1.4   Semivariable costs.

in the cost function, which might include a piecewise constant term. Sometimes, the term **semi-variable** cost is used to refer to such a case (see figure 1.4).

We stress again that we cannot really draw a thick line between the concepts above: A linear cost function can be a suitable approximation of a nonlinear one, and a fixed cost may be transformed (at least partially) into a variable one by suitable arrangements. So, we should just consider the above classifications as useful guidelines, which are best illustrated by a few examples.

We have seen that a supply chain is, from a physical point of view, a network of facilities on which goods are stocked and transported. A first set of costs is associated with building and maintaining facilities. These costs are sunk when we are operating the network, but they are a result of a decision when we are *designing* the network. The cost of a facility is a possibly complex function of its type, location, and capacity. A pure transit point is typically less expensive than a distribution warehouse. We need to find a suitable approximation of the cost associated with building and operating a facility, and this is certainly not a simple linear function. Some costs are fixed, such as those linked to the realization of basic infrastructures to get the facility working; other costs could be represented by a piecewise constant function depending on capacity, or by general nonlinear functions of the flows going through that node in the network. Recent trends tend to make some fixed costs variable, as we may lease warehouse space from a provider of logistic services; in a highly uncertain and dynamic setting, this may be an advantage.

Transportation costs present a similar structure, resulting from a mix of fixed and variable costs. When shipping a standard container from a certain point of the supply chain to another one, part of the cost is fixed and independent of the content. Transportation rates may be quite intricate, but again we may find a suitably accurate representation. If we want to compare two transportation modes, we are actually interested in figuring the best solution. If errors in the cost evaluation are not too large, they do not reverse the ranking of alternatives, and we make the correct decision anyway.

More often than not, there is a tradeoff between different cost components. For instance, transportation cost can be reduced by selecting a close supplier; however, this need not lead to the lowest overall cost because, when we order something from a supplier, several factors come into play besides transportation cost:

- ordering costs,

- the price charged by the supplier, which may also be affected by currency exchange rates,

- inventory holding costs.

Unlike transportation costs, ordering costs are *internal* costs, in the sense that they depend on the operations of the buyer firm, whereas transportation cost may depend on either the supplier, or the buyer, or a service provider. In the past, each order was associated with a procedure including some phone calls or fax messages. These costs were largely independent of the amount purchased. This is why we typically consider fixed ordering costs, i.e., associated with the order itself and not with the amount ordered. Electronic commerce has eased this burden considerably, but we may also consider receiving, inspecting, and handling incoming goods as components of the ordering cost. They can be partially captured by a fixed ordering cost. Sometimes, for the sake of simplicity, we aggregate all of the fixed cost components, including transportation, into a fixed ordering cost.

It is not uncommon to compare a geographically close supplier against a distant one who charges a lower price. The decision cannot be taken without specifying an ordering strategy, which is linked to the inventory control policy. The price can also depend on the purchased quantity, as quantity discount opportunities are sometimes offered. Should we take those opportunities? Reducing the purchase cost is certainly attractive, and the possibility of securing a known price might be too, if we fear an adverse movement in prices and/or exchange rates. However, ordering more materials also implies larger inventory holding costs. Inventory holding costs aggregate different cost components. To begin with, whenever we pay for some goods, and these stay in a warehouse for a possibly long time, we have an opportunity cost for the capital tied up in inventories, which we could have invested otherwise. From a financial point of view, too much capital sitting in inventories is bad news. More so, if we had to borrow money to purchase materials. Apart from financial issues, more inventory means more insurance charges, more material handling (with the possibility of wasting materials), larger expenses to heat or to refrigerate the warehouse, etc. If the goods are perishable or subject to obsolescence, we may also face the need of scrapping a significant amount of material: Cisco Systems was reported to take a staggering inventory write-off

($2.5 billion). All of these considerations lead to the idea that inventories should be kept low. Actually, inventory management is all about finding the right tradeoff; we will introduce the well-known Economic Order Quantity formula in section 1.4, to illustrate the tradeoff between cost of inventories and benefits of inventories (i.e., value of the functions performed by inventories).

We close this section by considering costs which may be very hard to quantify, i.e., stockout costs. We have a stockout whenever we run out of stock and we are not able to satisfy demand immediately; this may result in an unsatisfied customer or the stopping of downstream production. In the latter case, the stockout cost may be not too hard to estimate in terms of lost production, but when dealing with customers at a retail store, how much does an angry customer cost? To begin with, the loss of image associated with a stockout is an elusive concept, because it depends on consumer behavior. If we have a stockout and cannot meet an order from a customer, will she wait or go somewhere else? Assuming she is impatient, and the second case occurs, do we lose just this order or the customer altogether? This is very hard to tell; maybe we will never know, because she will just purchase a substitute item without telling anyone. As a further complication, the stockout cost can be linked to the *occurrence* of the stockout itself, or to the *size* of the stockout (e.g., number of customers that could not find the stocked out item). Even if we cannot quantify a stockout cost, we need to keep close control of the service level we offer, trading off other costs against this performance measure. We cover all of these considerations in chapter 5 on inventory management under uncertain demand.

1.3.3 Strategy

After this cursory look at competition factors and cost drivers in supply chain management, it should be clear that there is no way to find a single solution which is optimal from all of the conceivable points of view. In fact, firms adopt quite different strategies. The supply chain for a technologically mature product with a low profit margin must be efficient and inexpensive. In the case of an innovative product, with high margins and maybe a limited life, the overall strategy will be quite different:

- In the retail sector, the availability of goods on the shelves is essential. Still, an unsatisfied costumer might have a negligible impact, particularly for goods which have acceptable substitutes. However, a stockout for a whole product category (say, milk) or for products which are subject to strong brand loyalty (e.g., Coke in the soft drink industry or detergents for personal hygiene) may have serious consequences.

- In the "business-to-business" sector, we may have quite different priorities. Just think of managing the stock of spare parts to replace defective or failed ones in big industrial machining tools. Keeping such machines idle because of lack of spare parts may be extremely costly; indeed, this is a case where stockout costs may be easy to quantify, as there are contracts specifying penalties for lack of service. Quantifying the stockout cost of spare parts for life-critical equipments at hospitals is impossible, but we clearly see that in such a case we need to ensure immediate availability, either by suitable stock levels or, if the cost is too high, by very fast and expensive transportation.

In order to define a strategy, we must associate priorities to competition factors and find cost-effective ways to achieve a given performance target, possibly trading off performance against cost. Firms in different industries will probably define quite different strategies. It is no surprise that managing supply chains for high performance laptop computers requires a different approach than in the case of soap powder. However, even within the same given industry, we may observe quite different strategies.

Example 1.2 Personal computers are sold using different distribution channels, appealing to different consumers. Some consumers are quite sophisticated and want a very specific configuration; they are willing to wait relatively long lead times to get exactly the stuff they want. Others prefer a choice between a few well-defined alternatives, but fast delivery and cheap prices are essential to them. For similar reasons, some consumers do not mind ordering on a web site, whereas other consumers feel much safer buying from more traditional channels, because they want a personal contact in case of trouble with the product. In fact, different market segments can be dealt with by different marketing strategies.[15]   ⬚

Example 1.3 IKEA and MC are two dominant players in the Italian retail furniture business. They are both healthy and fast growing companies. However, they have fairly different strategies. IKEA basically designed a self-service environment where customer are asked to select the product they like, take note of the product code, and collect the selected item(s) at the warehouse. IKEA customers tend to transport goods by themselves. IKEA does not provide transportation services (though a business partner located near the counters sells transportation services). IKEA customers are even asked to design their own kitchen through the Internet or at do-it-yourself PC stands in the stores. Moreover, IKEA has a very wide number of product categories ranging from beds and chairs, to carpets and forks. However the range of product designs is rather limited and is dominated by the Swedish minimalistic design. The MC strategy is quite different. Though the prices are comparable, MC only sells furniture. In a MC store one cannot find carpets, forks, etc. However, in a MC store one can find furniture with very diverse designs ranging from classic, to modern, ethnic, romantic, etc. So

---

[15]See: V.K. Rangan and M. Bell, *Dell Online*, Harvard Business School Case No. 9-598-116, 1999.

the assortment offered by MC is very wide, though in a slightly different way: MC provides fewer product categories, but more styles than IKEA does. Also, while IKEA provides little sales assistance and delivery service, MC has a service intensive strategy. The vast majority of MC customers is attended by a salesperson. A salesperson can spend up to one hour designing the kitchen for a customer that then might simply walk away. Also, 90% of customers ask for the delivery of goods at their place (the cost of delivery is just 7% of the overall price). As we can see, the two companies have very different strategies (in many perspectives they have opposite strategies). However, they both are fairly successful. How can that be? Actually, the key idea is that the two companies appeal to two different segments of consumers and have two different value propositions. IKEA appeals mostly to youngsters (IKEA offers services such as day care for children), who can easily use technologies to design their own kitchen, can transport and assemble furniture on their own, and tend to appreciate the minimalistic Scandinavian style. MC tends to appeal to a more mature population that appreciates more traditional furniture and services such as sales assistance, delivery, and assembly of furniture.   ☐

Perhaps even more surprisingly, the same firm may pursue different operations strategies in space and/or time. In fact, operations may be diversified by geographic region, because alternative markets may require different approaches, depending on customers habits and cultural factors.

Example 1.4 Buying a car follows different patterns on the two sides of the Atlantic Ocean. In the USA, it is common to purchase a car on the spot, after having a look at what is available at the retailer. In Europe, it is more common to order a specific configuration, and possibly wait weeks for the desired model.   ☐

The level of market penetration and/or the potential entry of competitors may also contribute to the definition of a strategy. Finally, time is also essential, as a product at the beginning of its life cycle is typically not managed like an almost obsolete one. For example, a stockout late in the life cycle of a product is almost a desired outcome.

1.4 THE ROLE OF INVENTORIES

Much of what follows in this book deals with inventories; actually, three chapters (4, 5, and 6) are devoted to this topic. Keeping inventories implies a long array of costs, including less obvious ones such as an adverse effect on quality.[16] Indeed, given that many management philosophies are based on the

---

[16] Quality may be adversely affected because large amount of stocks typically require more material handling, which may result on accidental damage. High inventory levels also

idea of zero inventory, should we could consider inventory management a sort of more or less necessary evil?

Example 1.5 An intuitive consideration is that inventory availability has a positive effect on our ability to satisfy demand. What may be less obvious is that sometimes it is inventory itself that *generates* demand; just think of the allocation of shelf space at a big retail store. Even less obvious, inventory availability may be used to *sense* demand. Consider a large book store. Keeping an inventory of all possible titles is clearly out of the question. However, having some titles covering some discipline may be essential to check if there is potential demand for that kind of book (see case [11]). Otherwise, lack of inventory may imply lack of demand, which may be further motivation for not keeping stock; a perfect vicious circle.[17] Also, some companies keep deliberately large inventories if some staple products to show their dominance and as an implicit promise of product availability, which most customers tend to notice.   □

The example above does not imply that we should just increase stock availability. The message is that inventory has a purpose, and that we should understand its role and function in order to plan its level at a facility. The most complex problem is arguably the deployment of stock at the right installation of a large supply chain; on the one hand, we would like to place stock near customers, but this may be the worst place in terms of value of stock, as this is where we have the most added value; furthermore, stock near customers has been committed to a given retail region, potentially reducing flexibility in the allocation of goods. Generally, inventory reduction may be highly beneficial, provided that we eliminate the reasons for keeping it. In order to understand why we might need some inventory, a good starting point is the classical EOQ model.

1.4.1   A classical model: Economic order quantity

In this section we outline a sort of archetypal model for inventory management, the Economic Order Quantity (EOQ) model. Our purpose is just to illustrate how fixed ordering costs affect the need for some stock as well as to lay down some background which will be also used in chapter 2. Hence, the analysis is rather superficial, and much more detail is given in section 4.2.

Consider a distributor selling a good with a rather regular demand pattern. Taking it to the limit, we consider a perfectly constant demand over time. Let

---

imply longer waiting times on the shelves, which have an impact on perishable items. In manufacturing, high work in process levels are associated with longer flow times; if quality is checked at the end of the process, defects will be detected later, with a possibly significant increase in scrapped material.

[17]For a similar issue, related to phase-in/phase-out of products, see example 3.7 on page 100.

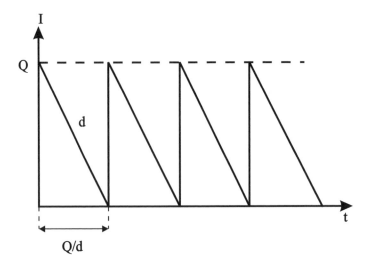

Fig. 1.5 Time evolution of inventory levels in the EOQ model.

d the demand per unit time; the specific time unit is not important, provided that we are consistent in specifying the remaining data (e.g., if demand in measured in units/daym the holding cost is unit of value – say euro – per unit, per day). The demand must be satisfied from stock, and goods are ordered from a supplier. A natural objective is finding an ordering strategy that allows the distributor to satisfy demand at minimum cost. Given that demand is constant, it is also reasonable to assume that the ordered quantity is always the same. Let Q be the lot size we choose and, without loss of generality, assume that we start with Q units on hand, as shown in figure 1.5. We will run out of stock after Q/d time units. Ideally, we would like to get a new lot of Q parts exactly when the inventory level drops to zero, as this will keep holding cost down. Such a perfect timing is possible if everything is certain and deterministic; this means not only demand, but also the supplier's delivery lead time. If the lead time is denoted by LT, it is easy to see that we should order Q whenever the inventory level[18] drops to a reorder point R given by the demand over the lead time : $R = d \cdot \mathrm{LT}$. If we repeat this cycle over and over, the time evolution of the inventory level will be periodic, as shown in figure 1.5, with cycles repeating every $T = Q/d$ time units.

Let c be the unit price charged by the supplier for each unit; we assume that whenever we order Q units from the supplier, we pay her an amount cQ. In other words, there are no discount opportunities we might take advantage of by ordering a larger amount. We see that cQ is a linearly variable cost. In

---

[18] In later chapters we will see that ordering decisions should not be simply based on *on-hand* inventory.

the cost c, we could also include a variable component of the transportation cost. Whenever we order, it is also reasonable to expect that a fixed cost has to be paid. This may be due to a fixed component of the transportation cost; or it could be a fixed ordering cost due to the need of issuing and tracking the order. Whatever the case, we denote this fixed ordering cost by A, which does not depend on Q. To summarize, whenever we order Q, the total cost of the order is $A + cQ$. This expression suggests the opportunity of not ordering too often a small amount. We have an economy of scale if we order a larger amount, because the fixed component is distributed on a larger number of parts.

However, there are good reasons to keep Q to a reasonable size. In this very simplified setting we do not consider the risk of obsolescence or perishability, nor physical space limitations in the warehouse. But the least we should do is to consider an inventory holding cost. The simplest reason for dealing with such a cost is the opportunity cost of capital tied up in inventory. There are many other factors which come into play here, but let us simply say that if we keep one part in inventory for a unit period, we face a cost h. Note that the dimensions of this unit inventory holding cost are money per part, per unit time. If we assume that this cost depends linearly on inventory, the total holding cost over some time period is h times the average inventory level.

Example 1.6 We should emphasize that using a linear inventory holding cost, as we will do in most of the book, can be a rather unsatisfactory approximation. To begin with, if we have discount opportunities, we should consider an explicit dependence $h(Q)$; clearly the total opportunity cost does depend on the price we pay per item, and this creates some dependence between the inventory holding cost and the average inventory level, which also depends on Q. Even if we assume that financial costs are more or less linear, other factors may have a nonlinear effect. For instance, consider a very perishable item, whose shelf life is just one day. If we keep inventory levels low, we will probably sell all of the available stock and no material will be scrapped. But if we raise inventory, under demand uncertainty, some leftover inventory will have to be occasionally disposed of. Hence, we see that cost linearity may be a debatable assumption. Still, all of these considerations point out some incentive to keep a low inventory level, maybe within a range such that a linear approximation is acceptable. Anyway, if demand is assumed deterministic, a limited shelf life would simply imply an upper bound on Q, which is easily dealt with. Hence, we will stick to linear holding costs in the following.   ▯

We see that we have two contrasting factors to account for in determining the order quantity Q. To spot the best compromise, we should quantify the total cost per unit time (say, one year) as a function of the decision variable Q. Since inventory level ranges between 0 and Q according to a linear pattern, we see that the average inventory level is $Q/2$. Hence, the holding cost component

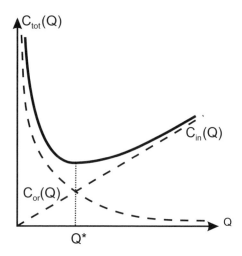

Fig. 1.6   Inventory holding and fixed cost components in the EOQ model.

is

$$C_{\text{in}} = h\frac{Q}{2}.$$

The contribution of the fixed cost component is A times the average number of orders issued per unit time. Since we have d/Q orders per unit time, this component is

$$C_{\text{or}} = \frac{Ad}{Q}.$$

Taking into account the purchasing cost of yearly demand, $C_{\text{pu}} = cd$, we have

$$C_{\text{tot}}(Q) = C_{\text{in}} + C_{\text{or}} + C_{\text{pu}} = h\frac{Q}{2} + \frac{Ad}{Q} + cd.$$

Leaving the last constant component aside, we may draw a qualitative picture of total cost in figure 1.6. We see that the objective function depends on a linearly increasing component C_{in} and a decreasing component C_{or}, displaying an economy of scale with respect to Q. The variable purchase cost plays no role really, as it does not depend on Q under our assumptions (but see example 1.7 below). Now we may find the optimal solution by equating the first-order derivative of the total cost to zero[19]:

$$Q^* = \sqrt{\frac{2Ad}{h}}. \tag{1.1}$$

---

[19] As we point out in appendix B, this first-order condition need not be sufficient for optimality, and we should also check the second-order derivative or, equivalently, show that the total cost is a convex function of Q. See example B.3 on page 547.

We have just derived the well-known EOQ (*Economic Order Quantity*) formula, which is valid only under a rather long list of limiting assumptions; nevertheless, it provides us with some useful insights. We see that the EOQ size increases with the fixed cost A and decreases with the inventory holding cost h. A simple calculation yields the optimal cost value for the optimal lot size:

$$C_{\text{tot}}(Q^*) = \sqrt{2Ahd} + cd. \tag{1.2}$$

This function shows that the total cost is a *concave* function of demand d; in other words, there is an economy of scale with respect to the demand a facility must face, and we will see in chapter 2 what impact this has on the design of a logistic network.

Example 1.7 (*A remark on relevant vs. irrelevant costs*) Expression (1.2) suggests that, unless discount opportunities are offered, the unit price c we pay for the stocked item is irrelevant in determining the optimal order size. Of course, it is very relevant for the bottom line, since it affects profitability, but we should observe that some costs may be irrelevant when making certain decisions. Actually, a closer look at the formula would suggest that probably c plays some role in determining h. Indeed, a common way to estimate inventory holding cost is to assume some opportunity cost of capital, that is a sort of interest rate i, say 15%, and setting $h = ic$. Nevertheless, the last term cd in (1.2) does look constant and irrelevant in determining Q^*. However, this holds only when we want to select the order quantity Q for a given supplier. If we change the problem at stake, things can change substantially. Suppose that we want to select a supplier and that there are two competitors, whose characteristics are represented by fixed and variable costs, c_1 and A_1, and c_2 and A_2, respectively. When comparing the two suppliers, in terms of the total cost as expressed in equation (1.2), we *cannot* overlook the last cost term. Hence, we see that cost elements and parameters may be irrelevant or not, fixed or not, and this depends on the decision at stake. Indeed, we can tell whether some kind of cost is relevant/irrelevant only with respect to a specific decision. □

Now suppose we wish to reduce the inventory level and/or the corresponding cost. A look at (1.1) suggests that unless we wish to reduce demand or we can reduce inventory holding cost (which increases Q^* but reduces the overall cost), we should reduce the fixed cost A. The fixed cost may depend on the ordering mechanism, the transportation cost, and possibly the setup cost. The setup cost is, within a production context, a fixed charge we pay whenever we start producing an item, independent of how many parts we make.[20] Clearly,

---

[20] We should note that, in a manufacturing context, setup cost might not be as relevant as the setup time, which reduces machine availability. When capacity is scarce, we cannot overlook interactions among products manufactured using a shared set of resources, and the EOQ model is not well-suited to this task.

if there is such a fixed charge, it is economical to buy or make a suitably large number of parts at once, and this is why inventories may be needed.

This reasoning points out a first function of inventories, which is linked to the need of adapting a relatively continuous and smooth consumption process to a replenishment mechanism, that on the contrary is very lumpy due to purchase, production, or distribution lots. The inventory we build because of this issue is called **cycle stock**. We cannot reduce cycle stock unless we reduce fixed charges, which create the need for a relatively large lots. Indeed, a mainstay of Japanese manufacturing philosophy has been the reduction of setup costs.

However, there are other reasons to build up inventory.

1. Stock is needed to decouple supply and demand, when one of them is subject to variability, and the other one is constrained and cannot follow such variability. In the next subsection we consider how transportation or capacity constraints generate stocks.

2. Stock is needed to hedge against demand uncertainty. The role of demand uncertainty is dealt with later in section 1.5.

We should also mention that there are many more factors that result in the creation of inventories. Raw material stock is sometime created in anticipation of unfavorable market conditions, such as increasing prices or uncertainty in the supply of a scarce commodity. We call this **speculative stock**.

Moreover, it is natural to think of stock as something sitting in a warehouse. However, inventory may be moving, as is the case of **in-transit** or **pipeline** stock. If transportation takes a few hours, in-transit inventory is actually negligible, but if long-distance transportation by ship is used, we may have a non-negligible impact. A similar consideration applies to manufacturing systems: The longer the flow time, the larger the work in process. We should note that while cycle stock depends on the order size, average in-transit stock only depends on average demand and the transportation delay, as illustrated by the following example.

Example 1.8 Consider an Italian firm importing a product from the Far East. The product is transported by ship, which takes one month, and the demand is constant and equal to 1000 pieces per month. If the firm issues a replenishment order once per month, each month it will issue an order for 1000 units, just when the previous one is being received. At each time instant, there is always a ship traveling with 1000 items. If the firm orders once per year, the order size is 12,000 pieces, and during the month following the order (say, January), there will be an in-transit stock of 12,000 items; for the remaining eleven months, in-transit inventory is zero, but its yearly average is still 1,000 anyway. □

Example 1.9 Let us consider a company from the Piedmont region that produces Barolo wine. Let us assume that the company sells 1,000 liters per year. Barolo wine needs to age for at the least three years before it can be sold. Two of these three years need to be spent in oak barrels. Given this demand and these technological constraints, at any point in time the company has at the very least 3,000 liters in stock. To cut this inventory investment either we reduce production volume, or change the technology in order to reduce the three year LT (i.e., find a way to make Barolo wine age more quickly), or simply decide to produce a different kind of wine. □

1.4.2 Capacity-induced stock

In the EOQ model we consider a constant and perfectly predictable demand. However, demand need not be constant to be perfectly predictable. In (very few) lucky cases we may have a time-varying demand which we know, as is the case if we make to order with a long delivery lead time. Ideally, we should be able to deliver all of the items just in time, with no need for stocking end items. As expected, cycle stock might be needed if there are fixed charges in making or buying the items. However, even if there is no fixed charge, we may have to resort to stock items in order to better match demand with capacity.

Example 1.10 Consider an item whose demand is strongly affected by seasonality. For instance, say that average demand is 100 per month, but the actual demand is 200 in spring and summer, and zero in autumn and winter. If items are produced by the firm, rather than purchased from an outside supplier, there are two extreme choices. It may size its manufacturing capacity to the maximum demand (200 units per month). In this case, there is no need for inventory, but capacity utilization is just 50%. At the other extreme, it could size the capacity to 100 items per month. In this case utilization is 100% but there is a considerable inventory buildup during the low-demand season. In this case we speak of **seasonal stock**. □

In figure 1.7 we illustrate a sample time evolution of seasonal stock when capacity is held constant and equal to average demand. It may also be the case that the mismatch is not between constant manufacturing capacity and time-varying demand, but between time-varying raw material availability and constant demand; this is the case for many food goods, such as canned tomatoes and olive oil. Sometimes, one can try to match capacity and demand by producing items with opposed seasonalities. For instance, winter and summer clothing can be produced in the same plants. In other cases, transportation capacity is used as a buffer: Apparently, for half a year kiwi fruit is imported from New Zealand to Italy, and vice versa in the other half.[21]

In a mathematical programming model illustrated on page 542 of appendix B, we illustrate how we might plan inventory buildup in order to match de-

---

[21] It remains to be seen whether eating kiwi twelve months per year is worth the resulting pollution.

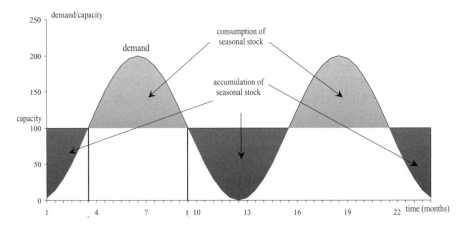

Fig. 1.7   Seasonal stock buildup and depletion.

mand and manufacturing capacity. Clearly, issues may compound with each other: in example B.12 on page 571, we consider a production planning problem where both capacity limitations and fixed charges call for the creation of stock. We should emphasize that besides fixed charges in the form of a fixed cost, we might have fixed consumption of capacity whenever we start production. If it takes a few hours to set up a machine to make an item type, we have to make a fairly large lot not to consume the capacity with setup times. These fairly large lots build up some inventories. This is again a form of cycle stock, even if the motivation is not strictly economical.

In the example above, we have considered production capacity, but in distribution logistics similar considerations apply to transportation capacity. Cycle stock may be necessary if transporting small orders is not economical, but inventory might also be required if the number of vehicles is limited and their capacities need to be fully utilized by full-truckload transportation.

1.5   DEALING WITH UNCERTAINTY

In distribution logistics there are many factors which are significantly affected by some form of uncertainty. For instance, we should extend the EOQ model to account for:

- uncertainty in demand, which may vary according to not perfectly predictable patterns;

- supplier lead time, which is affected both by transportation time and by possible material shortages.

At a different decision level, when tackling a long-period problem, we may have to face uncertainty in:

- prices, both in the sense of prices our suppliers charge and prices we may ask;

- exchange rates, which are relevant in an international context, on both the supply and demand side;

- changes in average demand; e.g., demand might simply fade away because of new emerging technologies.

Uncertain factors may be different in nature, depending on the length of the time horizon on which decisions must be made. Furthermore, different types of uncertainty may compound; for instance, demand uncertainty may be the result of short-term random variations in demand level, or of more systematic factors such as the success of a product and its market penetration, which also depends on the behavior of competitors.

In fact, we may consider different concepts of uncertainty. The probabilistic concept of uncertainty, which may be modeled by random variables following a given probability distribution, is the most common one. Other paradigms have been proposed, but we will essentially stick to a more familiar statistical framework (see chapter 5). If we know the relevant probability distribution, then we just have uncertainty in the realization of random variables. More often than not, properties of random variables must be inferred from available data, assuming they are available and reliable. In such a case, we have some uncertainty as far as the probability distribution itself, or its parameters, are concerned. Nevertheless, if data are available, we are still in the domain of probability and statistics and deal with a sort of "objective" uncertainty. In extreme cases, we deal with a brand new innovative product, and past information is simply unavailable, or its relevance might be questioned. In that context, we have to deal with subjective assessments of uncertainty (see section 3.12).

Whatever the nature of uncertainty, we must come up with some way to mitigate its effects. In the next two sections we consider two examples illustrating the role of safety stocks and proper product design.

1.5.1 Setting safety stocks

We have illustrated *how much* we should order according to the EOQ model, but we should also clarify *when* we should order (for a more detailed discussion see section 5.4). If both demand and supplier lead time are constant, it is easy to see that we should order an amount Q whenever the inventory level falls below a reorder point $R = d \cdot \text{LT}$ corresponding to the demand during lead time. In doing so, we should consider not only the physical (on hand) inventory, but also orders that we have already sent but have not been delivered yet, and backorders. Under deterministic assumptions, items will be delivered exactly when on-hand inventory reaches the zero level. When uncertainty is involved in either demand or lead time, or both, it is intuitive that we should raise the reorder level (in most situations, for a more detailed discussion see chapter 5). To do so rationally, we need two ingredients:

1. a description of the uncertainty of demand during lead time;

2. a suitable definition of the quality of service we want to offer our customers, in terms of our ability to meet demand immediately from stock.[22]

The uncertainty of demand during lead time depends on how the two basic uncertainties, in demand per unit time and in the lead time itself, are compounded. A typical assumption is that it can be modeled by a random variable D_{LT}, with normal distribution, expected value μ_{LT}, and standard deviation σ_{LT}.[23]

As far as the service quality is concerned, we will see in chapter 5 that different measures can be reasonably defined; we could also set up an optimization model, provided we may quantify the cost of a stockout. We consider here the simplest, not necessarily the best, alternative, which is to set a constraint on the probability of a stockout. This probability, denoted by α, should be suitably small; correspondingly, we define the quantity $1 - \alpha$ as our service level. Typical values of the service level could range between 90% and 99%. We have a stockout during lead time if demand in that time span exceeds the reorder point R. The probability of not having a stockout is

$$P\{D_{\mathrm{LT}} \leq R\} = 1 - \alpha.$$

We immediately see that R is the $1 - \alpha$ quantile of a normal distribution with parameters μ_{LT} and σ_{LT}.[24] As shown in appendix A, calculating the quantile of an arbitrary normal distribution boils down to finding the corresponding quantile for a standard normal distribution. Knowing the quantile $z_{1-\alpha}$ for a standard normal variable, we set

$$R = \mu_{\mathrm{LT}} + z_{1-\alpha}\sigma_{\mathrm{LT}}.$$

The idea is illustrated in figure 1.8: The shaded area, on the right of the quantile, corresponds to the stockout probability α. In the deterministic case, we simply set $R = \mu_{\mathrm{LT}}$; doing so when lead time demand is normally distributed would result in a 50% service level. In order to increase the service level, we add a **safety stock** given by $z_{1-\alpha}\sigma_{\mathrm{LT}}$. Clearly, safety stock increases the overall cost. On the average, we have an additional inventory of $z_{1-\alpha}\sigma_{\mathrm{LT}}$

[22] In a make-to-stock system or retail environment, this is the fundamental ability. In a make-to-order, quoting a reliable lead time may be more relevant, whereas in assembly-to-order, the customer should be allowed to customize her order in an easy and flexible way.

[23] Central limit theorem may justify such an assumption in the case of consumer goods; see page 470. This hypothesis should be checked by a suitable statistical procedure (see section A.9.1).

[24] See definition A.9 on page 456.

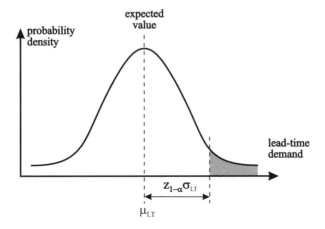

Fig. 1.8   Calculating safety stock based on lead-time demand uncertainty.

parts, with respect to the EOQ model; hence, if we set the order quantity according to the EOQ model[25] and we disregard variable purchasing cost, the average total cost (1.2) becomes

$$\sqrt{2Ahd} + hz_{1-\alpha}\sigma_{\mathrm{LT}}. \tag{1.3}$$

We should also mention that in this expression we are not considering stockout costs, which will be essential in chapter 5. Looking at equation (1.3), we see two sources of cost: cycle and safety stock. Setting a safety stock is, in some sense, a *passive* answer to the problem of uncertainty; we simply add some slack resources to reduce the effects of demand uncertainty. We could try to be proactive and prepare a set of actions to reduce the need for a large safety stock. Reducing safety stock, without reducing uncertainty is just not a good option, unless we want to give up service quality or ignore stockout costs. This calls for reducing uncertainty in lead time demand. On the one hand, lead time should be reduced; in a deterministic setting, the lead time might be irrelevant, because in that case we have just to anticipate the order timing (yet, it could be relevant in terms of in-transit inventory). In an uncertain setting, while the average lead time LT does not enter explicitly equation (1.3), it contributes to increasing σ_{LT}. We will see in chapter 5 that, if demands during different time periods are independent random variables, then σ_{LT} increases with the square root of lead time: $\sigma_{\mathrm{LT}} = \sigma\sqrt{\mathrm{LT}}$, where σ is the standard deviation of the demand per unit time. We will also see, in chapter 6, how demand and lead time uncertainty can be compounded.

---

[25] We will see later, in chapter 5, that this need not be the optimal choice. We should select the two parameters Q and R jointly, and perhaps consider an alternative definition of service level, taking into account the size of the stockout, and not only its occurrence.

It is tempting to believe that demand uncertainty is out of our control, and there is nothing we can do about it. Sometimes, this is true, but in many cases demand uncertainty is not exogenous. In many cases we can simply reduce demand variability that creates uncertainty. Demand spikes can be the result of unanticipated promotional sales; indeed, some large retail chains have decided to avoid promotional sales altogether, adopting an *every-day low prices* (EDLP) policy. In other cases, we may try to improve our forecasting procedures in order to (partially) transform unpredictable variability into predictable variability. This is basically the purpose of forecasting techniques described in 3.

In complex systems, many other actions can be attempted to reduce safety stock costs. In manufacturing systems, we may adopt preventive maintenance policies in order to reduce the occurrence of random machine breakdowns, and the related need for inventory buffers. In a multiechelon system, we may try to hold inventories in central warehouses where demand is more aggregate and thus more predictable, rather than at retail stores. In the next section, we illustrate the general idea of *risk pooling* by exploiting common components in assembling end items.

1.5.2 A two-stage decision process: Production planning in an assemble-to-order environment

When computing safety stocks, we do not plan orders in advance: We prescribe the structure of a policy (e.g., Q and R), and we let the system run and place orders when our policy suggests to do so (e.g., when we hit the reorder point R). In practice, the parameters are adjusted periodically. Furthermore, emergency actions are carried out when needed. All of these adjustments are carried out when additional information is obtained, but this is outside this formal model. The formal model is, in a sense, single-stage: We make some decisions and then see what happens. In some other cases, we want to include in a formal model the adjustments we might make at a later stage. In order to do so, we must formalize the dynamics of the decision process, whereby decisions are made and/or revised when new information is obtained. This may lead to very difficult stochastic models. We consider here a simple example of a two-stage model.

Consider an assemble-to-order (ATO) system. In such a system, we have to make (or buy) components, which are then assembled into some end item we sell. It would be nice to do everything after we receive a customer order, but we cannot afford this luxury if the customer is not willing to wait that much time. If the customer wants everything immediately, we have to keep a stock of end items; this may be difficult or impossible when end items come in a wide variety of configurations or when items cannot be stocked because of their cost. A compromise solution is feasible when making components requires a long lead time, but assembly is relatively fast. We can keep a stock of components, which are made or bought *before* we get customer orders. We assemble only on order, i.e., after we collect customer demand. Concrete examples of ATO processes are the automotive industry, at least in Europe, and the PC industry, where one can order a customized model and select among a number of feature/options.[26]

Let us build a simple but instructive model along with a small numerical example, under the following assumptions:

1. First, we decide how many units of each component we build, subject to manufacturing capacity constraints. This first decision sets the total production cost.

2. After receiving customer orders, we use components to assemble finished goods. The assembly plan in designed to maximize revenues; the cost term in the profit function is fixed by the previous decision (if we neglect assembly cost); if components are not enough to meet customer orders, we lose profit opportunities; if too many components are available, they are discarded with a possibly considerable loss of money.

The key point, apart from demand uncertainty, is that we have a limited time window for sales, after which components are no use. This is a limit assumption, typical of the classical newsvendor model (see section 5.2); in practice, components might have some salvage value, or they could be used in later time periods. In this setting, we have to make two decisions in sequence, in order to optimize profit. Literally, we cannot maximize profit, because it is a random variable depending on our decisions and on uncertain demand, but we may maximize its expected value.[27]

Since the main complicating factor is demand uncertainty, one possibility is to disregard it and just use expected values of demand in planning production of components. Another possibility is representing demand uncertainty by a set of scenarios. We will pursue both approaches and compare the decisions we make.

To set up a small toy example, say that we own a (very) small firm, producing just three end items (A_1, A_2, A_3), which are obtained by assembling components $(c_1, c_2, c_3, c_4, c_5)$. The components we use for each end item are described by a bill of materials, which is flat (just two levels: end items and components). The bill of materials is given in the left-hand side of table 1.1. From the bill of materials, we see that there are two common components, c_1 and c_2, while the remaining three are specific and characterize each end item. We assume that three resources (M_1, M_2, M_3) are used for production of components. On the right-hand side of the table we also see the bill of resources,

---

[26] A possibly more pleasing example is any pizzeria offering a wide array of pizzas; the pizza is made on order, but all of the components are prepared in advance.

[27] A more sophisticated approach would involve some considerations about risk, which is not fully captured by the expected value.

Table 1.1 Bill of materials for the assemble-to-order example

	c_1	c_2	c_3	c_4	c_5
A_1	1	1	1	0	0
A_2	1	1	0	1	0
A_3	1	1	0	0	1

Table 1.2 Bill of resources, cost of components, and available capacity

	M_1	M_2	M_3	Cost
c_1	1	2	1	20
c_2	1	2	2	30
c_3	2	2	0	10
c_4	1	2	0	10
c_5	3	2	0	10
Cap.	800	700	600	

Table 1.3   Demand scenarios, expected value of demand, and selling price of end items

	S_1	S_2	S_3	Exp. Demand	Selling Price
A_1	100	50	120	90	80
A_2	50	25	60	45	70
A_3	100	110	60	90	90

i.e., the time required on each resource to manufacture one component. In the table, we also give the available capacity for each resource type, and the cost of each component; this cost might include both direct variable production costs and material costs. We assume that assembly is not a bottleneck, hence its capacity is disregarded.

Other relevant data concern end items, demand, and the price at which end items are sold. They are given in table 1.3. Demand uncertainty is modeled by a set of three scenarios (S_1, S_2, S_3). If we have information about past sales, the three scenarios may result from the discretization of a continuous probability distribution (of course, more scenarios are needed in a practical setting to approximate the distribution); alternatively, they could result from an interview with three experts. Whatever the case, we assume that the three scenarios are equally likely, i.e., each probability is $1/3$.[28] We also give the expected value of demand, which is obtained by averaging the three scenarios for each end item. The last column displays the price at which end items are sold.[29] Also, note that the selling price is larger than 60, the total component

[28] When discretizing continuous distributions, we might use different probabilities to get a better approximation; see. e.g., [4, chapter 10] for an application of Gaussian quadrature. In the case of forecasts based on subjective judgment by experts, using the same probabilities means that we consider three equally reliable experts.

[29] If we do not want to disregard assembly cost, we may substitute selling price by contribution to profit from assembling and selling an item; this defines the second-stage cost, as it takes selling price and assembly cost into account, but not component costs.

cost, for all of the three end items, but A_3 looks more profitable, in a sense, because its profit margin including component costs is $90 - 60 = 30$, whereas A_2 is the least profitable; of course, this reasoning may be misleading in that it does not take into account resource consumption.[30]

We may tackle the problem of maximizing expected profit by the Linear Programming (LP) techniques described in appendix B. To build a simple model as a starting point, we could disregard uncertainty and deal with one scenario characterized by average demand. We get the following model:

$$\max \quad -\sum_{i=1}^{5} C_i x_i + \sum_{j=1}^{3} P_j y_j, \tag{1.4}$$

$$\text{s.t.} \quad \sum_{i=1}^{5} T_{im} x_i \le L_m, \qquad m = 1, 2, 3, \tag{1.5}$$

$$y_j \le \bar{d}_j, \qquad j = 1, 2, 3, \tag{1.6}$$

$$\sum_{j=1}^{3} G_{ij} y_j \le x_i, \qquad i = 1, 2, 3, 4, 5, \tag{1.7}$$

$$y_j, x_i \ge 0.$$

Here, subscript i refers to components, subscript j refers to end items, and subscript m refers to resource types. Input data correspond to those reported in the tables:

- the component cost C_i;

- the selling price P_j for each end item;

- the available capacity L_m for each resource type (measured in time units);

- the resource requirement (processing time) T_{im}, for component i on resource m;

- the number G_{ij} of components i going into an end item j (i.e., the bill of materials - BOM);

- the expected demand $\bar{d}_j$, which is assumed certain.

The decision variables are x_i, the number of components of type i that we produce, and y_j, the number of end items of type j that are assembled and sold; to be more precise, we *pretend* that we will really sell an amount y_j, because we disregard demand uncertainty. The model aims at maximizing profit, as expressed by the objective function (1.4), subject to capacity constraints (1.5). The inequality (1.6) states that we cannot sell more than what

---

[30]See example B.1 on page 537.

is demanded, whereas (1.7) says that we cannot assemble end items if the required components are not available. The decision variables are required to be non-negative. In fact, for the sake of simplicity, we consider a *continuous* LP model, which allows for fractional quantities; if we insist on requiring that produced and assembled quantities are integer, it is easy to incorporate this requirement (see section B.6).

Solving the model, e.g., by the simplex method (see appendix B), we get the following solution (rounded to two decimal digits):

$$x_1^* = 116.67, \quad x_2^* = 116.67,$$
$$x_3^* = 26.67, \quad x_4^* = 0.00, \quad x_5^* = 90.00,$$
$$y_1^* = 26.67, \quad y_2^* = 0.00, \quad y_3^* = 90.00.$$

In this very small example, we may easily interpret what this solution tries to accomplish. We assemble the maximum number of end items of type A_3, which is the most profitable one; this requires in turn the production of a corresponding number of common components c_1 and c_2, as well as the specific component c_5. Since demand limit is binding for A_3, there is some capacity left, which is used to produce a limited amount of the specific component c_3, which is needed to assemble end item A_1, plus common components. End item A_2 has the lowest selling price and is disregarded, as well as is its specific component c_4. It should be noted that, in general, one should not take for granted that the production of the highest profit item should be maximized; the consumption of available resources should be taken into account as well (see example B.1 on page 537 for a counterexample).

In this specific case, the solution is quite readable, but it is a bit "extreme." An expert planner would immediately see that it is a risky bet on high sales of the most profitable item. The optimal profit, according to this model, is 3233.33, but this is actually misleading. After planning production of components, we do *not* know the value of profit, but only its distribution (if we accept the validity of the demand scenarios). We cannot maximize optimal profit; what we can do is maximizing its *expected value*, and this requires a more sophisticated model that takes demand scenarios into account:

$$\max \quad -\sum_{i=1}^{5} C_i x_i + \sum_{s=1}^{3} \pi^s \left(\sum_{j=1}^{3} P_j y_j^s \right), \tag{1.8}$$

$$\text{s.t.} \quad \sum_{i=1}^{5} T_{im} x_i \leq L_m, \qquad m = 1, 2, 3, \tag{1.9}$$

$$y_j^s \leq d_j^s, \qquad j = 1, 2, 3, \quad s = 1, 2, 3, \tag{1.10}$$

$$\sum_{j=1}^{3} G_{ij} y_j^s \leq x_i \qquad i = 1, 2, 3, 4, 5, \quad s = 1, 2, 3, \tag{1.11}$$

$$y_j^s, x_i \geq 0.$$

The big change in this model, with respect to the expected demand model [(1.4)–(1.7)], is that demand uncertainty is taken into account explicitly. Here we consider demand d_j^s for item j in scenario s. Accordingly, the quantity assembled is now represented by scenario-dependent decision variables y_j^s; this is the amount of end item we assemble and sell, if and when scenario s is realized. Assembly decisions are not taken here and now, when we plan production of components, but they are *contingent plans*. The scenario-independent variables x_i are first-stage variables, whereas variables y_j^s are second-stage variables. So now we implement the production plan (i.e., first stage decisions x_i) and develop a contingency plan for the assembly operations (i.e., second stage decisions y_j^s). Only when demand is realized we choose among the contingency plans (y_j^1, y_j^2, y_j^3).[31] We should carefully notice the difference between a multiperiod model and a multistage model. We illustrate examples of multiperiod models in appendix B and in chapter 4. In such models, decisions will be implemented in later time periods, but they are all taken *now*, based on the currently available information. It is possible to revise such decisions by solving the model again according to a rolling horizon strategy, but this is outside the scope of the model itself. In a multistage model, we do not commit to one specific decision for the later stages; the decision that will actually be implemented depends on the realization of random variables, and it will be fixed only when the relevant information will be available in the future. Next-stage variables may also be used to "adjust" previous decisions, given current contingencies. This interpretation explains why models such as the one above are called stochastic programming models with recourse.

Going into details of the model above, the objective function (1.8) consists of a first-stage (deterministic) term accounting for the cost of components, along with a second-stage term, which is the expected revenue from selling end items (not including component cost); the expected value is computed by summing the revenues under the three possible decisions, times scenario probabilities π^s. The capacity constraint (1.9) is unchanged, because it pertains to first-stage only. The market constraint (1.10) is now scenario-dependent, as it considers the stochastic demand d_j^s. Finally, constraint (1.11) links the two stages, stating that assembly is constrained by component availability, for each end item and each scenario. Solving this model, we get the following solution:

$$x_1^* = 115.71, \qquad x_2^* = 115.71,$$

---

[31] Notice that this holds only when the three scenarios are actually the only three possible demand scenarios. In other cases, we can face a very large number of different scenarios (possibly an infinite number of different scenarios). In this case, the three scenarios are only meant to model demand uncertainty and make sure that first stage decisions account for demand uncertainty. The realized demand might differ from all three scenarios. In this case, once demand is realized we simply have to write a second model for assembly decisions, where we need to meet the realized demand with a limited quantity of components that was fixed through the above model.

$$x_3^* = 52.86, \qquad x_4^* = 2.86, \qquad x_5^* = 62.86,$$
$$y_1^{1*} = 52.86, \qquad y_2^{1*} = 0.00, \qquad y_3^{1*} = 62.86,$$
$$y_1^{2*} = 50.00, \qquad y_2^{2*} = 2.86, \qquad y_3^{2*} = 62.86,$$
$$y_1^{3*} = 52.86, \qquad y_2^{3*} = 2.86, \qquad y_3^{3*} = 60.00.$$

The real outcome of the model is the set of the first-stage decision variables x_i^*. Observing the component production plan, we immediately see a qualitative difference with respect to the model disregarding uncertainty: It is less extreme. We do not produce a large amount of component c_5, because we do not place a risky bet on high sales of A_3. In fact, scenario three would prove a disaster for the deterministic solution: In that scenario, sales are lower for A_3, but we could not react because we do not have enough specific components for the other end items. This also implies that many specific components[32] would be thrown away (according to our assumptions concerning the limited time window for sales and the lack of any salvage value of unused components). The stochastic model, instead, increases production of specific component c_3, which is needed to support assembly and sales of A_1; even a small amount of component c_4 is produced, in order to support the least profitable end item A_2, which helps in using common components when sales are low for other end items. While there is a big difference in terms of specific components, we see that as far as common components are concerned, the solutions of the deterministic and the stochastic solutions are essentially the same. There is a good reason for this, as common components are a flexible resource, which can be exploited to support different end items. Moreover, the demand for common components is the sum of the individual demands for the end items, and by aggregating demand we often reduce uncertainty. Indeed, this *risk pooling* effect is what we try to exploit in assemble-to-order systems. In chapter 6 we will see that the same mechanism is exploited in the management of distribution networks. However, it is also important to note that when end item demands are strongly correlated, the risk pooling effect is considerably reduced. In such a case, we should expect that even the produced quantities of common components differ in the deterministic and the stochastic model. Another relevant factor is capacity: If this is so tight that we may sell whatever we are able to produce, a simple deterministic model could be a viable option.

But how do the two solutions compare in terms of profit? The objective function from the solution of the second model is 2885.71; apparently, the stochastic solution is worse than the deterministic solution, whose objective value was 3233.33. But this comparison makes no sense. We are actually comparing two different situations rather than two different solutions. The above finding simply proves that we would rather face a certain demand rather than an uncertain one. The objective function of the first model is neither

---

[32] In the more general case even common components could be thrown away.

the true profit, which is uncertain, nor its expected value. It would be the optimal profit, if we knew that the average demand scenario is what will be realized. In the first model [1.4]–[1.7] we *pretend* to know the end item demand, and we get the illusion of higher profits. In order to compare the two solutions, we should fix the production plans for components suggested by the two models, and then we should solve a set of second-stage problems, where we optimize assembly of end items subject to component availability, for different demand scenarios. More formally, given a set of first-stage variables x_i^* for components, we should solve the following second-stage (recourse) problem for each scenario s:

$$R^s(\mathbf{x}^*) \equiv \max \sum_{j=1}^{3} P_j y_j^s,$$
$$y_j^s \leq d_j^s, \qquad j = 1, 2, 3,$$
$$\sum_{j=1}^{3} G_{ij} y_j^s \leq x_i^*, \qquad i = 1, 2, 3, 4, 5,$$
$$y_j^s \geq 0,$$

where $R^s(\mathbf{x}^*)$ is the optimal revenue we collect under scenario s, given the first-stage solution $\mathbf{x}^*$, and making optimal use of available components to meet demand. The first-stage solution can come from solving a stochastic or an expected-value model; whatever the case, its expected revenue is

$$\sum_s \pi^s R^s(\mathbf{x}^*).$$

Expected profit for an arbitrary solution can be obtained by subtracting its first-stage cost from its second-stage expected revenue.[33]   To evaluate the deterministic solution, we should plug it in this model; in case of scenario S_1, the optimal assembly and sales plan is

$$y_1^* = 26.67, \qquad y_2^* = 0.00, \qquad y_3^* = 90.00,$$

and the same holds for S_2. The bad news is that if scenario S_3 occurs, we are in trouble, because the high-risk solution does not fit demand very well. The optimal assembly and sales plan would be

$$y_1^* = 26.67, \qquad y_2^* = 0.00, \qquad y_3^* = 60.00.$$

This is a pretty bad scenario with low sales and corresponding low profit. We must compute revenue for each scenario, multiply it by its probability,

---

[33] We are evaluating expected profit *in-sample*, i.e., by using the same set of scenarios which are used in the stochastic model; we could use a much larger set of out-of-sample scenarios to get a more reliable estimate. The point is that solving a large number of small LP problems may take less CPU time than solving one large stochastic LP model.

sum everything to get the expected value, and subtract the component cost from the first stage. Doing so, we may see that the expected profit from the deterministic solution (2333.33) is much lower than what the objective function of the deterministic model [(1.4)–(1.7)] predicts (3233.33), based on one average-case scenario. The percentage improvement of the stochastic solution with respect to the deterministic one is

$$\frac{2885.71 - 2333.33}{2333.33} \approx 23.67\%.$$

Clearly, we cannot extrapolate general results from a small toy example. Indeed, the advantage of using a stochastic model is striking here, because specific components have a large impact. In a case featuring a lot more component commonality, the result would be less impressive. Furthermore, we have assumed that unused components are scrapped, which need not be the case. They could have some salvage value, and we could have a multistage problem so that they can be used in later stages. Nevertheless, the example is quite instructive in pointing out:

- the difference between decision *stages* and *time periods*,

- the role of risk pooling.

In this case, risk pooling is obtained by using common components and by deferring assembly decisions. To further illustrate the value of deferring decisions in a more specific distribution setting, some fashion retail chains send only a part of the items to retail stores at the beginning of a season; at a later stage, after observing sales at each retail store, the residual stock is sent downstream. Also in this case, the first decision, i.e., the purchase of items from suppliers, is often constrained by a budget assigned to each *buyer* in charge of a specific market segment. The second decision, inventory allocation, can be made by a different type of professional called *planner*.

A last important consideration, which applies to all models we describe in this book to deal with demand uncertainty, is that we have considered the maximization of expected profit as a suitable objective. We do not consider profit variability across scenarios, or what happens in extremely bad scenarios (the average smooths out single outcomes). This makes sense if we may repeat the game over and over (for various items or over multiple periods), so that what really matters is the average profit in the long run. However, in the short run we may take too many chances: If a single bad decision cannot be recovered, because we immediately go out of business, or get fired, a more careful approach should be taken to fully account for risks. An alternative view, for economically minded readers, is that considering expected profit is equivalent to assuming a risk-neutral attitude; risk-averse decision makers should consider different objective functions.

1.5.3 Inventory deployment

The previous section serves well to illustrate the role of commonality in order to reduce the impact of uncertainty. Common components mitigate uncertainty by providing flexibility and by allowing postponement of critical decisions. This is just one instance of the more general risk pooling concepts which are widely used in distribution logistics. When we consider an arborescent network like the one in figure 1.3, we should decide if and how much safety stock we should place at each node. We will see in section 2.1.1 that placing safety stocks upstream may reduce their aggregate level.[34] On the other hand, we should be careful to ensure suitable customer service, which would require locating stock downstream. We see that the inventory deployment decision is by no way trivial, and as usual there is no ready answer for all possible circumstances. As the following example shows, creative thinking may be required in peculiar cases.

Example 1.11 Consider the problem faced by a manufacturer of very expensive spare parts for some industrial equipment[35]; the manufacturer itself, or a firm providing maintenance services in its place, signs a contract requiring immediate replacement of faulty parts, say within a few hours. Where should spare parts be located, and how many of them are required? The second question requires possibly nontrivial probabilistic modeling. As far as the first question is concerned, allocating one part to each customer would certainly ensure satisfactory customer service, but it would be extremely costly. One alternative could be to place some stock at a facility which is more or less located in a barycentric position with respect to customers. However, if a customer is far, we should probably arrange for very fast transportation, maybe by air. With very fast transportation, the exact location of stock may be irrelevant. Hence, we could even consider placing spare parts at some customer location, reserving the right to collect the part for fast shipment to another customer in need of a spare part. This would save some warehouse cost, but it requires a shift in the paradigm prescribing that the owner of stock is the owner of the location where the inventory is placed. The spare part changes owner only when it is mounted on a machine. ⬚

The example illustrates a simple case of a more general strategy called Vendor Managed Inventory, which is later illustrated in example 1.12 on page 41. For reasons that will be later explained in chapters 6 and 7, it may be advantageous to have only one authority in charge of inventory management, since the interactions of different decision makers having limited information, and typically misaligned incentives, may generate unwanted spikes in demand; this phenomenon is known as the bullwhip (or Forrester) effect. In fact, it

---

[34] See section 2.1.1.
[35] See [8], page 611.

is important to keep in mind that managing a complex supply chain is not just a technical challenge, as human factors and different points of view may exacerbate difficulties (see chapter 7).

The possibility of postponing inventory allocation decisions and exploiting risk pooling depends on product design too. The supply chain of HP DeskJet printers was successfully reorganized by changing the assembly process,[36] in such a way to delay differentiation of products (e.g., according to destination country). For example, they assemble the printer with instruction manuals, cables, and plugs at the warehouse, rather than at the production site. This may result in an increase in the direct product cost, but the analysis must be carried out on a global level, taking into account the shorter and shorter life cycle of products, whose obsolescence may be very fast (indeed, this is the case in consumers' electronics). While this assembly process might add a few cents to the direct production cost, customizing products in the central warehouse might cut inventory investment and obsolescence cost by millions of euros. Generally, demand forecasting is easier whenever we may aggregate items by family. Consider clothing, which may differ in model, size, and color. If we may postpone dyeing items, in order to gain more reliable information about demand, considerable savings may be obtained. Indeed, a well-known case in this vein is Benetton, where cutting and dying operations were swapped in order to ease forecasting.[37]

1.6   PHYSICAL FLOWS AND TRANSPORTATION

In section 1.2 we have considered a network as a physical arrangement of facil-ities. An essential feature of any supply chain is the selection of a transporta-tion strategy and the management of physical flows, inbound and outbound from any node. Large organizations manage transportation by themselves, whereas in other cases this activity is outsourced; in general, we should de-cide between alternatives such as rail, ship, air, or trucks.

Restricting our attention to road transportation, we may arrange point-to-point transportation or route a vehicle to serve multiple destinations. For instance, referring to figure 1.3, we may have one vehicle for each link from node 3 to nodes 6,7, and 8; alternatively, the same vehicle may visit the three retail stores sequentially. A decision problem that may occur in the first case is the determination of a suitable transportation frequency; in section 2.1.2 we show that a simplified version of the problem, accounting for fixed and inventory holding costs, closely resembles the EOQ model. In the second case, we should find a suitable assignment of customers to vehicles, and a

---

[36] This case study is described, e.g., in [10].
[37] See S. Signorelli, J.L. Heskett. Benetton (A). Harvard University Business School Case, 1984.

customer sequence for each vehicle, in order to optimize a given performance measure; such a problem, known as the Vehicle Routing Problem, is dealt with in chapter 8.

When operating our own vehicles, we may try to utilize their capacity at best, according to a *full truckload* strategy (e.g., see the case [7]). Sometimes, the need for fast delivery requires *less-than-truckload* (LTL) transportation. For instance, fast mail couriers typically cannot easily exploit full transportation capacity (trucks and aircrafts), and they try to aggregate flows by proper design of the transportation network. In the LTL case, we may also consider the use of third-party transportation, leaving to our business partner the task of aggregating flows in order to better exploit capacity.

1.7 INFORMATION FLOWS AND DECISION RIGHTS

In figures 1.1, 1.2, and 1.3 we have illustrated the flow of goods, but the information flow is just as important. In principle, information pertaining to the whole supply chain can be collected and managed by a unique decision maker. This centralized manager, should be able to come up with globally optimal decisions. Information Technology (IT) might make all of this a concrete possibility, but there may be unsurmountable difficulties. To begin with, an all-encompassing decision model may be way too difficult to solve. A nastier difficulty is the reliability of information. All large retail stores use point-of-sale data acquisition, and we should be able to know exactly how much stock is available and where, for each item. In practice, such information need not be 100% reliable because of errors, theft, wrong deliveries on the part of suppliers, misplaced inventory, exceeded shelf-life, damage due to material handling, etc.

Even leaving the above difficulties aside, there are deeper difficulties with a fully centralized decision-making architecture:

- Actors in the supply chain may be unwilling to share information.

- Actors in the supply chain may be unwilling to relinquish decision rights to others.

Example 1.12 The **Vendor Managed Inventory** (VMI) approach is a good case to illustrate difficulties in information sharing and allocation of decision rights. Consider a supplier, who delivers goods to independently owned retail stores. Point of Sales (POS) information can easily be collected and sent to the supplier, who could plan inventory accordingly. By the same token, retailers should send timely information to the supplier in case of planned promotions; otherwise, unpredicted demand spikes may have both immediate consequences, such as stockouts, and long term ones, such as loss of customers to competitors, which further contribute to the difficulty in forecasting demand an planning inventory. In fact, a retailer who receives a reduced amount of stock, because of a shortage, may be tempted to order more than needed during the next replenishment cycles, in anticipation of rationing strategies on the part of the supplier. But if all of the demanded items are eventually delivered, a low-demand period will follow because the retailer must get rid of excessive stock. This contributes to an increase in demand volatility along the supply chain, as well as to an overall feeling of partner unreliability. These and other reasons contribute to the generation of the so-called **bullwhip effect**, which has been well-known since the 1960s (see section 6.3). One way to overcome this issue would be to centralize demand information from POSs, which can be collected by the supplier. While technically possible, this solution may be thwarted by retailers feeling that the supplier could share this information with their competitors. An even more radical approach is based on the idea that the supplier is not only the collector of all information in the supply chain, but also the only actor in charge of managing stocks. In VMI, goods are stocked at retail stores, but they are managed by the supplier and change owner only when goods are placed on the shelves. A very well-known case in this vein is Barilla,[38] a firm that had to work very hard to persuade retailers to adopt such a policy and give up authority on inventory.   ⬚

A general issue raised by VMI is: Assuming that a (maybe partially) centralized policy reduces the overall costs, who is going to enjoy the benefit? More generally, if multiple actors (different firms, or separate branches within the same firm) control different managerial levers along the supply chain, is there any guarantee that the overall strategy is optimal? There is no easy and general answer to such very delicate issues. In chapter 7 we clarify the related issues and outline the design of incentives to improve overall performance. Given the complexity of the involved issues, that chapter has more of a conceptual than operational nature.

1.8   TIME HORIZONS AND HIERARCHICAL LEVELS

In distribution logistics we have to tackle quite different problems in terms of time horizon, involved uncertainty, and impact of the decisions we make. Designing a new network of warehouse facilities, to be operated during the next few years, and organizing vehicle routes for the delivery of the next day are clearly two extreme examples of problems pertaining to different hierarchical levels.

- At the highest hierarchical we have **strategic** problems. The time horizon may be years or months. The longer the time horizon, the higher the level of uncertainty, which calls for suitable forecasting procedures

---

[38]See J.H. Hammond. Barilla SpA (A). Harvard University Business School Case, 1994. Alternatively, the case is described in [13].

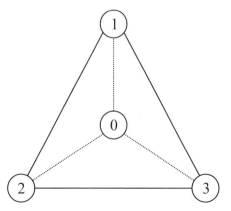

Fig. 1.9 Graphical illustration of the location problem in example 1.13.

and scenario analysis. Decisions made at the strategic level, such as warehouse capacity, will play the role of constraints at lower levels in the hierarchy.

- At an intermediate level we have **tactical** problems. Here, resource availability is usually fixed, but the time horizon (say, weeks) is long enough to require some form of forecasting. An example of tactical problem is the selection of an inventory management policy; changing such a decision is definitely easier than redesigning the structure of a distribution network.

- At an **operational** level, we have day-to-day decisions, where uncertainty is negligible, and we have to react to incoming information in a very short time span.

It is worth noting that the division between the three levels is not sharp at all. Third-party providers of logistic services allow us to enlarge warehouse space without building any new facility, and this makes the boundary between strategic and tactical problems less clear.

Furthermore, the hierarchical levels are interdependent. Of course higher-level decisions constrain lower-level management, but the link is two-way. Hierarchical decomposition is needed to tackle otherwise intractable problems, but when making a strategic decision we must somehow anticipate the effects on the tactical and operational levels. This must be done by some model simplification, resulting in a sort of "anticipation" function. We will see a few examples in the next chapter, but the next example illustrates how different decisions cannot always be taken disregarding their interactions.

Example 1.13 Consider a network consisting of three retail stores, located on the vertices of an equilateral triangle, as illustrated in figure 1.9. The position of nodes 1, 2, and 3 is given, and we should locate a distribution node (or a production plant) in such a way that the total transportation cost from the distribution to the retail stores is minimized. Assuming that the demand on the three retail stores is the same, an intuitive solution would be placing the distribution center in the barycenter of the triangle (node 0 in the figure; the resulting vehicle routes are drawn as dotted lines). However, this depends on the transportation mode. If demand is large with respect to vehicle capacities and we transport point-to-point, this solution is reasonable. However, if demand is low and distances are not too large, it could be much better to visit nodes 1, 2, and 3 in sequence with the same vehicle. In such a case, we could place the required node at the same location of any retail point on the perimeter and, in case it saves us some money, we can consider to place it in one of the three vertexes of the triangle (stores). Of course, a real-life problem should also account for fixed cost components in transportation, environmental issues, item perishability, etc. $\Box$

We close this section by stressing again the fundamental difference between *time periods* and *decision stages* (see section 1.5.2). In a long-term decisions, we may prepare plans which are implemented in successive time periods. This may result in dynamic problems. If the decisions are made here-and-now and are not changed later, we to have a multiperiod decision problem, but it is a single-stage one. We typically reserve the term "multistage" for problems in which future decisions are adapted as a function of additional information we gather and of the progressive resolution of uncertainty.

1.9   DECISION APPROACHES

Supply chain management strategies may differ according to priorities in objectives, information availability, and strategy of the firm (see section 1.3). There is wide array of possibilities, and confusion is sometimes added by ambiguous use of buzzwords, such as **push** vs. **pull** systems. Indeed, the manufacturing literature has largely contributed to this state of the matter, because of the confusion among different hierarchical levels, such as demand management (also known as master production schedule in a production environment) and shop-floor control.

The following classification criterion is suggested in [13, chapter 5]:

> In a push-based supply chain ... production decisions are based on long-term forecasts. (...) In a pull-based supply chain ... production is demand driven .... rather than..forecast (driven).

We may also substitute "purchasing" or "distribution" for "production," to make the definition more general. So the difference between a push and a pull strategy is the following:

- In a pull system, purchasing, production, or distribution orders are based on the consumption of a good in the downstream operation. For example, in a manufacturing environment, the production of a component might be triggered by the consumption of that component at an assembly plant. In a distribution environment, the distribution of a case-pack of canned tomatoes is triggered by the consumption of canned tomatoes at the stores. These policies are somehow based on minimum inventory levels and once they are reached the upstream stages start purchasing, producing or distributing the products.

- In a push system, purchasing, production, or distribution are based on a plan, which is based on a forecast of a future demand. For example, in a production environment we might decide to produce a component because our assembly plan for next week foresees the need for such a part. Also, in a distribution environment we might distribute a large quantity of a given product, because we foresee a peak in demand due to a promotion.

To be fair we shall say that somehow pull strategies are based on some sort of forecast as well. Indeed, while materials consumption triggers purchase, production, or distribution orders, these are governed by parameters that are based on some sort of forecast.

Example 1.14 Now, let us consider a simple EOQ-based policy, where a replenishment order is issued when we reach the reorder point. Is this a pull policy? From a certain point of view, it certainly is: We issue an order when inventory is pulled. However, some forecasting procedure is arguably used in setting the policy parameters, which depend on the expected value of demand and its standard deviation. We see that even a simple pull approach is based on a mix of demand forecast and materials consumption. The parameters are based on some sort of forecast, while the replenishment orders are triggered by actual demand and materials' consumption. □

Example 1.15 Kanban production control can give us another good example of a pull system. Kanbans are a means to control production at the shop floor level; they were invented in Japan and made famous by Toyota. Kanbans are basically a permission to start the production. These "permissions" to produce are released only when the components are actually consumed. So they are a very effective means to control the inventory level. If we only have permissions for 100 units (say we have 100 kanbans and each gives the permission to produce one unit), we never have more than 100 units of the component at stake. When a unit of the component is consumed in the assembly operation, we release the permission to produce one unit. On the other hand the manufacturing stage attaches one kanban to each unit manufactured. Therefore, while the consumption of the component releases "permissions" to produce, production consumes them. This process makes sure that the inventories of components do not get out of control, since the manufacturing stage can produce one unit if and only if one unit has been consumed. This makes the kanban production control the gold-standard for pull systems. However, one could wonder why we decided to have 100 kanbans rather than just 50? On the other hand, one might wonder whether 100 kanbans are enough? Actually, these decisions may be made by simple rules of thumb, simulation models, or even complex algorithms that lie outside the scope of this book that focuses on logistics. However, one can intuitively understand that the number of kanbans depends on the rate of consumption of the components, which in turn depends on the expected future demand for the finished good. Another relevant factor is uncertainty, which provides us with an incentive to raise the number of kanbans in order to add some safety stock.   ⬜

Quite often, "pull" is associated with a good and efficient policy, whereas "push" is associated with obsolete practice. Actually, in some contexts the pull strategy might perform very poorly whereas the push strategy might be very effective.

Also, there is nothing like a pure strategy, as real-life approaches are typically hybrid mixtures, and these terms should be associated to *features* of a solution approach, rather than to a specific choice. So the key issue is not choosing between one strategy and the other. We rather have to find the right blend at the various levels, as the two examples below show.

Example 1.16 For example, in many supply chains, we develop long-medium term plans to allocate resources, plan shifts, give suppliers advance notice of expected changes in demand, etc. For example, a company might sign a contract for the supply of 10,000 cans of Coke a month, based on the expected demand over the next 3 months. Nevertheless, the actual delivery-orders might be driven by the actual consumption of Coke at the stores; for example, stores might reorder a pallet of Coke when the inventory level of Coke drops below a given threshold. So we have a long term push strategy, whereby we commit to the overall quantity based on some sort of forecast of future demand. On the other hand, the short term replenishment process is driven by the consumption of Coke at the stores and thus can be considered to be pull. As we can see, push and pull are features of the solution rather than contrasting alternatives.   ⬜

Example 1.17 In a production environment we might have a master production schedule (that is the plan for production of the finished good) where we plan the production quantities over time, according to current inventory levels, future demand (either a demand forecast or firm orders or a mix of the two), setup costs, etc. This is actually a plan at the finished good level. So one would be led to think that companies that use the Master Production Schedule use a push strategy. Actually, at the shop floor level the replenishment of components to be assembled might be driven by their consumption and thus might fall under the "pull umbrella." The assembly of finished products is based on a schedule, whereas (some) components are produced and replenished as they are consumed by assembly operations. This example too shows that the push and the pull logics can coexist and very often are used by the same company. □

A second recurring theme in operations is the Make-to-Stock/Make-to-Order dilemma. First we should realize that **Make-to-Order** is not a synonym of pull system and **Make-to-Stock** is not a synonym of push system. An example will, hopefully, make the point clear.

Example 1.18 Let us go back to the car industry (see example 1.4). Both in the USA and in Europe the replenishment and production of components is based on a pull strategy, at the least in the short run as the kanban production control has become a sort of standard in this industry. Nevertheless, in the USA most cars are made to stock, while in Europe they are made to order. This clearly shows that push or pull can be associated with either Make-to-Stock or Make-to-Order. □

When one thinks carefully about it, the issue is actually fairly simple. The flow of components to the assembly line can be based on a pull or a push strategy, but the fact is that both strategies simply disregard whether a specific customer (say Mr. Brandimarte) is waiting for the blue car, with leather seats, and air conditioning, or the car is simply ordered by a retailer (or a commercial unit) that hopes to sell it sooner or later to a generic consumer.

Also, Make-to-Stock and Make-to-Order are not actually contrasting alternatives, but they should rather be considered as features of a strategy, and can be combined to design a reasonable solution. For example, in most good restaurants dishes are prepared to order, while raw materials are purchased to stock. Things are fairly easy for standard raw materials with a long shelf life such as flour or potatoes. Things are more tricky for very specific and short shelf life products such as mullets (a specific kind of fish that is used for very specific recipes). They are bought if and when we expect that on the same day (or the next day) a customer will ask for a very specific recipe.

Moreover, the assemble-to-order example of section 1.5.2 suggests the possibility of integrating different strategies. Components may be produced (or purchased) based on forecasts, whereas final assembly is made only when a customer order is received. As we pointed out, this is a necessary arrangement when the delivery lead time accepted by the customer is smaller than the overall lead time for producing the end item, but it is impossible to stock end items, because of their cost or their variety. We see that there is an *order decoupling point*[39] which separates subsystems governed by different policies. Finally, it is also worth noting that quite different approaches may be adopted within the same firm, depending on specific items (in terms of value, perishability, etc.) and customers.

---

[39]Sometimes, the term "push–pull boundary" is used in manufacturing.

Example 1.19 Consider for example a manufacturer of top-end watches. The basic models (maybe still worth a few thousand euros) are available at the stores. On the contrary, unique items such as top grand-complication items (that is items with an extremely complex mechanical movement that can account for lunar phases etc.) are made to customer order as demand is so sparse that it makes no sense to carry them over. Also these extremely expensive items are only bought by collectors that seem to enjoy the time they have to wait, as it testifies the product is really hard to make and is specifically made for them.   ◻

1.10   QUANTITATIVE MODELS AND METHODS

In this book we use quantitative models and methods extensively. Applying a quantitative approach means setting up a mathematical model and solving it by some appropriate method. The quantitative feature could be associated to some "scientific" or "objective" virtue, but this is a somewhat reductionist approach. As the saying goes, there is no such a thing like an exact model: All models are wrong, but some are useful. This is why modeling has been defined as the art of selectively simplifying reality. Choosing the right degree of simplification is indeed an art, which is subject to often contrasting views depending on personal taste and opinion. Since building and solving a model is done with some purpose in mind, different stakeholders may have quite different ideas about the right modeling approach. Whatever the case, there are many reasons making simplification necessary:

- Computational tractability: As we point out in appendix B, some optimization models may be hard to solve, and we must give up some modeling detail and/or resort to suboptimal solution methods.

- Uncertainty: In principle, we may use the machinery of probability theory and statistics (see appendix A) to represent uncertainty, but sometimes lack of data, or difficulty in the model, prevents an exact representation. We should also keep in mind that not all of the uncertainties can be formalized within the framework of probability theory.

- Complex dynamics may prevent elegant analytical modeling.

- There are often conflicting points of view, which cannot be analyzed objectively on a purely quantitative basis.

There are two wide classes of quantitative models:

1. Prescriptive models. Typical examples are optimization models, which are formulated with the aim of getting a decision directly. In principle, decision-making could be automated by gathering data, instantiating a mathematical programming model, and solving it by one of the many

commercial solvers. In practice, prescriptive models should just be used as a decision support tool.

2. Descriptive models. Unlike prescriptive models, modeling tools within this class do not aim at generating a decision. They just try to capture relationships between variables, shedding some light on key features of the problem at hand, which is then used by the decision maker.

While we will illustrate many quantitative models, we should emphasize that useful descriptive models may also be qualitative; their role is rationalizing a business process and reaching a common understanding, which is not to be taken for granted in large organizations or in contexts involving several firms with different views and incentives.

The descriptive models we consider in this book are mainly aimed at predicting something. Prominent examples that we will consider are time-series-based forecasting and regression models (see chapter 3). We might also consider performance evaluation models. The idea is predicting the performance of a real system, for a certain configuration and for a given setting of some parameters governing decision rules. To make this point a bit more concrete, let us denote by $f(\boldsymbol{\theta}; \omega)$ a performance measure depending on a set of decision variables $\boldsymbol{\theta}$, which are under our control, and a set of random variables, which are beyond our control; the dependence on random events is expressed by ω. A performance evaluation model aims at estimating the expected value of the selected performance measure:

$$H(\boldsymbol{\theta}) \equiv \mathrm{E}_\omega[f(\boldsymbol{\theta}; \omega)].$$

Performance evaluation models may further split into two subclasses:

1. Analytical models.

2. Simulation models.

Analytical models typically require some simplification. We will see some examples in chapter 5 when deriving approximations of expected cost as a function of inventory management policies under uncertainty. Analytical models in this domain may require some simplifying assumptions; for instance, we will assume that backordering is possible, i.e., customers are patient. But if customers are not necessarily willing to wait, demand can be lost, making modeling harder. Simulation models, on the contrary, are extremely flexible and powerful, at least in principle; however, they require much effort in data gathering, and maybe in solution time, and require a working knowledge of general-purpose programming languages or more specific simulation environments. While in other engineering-related problems we need continuous-time simulation models, in supply chain modeling we need *discrete-event* simulation models. By "discrete-event" we mean that the system state changes in correspondence with specific events: For instance, the inventory level changes abruptly when a supplier delivers an order, or when a customer asks for some material. Uncertainties are modeled by pseudo-random number generators, i.e., algorithms able to emulate random phenomena, such as customer demand. The simulation program includes event management and decision rules which allow us to emulate the time evolution of quantities of interest and to estimate required performance measures given the set of parameters $\boldsymbol{\theta}$. There are graphical description languages, which may make the modeling task easy in simple cases, as they require assembling and linking standard blocks with a graphical editor; still, nontrivial thinking may be required to fit a complex system within the bounds of the selected simulation environment.

Because of these reasons, we do not deal with simulation modeling in the book, but we want to point out that, as usual, we should not draw a very thick line separating prescriptive and descriptive models. For instance, many revenue management and dynamic pricing strategies use regression modeling (e.g., to capture the link between price and demand), as well as modern optimization software tools. Furthermore, modern simulation environments are integrated with optimization solvers able to manage simulation experiments in order to automatically search for the best setting of parameters with respect to a specified cost or profit function. Since there is randomness in any supply chain, we actually want to optimize (say, maximize) the *expected* value of some performance measure:

$$\max_{\boldsymbol{\theta} \in \boldsymbol{\Theta}} H(\boldsymbol{\theta}) \equiv \mathrm{E}_\omega[f(\boldsymbol{\theta}; \omega)],$$

where $\boldsymbol{\Theta}$ is the feasible set for the controlled parameters $\boldsymbol{\theta}$. The expected value is, when random variables are continuously distributed, a possibly high-dimensional integral. Then, we must resort to some sampling mechanism, yielding an approximation $\tilde{H}(\boldsymbol{\theta}) \approx \mathrm{E}_\omega[f(\boldsymbol{\theta}; \omega)]$. For simple systems, we may get an analytical approximation, which is suitable for optimization by mathematical programming, as we have seen in the stochastic optimization example of section 1.5.2. When simulation is needed, we have to resort to different optimization approaches. Typically, commercial software relies on some form of evolutionary computing able to deal both with noisy estimates of the performance measure and with usually nonconvex optimization problems.

1.11   FOR FURTHER READING

- In this book we will deal with problems which lie at the boundary between distribution logistics and production planning. An excellent book on manufacturing systems, including production planning and control, is [8].

- An excellent text covering supply chain management with a wider scope (and, necessarily, sometimes a more shallow level) is [5]. Among other things, the reader will find there some treatment of revenue management and electronic commerce. For a text very rich in references to practical cases, see also [13].

- We deal with distribution logistics from an *operations management* perspective, but we should keep in mind that this dimension must be linked to a financial perspective; models integrating the two sides of the coin are illustrated in [12].

- We have pointed out that there is no best supply chain management approach; the strategy must be adapted to the specific firm and market at hand, a point which is very well illustrated in [6].

- Readers interested in discrete-event simulation will find [9] very comprehensive and readable.

- A tutorial introduction to stochastic programming models in manufacturing can be found in [1]. For a comprehensive introduction to both models and solution methods, see, e.g., [3].

REFERENCES

1. A. Alfieri and P. Brandimarte. Stochastic Programming Models for Manufacturing Applications. In A. Matta and Q. Semeraro, editors, *Design of Advanced Manufacturing Systems*. Springer, Dordrecht, 2005.

2. C. Billington, G. Callioni, B. Crane, J.D. Ruark, J.U. Rapp, T. White, and S.P. Willems. Accelerating the Profitability of Hewlett-Packard's Supply Chains. *Interfaces*, 34:59–72, 2004.

3. J.R. Birge and F. Louveaux. *Introduction to Stochastic Programming*. Springer-Verlag, New York, 1997.

4. P. Brandimarte. *Numerical Methods in Finance and Economics: A MAT-LAB-Based Introduction (2nd Ed.)*. Wiley, New York, 2006.

5. S. Chopra and P. Meindl. *Supply Chain Management: Strategy, Planning, and Operation (2nd Ed.)*. Pearson Prentice Hall, Upper Saddle River, NJ, 2004.

6. M.L. Fisher. What Is the Right Supply Chain for your Products? *Harvard Business Review*, 75:105–116, 1997.

7. P. Ghemawat and J.L. Nueno. *Zara: Fast Fashion, case 9-703-497*. Harvard Business School Publishing, Boston, MA, 2003.

8. W. Hopp and M. Spearman. *Factory Physics (2nd Ed.)*. McGraw-Hill, New York, 2000.

9. A.M. Law and D.W. Kelton. *Simulation Modeling and Analysis (3rd Ed.)*. McGraw-Hill, New York, 1999.

10. H.L. Lee and C. Billington. Material Management in Decentralized Supply Chains. *Operations Research*, 41:835–847, 1993.

11. A. Raman and Z. Ton. *Borders Group Inc., case 9-601-037*. Harvard Business School Publishing, Boston, MA, 2003.

12. J.F. Shapiro. *Modeling the Supply Chain*. Duxbury/Thomson Learning, Pacific Grove, CA, 2001.

13. D. Simchi-Levi, P. Kaminsky, and E. Simchi-Levi. *Designing and Managing the Suppy Chain (2nd Ed.)*. McGraw-Hill/Irwin, New York, 2002.

2

Network Design and Transportation

In chapter 1 we have seen that logistic networks can be shaped according to several patterns; defining the structure of the network is a strategic task with a significant impact on the overall cost of the supply chain, and it results in constraints on its day-to-day operations. The main problem we deal with in this chapter is indeed the design of logistic networks. Actually, we should speak of network design *problems*, as there are many shades and nuances of this problem. In principle, designing a logistic network requires locating and sizing production plants, distribution centers, and retail stores. In practice, we typically face a subset of those decisions, since some part of the network is given. To begin with, we rarely design a network from scratch; we may have to redesign an existing network in order to adapt it to changing demand patterns or changing prices of inputs. Hence, we may have to relocate some facilities, to expand their capacities, or to locate a few new ones. Furthermore, (i) when locating plants or large distribution centers, retail store locations are taken as given; on the contrary, (ii) in retail management we often have to locate retail stores, i.e., the last nodes in the network (e.g., see [5]). The relevant criteria and constraints are quite different in the two problems. When locating retail stores, an important role is played by the logistic range, i.e., the maximum distance a potential customer is willing to travel to purchase a given item; hence, distance may drive sales rather than just contributing a cost term to the objective function. When locating a distribution center, the distance between the center and the retail stores is typically just an element to evaluate the total transportation cost. Moreover, in many location problems we take demand at final destination nodes as exogenously given. On the contrary, when locating retail stores, demand is a *result* of our decisions.

The design of a logistic network is typically considered a long-term, strategic problem. Indeed, building a large and expensive facility is certainly not a day-to-day decision. Nevertheless, recent trends, whereby third parties may offer logistic services, tend to make the problem a bit more tactical and shorter-term. Obviously, building a plant and renting shelf space for the next four months are different decisions. In the latter case, we are changing the nature of costs from fixed ones to (relatively) variable ones. Flexibility is a requirement dictated by the faster and faster introduction of new products and ever changing market conditions, which may call for the almost continuous redesign of the supply chain. In any case, even if we are making strategic decisions, we need to represent their consequences on tactical decisions, such as transportation optimization. We need a sort of "anticipation function" in order to estimate the costs of tactical decisions that we will make next, subject to constraints enforced by strategic decisions; this estimate need not be overly precise. In strategic models, we cannot take detailed issues, such as operational vehicle routing, into account; such decisions are the subject of chapter 8; by the same token, the optimal loading of a single vehicle is of no concern at this level. Still, a suitably aggregate representation of transportation flows and their costs is needed when designing a network.

An interesting feature of logistic networks is the presence of intermediate nodes, such as distribution warehouses or transit points, between production plants and retail stores. Since such facilities represent a cost, there must be some good reason to introduce them. We discuss their functions in section 2.1. In particular, we point out their potential role in reducing the impact of demand uncertainty in section 2.1.1, whereas in section 2.1.2 we consider their role in optimizing transportation and in managing assortment.

Section 2.2 deals with classical linear programming models to optimize transportation flows on a network, to locate facilities, and to choose their capacities. To keep computational effort limited, these models are static rather than dynamic, and we should wonder if such models are able to capture the interaction of flow routing and inventory management decisions. We cannot and should not mix detailed descriptions of both strategic and operational decisions in the same model; however, a suitable approximate model may be obtained by considering nonlinear cost functions. Then, to avoid the burden of solving a large nonlinear mixed-integer programming model, we may approximate nonlinear costs by piecewise-linear functions, as described in section 2.3. Since some model formulations may be tough to solve, a huge amount of literature has been produced, based on heuristic approaches to ease the computational burden or to make the solution process a bit more intuitive. We will not consider this literature, for which we point out a few references at the end of the chapter; by the same token, we refrain from describing complex models accounting for some additional issues. Indeed, the astonishing progress in both hardware and optimization software libraries has paved the way to the solution of large scale models. We believe that the main limitations of the modeling framework we describe here are not computational but, instead, lie in their limited ability to cope with demand uncertainty, as well as in the potential difficulty in understanding *why* we have obtained a certain optimal solution. Indeed, we should consider the models below as *one* tool within a complex decision support architecture; their role is to propose solutions, which could be modified in order to comply with some further requirements and should be thoroughly checked by detailed simulation.

The chapter is complemented by two web sections. Section W.2.4 deals with continuous-space location models. In fact, the previous sections assume that we have already identified sites for potential facilities, and we must make a choice between a discrete set of alternatives; in other cases, we would like to find ideal positions of facilities, in continuous space. This may be useful in the process of building alternatives. Section W.2.5 illustrates peculiarities of retail store location problems, compared with plant and distribution center location models. This topic is usually covered in books on marketing rather than in books on logistics. We believe it is actually a borderline issue as it defines the "last mile" (i.e., the last echelon) of the supply chain for consumer goods.

2.1 THE ROLE OF INTERMEDIATE NODES IN A DISTRIBUTION NETWORK

In section 1.2 we have considered the basic structures of logistic networks. In particular, figure 1.3 on page 8 illustrates a prototypical arborescent network; the network in the figure is an example of a distribution network consisting of three levels:

1. a first level, where production plants are located;

2. an intermediate distribution level;

3. a third level, where goods are finally routed to satisfy customer demand.

In practice, the network may be much more complex and it may feature more than three levels, but the basic question is: Since goods are not transformed at the intermediate nodes, why are they needed? Indeed, intermediate facilities are a cost, both for the structure itself and for the inventory they might carry. You may hear consultants stating that distribution nodes should be avoided. Indeed, in many industries/companies there is a need to rationalize the distribution network, and this may require the elimination of intermediate levels or the aggregation of distribution centers to cover a wider area.[1] Still, there may be good reasons to include intermediate levels, and we should understand them; if anything, removing a facility calls for the elimination

---

[1]In Italy, inefficiencies in distribution networks consisting of too many levels are often mentioned as a cause of higher prices of many goods, with respect to other countries.

of the reasons for its existence; i.e., the services it provides to the rest of the network. Furthermore, we should carefully consider the tradeoff between their cost and their benefit. In the next sections we illustrate simplistic models with the aim of pointing out what intermediate distribution centers try to accomplish. In the first case, we consider the impact of intermediate nodes on demand uncertainty; in the second case, we show how intermediate nodes can help manage product variety and transportation.

2.1.1   The risk pooling effect: reducing the uncertainty level

In section 1.5.1 we have introduced the concept of safety stock and we have obtained an expression for the total cost per unit time, in the case of a (Q, R) policy, with an economic order quantity Q, a reorder point R, and a stockout probability α:

$$TC = \sqrt{2Ahd} + hz_{1-\alpha}\sigma, \qquad (2.1)$$

where σ is the standard deviation of lead time demand, which is assumed to be normally distributed. We remind the reader that we have taken for granted the possibility of determining Q and R separately; as we shall see in chapter 5, this is actually an approximation. Anyway, this expression is useful to point out a factor that *may* make the inclusion of an intermediate distribution warehouse useful.

Let us consider a network with n retail stores, which could carry their own inventory. For each retail store $i = 1, \ldots, n$, let d_i be the expected value of lead time demand and σ_i its standard deviation. For the sake of simplicity, we also assume that the demands are independent random variables. Now we may use equation (2.1) to compare the cost we have when all of the inventory is allocated to the retail stores against the case in which inventory is centralized, i.e., it is kept at a central warehouse serving the retail stores.

If inventory is fully distributed to the retail stores, there is no central warehouse and the total cost is the sum of n terms, one per retail store:

$$TC_D = \left(\sqrt{2Ah}\sum_{i=1}^{n}\sqrt{d_i}\right) + \left(hz_{1-\alpha}\sum_{i=1}^{n}\sigma_i\right).$$

Note that in the expression above, as well as in the following, we do not consider stockout costs; we include such penalties in chapter 5. If we keep inventory at the central warehouse, this will see an aggregate demand with expected value

$$\tilde{d} = \sum_{i=1}^{n} d_i$$

and standard deviation

$$\tilde{\sigma} = \sqrt{\sum_{i=1}^{n} \sigma_i^2}.$$

Actually, these two relationships require some care and a few assumptions. To begin with, the second one is based on independence among the random demands. Furthermore, it is certainly true that the aggregate demand per unit time is the sum of the individual demands, but in computing safety stocks we must consider the demand during lead time. The lead times seen from the warehouse and from each retail store need not be the same, while here we are assuming that it takes the same amount of time to deliver directly to the stores, or through a distribution center. Also, demands could be subject to different uncertainties. The relationships are correct if the lead times are deterministic and the same for all of the network nodes. Under all of these assumptions, the overall cost is

$$TC_C = \sqrt{2Ah}\sqrt{\tilde{d}} + hz_{1-\alpha}\tilde{\sigma}.$$

In order to compare the two total costs, we may see that the following inequalities hold:

$$\sqrt{\tilde{d}} = \sqrt{\sum_{i=1}^{n} d_i} \leq \sum_{i=1}^{n} \sqrt{d_i}, \tag{2.2}$$

$$\tilde{\sigma} = \sqrt{\sum_{i=1}^{n} \sigma_i^2} \leq \sum_{i=1}^{n} \sigma_i. \tag{2.3}$$

To see why these inequalities hold, we may observe that $a^2 + b^2 \leq (a+b)^2$, for non-negative values of a and b; this can be generalized to the sum of n terms. As we have pointed out, these inequalities need not be very accurate in a realistic case, but they do suggest that centralization of inventories could yield some advantage.

On the one hand, inequality (2.2) suggests a possible economy of scale, essentially due to the concavity of the square root function which is involved in the EOQ cost also in the deterministic case. We can also provide an economic reading of this finding. A company enjoys economies of scale when it orders at a central warehouse rather than at n stores. Ordering at the central warehouse can cut the transportation cost and reduce cycle inventories.

Concept 2.1 *A central distribution center aggregates demand and thus enables the company to enjoy economies of scale in transportation and order processing.*

On the other hand, inequality (2.3) suggests that the uncertainty in the aggregate demand can be lower than the sum of uncertainties of the individual demands, and this results in a reduction of the safety stock. Here we see another example of the risk pooling effect, which we have already met in section 1.5.2. In other words, demand at the central warehouse is more stable than demand at single stores. Indeed, high demand at one store can be counterbalanced by low demand at another store. Thus we need less safety stocks when we carry inventories in the central warehouse rather than at stores.

Concept 2.2 *A central distribution center aggregates demand. Aggregate demand tends to be more stable, thus reducing the need for safety stocks.*

We see that centralization of inventories may be beneficial, but we cannot really draw a general conclusion, because our analysis is way too simplistic. The risk pooling effect may be reduced if there is a strong positive correlation among demands at different retail stores. To see this, let us consider the sum of two random variables D_1 and D_2, with standard deviations σ_1 and σ_2, respectively, and correlation coefficient[2] ρ:

$$\sigma_{1+2} = \sqrt{\text{Var}(D_1 + D_2)} = \sqrt{\sigma_1^2 + \sigma_2^2 + 2\rho\sigma_1\sigma_2} \leq \sigma_1 + \sigma_2,$$

where the inequality stems from the condition $\rho \leq 1$. If the two random variables are independent, then $\rho = 0$ and the previous analysis applies; if there is negative correlation, uncertainty is actually reduced. But if ρ is large (close to 1), then the inequality tends to an equality and there is little reduction in uncertainty. We may have such a positive correlation when demand volume depends on the success of a product, assuming this is homogeneous across the retail stores, or when it depends on general economic conditions.

Other very important points we have missed are transportation costs and the delivery lead time as seen from the customer. If we have inventory available at the retail stores, we may serve the customer immediately. Our analysis implicitly assumes that the customer will wait if we can guarantee that stock is available at the central warehouse. This may be true for certain products, but not always; there is wide spectrum of consumption goods for which lack of stock on the shelves simply kills demand.

To summarize, inequalities (2.2) and (2.3) cannot be used to conclude that it is always optimal to centralize stocks. However, they point out a potential tradeoff between the overall amount of safety stock, which is an argument for centralization, and the quality of customer service, which is an argument against centralization. It can well be the case that the optimal solution is a compromise between these extremes, depending on the product type (in terms of customer behavior and competition) and on the transportation times from warehouse to retail stores. A detailed analysis can be carried out, but it may be remarkably complex; we will consider such issues in chapter 6, which is dedicated to multiechelon inventory management.

---

[2]See section A.6.2 in appendix A.

2.1.2   The role of distribution centers and transit points in transportation optimization

In the previous example, we have seen that distribution centers may be helpful in mitigating the effects of uncertainty, but they can also be helpful in
exploiting economies of scale. In this section we disregard uncertainty, but we
illustrate the last point in a somewhat more realistic setting, where different
item types are transported on a simple network. Consider the network in
figure 2.1, consisting of two production plants and five retail stores. In the
two plants, items A and B are produced, respectively; both item types are
sold at the retail stores. We may look at the network of figure 2.1 as the
superimposition of two independent arborescent structures, featuring direct
shipment from factories to retail stores. A possible alternative is depicted
in figure 2.2, which features an intermediate distribution center. Note that
we are not considering uncertainty here, and the intermediate transshipment
point is not necessarily meant to be a warehouse. In order to compare the two
alternatives and to get a feeling for what might make the second one interesting, we will use a very simplified example, where both demand at retail stores
and production rates at factories are constant over time. In our analysis, some
inventory builds up at the intermediate node; but in a more practical setting,
shipments may be synchronized in order to operate the distribution center
as a pure cross-docking point, where items are received, fanned out, and are
immediately shipped to destination.

Let us analyze direct shipment first. We have a point-to-point transportation over ten links, consisting of a pair (factory, retail store). How should
we manage each transportation link? If we assume that demand is constant
over time, and that the production rate is perfectly synchronized with this
demand rate, what we need to find is an optimal transportation frequency, by
formulating and solving a model which is quite similar to the economic order
quantity.[3] To see this point clearly, let us focus our attention on one link, say
the transportation from factory A to retail store 1. We assume that transportation cost has a fixed and a linearly variable component, which could
be an approximation of an economy of scale. The fixed-charge component
induces a transportation batch, i.e., a quantity which is transported with a
fixed period (or frequency). The quantity should also take vehicle capacity
into account, but for the sake of simplicity we disregard such an issue, assuming that vehicles are large enough. In figure 2.3 we see how inventory levels
at the two nodes change over time. Note that the figure is drawn under the
assumption that the network consists only of these two nodes; inventory levels
in factory A are also affected by shipments to other retail stores. The figure
suggests that the relevant decision variable is the time T_c elapsing between

---

[3]What we present here is a rough-cut analysis, inspired by [4, chapter 1], where a larger-
scale problem is discussed more exhaustively.

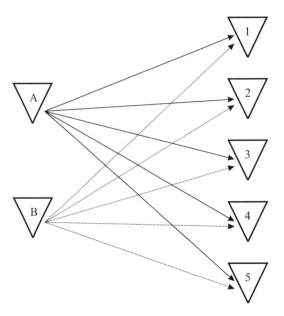

Fig. 2.1   Distribution network with direct shipments.

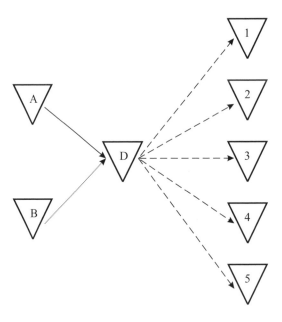

Fig. 2.2   Distribution network with an intermediate distribution center.

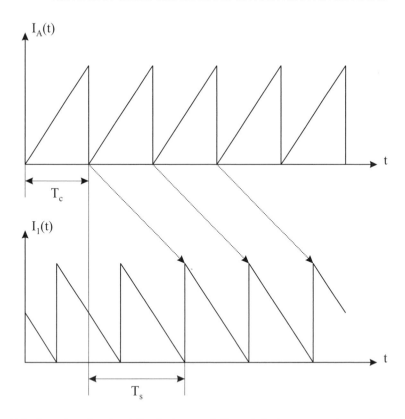

Fig. 2.3 Inventory levels at the factory and the retail store, assuming uniform production and demand rates.

two shipments, i.e., the transportation period or equivalently the frequency $1/T_c$. T_s is the time needed to transport items between the two nodes, which we consider deterministic and given; during this time interval, items on the vehicle are inventory in transit. Just like the EOQ case, the relevant data are the inventory holding cost h and the fixed transportation charge A. The variable transportation cost does not play any role, if it is linear, because all of the required items will be transported sooner or later; hence, the contribution of variable costs to the overall cost per unit time is constant with respect to the decision variable at stake.

We can write an expression of the total cost per unit time, which is similar to the EOQ objective function; the most notable difference is that the decision variable is a time period rather than a quantity, but if demand is constant, they actually boil down to the same decision. If we denote the demand rate by d, which is the same as the production rate according to our hypotheses, at each shipment the transported quantity is $Q = T_c d$. If vehicle capacity is not an issue, we also see that we have $1/T_c$ shipments per unit time; hence the fixed charge contribution per unit time is A/T_c. Now we must quantify the inventory holding cost. We should note that in this case we hold inventory in three stages: at the factory (there is an inventory buildup before each shipment), in transit, and at the retail store. We should figure out the inventory holding cost for the average piece, which waits somewhere between production and consumption. Consider the first piece produced at the factory after a shipment. It will wait in the outbound inventory for a time interval of length T_c; the last piece produced before the next shipment will not wait at all, because as soon it is available, the vehicle is ready for shipment. Hence, the average waiting time at the factory is $T_c/2$. The same occurs at the retail store, where inventory behaves just like in the EOQ model. The first unit sold right after the lot Q is received spends zero seconds in the retail stores, whereas the last unit of the lot Q spends T_c units of time in the store. The average unit spends $T_c/2$ units of time in the stores. This means that overall the average unit spends T_c units of time in inventories.[4]

We should also consider that there is inventory in transit waiting for a time interval T_s. Hence, the total waiting time is $T_c + T_s$ on the average, for each of the d pieces which are consumed per unit time. Therefore, the total cost per unit time is

$$TC(T_c) = \frac{A}{T_c} + hd(T_c + T_s). \tag{2.4}$$

Minimizing the objective function with respect to T_c, we get

$$T_c^* = \sqrt{\frac{A}{hd}}, \tag{2.5}$$

and the shipped quantity should be

$$Q^* = dT_c^* = \sqrt{\frac{dA}{h}}. \tag{2.6}$$

[4]Notice that here we implicitly made an assumption on the inventory build-up at the production plants. Indeed, the logic behind figure 2.3 is that the inventories for each and every store accumulate progressively. In other words, at any point in time each plant devotes a fraction of the capacity to each store (the fraction is proportional to the demand for the item at the store) and the inventories for the 5 stores build up in parallel. Figure 2.3 shows that there is a continuous production of inventories for the specific store at stake. However, there is a second, more efficient policy. One could allocate the production capacity in a slightly different way. We could produce at full speed for one store, prepare the distribution lot for the store, and then switch to the next store. In this case the inventory build-up for a given store is all but constant. The maximum inventory level at the production site is still Q but this quantity builds up over a shorter period of time, just before the goods are shipped. In this case, inventories at the central warehouse (and going to a specific store) are displayed in figure 2.4.

When we compare figure 2.4 with figure 2.3 we can see that the inventory level at the production site decreases because inventories are kept at zero for a fairly long period of time (on the average 80% of the cycle time T_c). So in our discussion we are cutting some corners; that is, we assume the company has a rather ineffective policy at the central warehouse.

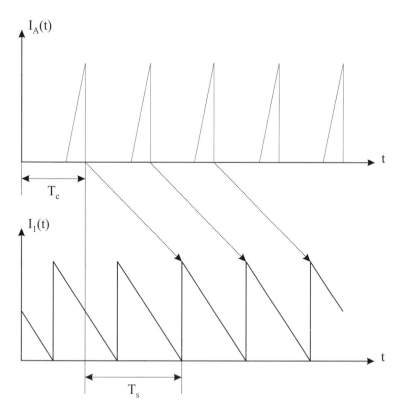

Fig. 2.4 Inventory levels at the factory assuming uniform production (and sequential allocation of capacity to stores); see footnote 4.

This is essentially the same as the EOQ formula, but for a missing 2 factor; this factor is missing because we are considering inventory holding at two locations rather than one. Also, the total cost for the optimal solution is similar to the EOQ case:

$$TC(T_c^*) = 2\sqrt{Ahd} + hdT_s. \tag{2.7}$$

As expected, the transportation lag T_s does not influence the optimal solution. Indeed, no matter what our policy is, all units spend T_s time on a truck (more generally, on some kind of transportation means). Nevertheless, T_s may be important when comparing different means of transportation with significantly different transportation lead times (e.g., ships vs. air freight). The same holds for variable transportation costs. In passing, we should also note that we are writing an objective function, while disregarding potentially thorny issues we face when *two* (or more) firms are involved; if we optimize the overall costs, who should reap the benefits? Actually, we do not need two separate firms to face this kind of difficulty; even two organizations within

Table 2.1   Coordinates of the facilities of a sample network

	A	B	D	1	2	3	4	5
x	100	100	300	200	400	600	300	700
y	300	700	400	400	200	100	300	300

the same firm may face conflicting incentives. We will get back to these issues in chapter 7 but, for the time being, let us assume that we live in a very idealized world where everyone is willing to improve the general welfare. Now we have analyzed a single link, and we should wonder how we can exploit this knowledge to compare the network with direct shipment against the network with the intermediate distribution center. In order to get a rough cut evaluation, we assume that the overall network cost can be estimated by summing the cost expression (2.7) over all of the links. This is not really correct, as we should pay close attention to the synchronization of transportation from a factory to the distribution center and from the distribution center to retail stores. This synchronization has an effect on the inventory at the distribution center, but we defer such issues to chapter 6.

Let us now tackle a small numerical case. We introduce subscript $i = A, B$ to denote production plants and the related items, and $j = 1, 2, 3, 4, 5$ to denote the retail stores; the distribution center is denoted by D. The coordinates of each location are given in table 2.1, with respect to some arbitrary point of reference; they can be thought of as miles or kilometers; we use these coordinates to evaluate distances and to quantify transportation costs. Transportation costs in practice depend on several factors, in a possibly complicated way. For our toy example, let us assume that there is a fixed transportation charge depending on distance traveled, and not on volume or weight; we have a cost q per unit distance (mile or kilometer), and the transportation cost is obtained by multiplying this factor times distance traveled. In a flat region, we may assume that the real road distance between two points is not too different from the Euclidean distance. This is actually an underestimate of the real distance, and we should take natural obstacles, such as mountains and lakes, into account; in practice, distances may be obtained by querying geographic information systems. To summarize, we may use the coordinates given in table 2.1 to compute fixed transportation charges A_{ij} between the plants and the retail stores, A_{iD} between the plants and the distribution center, and A_{Dj} between the center and the retail stores. Using plain Euclidean distance yields

$$A_{ij} = q\sqrt{(x_i - x_j)^2 + (y_i - y_j)^2}.$$

The remaining data we need to carry out the rough-cut analysis are, for each item i, its cost c_i and its demand d_{ij} per unit time at each retail store j; these data are given in table 2.3. The product cost c_i, multiplied by an interest rate

Table 2.2   Transportation costs for the network of figure 2.2

	D	1	2	3	4	5
A	223.61	141.42	316.23	538.52	200.00	600.00
B	360.56	316.23	583.10	781.02	447.21	721.11
D	—	100.00	223.61	424.26	100.00	412.31

Table 2.3   Product cost and demand per unit time at each retail store

Item	Cost	d_{i1}	d_{i2}	d_{i3}	d_{i4}	d_{i5}
A	100	1000	500	2000	300	2000
B	200	500	230	1300	120	1200

Table 2.4   Times between shipments and total cost per unit time on each link

	1	2	3	4	5
A	0.0752	0.1591	0.1038	0.1633	0.1095
B	0.1125	0.2252	0.1096	0.2730	0.1096
A	3760.60	3976.35	10378.02	2449.49	10954.45
B	5623.41	5179.03	14250.14	3276.15	13155.47

r, yields the unit inventory holding cost h_i for each item. If distances are short enough, the in-transit inventory holding cost is negligible. Let us carry out the calculations under the hypothesis that $q = 1$ and $r = 25\%$. The resulting costs are reported in table 2.2. Applying equations (2.5) and (2.7), for each link (i, j) in the network, we obtain the optimal periods between shipments and the total costs per unit time reported in table 2.4. Summing over all of the ten links, we get an estimated total cost per unit time of 73003.12.

Let us now analyze the network with the distribution center D. The problem is much more complex here, because the time evolution of the inventory levels, depicted in figure 2.3, need not apply to the new situation. To begin with, the inventory manager at the distribution center will see an aggregate demand for each item; this demand also depends, in a possibly intricate way, on the transportation pattern from distribution center to retail stores. Furthermore, we have the additional issue of the synchronization between inbound and outbound transportation from the distribution center, in terms of both timing and quantity. In practice, cross-docking transit points are operated in such a way that no inventory is held there. To keep the toy example simple, we will not optimize the overall transportation pattern, but we will estimate the total cost by applying equation (2.7) again to each link in the network. Even such a rough-cut estimate can be useful in showing if there is a definite advantage of one solution over the other one. Let us consider first the two links from plants to distribution centers. On these links, the two items travel separately and each plant "sees" an aggregate demand

$$d_i = \sum_j d_{ij}.$$

The time between shipments and the total cost on each link (i, D) are

$$T_{i\mathrm{D}}^* = \sqrt{\frac{A_{i\mathrm{D}}}{d_i h_i}}, \qquad TC_{i\mathrm{D}}^* = 2\sqrt{A_{i\mathrm{D}} d_i h_i}.$$

Carrying out the calculations, we get

$$T_{\mathrm{AD}}^* = 0.0393, \quad T_{\mathrm{BD}}^* = 0.0464, \quad TC_{\mathrm{AD}}^* = 11388.24, \quad TC_{\mathrm{BD}}^* = 15542.58,$$

yielding a total cost of 26930.82. Now we should add the cost related to the five links from the distribution center to the retail stores. Here we have the additional complication that items are shipped together. We may aggregate the two items together into a "virtual" product, which is the *bundle* of the two items.[5] A bundle is a virtual product that consists of a combination of items. Think of the bundle as a package containing some units of item A and some units of item B.[6] But how can we define such a bundle? Under deterministic conditions the mix of demand is fixed and so the mix of supply shall be fixed as well. So we can define these bundles (i.e., composite sets of products) and plan them rather than the finished products.

But how do we define the bundle? What is the demand for the bundle? On a closer look, we may see that the units we use to express the bundle are irrelevant; we have just to be consistent. For example, if at a retail store the demand per unit time is 300 for item A and 100 for item B, we may consider a bundle consisting of 300 pieces of A and 100 pieces of B, or a bundle consisting of 3 pieces of A and 1 piece of B, or finally a bundle consisting of 0.75 pieces of A and 0.25 pieces of B (as well as any other combination with a 3 to 1 ratio for products A and B). In the first case, demand for the bundle will be 1 unit per period, it will be 100 units per period in the second case, while it will be 400 units per period in the third case. Notice that the definition of the bundle implies a given level of demand and vice versa a level of demand for the bundle implies a definition of the bundle. Say we want to set the demand for the bundle to 10 units (per period). This means that we shall define the bundle in such a way that it consists of 30 units of A and 10 units of B. The inventory holding cost will be adjusted accordingly.

Once, the bundle is defined, what is the holding cost for the bundle? In the first case, holding one unit of the bundle means holding 300 units of A and 100

---

[5]We will use bundles again in section 4.6.1.
[6]Notice that we need not really assemble such packages. We simply refer to these bundles, sets, or packages as the unit we plan for.

units of B, so the holding cost for the bundle is $\tilde{h} = 300 \cdot h_A + 100 \cdot h_B$. In the latter case holding one unit of the bundle means holding 0.75 units of A and 0.25 units of B, so the holding cost for the bundle is $\tilde{h} = 0.75 \cdot h_A + 0.25 \cdot h_B$.

Now let us try to be more general. Say that we define the bundle in such a way that the bundle demand (per unit time) at retail store j is

$$\tilde{d}_j = \alpha,$$

where α is an arbitrary (positive) number. Then the number of units of item i contained in this bundle is

$$N_{ij} = \frac{1}{\alpha} \frac{d_{ij}}{\sum_k d_{kj}},$$

and the inventory holding cost for the bundle is

$$\tilde{h}_j = \sum_i h_i N_{ij} = \sum_i h_i \frac{1}{\alpha} \frac{d_{ij}}{\sum_k d_{kj}}.$$

Now we may see that α is actually irrelevant, because in the formulas giving the optimal period and the total cost for each link from the distribution center D to retail store j,

$$T^*_{Dj} = \sqrt{\frac{A_{Dl}}{\tilde{d}_j \tilde{h}_j}} =, \qquad TC^*_{Dj} = 2\sqrt{A_{Dl}\tilde{d}_j \tilde{h}_j},$$

the value of α gets canceled when multiplying $\tilde{d}_j$ and $\tilde{h}_j$. This is actually a rather obvious finding as the solution cannot possibly depend on an arbitrary parameter.

In order to carry out the calculations, we may assume

$$\tilde{d}_j = \sum_i d_{ij}.$$

In other words we assume that the number of bundle is equal to the total number of units over all products. But we must be careful and realize that this does *not* imply that we are summing demand for different items, which makes no sense. Each unit of the bundle consists of a percentage w_{ij} of each item i where

$$w_{ij} = \sum_i \frac{d_{ij}}{\sum_k d_{kj}}.$$

In the example above, we are simply assuming a demand for the bundle of 400 units per period and a bundle consisting of 0.75 units of A and 0.25 units of B. With this choice of bundle demand, the inventory holding cost for bundle j is a weighted combination of holding costs for single items

$$\tilde{h}_j = \sum_i h_i w_{ij},$$

Table 2.5 Demand, weights of single items in the bundle, inventory holding cost, optimal ordering period, and total cost per unit time for each bundle j and the corresponding link (D, j)

$store(j)$	1	2	3	4	5
$\tilde{d}_j$	1500	730	3300	420	3200
w_{Aj}	0.6667	0.6849	0.6061	0.7143	0.6250
w_{Bj}	0.3333	0.3151	0.3939	0.2857	0.3750
$\tilde{h}_j$	33.3333	32.8767	34.8485	32.1429	34.3750
T^*_{Dj}	0.0447	0.0965	0.0607	0.0861	0.0612
TC^*_{Dj}	4472.14	4633.17	13970.02	2323.79	13469.10

The resulting calculations are displayed in table 2.5. We see that the total cost of the second set of links is 38868.21; adding the cost of the first set of links (26930.82) we get an overall cost of 65799.04, to be compared against the cost of the solution with direct shipments (73003.12). According to the model, the percentage saving from the introduction of the distribution center is 9.87%.

We stress again that we have considered a toy model with a lot of debatable approximations, but the results suggest that the introduction of a distribution center may have some merits. The saving we have estimated might not be enough to justify the introduction of the distribution center, because we have not considered the cost of building and running it. To get a better idea of the effect of an intermediate distribution center, we may apply the same modeling approach to a network with more retail stores. Using data similar to those we have just used, generating retail store coordinates on a square where both edges are 1000 space units long, and placing the distribution center in the barycentre of retail stores, we get the following results: With 100 retail stores, the percentage saving is about 30%; if we add a third plant with a third product, the saving is 40%. The important message to get from this example is *why* a distribution center may reduce transportation and holding costs: The distribution center may improve the frequency of transportation from production to the retail stores. With a point to point transportation, in order to exploit economies of scale optimally, we may be forced to transport large amounts of goods, possibly exceeding warehousing capacities. If we mix different products at a distribution center, where we manage assortment, we are able to adapt transportation patterns on the two sets of links.

Concept 2.3 *A distribution center can consolidate flows of various goods, so that these share some fixed ordering and transportation costs. Thus, each single product is delivered in smaller quantities and more frequently. In other* *words the distribution center can create joint economies of scale since fixed costs are spread over a variety of products.*

Example 2.1 A somewhat paradoxical consequence of introducing a transit point is that goods travel longer distances than with direct shipment. This is particularly striking in the case of next-day shipments by logistic operators such as DHL or FedEx. Even considering a relatively limited region, such as one country, organizing a direct shipment network is out of the question. We have to introduce a hub to ensure economies of scale. FedEx, in its early days, had just one in Memphis: Hence, a parcel shipped from Oregon to California had to travel a long way.[7] In a large country, such as the USA, having multiple hubs may make sense; in a smaller country, such as Italy, this would be hardly justified. ☐

It is now important to close the section by listing all of the limitations of our exercise.

- We have considered one distribution center, in a given position. We have not considered the optimal location of single or multiple distribution centers, nor the problem of allocating retail stores to different distribution centers. We need optimization models to accomplish these tasks.

- In practice, we often see hybrid strategies. Retail stores with high sale volumes may justify direct shipments, whereas others need an intermediate point to exploit economies of scale in transportation.

- We have adopted a simplistic model of transportation costs and we have not accounted for vehicle capacities. Both may depend on the specific transportation link. It may also happen that different transportation means are used: In intermodal centers, we may have inbound transportation by rail or sea and have outbound transportation by trucks.

- The transportation period we get from the model above may not be practical, because it may take any value. From an organizational point of view, one might prefer a more meaningful and regular pattern. Imagine the difficulty in arranging a shipment every 3.57 days.

- We did not consider the costs of holding in-transit inventory. These costs may penalize the increased distance traveled when intermediate centers are used. We travel longer distances and goods spend more time traveling; thus the inventory holding costs increase. Actually, when

---

[7]See R.O. Mason, J.L. McKenney, W. Carlson, and D. Copeland. *Absolutely, Positively Operations Research: The Federal Express Story. Interfaces,* 27:17-36, 1997.

using transportation by truck, the traveling time is rather short and in-transit inventory cost may be neglected. However, when transporting perishable goods, time may be a very important issue.

- A similar consideration applies to variable transportation costs. If we approximate transportation costs by one fixed component and one variable component that varies linearly with the quantity transported, the latter does not affect the optimal transportation frequency. But if we are comparing different transportation patterns, variable costs may be relevant (think of different routes, one going through the distribution center and another one going straight into the stores)

- We have taken for granted that the distribution center had its own ware-house (without accurately modeling its dynamics). In fact, by carefully synchronizing shipments to and from a distribution center, we may avoid any inventory holding there; in this case, the distribution center works as a pure cross-docking point (i.e., a transit point). This means that the transit point does not reduce safety stocks by risk pooling; however, this does not imply that such a transit point has no role in dealing with uncertainty. As we will see in chapter 6, when transportation times are relevant, a transit point may help us in delaying the allocation of goods to specific final destinations; delaying the commitment of goods helps in reducing the impact of demand uncertainty. Another relevant consideration is that in the case of pure cross-docking, the facility is smaller and cheaper to build and manage (and we may also deal with highly perishable goods efficiently).

- Finally, we have neglected issues related to the increased material handling due to intermediate centers. Additional unloading/loading activities have a cost, and they may also increase the loss of material because of accidental damage.

All of these considerations illustrate adequately the extreme complexity of network design, which calls for the development of suitable models to take relatively strategic decisions. These models have to be reasonably simplified, yet we must anticipate the effect of strategic decisions on the costs associated to tactical and operational management. All we can hope to do is to approximate these costs; hence, optimization models have the role of generating a restricted set of candidate solutions, which must be fully evaluated by an accurate simulation model. It should be emphasized that a simulation model *per se* need not be an effective or efficient way to generate solutions. This is why in this chapter we deal with optimization models. We should also mention that heuristic solution approaches have been developed over the years. We refer the reader to [3] and to our web supplements.

2.2 LOCATION AND FLOW OPTIMIZATION MODELS

In this section with deal with a subclass of linear programming models, possibly mixed-integer ones, linked to classical network optimization problems; they include plant location and optimal flow models. Plant location is relevant at a strategic level, whereas flow optimization is of a more tactical or operational nature, but it occurs as a subcomponent of strategic models. Indeed, when locating plants we should account for the impact of the logistic infrastructure on the transportation cost. Hence, in illustrating models, we will not move from strategic down to tactical levels; rather, we will present models in increasing order of complexity. The simplest model is the classical transportation problem, and we will proceed to more realistic modeling frameworks including demand uncertainty and nonlinear costs. Our aim is not to propose a general, all-encompassing model, since there is no such thing. We want to present building blocks that can be assembled when needed; the focus is on modeling frameworks, and *not* on solution methods. We take for granted that a good commercial solver is available, to solve the models by standard methods such as simplex or branch and bound algorithms (see appendix B). This need not be the case, as some really large-scale models may require specific solution methods; however, we feel that algorithmic finesse is beyond the scope of the book; furthermore, the astonishing progress in optimization software is pushing the intractability frontier further and further.

All of the models below are based on the mathematical concept of a network, which is essentially a graph with some additional information. A **graph** consists of two components: nodes and arcs (see, e.g., figure 2.5). **Nodes** are, in our case, facilities along a supply chain; **arcs**, connecting nodes, represent the flow of goods along a certain transportation link. Formally, an arc is just a pair of nodes. Not every pair of nodes is directly connected by an arc; to reach a destination node starting from a source node, we may need to traverse several arcs, modeling a sequence of transportation activities. Arcs may be directed (oriented) or not; formally this depends on the type of node pair, which can be ordered or not. The arc orientation is associated with the sense of an arrow, and it represents the direction of goods flowing along the arc. A directed graph consists of nodes and oriented arcs. In this chapter, goods flow along a specific direction; hence, we will deal with directed graphs. We will meet undirected graphs in chapter 8 on vehicle routing, but we may also think of an undirected arc as a pair of directed ones.[8] We have a **network** when elements of the graph are associated with additional information related to costs or constraints. For instance, we may give the maximum amount of flow that can go through a node per unit time, i.e., the material handling

---

[8]Strictly speaking, in an undirected graph we should speak of vertices, rather than nodes, and edges, rather than arcs; we will use terms rather loosely.

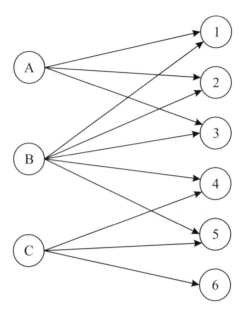

Fig. 2.5   A graph corresponding to a transportation problem.

capacity of a transit point, or the cost associated to an arc, representing unit transportation costs.

2.2.1   The transportation problem

The classical transportation problem is actually a very simplified view of a real-life transportation problem. It is a linear programming model dealing with a two-level network, on which a single type of good flows. We have two disjoint sets of nodes: the set $\mathcal{S}$ of source nodes and the set $\mathcal{D}$ of destination nodes. Referring to figure 2.5, we have $\mathcal{S} = \{A, B, C\}$ and $\mathcal{D} = \{1, 2, 3, 4, 5, 6\}$. Examples of (directed) arcs are $(A, 1)$ and $(C, 4)$; there is no $(B, 6)$ arc. Also, when there is no arc connecting two sources or two destinations; we say that the graph is bipartite. Destination nodes represent retail stores, which are characterized by a demand d_j $(j \in \mathcal{D})$, which may be given per unit time or over a time span of interest. Source nodes might represent production plants, with a given limited capacity R_i $(i \in \mathcal{S})$, measured over the same time span as demand. For each source–destination pair, i.e., for each arc, we have a unit transportation cost c_{ij}. This is considered as a variable linear cost; clearly, this is just a very rough-cut approximation of a real-life transportation cost. The problem consists of finding the minimum-cost set of flows, over all of the links (i, j), such that demand is met and plant capacities are not exceeded.

To represent the transportation problem as a mathematical programming model, we have to find suitable decision variables first. In this case, it is fairly evident that we need one decision variable for each link (i, j); let x_{ij} be the flow on each arc. We will assume that the only requirement on flows is that they be non-negative; we do not require that they be integers, which makes sense if we consider large flows, so that possible issues with a continuous approximation are of no concern.[9] The resulting linear programming model is

$$\min \quad \sum_{i \in \mathcal{S}} \sum_{j \in \mathcal{D}} c_{ij} x_{ij}, \tag{2.8}$$

$$\text{s.t.} \quad \sum_{i \in \mathcal{S}} x_{ij} = d_j \qquad \forall j \in \mathcal{D}, \tag{2.9}$$

$$\sum_{j \in \mathcal{D}} x_{ij} \leq R_i \qquad \forall i \in \mathcal{S}, \tag{2.10}$$

$$x_{ij} \geq 0.$$

The objective function (2.8) is a sum over all the pairs of nodes, and it amounts to the total transportation cost. The expression above assumes that there is an arc for any source–destination pair, which need not be the case. We could think of associating a suitably high cost c_{ij} to nonexistent arcs, so that they are never used. A possibly more elegant solution is to represent explicitly the arcs in the network by a set $\mathcal{N}$, and writing the double sum as $\sum_{(i,j) \in \mathcal{N}} c_{ij} x_{ij}$. The constraint (2.9) makes sure that demand is met at each destination node, by summing inflows from plants. The capacity constraint, limiting outflows from any source, is represented by (2.10). Notice the reversal of roles between subscripts i and j in constraints (2.9) and (2.10).

This model is extremely simplistic and it just provides us with a starting point for further modeling. To begin with, it is a static model ignoring time patterns in demand (demand variability). In principle, it is easy to extend the model to a multiperiod one; we need to introduce a time-varying demand d_{jt} and inventory variables at nodes, along the lines of section B.1. By the same token, we could consider diversified production costs across the plants, different items or families, and a more realistic transportation cost structure, possibly including fixed charges and economies of scale. Pursuing this line, we may come up with an integrated production–distribution model. Obviously, the computational requirements would grow, but this need not be the main difficulty of such an exercise; additional critical points are the following:

---

[9]In fact, if all problem data in the transportation problem are integers, it can be shown that there is no need to use branch and bound methods to get an integer solution. The simplex algorithm will always yield an integer solution because of the structure of the problem. Actually, this structure is so peculiar that we may use more specialized and efficient algorithms. We refer the reader to the available literature on model solving, which is not too relevant for our purposes. Anyway, good commercial solvers are able to spot network structures in linear programming models and to exploit them properly.

1. Any model taking dynamic demand patterns into account may be flawed if demand is highly uncertain. The longer the time horizon, the higher demand uncertainty. We could build scenario-based optimization models where demand uncertainty is suitably modeled. We have seen how this can be accomplished in the two-stage case in section 1.5.2, but the solution effort grows very quickly for the multistage case. Nevertheless, scenario-based models have been proposed for strategic level decisions; we will illustrate an example in section 2.2.3.

2. Another fundamental limitation is that transportation costs are assumed linear, which rules out economies of scale. In a strategic or tactical model, it may be difficult to account too accurately for transportation costs that also depend on operational decisions; however, we may approximate such costs by a nonlinear function, which in turn can be approximated by a piecewise linear function, as we illustrate in section 2.3, with a corresponding increase in the computational effort.

2.2.2   The minimum cost flow problem

In the classical transportation problem we have a two-layer network. If we generalize the network to an arbitrary structure, we obtain the minimum cost flow problem. In the classical version of the problem, we have one source node, which has to send a given flow, and a destination node, where the flow must ultimately be routed; our task is to find a minimum cost transportation plan. Since source and destination nodes need not be connected directly, we must use intermediate transshipment points. A complicating factor is represented by arc capacities, which limit the amount of flow we may transport on each link. The optimal flow might be split over multiple routes. This may actually be the case in telecommunication networks, but it is uncommon in distribution logistics. With respect to the basic network flow problem, we have other complicating factors, such as multiple destinations and sources, different commodities, nodes' capacities expressing limitation of transshipment nodes, etc. For the sake of simplicity, we will consider a three-layer network, of the type we have already seen in figure 1.3. Actually, arbitrary flow structures can be modeled. As in the transportation model, we denote source and destination nodes by $\mathcal{S}$ and $\mathcal{D}$, respectively, and denote by $\mathcal{C}$ the set of intermediate transshipment nodes. We consider multiple items, whose set is denoted by $\mathcal{L}$. Production plants, i.e., flow sources, need not be able to produce the whole spectrum of products; furthermore, capacities and production costs may vary across source nodes. We will also consider transportation capacities, e.g., linked to either volume or weight.

Given the increased complexity of this model, we proceed step by step. As we have already pointed out, a good starting point is figuring out which decision variables we need. In network flow problems, we always need to represent the amount of goods shipped along each link, with reference to some unit of time. Let us denote by x_{ikl} the amount of product type $l \in \mathcal{L}$, produced on plant $i \in \mathcal{S}$ and shipped to distribution center $k \in \mathcal{C}$; similarly, we denote by y_{kjl} the amount of product l shipped from distribution center k to destination node $j \in \mathcal{D}$. To be precise, we should define variables x_{ikl} only for items l which can be produced in plant i; by the same token, we should only define both variables for arcs which are included in the network. To ease the notation, we refrain from doing so, but we could easily add the required subsets.

Now we should write the constraints that must be enforced on flows. We start from downstream nodes and then move upstream. To begin with, we want to meet the demand for any destination node:

$$\sum_{k \in \mathcal{C}} y_{kjl} = d_{jl} \qquad \forall l \in \mathcal{I}, \, \forall j \in \mathcal{D},$$

where d_{jl} is the demand for item l at destination j.

We also have limited transportation capacities on each link, in terms of volume or weight. We have to aggregate different items according to these dimensions. For the sake of simplicity, we consider only volume, and we let v_l be the unit volume of item l and let V_{ik} and W_{kj} be the maximum volumes per unit of time which can be shipped on arcs (i, k) and (k, j), respectively. Then we may express capacity constraints on each arc:

$$\sum_{l \in \mathcal{I}} v_l x_{ikl} \leq V_{ik} \qquad \forall i \in \mathcal{S}, \forall k \in \mathcal{C},$$

$$\sum_{l \in \mathcal{I}} v_l y_{kjl} \leq W_{kj} \qquad \forall k \in \mathcal{C}, \forall j \in \mathcal{D}.$$

If some links are not really available, we may think of setting their capacities to zero. We also associate unit transportation costs c_{ik} and g_{kj}, which are related to the volume shipped on each link.

Then we also have to formalize constraints on activities at transshipment nodes. A typical constraint we need for multihop flow routing is the conservation of flows. The amount of goods flowing into a node must be equal to the amount flowing out of that node:

$$\sum_{i \in \mathcal{S}} x_{ikl} = \sum_{j \in \mathcal{D}} y_{kjl} \qquad \forall k \in \mathcal{C}, \forall l \in \mathcal{I}.$$

We should also consider the material handling capacity of distribution centers. If H_k the maximum amount of volume that can be handled at node k per unit time, we have to enforce the inequality

$$\sum_{i \in \mathcal{S}} \sum_{l \in \mathcal{I}} v_l x_{ikl} \leq H_k \qquad \forall k \in \mathcal{C}.$$

Finally, we must express capacity constraints on each production plant. To this aim, we need resource requirements for each item. There may be several relevant resources, but if there is a clear bottleneck, e.g., labor, we may collect the unit resource requirements r_{il} to produce one unit of item l on plant i, along with the resource availability R_i on that plant. Note that, because of possibly different technologies, we allow for plant-dependent resource requirements; for the same reason, we allow for different unit production costs p_{il}. The following production capacity constraint must be written for each plant:

$$\sum_{k \in \mathcal{C}} \sum_{l \in \mathcal{I}} r_{il} x_{ikl} \leq R_i \qquad \forall i \in \mathcal{S}.$$

Note that we are summing over both item types and distribution centers as we consider the overall workload of the plant, regardless of what the item is and where it is shipped to. In fact, we have not used production variables, because they are directly disaggregated into shipments to distribution centers. In a multiperiod model, where inventories are introduced, we would need to make production variables explicit.

Wrapping it up, and including the objective of minimizing the total cost, production plus transportation both to the warehouses and the stores, we obtain the following linear programming model:

$$\min \quad \sum_{l \in \mathcal{I}} \sum_{i \in \mathcal{S}} \sum_{k \in \mathcal{C}} (p_{il} + c_{ik} v_l) x_{ikl} + \sum_{l \in \mathcal{I}} \sum_{k \in \mathcal{C}} \sum_{j \in \mathcal{D}} g_{kj} v_l y_{kjl},$$

$$\text{s.t.} \quad \sum_{k \in \mathcal{C}} \sum_{l \in \mathcal{I}} r_{il} x_{ikl} \leq R_i \qquad \forall i \in \mathcal{S},$$

$$\sum_{l \in \mathcal{I}} v_l x_{ikl} \leq V_{ik} \qquad \forall i \in \mathcal{S}, \forall k \in \mathcal{C},$$

$$\sum_{l \in \mathcal{I}} v_l y_{kjl} \leq W_{kj} \qquad \forall k \in \mathcal{C}, \forall j \in \mathcal{D},$$

$$\sum_{i \in \mathcal{S}} x_{ikl} = \sum_{j \in \mathcal{D}} y_{kjl} \qquad \forall k \in \mathcal{C}, \forall l \in \mathcal{I},$$

$$\sum_{i \in \mathcal{S}} \sum_{l \in \mathcal{I}} v_l x_{ikl} \leq H_k \qquad \forall k \in \mathcal{C},$$

$$\sum_{k \in \mathcal{C}} y_{kjl} = d_{jl} \qquad \forall l \in \mathcal{I}, \forall j \in \mathcal{D},$$

$$x_{ikl}, y_{kjl} \geq 0.$$

For the sake of brevity, we will not repeat the cautionary remarks we made for the basic transportation problem.

2.2.3   The plant location problem

In the two models above we have taken the network structure as given. Hence, the decisions we had to make were tactical or operational, and just linked to flow routing. However, at a more strategic level, we have to make decisions concerning:

- the location (or relocation) of production plants;

- the sizing (or the expansion) of production capacities;

- the capacity and location planning for distribution centers;

- the allocation of retail stores to distribution centers.

As far as the last point is concerned, we may consider a purely exogenous demand, which we have to satisfy, say, at minimum cost. However, there are problems, such as the choice of the location for retail stores, in which the demand is a *result* of our decisions.

What we describe here is a straightforward extension of the transportation problem, whereby source nodes are just *potential* locations of plants. We should decide where (in the set of predefined options) a plant must be opened, taking into account the related costs. Such decisions (and the related variables) are logical (i.e., binary) in nature: Either we open a plant, or we do not. This is a typical setting in which binary decision variables are used:

$$y_i = \begin{cases} 1 & \text{if source node } i \text{ is opened,} \\ 0 & \text{otherwise.} \end{cases}$$

When opening a plant, the related costs include a fixed component, linked to the binary decision variables y_i. Finding a good solution calls for trading off the cost of opening a plant against transportation costs. Even if we consider only a fixed charge for opening a plant, we must be careful in making it comparable with transportation costs (basically we turn a one-time-only cost into a kind of per-unit-of-time fee, say a monthly fee). If demand is given per unit time, and we measure transportation cost on the same basis, we must somehow amortize opening costs to make all of them comparable. If we do this, we end up with a fixed charge for operating plant i, which we denote by f_i. The classical plant location model, where one item type is considered, has the following form:

$$\min \quad \sum_{i \in \mathcal{S}} f_i y_i + \sum_{i \in \mathcal{S}} \sum_{j \in \mathcal{D}} c_{ij} x_{ij}, \tag{2.11}$$

$$\text{s.t.} \quad \sum_{i \in \mathcal{S}} x_{ij} = d_j \qquad \forall j \in \mathcal{D},$$

$$\sum_{j \in \mathcal{D}} x_{ij} \leq R_i y_i \qquad \forall i \in \mathcal{S}, \tag{2.12}$$

$$x_{ij} \geq 0, \ y_i \in \{0, 1\}.$$

Comparing this model against the transportation problem, we see two basic differences:

1. There is an additional term in the objective function (2.11), which is typical of models including fixed charges.

2. The capacity constraint (2.12) does not include a given capacity, but a capacity depending on our strategic decisions. If a plant is not opened ($y_i = 0$), there can be no flow going out of the corresponding node. This way of linking continuous decision variables, in our case the flows, to binary variables is quite common.[10]

Since the model includes binary decision variables, it must be solved by mixed-integer programming methods such as branch and bound. Leaving solution issues aside, it is important to realize that the main difference between the two sets of decision variables is not due to integrality requirements. One set of variables is related to strategic decisions, which are not easy to change on a short time scale. Another set of variables is related to tactical decisions: Transportation decisions, should the demand pattern change, can be adapted on a short notice, subject to plant capacity constraints. The role of the flow variables x_{ij} is to "anticipate" in a strategic model the effects of tactical decisions which will be made later; in the model below, they define an anticipation function in the form of a linear transportation cost.[11] .

This difference in the nature of decision variables gets clearer if we extend the model to account for demand uncertainty. To do so, we may exploit the same concepts we introduced in section 1.5.2, where we illustrated a two-stage stochastic programming model for optimization under uncertainty. Like we did there, we represent demand uncertainty by a set of scenarios, indexed by s and associated with a probability π^s. Let d_j^s be the demand at retail store j under scenario s; for the moment, let us assume that demand must be satisfied anyway. The decision of opening a plant is a first-stage decision, which must be taken here and now; production and transportation decisions will be taken later, once demand is known. Hence, we have a set of second-stage decision variables x_{ij}^s, which are contingent on the realization of scenario s. The minimization of the total plant cost plus the expected transportation cost is obtained by solving the following model:

$$\min \quad \sum_{i \in \mathcal{S}} f_i y_i + \sum_s \pi^s \left(\sum_{i \in \mathcal{S}} \sum_{j \in \mathcal{D}} c_{ij} x_{ij}^s \right),$$

$$\text{s.t.} \quad \sum_{i \in \mathcal{S}} x_{ij}^s = d_j^s \qquad \forall s, \ \forall j \in \mathcal{D},$$

[10] See also the lot-sizing model (B.15) on page 573.
[11] An anticipation function actually anticipates the effects of the decisions at stake in the model on future performance. In this case, the function anticipates the effects of location decisions on future transportation costs. We could say that the model encompasses *design* variables y_i and *control* variables x_{ij} that are used to capture the effect of design variables on future costs. Also, arguably a more accurate anticipation function should be nonlinear.

$$\sum_{j \in \mathcal{D}} x_{ij}^s \leq R_i y_i \qquad \forall s, \ \forall i \in \mathcal{S},$$

$$x_{ij}^s \geq 0, \ y_i \in \{0, 1\}.$$

We see that capacity constraints link the first-stage variables y_i with the second-stage variables x_{ij}^s, which must adapt to contingent demand d_j^s, subject to available capacity. So, first stage decision variables generate the capacity that we then use in the second stage to meet demand. Solving this model could yield a very costly solution, if extreme but unlikely high-demand scenarios are included, since we would be forced to buy a lot of capacity just in case that odd scenario comes true. Hence, we could also consider a more "elastic" formulation allowing for the possibility of leaving some demand unsatisfied (at least in some high-demand scenarios). Let $z_j^s \geq 0$ be the amount of unmet demand at node j under scenario s; these decision variables are included in the objective function multiplied by a penalty coefficient β_j, yielding the elastic model formulation:

$$\min \quad \sum_{i \in \mathcal{S}} f_i y_i + \sum_s \pi^s \left(\sum_{i \in \mathcal{S}} \sum_{j \in \mathcal{D}} c_{ij} x_{ij}^s \right) + \sum_s \pi^s \left(\sum_{j \in \mathcal{D}} \beta_j z_j^s \right),$$

$$\text{s.t.} \quad \sum_{i \in \mathcal{S}} x_{ij}^s + z_j^s = d_j^s \qquad \forall s, \ \forall j \in \mathcal{D},$$

$$\sum_{j \in \mathcal{D}} x_{ij}^s \leq R_i y_i \qquad \forall s, \ \forall i \in \mathcal{S},$$

$$x_{ij}^s, z_j^s \geq 0, \ y_i \in \{0, 1\}.$$

The penalty coefficient β_j could be quantified by taking the relative importance of different markets into account; alternatively, it could be related to the cost of meeting demand by resorting to external suppliers.

It is important to really understand the meaning of the model above. The second-stage cost term is just an anticipation function: Transportation plans will be determined by possibly complex strategies, and in a real setting we could have inventories at destination nodes. The meaning of the model above is the minimization of the long-run average cost, assuming that similar demands are observed over multiple periods (in modeling terms, independent experiments are repeated taking independent and identically distributed demand samples). If we anticipate possible trends in demand and we foresee significant changes on top of random fluctuations (say we expect demand to increase), we need to build a multiperiod model that accounts for demand variability, with considerable complications. An advantage of a multistage formulation would be the ability of including the redesign of the network. In real life, we typically have to redesign the network by closing facilities, building brand new ones, or expanding the capacity of existing facilities. The next model shows, in a deterministic setting, how this could be accomplished.

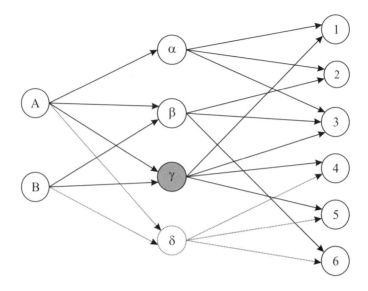

Fig. 2.6 Capacity expansion and relocation in a distribution network.

Concept 2.4 *When we design a model to set the infrastructure of a supply chain and locate warehouses, we shall anticipate the effects of such structural decisions on the ongoing operational performance. In other words, when we make strategic decisions we shall anticipate their effect on operational performance.*

2.2.4   Putting it all together

In this section we discuss a model which summarizes modeling elements we have introduced before.[12] For the sake of simplicity, we deal with one item type and disregard uncertainty; yet, the model is a good example of how we can improve an existing network to adapt it to changing demand patterns. The network illustrated in figure 2.6 consists of two production plants, *A* and *B*, of given capacity, which can manufacture a product which must be ultimately be shipped to six final destinations. The product is shipped through three distribution centers, *α*, *β*, and *γ*, with some current capacity level. The problem calls for the redesign of the network, pursuing the following options:

1. We could expand the capacity of distribution center *γ*, which is drawn as a shaded area to point out this opportunity.

2. We could build a new center *δ*, which is drawn using a dashed line, endowing it with a relatively low capacity.

---

[12]The example described here is a simplified version of a similar model considered in [9].

3. We could build center δ, with a relatively high capacity.

Also, the transportation links going into and out of center δ are dashed to point out that they are potential transportation links. For organizational or budget constraints, if we open the new center δ, we must close (at least) one between centers α and β. In other words, we do not want to use more than three centers. Center γ might be expanded, but we do not consider it a candidate for dismissal. Every decision has a given cost, or a benefit, as is the case of the savings associated with closing centers α or β; all of these quantities are expressed on a per unit of time basis (i.e., are amortized), in order to make them compatible with transportation costs and flows per unit time. The capacity of all transportation links is assumed unbounded; it is node capacity that constrains goods flow.

To write the model, we will use subscripts $i = A, B$ for production plants, $k = \alpha, \beta, \gamma, \delta$ for distribution centers, and $j = 1, 2, \ldots, 6$ for retail stores. The available data, which may be referred to time units when necessary, are:

- demand per unit time d_j at retail stores;

- unit transportation costs c_{ik} and g_{kj}, between the different network layers;

- the current handling capacity T_k for the three active distribution centers $k = \alpha, \beta, \gamma$;

- the possible capacity expansion U_γ for center γ, along with its cost per unit time q_γ;

- the two possible capacity levels, high and low, U_δ^l e U_δ^h, for the potential center δ, along with the related fixed charges q_δ^l and q_δ^h;

- the saving r_k for the potential closure of centers $k = \alpha, \beta$.

- production capacity R_i, per unit time, at the production plants, which we assume have the same technology, hence the same unit production costs;

The decision variables are the material flows per unit time, x_{ik} and y_{kj}, on the two sets of links, along with the logical variables:

$$w_\gamma = \begin{cases} 1 & \text{if the capacity of center } \gamma \text{ is expanded,} \\ 0 & \text{otherwise;} \end{cases}$$

$$z_k = \begin{cases} 1 & \text{if center } k = \alpha, \beta \text{ is kept open,} \\ 0 & \text{otherwise;} \end{cases}$$

$$s_\delta^l = \begin{cases} 1 & \text{if we open center } \delta \text{ with low capacity,} \\ 0 & \text{otherwise;} \end{cases}$$

$$s_\delta^h = \begin{cases} 1 & \text{if we open center } \delta \text{ with high capacity,} \\ 0 & \text{otherwise.} \end{cases}$$

We obtain a model which is a hybrid between the minimum cost flow problem of section 2.2.2 and the plant location model of section 2.2.3:

$$\min \quad q_\gamma w_\gamma + q_\delta^l s_\delta^l + q_\delta^h s_\delta^h - \sum_{k \in \{\alpha, \beta\}} r_k (1 - z_k)$$

$$+ \sum_i \sum_k c_{ik} x_{ik} + \sum_k \sum_j g_{kj} y_{kj} \qquad (2.13)$$

$$\text{s.t.} \quad \sum_k y_{kj} = d_j \qquad \forall j, \qquad (2.14)$$

$$\sum_i x_{ik} = \sum_j y_{kj} \qquad \forall k, \qquad (2.15)$$

$$\sum_i x_{ik} \leq T_k z_k, \qquad k = \alpha, \beta, \qquad (2.16)$$

$$\sum_i x_{i\gamma} \leq T_\gamma + U_\gamma w_\gamma, \qquad (2.17)$$

$$\sum_i x_{i\delta} \leq U_\delta^l s_\delta^l + U_\delta^h s_\delta^h, \qquad (2.18)$$

$$s_\delta^l + s_\delta^h \leq 1, \qquad (2.19)$$

$$z_\alpha + z_\beta + s_\delta^l + s_\delta^h \leq 2, \qquad (2.20)$$

$$\sum_k x_{ik} \leq R_i \qquad \forall i, \qquad (2.21)$$

$$w_\gamma, s_\delta^l, s_\delta^h, z_\alpha, z_\beta \in \{0, 1\},$$

$$x_{ik}, y_{kj} \geq 0.$$

The objective function (2.13) consists of two terms: The first one is linked to capacity modifications; the second one (an anticipation function) is linked to transportation costs. The only point worth noting is the negative sign of the term associated with decisions pertaining to centers α and β: It is a saving, and the binary variables are complemented to one, because we have a saving if we do *not* keep the center open. Notice that implicitly we consider the current situation where α, β and γ are currently open as our base-case scenario. Obviously, any other base case scenario would work as well (we suggest the reader to restate the model with other base cases). The constraint (2.14) says that demand must be satisfied. The flow equilibrium on distribution centers is expressed by (2.15). Constraint (2.16) says that it is possible to have transshipment through centers α and β only if they are kept open, in which case total flow is limited by handling capacity. The constraint (2.17) is also a node capacity constraint, but in this case we include a potential expansion of capacity. The capacity of center δ can take one of three values: zero, low, or high, depending on our decisions; constraint (2.18) takes care of this. We should note that capacity in γ cannot be the sum of low and high capacity, because the related decision variables are mutually exclusive, courtesy of constraint (2.19). Finally, inequality (2.20) has the effect that at most two centers among α, β, and δ are active in the new network, whereas (2.21) is the capacity constraint on production plants.

2.3 MODELS INVOLVING NONLINEAR COSTS

The careful reader has certainly noticed something strange in the last model we have considered: Why didn't we consider the possibility of shipping goods directly from a factory to a retail store? Extending the model by the inclusion of a new set of decision variables, say z_{ij}, to model the direct flow from factory i to retail store j would be rather trivial; we just have to adjust constraints on outflows from factories and on inflows to retail stores. The real issue is that this linear model is not able to capture economies of scale. Since the objective function includes linearly variable transportation costs, if it is convenient to ship a large amount directly rather than through the distribution center, this will also be the case for a small amount. What we observe in practice is that direct shipments are used only for large demand volumes at destination. The reason, as we have hinted at in section 2.1.2, is that we must achieve economies of scale in transportation. We should better represent costs, which are actually nonlinear. In fact, equation (2.7) on page 63, despite its limitations, suggests that the cost associated to a transportation link is not only a nonlinear function of flow, which in this case is essentially given by demand d per unit time; this function is also concave, because it includes the square root of d.[13] Concave cost functions model economies of scale. The total cost function of the Economic Order Quantity model offers a similar suggestion.

The actual cost associated with transportation flows on a link, with inventory holding, and with material handling at a facility is a complicated function depending on dynamic system behavior. At the network design level, we must settle for a suitable approximation by some anticipation function, aggregating costs on a relatively long-term horizon; then we may validate the solution we have obtained by simulating operational decisions. One possible approach is to postulate some functional form, like

$$C = \alpha \cdot V^{\beta},$$

where C is the cost per unit time, V is the flow volume, and α and β are coefficients we should estimate. For values of β such as 0.5 or 0.8, this cost function is concave. One possible way of finding suitable values for the coefficients in the assumed functional form is by analyzing approximate models. An interesting alternative is carrying out simulation experiments and then

---

[13]See section B.3 for the definition of convex and concave functions.

fitting a functional form against experimental results, e.g., by least-squares methods (see section A.10.6).[14]

Introducing a nonlinear cost function in a network optimization model may significantly change the nature of the model, the nature of the solution, and the computational effort to get this solution by solving the model:

1. In general, nonlinear programming models are harder to solve than their linear counterparts, and this may limit the size of the model we can afford to tackle.

2. Location models involve binary decision variables; solving a nonlinear mixed-integer model may be difficult.

3. Recent research has spawned a host of efficient algorithms for general convex optimization, and solvers have been introduced for nonlinear mixed-integer programming. Regrettably, minimizing a concave function is not a convex problem. For a model with a nonconvex objective, even solving a continuous relaxation within a branch and bound strategy (see section B.6.1) may be difficult because of potential local optima. We should use possibly demanding *global* optimization methods.

Even if we refrain from dwelling too deeply in algorithmic details, we immediately see that solving a suitably accurate network optimization model may be a time-consuming task. Network design is not a real-time decision-making task and that possibly significant savings are involved by proper analysis; hence, much CPU time can be afforded, but if we want to play with alternative scenarios to get a robust solution, we should try to keep computational requirements as low as possible.

One way out of this difficulty is approximating a nonlinear cost function by a piecewise-linear function, like those illustrated in figure 2.7. Given a function $f(x)$, we can define a set of "knots" $x^{(i)}$ which separate intervals over which the function is approximated by a linear piece. Determining how many linear pieces are needed and how knots should be placed requires some skill and experience, but we see that we may boil down a possibly complex model to a linear programming model. The nature of the function dictates if this may be solved as a continuous linear programming model or if mixed-integer modeling is necessary. We have the first case when the function we approximate is convex, so that its approximation may be convex too. For instance, let us consider a function like

$$f(x) = \begin{cases} c_1 x, & 0 \leq x \leq x^{(1)}, \\ c_2\left(x - x^{(1)}\right) + c_1 x^{(1)}, & x^{(1)} \leq x \leq x^{(2)}, \\ c_3\left(x - x^{(2)}\right) + c_2\left(x^{(2)} - x^{(1)}\right) + c_1 x^{(1)}, & x^{(2)} \leq x \leq x^{(3)} \end{cases}$$

[14] The approach of using a simulation model to build an approximate analytical model is called *meta-modeling*.

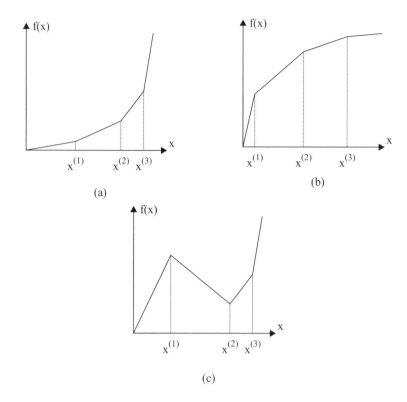

Fig. 2.7 Piecewise approximations of nonlinear functions $f(x)$: (a) convex, (b) concave, (c) neither convex nor concave.

modeling a cost depending on the level x of some activity. If $c_1 < c_2 < c_3$, like the case of figure 2.7a, then marginal costs are increasing; in other words we have a diseconomy of scale, which is represented by a convex cost function. If, on the contrary, $c_1 > c_2 > c_3$, marginal costs are decreasing and we have a concave function displaying economies of scale, as depicted in figure 2.7b. In our applications, this is the case most likely to occur, but in principle we might have the case of a generic function, like in figure 2.7c.

If the piecewise linear function is convex, its minimization is easily recast as a continuous linear program which can be efficiently solved. We have to transform the function $f(x)$ into the sum of linear terms, depending on auxiliary variables, say y_1, y_2, and y_3 if the function consists of three pieces:

$$f(x) = c_1 y_1 + c_2 y_2 + c_3 y_3,$$
$$x = y_1 + y_2 + y_3,$$
$$0 \le y_1 \le x^{(1)},$$
$$0 \le y_2 \le \left(x^{(2)} - x^{(1)} \right),$$

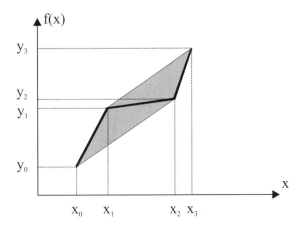

Fig. 2.8   Modeling a nonconvex piecewise linear function.

$$0 \leq y_3 \leq \left(x^{(3)} - x^{(2)} \right).$$

In practice, each variable y_i is associated with an interval, and the original variable is expressed as the sum of auxiliary variables. In order for this approximation to work properly, auxiliary variables should be "activated" in the correct order. First we use y_1, and we should activate further variables only if $x > x^{(1)}$; in other words, each "subinterval" must be saturated before using the next one. But since in the convex case we have $c_1 < c_2$, y_2 will be positive in the optimal solution only if y_1 reaches its upper bound $x^{(1)}$. We will not use y_2 in place of y_1, unless strictly necessary, because y_2 is more expensive to use. By the same token, y_3 is activated only if both y_1 and y_2 reach their upper bounds.

This reasoning applies for the minimization of a convex function, or the maximization of a concave one. But if we are minimizing a concave function, due to decreasing marginal costs, the solution algorithm would find it advantageous to use variable y_3 first, because it is the cheapest one. Of course this is no surprise, because we cannot expect to recast a nonconvex problem into a convex one. However, we may trade one nonconvexity for another one, by transforming the model into a mixed-integer linear programming model. The trick is associating a binary decision variable with each interval, making sure that only one interval is used. To see the idea, let us refer to figure 2.8, in which the piecewise linear approximation is encoded by a set of points of coordinates (x_i, y_i), where $y_i = f(x_i)$, $i = 0, 1, 2, 3$. Each point on the segment from (x_i, y_i) to (x_{i+1}, y_{i+1}) can be expressed as a convex combination[15] of

---

[15] A convex combination of an arbitrary number points in $\mathbb{R}^n$ is just a linear combination of those points, such that weights are non-negative and their sum is 1. Given a set of points S, the set of all of the possible convex combinations of them is called the *convex hull* of S.

those two extreme points:

$$x = \lambda x_i + (1 - \lambda)x_{i+1},$$
$$y = \lambda y_i + (1 - \lambda)y_{i+1},$$

where $0 \le \lambda \le 1$. Now, let us see what happens if we form a convex combination of all four points:

$$x = \sum_{i=0}^{3} \lambda_i x_i,$$

$$y = \sum_{i=0}^{3} \lambda_i y_i,$$

$$\sum_{i=0}^{3} \lambda_i = 1, \qquad \lambda_i \ge 0.$$

What we get is not really the piecewise linear function, but rather the convex hull of the four knots, which is depicted as the shaded area in figure 2.8. Nevertheless, we are close to our aim. We should find a way to enforce the use of only pairs of adjacent points in forming the convex combinations. In other words, only pairs of adjacent coefficients λ_i can be positive. For instance, if λ_0 and λ_1 are allowed to take positive values, whereas λ_2 and λ_3 are stuck at zero, we get the first line segment; if only λ_1 and λ_2 are free, we get the second one, and so on. To accomplish this, we may introduce a set of three binary variables, s_i, $i = 1, 2, 3$, one for each segment $(i - 1, i)$, and link these variables to the weights λ_i by the following constraints:

$$0 \le \lambda_0 \le s_1,$$
$$0 \le \lambda_1 \le s_1 + s_2,$$
$$0 \le \lambda_2 \le s_2 + s_3,$$
$$0 \le \lambda_3 \le s_3,$$
$$\sum_{i=1}^{3} s_i = 1, \qquad s_i \in \{0, 1\}.$$

This may look like a rather involved trick, but many software packages for mathematical programming ease the burden of introducing binary decision variables by just requiring the knots of the approximation, and automating the generation of auxiliary variables. However, it is important to realize what's happening behind the scenes; when we approximate a nonconvex function in a minimization problem, we introduce binary decision variables, and the resulting model may be hard to solve. We must be careful in striking a suitable compromise between accuracy of the approximation and the computational effort for solving the resulting model.

While we are discussing modeling by binary variables, we should also mention another typical modeling trick. To motivate it, suppose that we are considering the use of some transportation link, which should not be used if flow traveling on it is below a certain minimal threshold. Note that we are not saying that a certain activity level x must lie in the range $[L, U]$, where L and U are lower and upper bounds, respectively. Doing so would enforce a strictly positive value of x; however, what we want to express is that *if x is positive*, then it must stay within that interval. More formally, the feasible region for x is $\{0\} \cup [L, U]$. Since this set is not convex,[16] we cannot just resort to continuous linear programming. Yet, we may express the requirement within the mixed-integer linear programming framework, by introducing a binary decision variable s, set to 1 if the service is activated ($x > 0$), and set to 0 otherwise. Our aim is easily accomplished by the following constraints:

$$x \geq Ls, \qquad x \leq Us.$$

We see that if $s = 0$, then $x = 0$; if $s = 1$, then $x \in [L, U]$.

W.2.4   CONTINUOUS-SPACE LOCATION MODELS

In the last section, we considered a location–routing model in which potential sites for distribution centers have already been selected. Continuous-space location models are relevant when we want to generate alternatives. In the web section we describe "minsum" models, in which the aim is to minimize the sum of the distances between the new facility and, say, the retail centers; alternatively, we might wish to minimize the maximum distance, which leads to "minmax" models. By solving such a model, a new facility might well be located in the middle of a lake; nevertheless, the solution is useful in order to spot a neighborhood which could be searched for real location opportunities.

W.2.5   RETAIL-STORE LOCATION MODELS

In this chapter we have considered models in which demand was given exogenously. However, consider a consumer who has to travel a long distance to get to a retail store we want to build. If she is offered alternatives, she is not likely to become our most loyal customer. Indeed, an important concept in retailing is the logistical range, i.e., the maximum distance a customer is willing to travel. This depends on many factors, including the type of product and the level of competition, but it is easy to see that demand is endogenously generated by our choice of retail store location. In the web section we outline a few models which are suited to this task.

---

[16]In general, the union of convex sets need not be convex.

2.6   FOR FURTHER READING

- Background references which are relevant to this chapter are [1] and [2].

- The analysis outlined in section 2.1.2 is a simplified version of what is proposed in [4], to which we refer the reader for more details and further justification. In practice, organizational constraints may dictate that transportation frequencies are restricted to discrete values; a model to cope with this case is described in [8].

- The example described in section 2.2.4 is a simplified version of a similar model considered in [9].

- In [7] the reader may find approaches to reflect uncertainties, safety stocks, and alternative transportation modes within a static modeling framework.

- The reader interested in further information on location models can have a look, e.g., at [6], while [5] is useful to those working in the retail sector.

- Commercially available optimization solvers and languages are described, e.g., in `http://www.ilog.com` and `http://www.ampl.com`

- To have an idea of what software is commercially available for the logistic network design, we suggest visiting `http://www.slimcorp.com` and `http://www.bestroutes.com`

REFERENCES

1. R.K. Ahuja, T.L. Magnanti, and J.B. Orlin. *Network Flows: Theory, Algorithms, and Applications.* Prentice Hall, Upper Saddle River, NJ, 1993.

2. M.O. Ball, T.L. Magnanti, C.L. Monma, and G.L. Nemhauser, editors. *Network Routing (Handbooks in Operations Research and Management Science, Vol. 8).* Elsevier Science, Amsterdam, 1995.

3. R.H. Ballou. *Business Logistics Management (4th Ed.).* Prentice Hall, Upper Saddle River, NJ, 1999.

4. C.F. Daganzo. *Logistics Systems Analysis (3rd Ed.).* Springer-Verlag, Berlin, 1999.

5. M. Levy and B.A. Weitz. *Retailing Management (5th Ed.).* McGraw-Hill/Irwin, New York, 2003.

6. P.B. Mirchandani and R.L. Francis, editors. *Discrete Location Theory.* Wiley, Chichester, 1990.

7. J.F. Shapiro. *Modeling the Supply Chain.* Duxbury/Thomson Learning, Pacific Grove, CA, 2001.

8. M.G. Speranza and W. Ukovich. Minimizing Transportation and Inventory Costs for Several Products on a Single Link. *Operations Research,* 42:879–894, 1994.

9. H.P. Williams. *Model Building in Mathematical Programming (4th Ed.).* Wiley, Chichester, 1999.

3

Forecasting

3.1 INTRODUCTION

Before we discuss how to forecast, we shall wonder whether we should do so and why. Over the last few years many managers and academics have been supporting the drive towards lead time reduction and *Make to Order* (MTO). A basic truth about forecasts is that they turn out to be wrong. Hence, some managerial theories suggest that you would better not forecast; and actually, if a company is quick enough, it does not need forecasting. But what does "quick enough" mean? And is lead time reduction free?

Certainly, cutting lead times is a fruitful endeavor (e.g., see [12]). However, reality is a little bit more complex than these theories suggest. First, while these theories contrast *Make to Order* and *Make to Stock* (MTS) supply chains, almost all supply chains are partially driven by customers' orders (think about the assembly of a car that, at the least in Europe, is almost always custom-built) and partially driven by demand forecasts (think of components or raw materials purchases).

Example 3.1 Dell computers is today one of the largest PC manufacturers in the world and is considered to be the champion of Make to Order supply chains. Dell assembles PCs to customers' order. However, not the whole Dell supply chain is order-driven. Components' inventories are set according to demand forecasts. Thus, a more appropriate description of the Dell supply chain is: Distribution and production are order-driven (MTO) while components are made to stock (MTS). This is a significant advantage over other competitors, as Dell carries inventories where consumption is more predictable (component level) rather than where it is less predictable (single product configuration/single store). This redesign of the supply chain makes Dell a very efficient manufacturer and a very successful competitor in the tough PC business.

Also, Dell provides a very interesting answer to the question, what is "quick enough?" Dell significantly reduced the production lead time and can deliver in 23 days. Is that enough? The answer is that for PCs it is enough for most customers that do not need the computer they bought immediately. However, it is not enough for all users. Think of a situation where you lost your PC and need to make an important presentation tomorrow. Dell is not your favorite supplier. Also, this depends on the product a company is selling. While 2-3 days is fair enough for most customers for a PC, it is definitely too long if we are speaking about drugs for acute diseases (for further information on these examples see [17] and [18]). □

Moreover, many companies forecast demand implicitly. For example, in the grocery business many companies state that they do not generate any forecast (especially at the store/item level). However, when one digs into the planning systems, he/she can see that one key input to the purchase/delivery plan is a demand forecast, though it is often fairly rudimentary. For example, at a couple of grocery retailers in Italy, the target inventory level for the next week depends on the demand during the previous week. Thus, these companies implicitly assume a stationary demand and use the so-called "naïve approach"; that is, demand forecast for the next period (read "week" in the example) is equal to the demand in the previous one.

Generally speaking, when the Delivery Lead Time that customers want is shorter than purchasing, production, and distribution lead time, one needs to perform some sort of forecast to execute some activities before customers' orders are collected.

Example 3.2 In most retail outlets customers expect to collect immediately the goods they are looking for. This means that most retail companies shall somehow forecast demand to plan inventories for the finished products carried at each single store.

However, for some product categories the situation is rather different. For food products such as pizza, we might not need to carry all possible product variants, as customers might be willing to wait while their pizza is being cooked. Does this mean that all operations in a pizza restaurant are made to order? Actually, in Italy the average customer is just willing to wait while the raw materials are "assembled" and cooked. Most customers are not willing to wait while the cook looks for and buys the topping(s) they have ordered. Thus, even in a simple pizza restaurant we need to forecast the consumption of raw materials to purchase them in advance (pre-position raw materials). Quite interestingly in this case too we can see that different customers have different needs. While in traditional pizza restaurants pizzas are Made to Order, in fast-food and most US pizza restaurants the basic cheese pizza is cooked and then toppings are added according to customers' orders, as customers are not much willing to wait. In this case, customers are willing to give up a bit of product quality to reduce Delivery Lead Time (DLT). For them, 15 minutes is just not "quick enough."   □

Forecasting needs to cover and guide the portion of the supply chain operations that cannot be driven by customers' orders (see the order decoupling point concept in chapter 1). Let us consider a specific activity I, and let us use i as the index for activities performed by the supply chain starting from the delivery to customers (we number activities starting from downstream and move upstream). If the lead time of all downstream activities $\sum_{i=1}^{I} LT_i$ is greater than the DLT, then activity i cannot be driven by orders and we shall perform some sort of forecast to plan it.

Concept 3.1 *Forecasting is required when customers are not willing to wait long enough for all activities (purchasing, production, and distribution) in the supply chain to be performed based on firm customers' orders. So, the relevance of forecasting also depends on the strategy of the firm. In particular, it is very relevant for companies that rely on quick delivery and high service levels to gain a competitive advantage.*

Before we get into the details of algorithms to generate a forecast and measure forecasting errors, in section 3.2 we investigate what we mean by forecasting and how to choose a forecasting technique; in particular, we describe a forecasting process in section 3.2.1. Then, in section 3.3 we analyze how to measure forecast quality by means of accuracy and bias metrics. The remainder of the chapter discusses forecasting techniques. Section 3.4 classifies forecasting techniques. Sections 3.5–3.10 discuss several techniques starting from simple ones such as moving average to slightly more complex ones such as simple linear regression and exponential smoothing with trend and seasonality. In section W.3.11 we describe an example of how multiple linear regression (see section W.A.11) can be used in forecasting. Finally, forecasting techniques for new products are covered in 3.12–3.13.

3.2   THE VARIABLE TO BE PREDICTED

Before we move on to the forecasting techniques ("how should we forecast?") we shall introduce some parameters that help us define the variable we want to forecast ("what should we forecast?").[1] We need to define this concept carefully in order to set the forecasting problem properly.

---

[1]Often we just say that we want to forecast demand; as we shall see, this answer is just way too broad and fuzzy. In the remainder of this chapter we will refer to demand as the variable we want to forecast. This is just an example, as one might want to forecast other variables. We use demand instead of "variable we want to forecast" for the sake of clarity. However, the concepts we investigate apply to the more general problem of forecasting.

The time bucket   First, to properly set the forecasting problem we shall choose the *time bucket*, that is, the unit of time. The time bucket is a quantum of time, that is, a minimum amount of time we use for our analysis of demand. When we choose months as the time bucket, we never look at demand at the week, day, or hour level.[2] We basically choose to look at time as a set of months. Clearly, forecasting demand at the day level can be harder than forecasting it at the week or month level. So the forecasting problem is not properly set until we define the time bucket.

Example 3.3 A retailing company in Italy has historically recorded demand and delivery data during periods consisting of ten days. The logic behind this choice is that tens of days is a convenient metric to split monthly data.[3] However, such a logic has significant drawbacks for a retailer. In retailing, sales peak on Saturday and Sunday (in case stores are open). For this specific company, sales on Saturday and Sunday are more than twice the sales of the average weekday. A time bucket of 10 (or 11 days) is a poor choice, as demand data become very bumpy. Indeed, when we have only one weekend in ten days, demand is relatively low, while when we have two weekends in ten days, demand is substantially higher (on the average, by roughly 20%). Once the problem was identified, the company switched to weekly time buckets for all operational processes, from inventory planning, to sales force and delivery scheduling. The monthly data are used only for sales reporting and budgeting purposes and tens of days are no longer used.   ⬚

The forecasting horizon.   Second, we shall set the *forecasting horizon*, that is, how far into the future we want to foresee demand. For example, given a time bucket of one week, we shall wonder whether we want to forecast demand for the next week rather than 52 weeks into the future. In many instances we have to forecast demand over a variety of forecasting horizons rather than just one. For example, we might need a demand forecast for each of the next 4 weeks. Thus we might forecast demand over multiple forecasting horizons rather than a single one.

The forecast frequency.   The third relevant parameter is the frequency of forecasting updates. For example, let us assume we have to forecast demand for each of the next 52 weeks. On the one hand, such forecasts can be updated each and every week; we call such practice *rolling forecast*. On the other hand,

---

[2]Note that one can also try to predict when a given event is going to occur. For example, one can try to predict when a given customer is going to place his/her next order. In other words, in this book we try to answer the question: How many units are going to be requested in a given time bucket? Another question is: In which time bucket is a given event (say an order) going to occur? The former question is much more common and, generally speaking, more relevant.

[3]Actually, each month is split in three sections: the first ten days of the month, the second ten days of the month, and the rest of the month.

we can update the forecast at the end of these 52 weeks (so called *fixed horizon*). In the former case, the company can always foresee 52 weeks into the future. In the latter case, the company can foresee 52 weeks into the future, just after the forecast is generated, but the forecasting horizon progressively decreases down to just a single week. However, in the former case the cost of forecasting is substantially higher than in the latter one, as 52 different forecasts rather than a single one are generated in a year. Also, it makes sense to update demand forecast only when additional pieces of information are available. For example, let us consider a retailer that uploads demand information from each single stores once a month. For retailer like this it does not make sense to update demand forecasts weekly. Thus the appropriate forecast frequency depends on the cost of the forecasting process, on the availability of additional information, and on the potential benefits of a fresher (and thus usually "better") forecast.

The product.   A fourth relevant parameter is the definition of the *product* or set of products we refer to. Forecasting demand for a specific model of shoes in a given size (for example, Clark's Desert Boots, brown, size 43) is definitely more complex than forecasting the aggregate demand for all shoes in a given market.

The market.   The last relevant issue is the *market* or *geographical area* we refer to. Forecasting aggregate demand for shoes in Italy is relatively simple, whereas forecasting it at the single store level can be all but trivial. First, at the single store level, demand is lower and thus it tends to be (relatively) more variable (i.e., the coefficientof variation is larger). Also, exogenous factors such as local weather or even simple road-works can change the demand pattern significantly.

We have introduced the five dimensions that identify the object of forecasting, i.e., the variable we want to forecast; but we still have to answer a key question: What is the right choice for these five dimensions? What is the right set of products? Should we forecast at item of family level? What is the right time bucket? Should we forecast at the day or year level?

Forecasting the aggregate demand for a whole country in a year is definitely simpler than forecasting demand for a single model, in specific color in a specific size at a given store in week 4 of year 2007. This simplistic analysis can lead us to believe that we should always try to aggregate demand over product, time, and locations to reduce the forecasting error. Indeed, as the object of forecasting is more aggregate, the demand pattern tends to be more stable. Thus, it is easier to read past history and predict future one. This view of forecasting overlooks the relationship between the *forecasting process* and the *decision-making process*. The forecasting process is part of a broader decision-making process. To properly set the parameters of the forecasting process, we should first understand the decision-making problem(s) we want to support through a better forecast. In the specific case of logistics, the identification of the appropriate variable to forecast depends on the features of the planning problem(s) we're facing. For example, if we want to plan deliveries of drugs to a chain of drugstores that are replenished weekly, we must forecast the weekly demand for each single drug in each single store.

Example 3.4 Often these basic concepts are overlooked by many companies. Company Gamma is a market leader in the US office supply retail sector and operates hundreds of stores. Gamma wants to forecast promotional demand. Promotions last two weeks. Gamma thought they had found a great forecasting tool, as the forecasting error was apparently just 2%. This would be a very impressive result by any standard, as a 50% error is rather common for promotional items. However, this error was measured on the overall turnover for all promoted items in the whole chain. The metric of accuracy was totally inconsistent with planning problem the company was facing. Gamma corporation needs to plan how many units of each item (tens of items are promoted in any given week) shall be sent to each single store (the chain consists of hundreds of stores). Thus, the aggregate metric of forecasting performance has nothing to do with the very detailed decision problem the company is facing.   ⬜

Example 3.5 THREE is a company that sells furniture in Italy. Each three months they place orders to their Asian suppliers. Suppliers deliver in three months, so the lead time is three months.

   THREE has hired a new employee to improve the forecasting and planning process Table 3.1 shows demand data downloaded from the company's IT systems. The employee needs to forecast demand and measure its variability to properly set safety stocks. One might be tempted to use monthly demand data. Nevertheless, the company does not need really such a detailed forecast. The company only needs to forecast demand at the quarter level, as the frequency of orders (and thus the frequency of deliveries) and the lead time are three months. So, for any practical purpose the decisions of the company do not depend on whether the demand for April is high and for May is low or vice versa. Indeed, the company needs to place an order in early January for delivery in early April. Such an order shall meet demand for April, May, and June regardless of the distribution of demand among the three months. Thus, we actually need the demand during each quarter. In other words, we should restructure the demand database as table 3.2 shows.   ⬜

Concept 3.2 *The forecasting problem is properly set only when we have set the time bucket, the forecasting horizon and frequency, and selected the appropriate aggregation of products and markets. Also, these choices really depend on the decision-making process forecasting is supposed to support. Indeed, we forecast to make better decisions.*

Table 3.1   Forecasting example: demand data

Month/Year	2000	2001	2002	2003
1	127	111	111	119
2	130	131	132	136
3	134	131	124	136
4	134	137	134	130
5	126	119	111	118
6	103	103	105	119
7	91	96	94	92
8	88	96	98	100
9	90	91	96	99
10	93	84	101	86
11	103	96	95	98
12	115	101	108	108

Table 3.2   Forecasting example: aggregate demand data

Month/Year	2000	2001	2002	2003
First Quarter	391	373	367	391
Second Quarter	363	359	350	367
Third Quarter	269	283	288	291
Fourth Quarter	311	281	304	292

3.2.1   The forecasting process

When one says "forecasting," most people tend to think about algorithms. Indeed, in some instances algorithms can be used to forecast. However, *forecasting is a process* rather than an algorithm or a set of algorithms. Algorithms are just part of the broader process that consists of various phases presented in the following sections.

Analysis of decision-making processes.   The first step of a forecasting process is to analyze the decision making process one wants to support. This sets the basic output of the forecasting process (definition of product, time bucket, and market demand refers to, and choice of forecasting horizon(s) and frequency of updates). It is actually fairly hard to prescribe how this task shall be performed. However, we have to realize that any mistake in this initial phase has substantial consequences. A guiding principle is to look at the information one needs to make decisions and make sure that the forecasting process provides it.

If the forecasting process is too detailed, the output is too inaccurate (see previous section). On the other hand, if the forecasting process is too aggregate the output is generic and hardly helps the decision maker. For example, consider company Gamma from example 3.4. Probably, aggregate figures on consumption of paper in the USA are hardly the input that inventory planners expect in order to decide how many reams of paper should be sent to store 346 tomorrow.

Gathering information.   This is the second phase of the forecasting process. Once the output of the forecasting process is properly defined, we shall investigate what pieces of information are available to generate it. Forecasting, like any other statistic, is conditioned upon (i.e., depends on) an information set. In other words, the quality of the final forecast depends, among other things, on the quality and quantity of data and information used to generate such a forecast. Thus finding the right set of information to forecast demand can be as important or even more important than the selection of the appropriate forecasting algorithm. Indeed, even the best algorithm cannot possibly operate successfully without key pieces of information.

Example 3.6 Figure 3.1 shows the demand pattern of a food product in a large Italian grocery chain. The graph shows wide variations as demand jumps from 10 to 240. The root cause of such bumps are *trade promotions*.

It is rather apparent that the manufacturer of this product cannot possibly forecast demand accurately with no information on trade promotions, no matter what the forecasting algorithm is. Indeed, there is no clear pattern in promotions and thus an algorithm cannot predict when they will occur in the future and forecast their impact on demand. However, the retailer and the manufacturer agree on the promotions well in advance of their start. Indeed, both the retailer and the manufacturer enjoy the beneficial increase of demand.

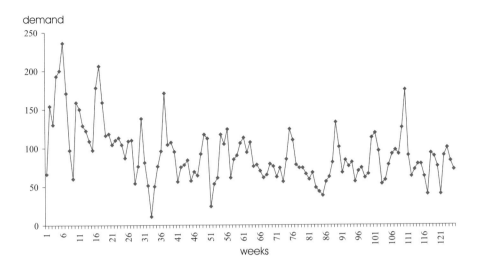

Fig. 3.1   Demand time series for a fresh product in a grocery chain.

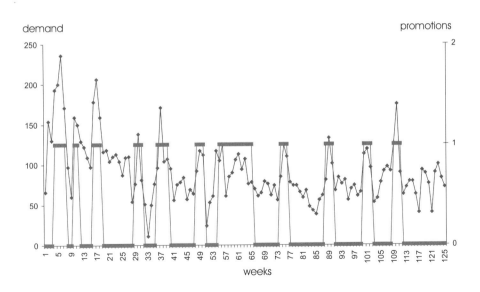

Fig. 3.2   Demand and promotions time series for a fresh product in a grocery chain.

Thus the retailer asks the manufacturer to cut the wholesale price (i.e., the price the manufacturer charges the retailer) temporarily. So the manufacturing company knows when the promotions are going to occur few a weeks before they start. The manufacturer needs to make this precious information available to the forecasters. Unfortunately, those that collect the information from the retailers typically belong to the sales departments, while the persons in charge of forecasting belong to other departments (e.g., logistics or manufacturing), and in many companies information does not flow smoothly across departmental boundaries. The benefits of such an information can be appreciated by looking at Figure 3.2. 〇

The key pieces of information to predict future demand depend on the specific forecasting problem one faces. Thus we cannot provide an exhaustive list of variables one might want to consider. However, we can discuss issues and variables that are often overlooked and do require some careful attention.

Forecasting tries to predict the future behavior of an exogenous variable, in our case, future demand.[4] Hence, it is very important to use *demand rather than sales* as the input to the forecasting process. Actually, sales depend on true customer demand (that is a truly exogenous variable one tries to predict) and on the availability of products (that is a lever for the supply chain manager). Product availability censors demand. In most situations a company can only sell the products that are currently available in the warehouse or in the store. When 30 cans of beer are available in a supermarket, we cannot sell more than 30 cans. If sales are used to forecast future demand, a low demand forecast might turn out to be a self-fulfilling prophecy. Low sales might reduce the forecast, which then leads planners to reduce inventories. Finally, low inventories might further reduce sales.[5]

Example 3.7 For example, a leader in the production of dry pasta in Italy uses time-series models (see section 3.5 in this chapter). When a new kind of pasta was launched, the company decided to postpone the launch in a given region because the company wanted to consume inventories of a preexisting item that the new one was going to cannibalize. The automatic forecasting and replenishment system immediately started to record zero sales for the new product in that region, thus predicting no demand and suggesting to ship zero units of the new kind of pasta. The vicious circle was interrupted only when the product manager spotted the anomaly in sales, investigated the issue, and finally discovered what was going on. 〇

---

[4]Please note that demand is not completely exogenous for a company, since many levers such as price can influence it. However, in our context we can assume demand to be exogenous, as logistics and supply chain managers are supposed to meet demand. In other functions such as marketing and sales, demand is actually the variable that one tries to control through pricing, promotions, new products, etc.

[5]Note that this process might be very dangerous in case of products with low margin, as companies tend to provide low service levels and a relatively large portion of demand can be lost (see section 5.2).

Moreover, a stockout of a given product can perturb the sales pattern of other products since some customers might be willing to substitute the product they were looking for with a surrogate. In some industries such as business to business, e-commerce, or catalogue sales, it is relatively easy to capture the gap between sales and demand as one can keep track of customers orders. In other instances, like "brick and mortar" retail chains, this is more complex, as customers do not formally place orders. In this case too, though, one can use statistics to estimate the potential customer demand out of censored sales data (among others see [15] and [20]).

Analysis of demand.   The third phase of the process is the analysis of demand. In this phase one shall study and identify demand patterns. As we further discuss in the next sections, all quantitative forecasting techniques make some assumptions on demand behavior and pattern. Thus one should first analyze demand to figure out its actual behavior and then look for a forecasting technique that fits it. For example, we might investigate the demand to check whether it is stationary, it shows seasonal fluctuations, or it is influenced by phenomena such as weather conditions, promotions, or fashion. We should understand first the drivers of demand, and then we can design (or choose) an appropriate forecasting model that is able to read past demand behavior and predict the future one.

Selection of forecasting technique and fine tuning of parameters.   The fourth phase of the process consists of (i) the selection of the appropriate forecasting model and (ii) the fine tuning of its parameters. In simple cases, one can just select a forecasting model off the shelf, i.e., adopt an existing model as it fits very well. Commercial software provides several standard forecasting techniques to choose from.[6] Very often, though, real-life problems require more complex or at the least "ad hoc" solutions. This is the reason why one shall fully understand assumptions, mechanics, and applicability of standard forecasting techniques. If one does not fully understand the details of standard techniques, he/she is bound to use them as they are and cannot adapt them to the unique features of any given demand. Moreover, the effectiveness of many models depends on the selection of proper values of the parameters.

Usually, forecasters judge the quality of a model or a set of parameters by looking at their ability to generate small errors. In the next section we discuss several metrics for forecasting errors. Notice that the selection of a model (or set of parameters) should be based on its ability to forecast future demand. Unfortunately, future demand is not known yet. This makes the selection of the "best" model tricky. Often one looks at what would have been the performance of the forecasting model (or set of parameters), had it been used in the past. This is typically the only way out, but we are implicitly

---

[6]For a list of software providers see www.forecastingeducation.com.

assuming that the basic demand pattern will not change: The best model to predict past demand will still be the best model for future demand as well. In case we expect a significant change in demand – say we expect a stationary demand to start growing – this approach might lead us to poor performance. In these cases, we might want to select a model simply because it logically fits the demand pattern we expect to observe in the future.

Forecast generation   Once the model is selected and parameters are set, we can start using them to generate demand forecasts. During this phase, data are processed and forecasts are used to make decisions.

Measuring forecasting errors   While we continuously generate demand forecasts, we shall keep track of errors. By doing so, one can spot any inconsistency between the model and current demand behavior, which in real contexts is dynamic and thus requires periodic tunings. Moreover, the quality of forecasts is a relevant input for the distribution and production planning process. As chapters 5 and 6 discuss in detail, uncertainty (as measured by forecasting error) changes the very nature of decision-making and planning problems. Under uncertain conditions we shall deliberately acknowledge that very different scenarios might come true. Also, forecasting errors can be used to judge the quality of a forecasters' job and, through appropriate incentives, lead him/her to improve it over time.

Often this phase of the forecasting process is overlooked. The basic logic is that right or wrong, the story is over once we have observed demand. Many companies do not record forecasts in their systems. They simply record the purchase, production, or distribution plans. Some companies think that if 200 units were manufactured and 200 units were sold, the forecast quality was good. This simplistic vision overlooks a basic difference between a forecast and a plan. The forecast is the expectation of the future behavior of a variable which is at least partially exogenous. A plan is the response the company believes to be optimal in the face of all possible future levels of demand. Thus the demand forecast and the plans to meet it are logically very different and should be treated as such. As we discuss in further detail in chapter 5, producing 100 units while we expect a demand for 100 units can be a very bad decision, though an apparently reasonable one.

Also, even when forecasts are recorded, they are often overwritten as they are updated. Thus, only the most recent, and usually most accurate, forecasts are left in the databases. The following example shall make the concept clearer.

Example 3.8 Let us assume that a company forecasts demand and plans inventories with a monthly time bucket. Also, let us assume that the company forecasts and plans 12 months into the future with a rolling horizon, i.e., every month it forecasts demand and plans inventories for each of the next 12 months. At the end of year 2006 the company updates forecasts for January–

November 2007 and creates a brand new one for the month of December 2007. The forecast for December is going to be updated in January 2007, February 2007, and so on. Often companies tend to overwrite the original forecast for December 2007 with more recent ones. Thus in databases we tend to find forecasts with very short horizons and thus relatively small errors. This often leads companies to overestimate their ability to forecast demand and underestimate the uncertainties they face. For example, consider a company that wants to forecast the total turnover for a fiscal year and during the year constantly keeps on updating the forecast to get an accurate figure. By the end of the year the figure is going to get very accurate by definition, as we are basically looking back at past sales rather than predicting future ones.   ⬜

3.3 METRICS FOR FORECAST ERRORS

To properly define a metric for forecasting quality, we must first understand the nature and objectives of the forecasting process. In case of point forecasts, the relevant performance is the percentage of correct forecasts. For example, in the case of sport bets, what matters is the number of correct predictions. The extent to which a forecast was wrong does not actually matter. If you predict soccer teams I and M will draw, no matter whether M won 3 to 2 or 6 to 0, you still made an error. In general, a point forecast is relevant when any difference between the forecast and the actual event is equally damaging (in the case of sport bets, no matter how close to the final outcome your prediction was, you still lose your money).

In most circumstances, though, we do not use point forecasts. When we say that we expect a demand of 1,000 units, we really mean that we expect demand to be around 1,000 units rather than exactly 1,000 units. Thus, we do not really care about the frequency of perfect forecasts. If demand is a continuous variable (think of demand for energy or demand for cheese over the counter at a supermarket), the probability that demand will equal the point forecast is zero (see appendix A). Hence, we do not care about the frequency of perfect forecasts, but we should rather capture the differences between our predictions and actual demand.

Measuring the quality of a forecast for a single product, in a single market, for a single time bucket is relatively straightforward, as you only need to compare actual demand with your forecast. Often one needs more aggregate figures to judge the performance of a forecasting tool (or a forecaster) over multiple periods of time, multiple items, or multiple markets. In this book we investigate in detail the case of a single product in a single market over multiple time buckets. The final section of this chapter presents some extensions to the multi-item or multi-market case.

To measure the forecast error, we need to introduce some notation:

- $F_{t,h}$ is the forecast generated in period t with an horizon h; thus $F_{t,h}$ is a prediction of demand at time $t + h$, where $h = 1, 2, 3, \ldots$

- Y_t is the realization at time t of the variable we try to forecast; in our examples it is the demand at time t.

- $e_t = Y_t - F_t$ is the forecast error at time t.

Notice that F_t is the forecast of *demand in period t* regardless of when such forecast was generated. We look back *ex post* and compare it with demand at time t to judge its quality. The time at which the forecast was generated depends on the decision process we have to support, and it is irrelevant if we are evaluating the forecasting process. On the contrary, $F_{t,h}$ is the forecast generated *in* time t *for* time $t + h$.

Also, notice that in our definition the error is positive when demand is larger than the forecast (i.e., we under-forecasted), whereas the error is negative when demand is smaller than the forecast (i.e., we over-forecasted).

3.3.1   The Mean Error

A first metric of "forecasting quality" is the simple average of past errors, that is, the mean error (ME):

$$ME = \frac{1}{n} \sum_{t=1}^{n} e_t. \tag{3.1}$$

As equation (3.1) clearly shows, with this metric, positive errors counterbalance negative ones. In other words a forecasting method that generates no error in each of the n periods in our sample and a forecasting method that generates a +10 units error in 50% of the periods and generates a –10 units error in 50% of the periods are just as good, from the ME standpoint (see table 3.3). In fact, ME is just a metric of *bias*, since it just captures whether our forecasting process is on the average pessimistic (it tends to under-forecast and thus ME is positive) or optimistic (it tends to over-forecast and thus ME is negative).

Therefore, we need to consider other metrics that can capture *accuracy*, that is, the ability to generate a forecast that is close to actual demand in *each* period. Metrics of accuracy differ from ME (and more generally metrics of bias) as positive errors do not cancel negative ones; rather, they add up.

3.3.2   Mean Absolute Deviation

A first metric for accuracy is MAD (Mean Absolute Deviation), which basically uses the absolute error to make sure negative and positive errors add up:

$$MAD = \frac{1}{n} \sum_{t=1}^{n} |e_t|. \tag{3.2}$$

Table 3.3   Mean Error a metric for bias

Period	1	2	3	4	5	6	ME
Demand	90	110	110	90	110	90	
Forecast 1	100	100	100	100	100	100	0
Forecast 2	90	110	110	90	110	90	0

Table 3.4   Comparison between ME and MAD

Period	1	2	3	4	5	6	ME	MAD
Demand	7	13	9	12	8	11		
Forecast 1	10	10	10	10	10	10	0	2
Forecast 2	6	12	8	11	7	10	1	1

The example in table 3.4 tells the difference between ME and MAD. The first forecast is not biased, as the mean demand equals the mean forecast. On the contrary, the second series of forecasts is biased, as it is always conservative: The forecast is always one unit below the demand. ME actually tells us that the first series of forecasts is unbiased while the second one under-forecasts. However, the second forecast captures and follows demand fluctuations more accurately than the first one. Thus, in each single time bucket the second forecast tends to be closer to demand than the first one. MAD catches such a difference as it tells that the second forecast is more accurate than the first one.

Finally, which forecast is the best option? Should we care more about accuracy or bias?

Actually, we cannot tell whether one forecast is better than the other. One is better for bias, the other for accuracy. In some contexts bias might matter more than accuracy and vice versa. However, we may see that correcting for bias is relatively easier than correcting for inaccuracy. If a forecasting process is consistently conservative, but it follows demand fluctuations very closely (see the example in table 3.4), we can improve the forecast by adding the average bias to the forecast. For example, if a forecaster is conservative and consistently underestimates demand by 10 units, when he/she generates a demand forecast of 110 units for next period, we might expect demand to be around 120 units (110 units + 10 units). In the example above, if the second forecaster predicts a demand of 12 units for the next period, we might add one extra unit to it since in the past we have noticed that he/she tends to under-forecast by one unit. Thus we might expect demand for 13 units. Such an adjustment to the forecast improves both bias and accuracy, thus reducing both the ME and the MAD. On the contrary, there is no obvious solution to inaccuracy. Say you want to improve the quality of the first forecast in table 3.4. What would you do? Actually there is no easy fix with regard to inaccuracy.

Concept 3.3 *A good forecast is both accurate and unbiased. Both are very relevant performance metrics, but while there is a fairly easy fix for a consistently biased but accurate forecast, there is no such easy fix for an unbiased and inaccurate one.*

3.3.3   Root Mean Square Error

A second metric for accuracy is Root Mean Square Error (RMSE). This metric squares errors to sum positive and negative ones.

$$\text{RMSE} = \sqrt{\frac{1}{n} \sum_{t=1}^{n} e_t^2}. \tag{3.3}$$

RMSE is a very commonly used metric, as in statistics squared errors are often used instead of absolute ones (they result in a differentiable function, whereas the absolute value function is kinky). Thus, a quadratic error provides estimates that are more directly linked to the variance and standard deviation (see appendix A) of the demand distribution. Often we use the forecast that an algorithm generates as an estimate for the expected level of demand while we use RMSE as an estimate of standard deviation.

Table 3.5 shows the differences among ME, MAD, and RMSE. Forecast 2 differs from Forecast 3, as errors are more frequent but they tend to be smaller. This is why RMSE considers Forecast 2 to be more accurate than Forecast 3. This finding can be generalized by saying that RMSE is a quadratic metric for error and thus it tends to overweight large errors. So RMSE "prefers" forecasting algorithms that generate constant errors, rather than algorithms that are very accurate in some periods but can generate significant errors in others. MAD is a linear metric for error and thus gives the same weight to all errors, small or large.

ME, RMSE, and MAD measure the forecast error using the same units of measurement as demand. For example, if demand is measured in units or kg, then ME, RMSE, and MAD are measured in units or kg as well. This can be a drawback: When reading the performance of any forecast, we should carefully consider the scale that is adopted. If one decides to use kg rather than hg to measure demand for cheese, ME, MAD, and RMSE drop by a factor of 10.

Moreover, these metrics make the comparison of performances across products very hard. As table 3.6 shows, the metrics presented so far might lead us to believe that the forecast for item A is more accurate than the forecast for item B. However, an error of one unit out of an average demand of 10 units

Table 3.5 Comparison between accuracy metrics: MAD and RMSE

Period	1	2	3	4	5	6	BIAS	MAD	RMSE
Demand	7	13	9	12	8	11			
Forecast 2	6	12	8	11	7	10	1	1	1
Forecast 3	7	10	9	9	8	11	1	1	1.73
Error 2	+1	+1	+1	+1	+1	+1	1	1	1
Error 3	0	+3	0	+3	0	0	1	1	1.73

Table 3.6 Comparison between accuracy metrics: MAD and RMSE

Period	1	2	3	4	5	6	ME	MAD	RMSE
Demand A	7	13	9	12	8	11			
Forecast A	8	12	10	11	7	12	0	1	1
Error A	−1	+1	−1	+1	+1	−1	0	1	1
Demand B	70	130	90	120	80	110			
Forecast B	75	125	95	115	75	115	0	5	5
Error B	−5	+5	−5	+5	+5	−5	0	5	5

is "worse" than an error of 5 units out of a demand of 100 units. Thus, often one wants to look at percentage error metrics.

3.3.4 Mean Percentage Error and Mean Absolute Percentage Error

The drawbacks of metrics such as ME, MAD, and RMSE lead us to introduce percentage errors that basically try to compare the forecasting error with demand. The most classic metrics in this vein are Mean Percentage Error (MPE) and Mean Absolute Percentage Error (MAPE), which measure percentage bias and percentage accuracy, respectively. Notice that, as following equations show, these metrics compare the error in period t with the demand in the same period:

$$\text{MPE} = \frac{1}{n} \sum_{t=1}^{n} \frac{e_t}{Y_t}, \tag{3.4}$$

$$\text{MAPE} = \frac{1}{n} \sum_{t=1}^{n} \frac{|e_t|}{Y_t}. \tag{3.5}$$

These metrics are pure numbers and thus do not depend on the scale one uses to measure demand. Hence, one can easily compare the accuracy and bias across various product or markets.[7]

Example 3.9 Some European Fortune 500 companies have adopted different percentage errors metrics. They basically divide the error by the forecast rather than by the demand; hence, they use the metrics below, which are modified versions of MPE and MAPE:

$$\mathrm{MPEM} = \frac{1}{n} \sum_{t=1}^{n} \frac{e_t}{F_t}, \tag{3.6}$$

$$\mathrm{MAPEM} = \frac{1}{n} \sum_{t=1}^{n} \frac{|e_t|}{F_t}. \tag{3.7}$$

This might be a tempting solution but is actually an awful one. Indeed, this definition of percentage error provides the forecasters (whose reward may depend on these metrics) with two means to improve their performance:

- First, they can reduce the numerator, that is reduce the forecasting error.

- Second, they can increase the denominator, that is increase the forecast.

This gives the forecasters an incentive to overstate their forecast. Not surprisingly the companies noticed that the predicted demand was on the average above the actual one.

   These metrics are particularly dangerous in the case of low or highly variable demand. Let us consider the case of a demand that in $1/3$ of the cases is zero, in $1/3$ of the cases is one, and in $1/3$ of the cases is two. Let us assume that the forecaster is judged and rewarded on the basis of MAPEM. Also, let us assume that he/she has no specific idea about what is going to happen in the next period. So he/she basically faces the long term demand distribution. He/she has two options. The more reasonable one is to forecast one unit for all future periods. In this case, in $2/3$ of the cases the absolute error is 1 and in $1/3$ of the case it is zero. Given the forecast of one, the MAPEM is going to be 0.66. The other apparently less reasonable option is to forecast two units for all future periods. In $1/3$ of the cases, demand is going to be zero and the error is going to be 2. In $1/3$ of the cases demand is going to be one and error is going to be one, and finally in $1/3$ of the cases the forecast is going to be correct. This really means that the MAPEM is just 0.5 $(33.33\% \cdot 2 + 33.33\% \cdot 1 + 33.33\% \cdot 0)/2$. As this example clearly shows,

---

[7]Note that, in general, we expect products/markets with higher demand to have less variability. Thus, in general, we also expect that the higher the demand, the lower the percentage error, as the forecasting problem is simpler.

Table 3.7   Percentage error metrics: MPE and MAPE

Period	1	2	3	4	5	6
Demand A	7	13	9	12	8	11
Forecast A	8	12	10	11	7	12
Error A	−14.3%	+7.7%	−11.1%	+8.3%	+12.5%	−9.1%
Demand B	70	130	90	120	80	110
Forecast B	75	125	95	115	75	115
Error B	−7.1%	+3.8%	−5.6%	+4.2%	+6.3%	−4.5%

Table 3.8   Comparison between absolute and percentage error metrics

	ME	MAD	MPE	MAPE
Forecast A	0	1	−1%	10.5%
Forecast B	0	5	−0.5%	5.3%

these metrics, which are apparently very similar to MPE and MAPE and are commonly used, provide very odd incentives to overstate the forecast.   ⬜

We can reconsider the data in table 3.6 and calculate the percentage errors displayed in table 3.8. Data show that the forecast for demand B is actually more accurate than for demand A.

The use of MPE and MAPE as performance evaluation measure is suggested in the literature (see, e.g., [13]), but these metrics have several drawbacks and weaknesses:

- They cannot be adopted when demand during a time bucket can be zero. Indeed, when demand is zero we cannot compute the percentage error. In real applications, such a case is relatively frequent. For example, in the case of retail chains, replenishments are so quick and frequent that one needs to forecast demand down to the single day or single week. Also, assortments tend to be very wide and thus many products have relatively low demand rates. These trends make the likelihood of a zero demand for a single product, in a single store, in a given day quite sizeable. Understandably, the extent of this problem depends on the definition of the demand one wants to forecast: The longer the time bucket, the larger the market (nation vs. single store), and the broader the set of product variants (single SKU or product family), the

Table 3.9   Percentage error metrics in case of variable demand

Period	1	2	3	4	5	6	7	8	9	10
Demand	10	10	10	10	1	10	10	10	10	10
Forecast 1	10	10	10	10	10	10	10	10	10	10
Error 1	0	0	0	0	-9	0	0	0	0	0
Forecast 2	12	12	12	12	1	12	12	12	12	12
Error 2	-2	-2	-2	-2	0	-2	-2	-2	-2	-2

	ME	MAD	MPE	MAPE
Forecast 1	-0.9	0.9	-90%	90%
Forecast 2	-1.8	1.8	-18%	1 8%

higher the expected demand and thus the lower the probability of a zero demand.

- Even in cases of nonzero demand, these indexes can give really odd results when demand shows wide variations. Indeed, as the example in table 3.9 shows, MPE and MAPE tend to overweight errors in low demand periods. In the example, the error of the first forecasting method in period five is so large (in percentage) that it more than counterbalances the greater accuracy that this method achieves in other periods.

Thus, these metrics cannot possibly be computed when demand is zero, and when demand varies substantially they might provide misleading insights. For example, in table 3.9 the first forecast seems to be more accurate and less biased than the second one, while MPE and MAPE seem to suggest just the opposite. Thus these metrics might lead us to erroneous conclusions. Indeed, in most circumstances the cost due to a forecast error of 2 units in a low demand period is quite similar to the cost of a 2 units error in a high demand one. Finally, these metrics actually build strange incentive schemes for the forecasters. If a forecaster is to allocate his/her efforts among different products or over time, he/she might end up focusing on items in periods of low demand since a unit of error is more heavily penalized by the error metric.[8]

3.3.5   ME%, MAD%, RMSE%

The problems discussed in the previous section lead us to design new performance metrics that

---

[8]In this case, we clearly overlook the fact that the effort required to cut the error by one unit might be different for different products/periods.

- consider errors in low and high demand periods equally damaging and

- allow us to compare the performance across products and markets with different mean demand.

Such metrics are ME%, MAD%, and RMSE%.

These performance measures compare the ME, MAD, and RMSE to the mean demand for the product/market combination:

$$ME\% = \frac{ME}{\overline{Y}},$$

$$MAD\% = \frac{MAD}{\overline{Y}},$$

$$RMSE\% = \frac{RMSE}{\overline{Y}},$$

where

$$\overline{Y} = \frac{1}{n}\sum_{t=1}^{n} Y_t.$$

These metrics still retain the good features of MPE and MAPE. Indeed, if we apply them to the data in table 3.6, they suggest that forecast B is more accurate than forecast A: MAD% and RMSE% are 5% (5/100) for B, while they are 10% (1/10) for A; ME% is zero in both cases.

Moreover, they avoid some of the drawbacks of MPE and MAPE as they can properly judge the quality of the forecasts in table 3.9. MAD% for forecast 1 is 9.9% (0.9/9.1) while it is 19.8% (1.8/9.1) in case of forecast 2.

These metrics can measure the quality of a forecast and compare it with the average demand.[9] However, predicting an extremely variable demand can be more complex than predicting a very stable one. In other words, a given forecasting error might be very good in the case of an extremely variable demand, whereas it might be very poor in the case of a flat one. Thus we might not want to look at the forecasting error *per se*, but we might want to put it in the right perspective and analyze the complexity of the forecasting task.

---

[9]Note that in this case the denominator depends on the sample we choose. Thus, if we consider the accuracy of the forecast for May 2006 and look at the demand over the first five months of 2006 or over the last 12 months, we are going to get two different figures. Therefore, to make sure metrics for accuracy and bias do not change over time, we shall define sampling policies. For example, a company that generates forecasts at the day level might want to record accuracy and bias at the month level to properly define the sample and thus the average demand in each sample.

Table 3.10   The impact of demand variability on forecasting performance.

Period	1	2	3	4	5	6	ME%	MAD%	RMSE%
Demand A	10	9	10	11	10	10			
Forecast A	9	10	11	10	9	11	0		
Error A	+1	−1	−1	+1	+1	−1	0	10%	10%
Demand B	15	8	5	12	13	7			
Forecast B	14	9	7	10	12	8	0		
Error B	+1	−1	−2	+2	+1	−1	0	13.3%	14.3%

3.3.6   Theil's U statistic

Often, the Theil's U statistic is used to put the accuracy of a forecast in perspective. This statistic is defined as

$$
U = \sqrt{\frac{\sum_{t=1}^{n-1}\left(\dfrac{F_{t+1} - Y_{t+1}}{Y_t}\right)^2}{\sum_{t=1}^{n-1}\left(\dfrac{Y_t - Y_{t+1}}{Y_t}\right)^2}}. \tag{3.8}
$$

We can interpret the U statistic by looking at the numerator terms within the squared ratios. The term $F_{t+1} - Y_{t+1}$ is the error at time $t + 1$. $Y_t - Y_{t+1}$ is the error we would have made, had we adopted a *naïve* forecasting technique where the forecast for next period $t + 1$ is equal to the demand in the last period t ($F_{t,1} = Y_t$). Thus, the U statistic compares the error of the method we have adopted with the error that a simplistic model would generate. In case our model generates an error that is larger than the error of the naïve one, the U statistic is greater than 1. If, on the contrary, the forecasting model used is just as accurate as the naïve one, the U statistic is equal to 1. Finally, if the forecasting model adopted is much more accurate than the naïve one, the U statistic is close to 0.

Thus the U statistic gauges the ability of the model adopted to be more accurate than a naïve technique that is somehow considered a point of reference. In other words, the U statistic does not measure the accuracy of the forecast, but rather relates the error to the complexity of the forecasting task.

In the example of table 3.10, the U statistic is 1.11 for forecast A and 0.31 for forecast B. Thus the statistic captures the fact that forecast B is less accurate than forecast A simply because time series B is harder to predict than time series A. Actually, the U statistic uses the naïve method as a point of reference and tells that forecast A is 11% worse than the naïve one, while forecast B is 69% better than the naïve one. Although MAD% and RMSE% suggest that forecast A is more accurate than forecast B, the latter is actually a more appropriate model than forecast A.

The definition of the U statistic shows that the forecast error at time $t + 1$ (of both the forecasting model adopted and the naïve one) is divided by the demand at time t. This makes the reading of the statistic less than intuitive. Thus, we often use the simple ratio between the performance (e.g., MAD% and RMSE%) of the forecasting model adopted by the company and the performance of the naïve method ($F_{t+1} = Y_t$).

Example 3.10 A grocery retail company in northern Europe has a rather *heterodox* and interesting view of forecasting accuracy. Basically, this company theorizes that classic measures of forecasting accuracy are simply not relevant. Actually, they think that a 2% error is simply not relevant. A forecast error of 2% is basically as good as a perfect forecast, from their standpoint. Their idea is that a forecast error matters simply because it can increase the costs of the company. Also, they noticed that the flexibility in the supply chain enables them to recover, say, a 30% forecast error during promotions. This really means that any error below 30% has basically no consequence whatsoever. A 20% error and a 7% error are just as good. So, their measure of forecast accuracy is the percentage of forecasts that are within 30% of the actual demand. So a 98% accuracy means that in 98% of the cases demand is within 30% of demand. The idea behind this is to use a metric for accuracy that is a good proxy for the cost function of the company. The cost of errors below 30% is limited and is assumed to be zero in the metric developed by this company. Errors above 30% are expensive for the company. Though understandably a 70% error can be more expensive than a 35% one, the simple metric catches the fact that they are both expensive. In other words, though the cost function can be more complex, the metric adopted by the company assumes that it resembles a step function that is 0 if the (absolute) error is lower than 30% and is 1 if the error is above this threshold.

Actually this uncommon, though fairly interesting, practice is consistent with a stream of research that investigates whether costs are somehow related to any specific metric of forecasting performance.   □

3.3.7   Using metrics of forecasting accuracy

The metrics for forecasting accuracy presented in the previous section can be used for various purposes.

1. First, the metrics can be used to monitor performance over time. Measures of accuracy are used to gauge demand uncertainty (i.e., our ability to predict demand) that is a key input to the planning process (see chapters 5 and 6).

2. The error can be used to set forecasters' incentives, focus their efforts over time, and hopefully improve performance.

3. Also, the control of performance over time can be used to judge whether the forecasting method currently adopted fits the current demand pattern.

In case performance is unsatisfactory, one can (i) change the parameters, (ii) adapt the technique, or (iii) adopt a completely different forecasting approach. But how can we judge what is the most appropriate forecasting method or the most appropriate set of parameters? Obviously, the best method/set of parameters is the one that generates the best performance. But how can we estimate the performance that the forecasting process would generate if we choose to adopt it in the future? Basically, this is a very hard exercise, as we do not know how demand will behave in the future. We basically have two options.

A first option is to actually try the forecasting process and measure the performance it actually generates over a trial period. This approach selects the forecasting method, based on actual performance. However, it is very expensive, since several forecasting tools and processes (including human interactions and corrections) must run in parallel for a period of time that shall be long enough to draw statistically significant conclusions. Also, in the trial period we might be using data from a forecasting process that actually generates very poor predictions and thus we might make poor decisions and experience poor operational performance and high costs.

A second, widely adopted approach is to use past history to test the performance that the various alternative methods would have generated had they been adopted in the past. This selection process makes an implicit assumption. It assumes that the method that would have worked best in the past will be the best option for the future.[10]

Then the question becomes: How can we judge the performance a that forecasting process would have generated in the past?

To do so, we must use past demand data both to generate a forecast and to test its quality. When we do such analysis we must be extremely careful and avoid a frequent conceptual error. No data about any period after t shall be used to generate the forecast $F_{t,h}$. In other words, we want to make sure we appropriately simulate the forecasting process. While we forecast demand in period t (for period $t + h$), only information about demand (as well as other variables) in periods $\tau \leq t$ is available. In particular, demand in period $t + h$ shall not be used in any even indirect way to generate $F_{t,h}$.

Several forecasting methods depend upon some parameters that influence their behavior and performance. These parameters are set by using a sub-

---

[10] Actually even the first option makes a similar assumption. Indeed, it assumes that the method that performs better in the near future (trial period) works best in the long term as well.

sample of the demand data that we call **fit sample**. Thus, when one wants to use past demand data to judge the quality of various forecasting methods, he/she shall identify a fit sample to set the parameters of the forecasting models and a second subsample (often called **test sample**) to judge their performance.[11] The larger the fit sample the better the choice of the parameters of the models and thus their performance. On the other hand, a large fit sample implies a small test sample (given the limited amount of relevant data available). Hence, we face a tradeoff between the choice of the appropriate parameters for each of the competing models and the ability to properly judge the quality of the forecasts they generate.

Example 3.11 Let us assume that 100 demand observations are available and we are considering two alternative forecasting algorithms. Also let us assume that the forecasting horizon is 1.

A first choice is to use 99 demand observations to set the parameters of the two models and compare them on their ability to predict the demand in the 100th period. In this case, the parameters of the two forecasting models are set very effectively and thus we compare the two forecasting models at their full potential. However, we are judging the quality of the two options on their errors in a single period. Thus, our conclusions have little statistical significance and might be wrong. In other words, we might choose the forecasting method with an higher error simply because it was "lucky" in the one period we used to compare our two alternatives.

On the other hand, we might be tempted to use very few demand observations (in the extreme case, just 1) to set the parameters of the two models so that we can enjoy a fairly large test sample (in the extreme case, 99 periods). In this case, we compare the performance of the two forecasting methods over multiple periods, and thus conclusions might seem statistically reliable. However, in this case the parameters of both methods would be set poorly. Thus we might choose the method that requires less data to set up the parameters (often the simpler method) or the method that by pure chance got better parameters. Clearly, in both cases there is little guarantee that the best forecasting method is selected. □

---

[11] It is interesting to notice that more complex models tend to have more parameters and thus more degrees of freedom. This greater degree of flexibility makes them the perfect candidate to fit the past demand data. In principle, a model with 100 degrees of freedom can perfectly fit 100 demand observations. However, this does not necessarily mean that it will generate better forecasts. Actually, literature (e.g., [1]) shows that often more complex models have little or no advantage over simpler ones. Indeed, simple models can be crude, but from a statistical point of view they are actually more solid than complex ones. The latter, under perfect circumstances and with a lot of information, might perform better than simple ones, but in real-life situations they tend to perform rather poorly.

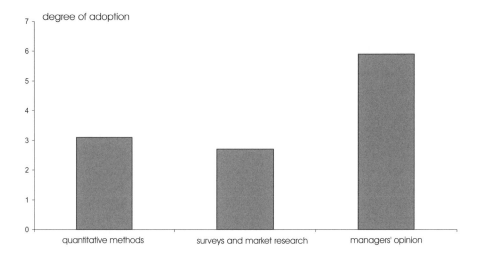

Fig. 3.3 Adoption of various categories of forecasting methods in the Italian mechanical machinery industry.

3.4 A CLASSIFICATION OF FORECASTING METHODS

Forecasting methods can be classified in two broad categories:

- quantitative methods

- qualitative methods

A large portion of this chapter is devoted to quantitative methods, as they can be properly described through formulas and equations, and this is one of the key features of this book. This does not mean that, from any practical standpoint, qualitative methods are less important or less performing than quantitative ones.[12] On the contrary, qualitative methods are widely adopted as figure 3.3 shows.

In general, we cannot say that one approach works better than the other. Rather, they have contrasting pros and cons. So there is no one-best-way but rather one shall choose the right blend of quantitative and qualitative methods according to the specific forecasting problem one is confronted with.

Qualitative forecasting methods are very flexible, since they do not require any explicit assumption on the relationship between the pieces of information

---

[12]Some articles (e.g., [19]) try to compare the performance of the two groups of forecasting methods. Often they find that quantitative methods are more accurate than qualitative ones. However, these research studies fail to account for the different demand patterns they try to predict. Often qualitative methods are used in more complex situations. Thus they can generate larger errors either because they face harder forecasting problems or because they are less accurate.

that are believed to be relevant and the forecast (i.e., inputs and outputs of the forecasting process). Basically, they are as flexible as the minds of human experts. Thus they can fit rather complex situations such as new product launches and/or long-term forecasts. However, these methods can accurately predict future demand only if the forecasters are true experts. So these methods must be expensive (i.e, they have to exploit a lot of very scarce and precious resources) to be effective (i.e., generate rather accurate forecasts). Thus these methods can be deployed only when the relevance of the issue at stake justifies the usage of such precious resources. Furthermore, qualitative methods can capture changes in the demand pattern, as human beings can capture a variety of variables, adding new ones as they become relevant. Often though, experts' forecasts are inconsistent: Human beings are unable to provide consistent estimates. This means that the same person facing the same evidence at different points in time might generate very different forecasts. Also, when experts are asked to forecast demand for each single item in a product family, the single numbers might be very well thought out, but the overall demand forecast for the product family might immediately sound unreasonable, let alone accurate.

Often incentives can push experts to overestimate/underestimate demand. For example, think about the incentive of the sales force to underestimate the demand in case it is used to set sales targets. On the other hand, think about the incentive of the sales force to overstate the demand forecast in case the forecast is used to set inventory targets: The higher the demand forecast, the higher the inventory level; this in turn implies more available products and easier sales.

Example 3.12 A large manufacturer of white goods has a 6 weeks rolling forecast. The total lead time for its products is roughly 3 weeks, so the most relevant forecasts are +1, +2, and +3 weeks. Other forecasts (+4, +5, and +6) are basically an advance information for the purchasing department and suppliers. The company has a team of forecasters that update the system forecast (i.e., a forecast generated by the company's IT systems) through their personal reading of demand trends (so-called "experience"). To drive their behavior, the company has designed an incentive scheme that rewards them on forecasting accuracy. The company has decided to reward them just on the accuracy of the +3 weeks forecast (i.e., the forecast three weeks into the future), to make the incentive scheme simple. The forecast for week +3 tends to be more accurate than the forecasts for longer horizons (+4, +5 and +6). Quite interestingly, though, forecasts for weeks +2 and +1 are just as accurate as the +3 weeks forecast. Managers were surprised to see such a pattern, as they expected the forecasters to collect more information and thus be more accurate. Actually, this odd result has more to do with incentives than with information or forecasting. Indeed, the forecasters did exactly what managers, through the incentive scheme, told them to do: improve accuracy of the forecast three weeks into the future and disregard any further potential improvement of the forecast. Interestingly though, this was not a deliberate choice. Indeed, the purchase plan is driven by the +3 weeks forecast, but the assembly and distribution operations are driven by the +1 and +2 weeks forecast. So a relatively poor short-term forecast (i.e., a forecast that is less accurate than it could be for weeks +1 and +2) might be quite expensive for the company. □

On the contrary, quantitative methods require an explicit assumption on the demand behavior (e.g., a seasonal, rather than stable, or linearly increasing demand). This makes them less flexible. If the demand behavior changes, the forecasting method performs very poorly. Nonetheless, these methods are more efficient, as a fairly large number of products and markets can be managed with very limited resources. Also, these methods provide very consistent results since computers will do the same task over and over again and are not influenced by any kind of incentive scheme. So, however wrong they might be, one can track their performance, spot their weaknesses, and hopefully correct them over time.

Example 3.13 A retail company in the furniture business used to adopt qualitative methods to forecast demand (and plan inventories). When two forecasters faced with the same demand pattern were asked to predict future demand, they could provide fairly different forecasts. What is even more interesting is that the same person would generate different figures on different days. When the company switched to quantitative methods, it was able to double the store/items combinations each forecaster could manage. Also, when four forecasters out of four left the company, the company could still operate normally, as the four new employees now in charge of forecasting could leverage on the demand knowledge built into the company's systems. We cannot tell what would have happened had the forecast been completely qualitative. Still, we can argue that it would have been harder to survive the change. □

Broadly speaking, quantitative methods consist of two subfamilies.

- **Time-series models** basically look at the past demand pattern over time and extrapolate future demand levels. In time-series models, we only look at demand data over time and thus do not account for variables that might influence them such as price, weather, competition, distribution, promotions, advertising etc. In these models we only have two variables, demand and time. Therefore, they are effective only when demand changes depend on time. Demand can be stable, might be growing over time, and can show periodic fluctuations (weekly, monthly quarterly or yearly fluctuations). These models, however, fail to work properly when other variables play a major role and determine significant changes in demand. Time-series models are the most widely adopted quantitative forecasting technique and a wide array of algorithms are investigated in sections 3.5–3.9.

- **Explanatory models** try to find a relationship between demand and some explanatory variables such as price, promotion, time, etc., that drive it. These models are often called **causal models**, as most variables they use might cause changes in demand. We call them explanatory models, since actually the statistics behind the models hardly provide any causal relationship. Rather, these models simply observe that when price goes down, demand goes up, and thus they predict that if in the future the company will reduce price, demand will go up again. So *our reading* of the models is causal, while they simply observe that low prices go together with high demand. The most basic explanatory model, simple linear regression, is discussed in sections 3.10 and A.10.

Qualitative and quantitative methods are often presented as alternative solutions. On the contrary, in many contexts they can be integrated to exploit the respective strengths. A blend of the two approaches can enjoy the flexibility and reactivity of humans and the consistency of an algorithm. Actually, there is a growing body of evidence that a combination of quantitative methods and qualitative ones can outperform both purely quantitative and purely qualitative methods (e.g., see [19]). Quantitative methods can generate a forecast for a large number of product/market combinations. The quantitative method might be based on a simple assumption of demand behavior, but still it provides very consistent forecasts. The outcomes of this first forecasting process can then be controlled and, eventually, adjusted by human experts to account for all the variables and phenomena the quantitative systems fails to account for properly.[13]

For example, a number of quantitative methods analyze time series of demand and extrapolate some sort of demand pattern (steady, linear, seasonal etc.) from past observations. These methods, by their very nature, fail to capture the effect of variables that change the demand pattern, such as the launch of competitive products. For example, a quantitative method could generate a demand forecast of 100 tons for a given kind of fresh filled pasta. However, an expert might reduce this forecast as a new competing product is being launched and it might be expected to partially cannibalize the existing one. Notice that the role of the quantitative method is to (i) provide a point of reference so that the expert can just focus on the net effect of the launch of the new product on the demand for the existing one and (ii) take

---

[13] Notice that this is one of the basic problems with neural networks. Neural networks are a forecasting technique that tries to simulate the learning process of human brain. The good thing about this forecasting technique is that the user is not forced to make any assumption about the demand pattern. Basically, we let the neural network observe past data and try to figure out a pattern. One of the flipsides of this model is that under these circumstances a human being trying to improve the system forecast would not know the assumptions behind the forecast, and thus he/she would hardly be in a position to properly modify and improve the forecast to account for other phenomena the system might have overlooked (e.g., because no quantitative data are available or because some new trend or relevant variable is emerging).

off the forecaster's table all the items that do not require specific attention, as demand is relatively stable, so that the forecaster can devote as much time as he/she needs to understand what is going on with products that face unusual conditions (that is, conditions that do not fit the assumptions behind the quantitative model).

Also, the integration between qualitative and quantitative models can be a sort of weighted average of qualitative and quantitative methods. This, say parallel, method of integration is actually seldom used though its effectiveness has been often proven in literature.

Finally, qualitative forecasts can be used as an input to a quantitative forecast. For example, experts' opinions can be the independent variables of a linear regression. Also, market research can be one of the key inputs to estimate the market potential of a new product. In this case as well, the quantitative methods can read the signal in experts' opinions, but at the same time they can account for any bias they might have. So, they can generate a consistent forecast as they correct for bias, but they still can be accurate since they exploit experts' knowledge.

The bottom line is that one can generate a consistent but still flexible forecasts by blending the qualitative and quantitative forecast in various ways.

3.5 MOVING AVERAGE

Moving average is the simplest time-series model. In this class of models we analyze past demand patterns to extrapolate a future forecast. All these models make an assumption about the pattern of demand: They try to identify it in past data to project it into the future. Hence, the performance of these forecasting techniques really depends on whether the underlying assumptions fit the actual demand pattern. This is why we devote a specific section to describe the basic assumptions on demand that each model in this class relies on.

3.5.1 The demand model

A first forecasting model is moving average. The assumption behind this approach is that demand is steady, as we expect neither major trend (neither downward nor upward) nor periodic fluctuations (seasonal patterns).

More formally, we assume that demand data are generated by a process like

$$Y_t = \mathrm{E}(d_t) + \epsilon_t, \tag{3.9}$$

where $\mathrm{E}(d_t)$ is the expected demand, which is an unknown parameter we want to estimate, and ϵ_t is a noise term such that $\mathrm{E}(\epsilon_t) = 0$. Actually, we do not expect $\mathrm{E}(d)$ to be truly steady but we expect smooth and random fluctuations of the expected value over time:

$$\mathrm{E}(d_t) = \mathrm{E}(d_{t-1}) + e_t, \tag{3.10}$$

where $E(e_t) = 0$.

Given these assumptions, at any given point in time t the demand forecast is the same for all future periods ($F_{t,h}$ does not really depend on h). Indeed, we assume demand to be statistically stationary and thus have no reason whatsoever to expect an increase or a decrease in demand. This does not mean that the forecast cannot be updated. Actually, as more recent observations of demand are collected, the demand forecast is updated ($F_{t,h}$ does depend on t, but not on h).

3.5.2   The algorithm

The moving average algorithm estimates the level demand (so called baseline demand) B_t for the future as the average of the last k demand observations.

$$B_t = \sum_{i=t-k+1}^{t} \frac{Y_i}{k}. \tag{3.11}$$

We can think of k as a "time window" which we apply to past data to include only the most recent ones. Also, given the assumptions of this model, we predict a flat demand for any future period:

$$F_{t,h} = B_t \qquad \forall h. \tag{3.12}$$

3.5.3   Setting the parameter

To use the moving average method, we shall set the parameter k, that is, the number of demand observations we want to use to generate the forecast. To select this parameter, we face a tradeoff between:

- The ability of the model to filter noise, that is, to avoid overreactions to demand observations that are significantly above or below the average.

- The ability of the model to promptly react to changes in demand such as a sudden increase or decrease in expected demand.

If a *large* value of k is chosen, the moving average method shows a strong inertia. On the one hand, a single observation significantly above (or below) the average has little consequence. On the other hand, it takes time for the model to adapt to any significant change in average demand. So, in this case the moving average *filters noise* very effectively, but it *adapts* to changes in demand *slowly*.

On the contrary, if a small value k is chosen, a single demand observation has a great deal of influence on the future forecast (to an extreme, if $k = 1$ the forecast just equals last demand observation). Thus a small k makes the moving average very *reactive* but at the same time very *sensitive to noise*. In other words, demand observations significantly above or below the average lead to bumps in the demand forecast, which turn into a larger forecasting error.[14]

Figures 3.4–3.7 show the behavior of moving average with various values of k. The examples consider the moving average with $k = 2$ and $k = 6$, and a forecasting horizon of one period ($h = 1$).

When we analyze the performance of the moving average with a statistically stationary demand (figure 3.4 e 3.5) we can see that:

- The moving average with time window 6 ($k = 6$) requires a longer initialization;

- The moving average with $k = 6$ is more stable than with $k = 2$. This leads to more accurate forecasts, if the expected demand is stable (and the random part of the demand is not auto-correlated, that is variables ϵ_t in equation (3.9) are independent—see definition A.11). In these cases, stable forecasts are more effective simply because fluctuations in forecasts add to the fluctuations in demand and tend to increase the gap between the two variables, that is, the forecast error.[15]

In the case of the demand patterns displayed in figures 3.4 and 3.5, $k = 6$ guarantees more accuracy than $k = 2$ (RMSE is 7.67 and 10.11, respectively, while MAD is 6.96 and 8.26, respectively).[16]

Figures 3.6 and 3.7 show how the moving average reacts to an odd demand observation that significantly differs from the mean. The figures show that the reaction to the anomaly is definitely larger in the case of $k = 2$ than in the case of $k = 6$. However, the effect of the odd observation lasts longer in the case of $k = 6$. Indeed, in the case of $k = 2$ the anomaly in period 15 quickly exits the sample we consider to generate the new forecast. This really means that if $k = 2$ the effects of the outlier are not larger but simply more concentrated over a shorter period of time. While the differences in MAD are negligible (MAD is 38.7 and 37.2 for $k = 6$ and $k = 2$ respectively), the differences in RMSE are sizable (RMSE is 82.6 and 71.8, respectively) since RMSE penalizes larger errors (see section 3.3).

---

[14]We basically add the fluctuations of demand to the fluctuations of forecast in a scenario where expected demand is stable.

[15]If the process is truly stationary, there is no reason whatsoever to consider only the last k demand observations. If demand is truly stationary, we should simply take the average of all demand observations we have. However, in real-life contexts, this situation is hardly the rule. So we only consider the last k demand observations, as we believe them to be the only relevant ones to estimate future demand. Adding an extra observation from the past adds more information on the one hand, and thus should increase accuracy, but on the other hand it reduces the quality of our inputs, as the older the data, the least significant they are to predict future demand.

[16]Notice that we only use periods 7 to 30 to measure accuracy so that performance of both alternatives are measured on the same sample.

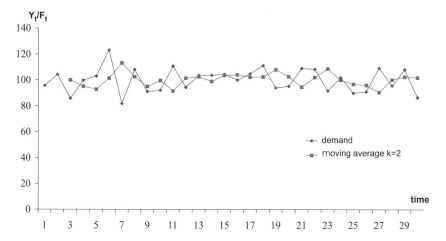

Fig. 3.4 Behavior of moving average: case of $k = 2$, stationary demand.

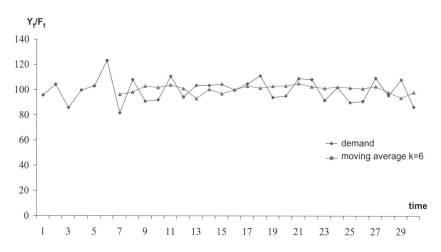

Fig. 3.5 Behavior of moving average: case of $k = 6$, stationary demand.

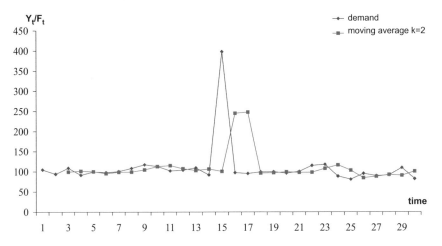

Fig. 3.6 Behavior of moving average: $k = 2$, demand featuring a pulse.

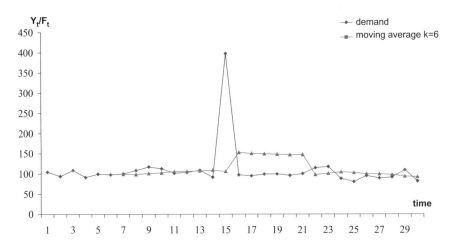

Fig. 3.7 Behavior of moving average: $k = 6$, demand featuring a pulse.

Also, figures 3.6 and 3.7 show that when one uses time-series models, outliers cause forecasting errors both when they occur (as they are unpredictable for time series models) and in successive periods as they bias forecasts.

The previous examples show that, for "large" values of k, the moving average "filters noise" very well, that is, it effectively tells the average behavior of demand from random short-term fluctuations. The example of figures 3.8 and 3.9 shows that large values of k entail a poor reactivity of the model, that is, they limit the ability to adapt to changes in expected demand. In the case of $k = 2$ the moving average completely "forgets" the previous behavior of

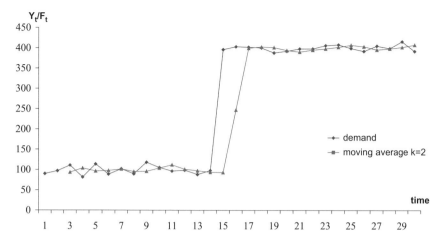

Fig. 3.8   Behavior of moving average: $k = 2$, demand is a step function

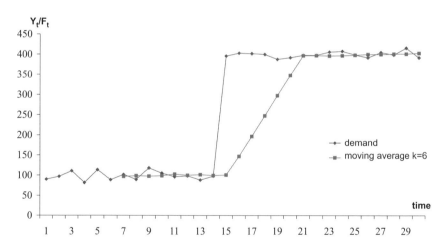

Fig. 3.9   Behavior of moving average: $k = 6$, demand is a step function

demand while in the case of moving average with step 6 ($k = 6$) the transient state is much longer and thus the accuracy is worse (MAD is 29.0 and 56.2 while RMSE is 76.7 and 107.1, respectively).

3.5.4   Drawbacks and limitations

The moving average is a rather simple forecasting method that is widely used. However, it has drawbacks and limitations. This method gives an equal weight $1/k$ to the last k demand observations, while it totally neglects previous ones.

Table 3.11 Demand data for vanilla ice cream

period	1	2	3	4	5	6	7	8
Y_t	116.36	96.30	109.64	99.92	110.31	99.88	89.07	107.38

period	9	10	11	12	13	14	15	16
Y_t	121.21	100.99	89.63	88.43	83.83	95.87	102.17	103.43

period	17	18	19	20	21	22	23	24
Y_t	104.55	88.19	98.53	103.58	87.95	110.83	103.87	115.57

One could think that it might be more reasonable

- To give more recent observations a greater weight than more remote ones; for example, one might want to give more weight to observation t than to observation $t - 1$;

- To give even more remote demand observations a nonzero weight.

Example 3.14 Let us consider a store that sells ice cream on a beach. The demand for vanilla ice cream over the last 24 days is shown in table 3.11. Demand is rather stationary with some minor variations.

The lead time is two days and deliveries are daily. This means that the time bucket is the single day and the forecasting horizon is two days ($h = 2$) The manager of the store is trying to predict future demand with the moving average algorithm. He wonders whether he shall be using moving average with $k = 2$ or $k = 5$.

To choose between the two options, we can measure which one would have performed better in the past, assuming that the option that would have worked better in the past is going to be the better performer in the future as well. The moving average with step 5 ($k = 5$) can generate the first forecast only in period 5. Our horizon consists of two periods; hence, in order to get a fair comparison, we are going to compare the accuracy of the two parameters in periods 7 to 24.

Let us take you through the forecast generated in period 5 for period 7, i.e., $F_{5,2} = F_7$:

- If $k = 2$, the forecast generated in period 5 is the average of demand in period 4 and in period 5. So, $F_{5,2} = F_7 = (99.92 + 100.31)/2 = 105.12$. Given the demand in period 7 $Y_7 = 89.07$, the error is $e_7 = 89.07 - 105.12 = -16.05$.

- If $k = 5$, the forecast generated in period 5 is the average of the demand in the first 5 periods. So $F_{5,2} = F_7 = (116.36 + 96.30 + 109.64 + 99.92 + 100.31)/5 = 106.51$. So the error in period 7 is $e_7 = 89.07 - 106.51 = -17.44$.

Table 3.12   Forecast with step 2 $(k = 2)$

period	7	8	9	10	11	12	13	14	15
F_t	105.12	105.10	94.48	98.23	114.30	111.10	95.31	89.03	86.13

period	16	17	18	19	20	21	22	23	24
F_t	89.85	99.02	102.80	103.99	96.37	93.36	101.06	95.77	99.39

Table 3.13   Error with step 2 $(k = 2)$

period	7	8	9	10	11	12	13	14	15
e_t	-16.045	2.285	26.735	2.765	-24.665	-22.67	-11.48	6.84	16.04

period	16	17	18	19	20	21	22	23	24
e_t	13.58	5.53	-14.61	-5.46	7.21	-5.41	9.775	8.105	16.18

Table 3.14   Forecast with step 5 $(k = 5)$

period	7	8	9	10	11	12	13	14	15
F_t	106.51	103.21	101.76	101.31	105.57	103.71	101.66	101.53	96.82

period	16	17	18	19	20	21	22	23	24
F_t	91.75	91.99	94.75	97.97	98.84	99.37	99.66	96.56	97.82

Table 3.15   Error with step 5 $(k = 5)$

period	7	8	9	10	11	12	13	14	15
e_t	-17.44	4.17	19.45	-0.32	-15.94	-15.28	-17.83	-5.66	5.35

period	16	17	18	19	20	21	22	23	24
e_t	11.68	12.56	-6.56	0.56	4.74	-11.42	11.17	7.31	17.75

We can repeat this process for $t = 8, ..., 24$ and obtain tables 3.12 and 3.13, which show the forecasts and errors, respectively, in the case of $k = 2$, and tables 3.14 and 3.15 that show the forecasts and errors, respectively, in the case of $k = 5$.

Finally, with the error data we can compute accuracy metrics. For example, the RMSE is 13.95 for $k = 2$ and 11.90 for $k = 5$. Thus we draw the conclusion that we would rather select $k = 5$.   ⬜

3.6   SIMPLE EXPONENTIAL SMOOTHING

3.6.1   The demand model

The drawbacks discussed in the previous section suggest replacing the moving average with the simple exponential smoothing method. This method assumes the very same demand behavior and pattern as the moving average, that is, a stochastic but stationary demand or a demand with very smooth changes in the expected demand.[17]

3.6.2   The algorithm

In the simple exponential smoothing technique, the current level of demand is estimated through a weighted average of the last demand observation Y_t and the previous estimate of the demand level B_{t-1}. This method increases the previous estimate B_{t-1} when actual demand Y_t is greater than we had estimated, while it reduces the previous estimate B_{t-1} when demand Y_t turns out to be lower than we thought:[18]

$$B_t = \alpha Y_t + (1 - \alpha)B_{t-1}, \qquad 0 \le \alpha \le 1. \tag{3.13}$$

Also, given the assumption of a stationary demand over time, the forecast generated at a given point in time t is the same for all forecasting horizons h:

$$F_{t,h} = B_t \qquad \forall h.$$

Notice that, just like in the case of moving average, we keep on updating the estimate of the demand level B_t. So $F_{t,h}$ really depends on t but does not depend on h. Given equations (3.13) and (3.6.2), we can also write

$$F_{t,h} = \alpha Y_t + (1 - \alpha)F_{t-1,h}, \qquad 0 \le \alpha \le 1 \qquad \forall h. \tag{3.14}$$

In this forecasting model, α is a parameter between 0 and 1 that determines the reactivity (i.e., promptness) of the model. Indeed, as α changes we change the weight of the most recent demand observation Y_t and of the previous expectation of demand B_{t-1}. If α is 1, the smoothing algorithm behaves just like a moving average with a unit time window ($k = 1$) and thus reacts very promptly to any change in demand.

If α is set to zero, then the previous estimate B_{t-1} is not affected by the last demand observation Y_t and thus $B_t = B_{t-1}$. This clearly makes the forecasting technique extremely stable. Also, noise has no influence whatsoever on future forecasts. However, this brings the forecasting technique to a standstill and the model cannot adapt to any change in demand.

---

[17] As we already discussed, for a truly stationary process the best estimate of the expected demand is the simple mean of all observations. If this was the case, taking only the last k observations would not make sense. Also, it would not make sense to give more recent observations a greater weight. If demand is really stationary, all observations are equally relevant and thus have the same weight.

[18] Notice that we assume that we update our forecast at each period. If the forecast is reviewed less frequently (say every j periods), we simply take the weighted average of the average of the last j demand observations and the level of demand at time $t - j$.

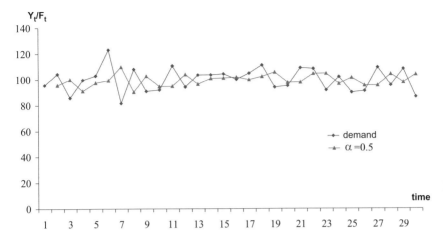

Fig. 3.10   Behavior of exponential smoothing: $\alpha = 0.5$, stationary demand.

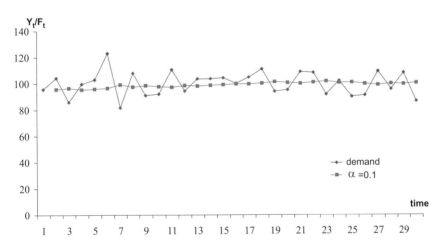

Fig. 3.11   Behavior of exponential smoothing: $\alpha = 0.1$, stationary demand.

The parameter α plays a role that is very similar to the role of k in the moving average technique. Figures 3.10–3.15 show that the pattern we get with a low α resembles the one we get with a large k and vice versa. Figures with $\alpha = 0.1$ show a rather inertial behavior, but also a great ability to filter noise, just like in the case of $k = 6$ for the moving average. Figures with $\alpha = 0.5$ resemble the ones with $k = 2$, as both techniques show a good reactivity, but a poor ability to filter noise.

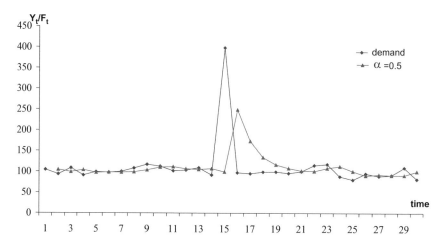

Fig. 3.12   Behavior of exponential smoothing: $\alpha = 0.5$, demand featuring a pulse.

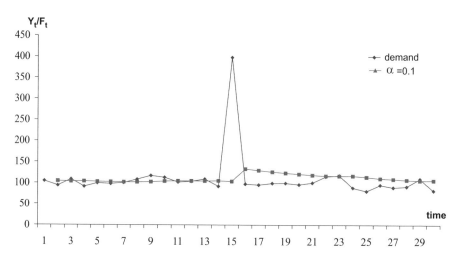

Fig. 3.13   Behavior of exponential smoothing: $\alpha = 0.1$, demand featuring a pulse.

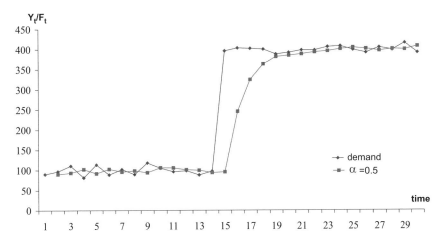

Fig. 3.14   Behavior of exponential smoothing: $\alpha = 0.5$, step demand.

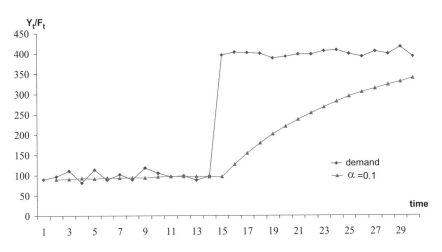

Fig. 3.15   Behavior of exponential smoothing: $\alpha = 0.1$, step demand.

We can elaborate on equations (3.13) and (3.14) to derive two formulations that provide interesting insights. A first reformulation is

$$F_{t,h} = B_t = \alpha Y_t + (1 - \alpha)B_{t-1} = B_{t-1} + \alpha(Y_t - B_{t-1}) \qquad \forall h. \qquad (3.15)$$

In other words, the new forecast generated at time t is equal to the previous one generated at time $t-1$ ($F_{t-1,h} = B_{t-1}$) plus a term smoothed through the parameter α, which can be interpreted as the error we made while attempting to forecast demand Y_t at time $t - 1$. Indeed, as the formula holds for all h we can set $h = 1$ and read the smoothed factor as $Y_t - F_{t-1,1}$. Thus exponential smoothing can be interpreted as a method that tends to correct the error by reducing the forecast when errors are positive and by increasing it when errors are negative.

We can provide a second reading by exploiting the recursiveness of equation (3.13):

$$B_{t-1} = \alpha Y_{t-1} + (1 - \alpha)B_{t-2}.$$

By substituting in equation (3.13), we obtain

$$
\begin{aligned}
B_t &= \alpha Y_t + \alpha(1 - \alpha)Y_{t-1} + (1 - \alpha)^2 B_{t-2} \\
&= \alpha Y_t + \alpha(1 - \alpha)Y_{t-1} + \alpha(1 - \alpha)^2 Y_{t-2} + (1 - \alpha)^3 B_{t-3} \\
&= \alpha Y_t + \alpha(1 - \alpha)Y_{t-1} + \alpha(1 - \alpha)^2 Y_{t-2} + \alpha(1 - \alpha)^3 Y_{t-3} + (1 - \alpha)^4 B_{t-4}
\end{aligned}
$$

$$\cdots$$

This formulation shows that exponential smoothing gives past demands a weight that decreases with the time elapsed since the demand observation. The weight of the demand observation at time $t - i$ is a decreasing function of i. Figure 3.16 shows the pattern of these weights with various levels of α.

For low α the weight of observation Y_t is very similar to the weight of observation Y_{t-1}, and so on. On the contrary, for high values of α the weight of observation Y_{t-1} is significantly lower than for the latest observation Y_t.

Also, we can use the properties of geometric series to show that the sum of all weights is just 1, as one would intuitively expect.[19] This property also suggests that all demand observations prior to $t - 20$ have an overall weight that is equal to 1 minus the sum of weights of all demands from period $t - 20$ to t. Figure 3.16 shows that in case of a very small α, the weight of the "remote past" is fairly relevant (see "other periods" in the figure).

3.6.3   Setting the parameter

The above analysis suggests that high values of α enjoy reactivity, that is the ability to promptly react to changes in average demand, whereas low values of α filter noise very effectively. This is why in real-life situations the choice

---

[19]In case we sum the weights of an infinite number of demand observations.

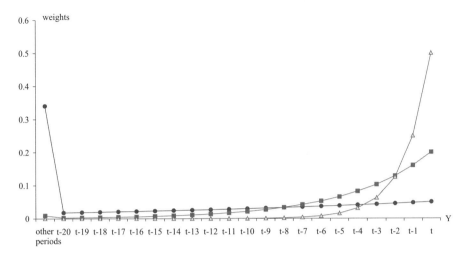

Fig. 3.16 Weights of the demand observations with various levels of α.

of α (and more generally all smoothing parameters that are presented in this chapter) should be dynamically adapted to the changes in demand. We shall increase α, as demand is going through a period of changes, while we shall reduce it when we expect demand to be rather stable and we only observe random fluctuations around the mean demand.

To support the choice of the appropriate level of α, we can use the *tracking signal* (TS_t):

$$TS_t = \alpha' \frac{e_t}{Y_t} + (1 - \alpha') TS_{t-1}, \qquad 0 \le \alpha' \le 1.$$

The tracking signal is basically a smoothed average of most recent errors. The logic behind this tool is that if expected demand is relatively stable, the demand forecast is unbiased, however inaccurate it might be. Thus, errors are positive in some periods and negative in other periods: They tend to cancel out and the tracking signal tends to be close to zero. On the contrary, if demand starts growing (or decreasing), exponential smoothing generates conservative (optimistic) forecasts and errors tend to be positive (negative). Thus errors tend to add up rather than cancel out, and the tracking signal (TS_t) significantly differs from zero.

TS_t signals the tendency of demand to increase (decrease) as it significantly differs from zero. So it can be used to decide when to choose large values of α (tracking signal differs from zero) and when to choose small ones (tracking signal close to zero).

The choice of the appropriate values of α (and more generally the parameters of the smoothing algorithms) is a key lever to control and improve the forecasting process. So, in general, it requires some managerial attention. Most software (and even Excel, if managed properly) can automatically search for optimal values of α that can then be used for a while. In the case of a very large numbers of time series (e.g., cheap products sold in various markets) the continuous control of a lot of parameters can be fairly expensive and not worth the effort. This is why one might consider the so-called *self-adaptive* methods that self-select the parameters according to the demand patterns.

In general, in a self-adaptive method, the value of α depends on the tracking signal, that measures the rate of change of demand. A possible choice is to set $\alpha = a \cdot |TS_t|$, that is α changes proportionally to the absolute value of TS_t. The parameter a is often set to 1.

3.6.4   Initialization

Equation (3.15) highlights one of the key issues for this forecasting algorithm: It is recursive and generates a new estimate of the expected demand in period t with a previous expectation from period $t-1$. However, this method needs a starting point, i.e., an estimate to start from. We call this the initialization of the smoothing algorithm, that is, the generation of the first estimate B_0.

Before we get into the details of how one can initialize the forecast, we notice that this can be a fairly important issue. Indeed, we can show that the initial estimate B_{t-I} can have a significant impact on the forecast we generate at current time t. Let us assume we have a set of I demand observations, $Y_{t-I+1}, \ldots, Y_t$. It is easy to show that the initial estimate (that refers to period $t-I$, that is the period before the start of our sample) is updated I times and thus has a weight of $(1-\alpha)^I$ (this means that a percentage $(1-\alpha)^I$ of the current estimate B_t depends on the initial one B_{t-I}). Thus for high values of α and low values of I, the initialization plays a key role, as a large portion of the current estimate B_t depends on the initial one (B_0). When α is low (thus, the estimate remains stable over time) and the initial estimate is updated a limited number of times I, the initial estimate B_{t-I} can be the single most relevant "ingredient" of the final estimate B_t and of the forecast for future demands. Oddly, it can be even more important than the last demand observation (see figure 3.16, case of $\alpha = 0.05$).

There are several approaches to set B_{t-I}:

1. A first option is to start with a zero estimate ($B_{t-I} = 0$). This makes the initial estimate biased. In case of low values of α and I, this makes the current estimate B_t and future forecasts $F_{t,h}$ significantly biased as well (see figure 3.17).

2. A second option is to set $B_{t-I} = Y_{t-I+1}$, that is, we set the first estimate of demand level equal to the first demand observation. Apparently this is unfair cheating. It seems we are using demand in period $t - I + 1$ to predict the demand itself, since by setting $h = 1$ we have $B_{t-I} = F_{t-I,h} = F_{t-I,1} = Y_{t-I+1}$, that is we are guaranteed not to make any error in the first forecast. In other words, we are violating the basic principle of non-anticipation. However, we must keep in mind that we

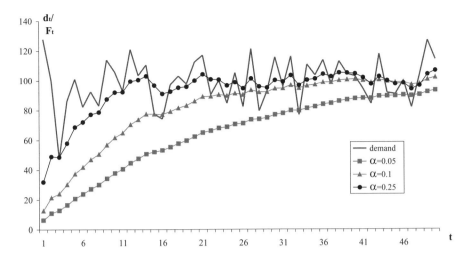

Fig. 3.17   Case of initial forecast equal to 0 with various levels of α.

are using a *fit sample* (see section 3.3.7); that is, we are using demand data to initialize the forecasting process. We shall simply be careful and fair when we judge the performance of our forecasting method. When we measure the performance of our method, we shall use a *test sample* that does not contain any data we have used to initialize the exponential smoothing technique.

Using demand Y_{t-I+1} to set the initial estimate of the baseline demand B_{t-I} is totally acceptable. We must be careful not to use it to judge the quality of our forecast. Therefore, when we use this approach the first demand observation Y_{t-I+1} cannot be used to measure the accuracy and bias of our forecasting process. To put it in a different way, we use the initial value to initialize the estimate of demand but we do not use it to forecast.

This second approach provides an initial estimate B_{t-I} that is not blatantly biased like in the former case. Nevertheless, it might significantly differ from the average demand since it is based on a single demand observation that might be affected by noise (see figure 3.18).

3. A third approach is designed to partially fix the problems we have just highlighted. We can use the average of the first l periods to initialize

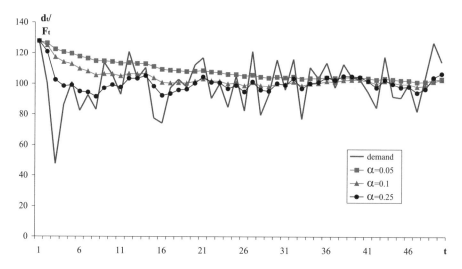

Fig. 3.18 Case of initial forecast equal to the first demand observation with various levels of α.

the estimate of demand level[20]:

$$B_{t-I} = \sum_{i=t-I+1}^{t-I+l} \frac{Y_i}{l}.$$

In this case, the initialization is based on l periods rather than a single one. Thus it can capture more accurately the long run average demand (see figure 3.19). However, this approach too has a side effect: We cannot use l periods to judge the quality of the forecasting process. For these periods the demand forecast depends on (i.e., exploits the information about) the demand itself (the forecast depends on the initialization that in turn depends on the demand during the first l periods).

This is actually a minor problem, when one just wants to generate a demand forecast in current period t. However, when one wants to in-

---

[20]Notice that we use l periods but still initialize at period $t-I$, that is we initialize as far back into the past as possible. Indeed, one could be tempted to set $B_{t-I+l} = \sum_{i=t-I+1}^{t-I+l} Y_i/l$, or even worse $B_t = \sum_{i=t-I+1}^{t-I+l} Y_i/l$. Actually, the initialization procedure is just a violation of the basic mechanics of this forecasting process that is based on progressive updates of previous estimates of demand. The more the initialization is set far into the past, the more time the exponential smoothing has to actually update demand and to limit the effect of the initialization. On the contrary, if we set $B_{t-I+l} = \sum_{i=t-I+1}^{t-I+l} Y_i/l$, we increase the weight of the initialization by a factor $1/\alpha^I$. Finally, if we set $B_t = \sum_{i=t-I+1}^{t-I+l} Y_i/l$. Basically, the first forecast $F_{t,h} = B_t$ is not based on any sort of exponential moving average, but rather on a simple average of demand observations that might not even be recent. Thus, we would simply be using a different forecasting method rather than the one we believe is appropriate for our forecasting problem.

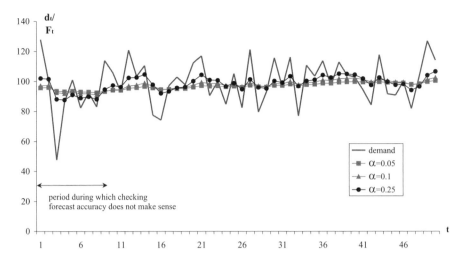

Fig. 3.19 Case of initialization equal to the average of the first 10 periods with various levels of α.

vestigate the performance of various methods (or various sets of parameters) to select the best one, we have to set aside a "test sample" to measure the forecasting errors. Thus we face a tradeoff between (i) the quality of the initialization and (ii) our ability to judge what is the best forecasting method (or the best set of parameters).

Figures 3.17–3.19 show the behavior of the smoothing algorithm under the three initialization policies. Figure 3.17 shows that setting the initial forecast to zero leads to a biased forecast during the first periods. The duration of this transient state depends on α: The higher the value of α, the more quickly the initial forecast loses weight and the forecast reaches steady state values.

Figure 3.18 shows that, in the second case, initialization is no longer biased, but it can be fairly inaccurate, as it is based on a single demand observation. So, also this method for initialization can generate fairly inaccurate forecasts for the first few periods, especially in the case of low α.

Figure 3.19 shows that the third choice usually guarantees a better initialization. This is particularly important in the case of low α. However, we also notice that we can measure the accuracy of the forecasting algorithm from period 11 onward, as the first 10 observations were used to initialize the forecasting algorithm. In the second case, instead, one can start measuring accuracy in the second period.[21] Hence, the third option tends to provide a better initialization, but it "consumes" a lot of data and we might be left

---

[21] Notice that we shall be expecting larger errors in the early periods as the forecasting technique is basically drawing conclusions on very small samples. So in an odd way even the second option might be misleading, as it might lead us to prefer the option with lower data

with a small test sample. So, in an odd way the error might be smaller but we have a limited ability to properly quantify it.

3.6.5   Drawbacks and limitations

Both forecasting methods presented so far are designed to manage a rather simple demand pattern, that is, a basically stable demand with random fluctuations. These simple methods are effective to the extent that their underlying assumptions hold. For example, when demand is expected to grow or vary according to season, these forecasting methods are inaccurate. The next three sections present forecasting techniques that are suited for these more complex and nonstationary demand scenarios.

3.7   EXPONENTIAL SMOOTHING WITH TREND

3.7.1   The demand model

Trend can be interpreted as the consistent change (growth or decrease) of expected demand over time. Demand growth (decrease) can be either:

- Linear, that is, we observe a constant increase; e.g., demand increases by 20 units per period, or

- Exponential, that is, a constant percentage change; e.g., demand grows by +20% per period.

In this book we discuss in full depth the linear trend model, while exponential trend is just presented briefly. The logic of, drawbacks of, limitations of, and comments about the linear trend model apply, *mutatis mutandis*, to the exponential trend model as well.

3.7.2   The algorithm

The exponential smoothing with trend algorithm uses two parameters, as the demand model is more complex. The two parameters are:

- B_t, the baseline demand (or level of demand) in period t;

- T_t, the trend of demand in period t; demand is growing when T_t is positive and is decreasing when T_t is negative.

The demand model is nonstationary and thus the forecast depends on the forecasting horizon. For example, if we expect a growth of 10 units per month,

---

requirements, which more quickly "forgets" an erroneous initialization, or simply performs better by pure chance.

demand for the next month is lower than the forecast for the following months. In particular, given the linear trend assumption, we obtain:

$$F_{t,h} = B_t + hT_t. \tag{3.16}$$

In other words, the demand forecast for period $t + h$ equals the baseline level in period t plus h times the growth we expect in a single period.

This method uses the exponential smoothing logic to update the two parameters B_t and T_t. As to the baseline level B_t, we use the last demand observation to update previous estimates, just like in the case of simple exponential smoothing. However, in this case we cannot just average the last demand observation Y_t with the last estimate of the baseline demand B_{t-1}, as they are actually hardly comparable numbers. In an odd way, they are apples and oranges. Indeed, we are facing a nonstationary process and thus the baseline at time $t - 1$ and the demand at time t are not directly comparable figures. Actually, demand at time t can be compared with the most recent forecast of demand at time t, that is, the one generated at time $t - 1$ ($F_{t-1,1} = B_{t-1} + 1 \cdot T_{t-1}$). Thus we can take a weighted average between actual demand and our latest expectation about it. Basically, we increase our expectations if demand has exceeded them and reduce them in case demand was lower then we thought:

$$B_t = \alpha Y_t + (1 - \alpha)(B_{t-1} + T_{t-1}), \qquad 0 \leq \alpha \leq 1. \tag{3.17}$$

As to the trend factor, we shall update the last period's estimate with the latest observation of the demand growth (decrease). We can measure the last growth of demand through $B_t - B_{t-1}$ and thus update the trend factor as follows:

$$T_t = \beta(B_t - B_{t-1}) + (1 - \beta)(T_{t-1}), \qquad 0 \leq \beta \leq 1, \tag{3.18}$$

where β is a second smoothing factor that is used to update the trend parameter.

3.7.3   Setting the parameters

This forecasting method uses two parameters α and β so that we can differentiate the speeds at which estimates B_t and T_t are updated. The effects of a high rather than low value of β resemble those of α. However, the trend factor T_t depends on α as well as on β. If α is low, then the difference $B_t - B_{t-1}$ is basically equal to T_{t-1} and the trend factor changes very slowly.[22]

---

[22]Notice that one could also update the trend through the difference between the last two demand observations.

$$T_t = \beta(Y_t - Y_{t-1}) + (1 - \beta)(T_{t-1}), \qquad 0 \leq \beta \leq 1. \tag{3.19}$$

3.7.4   Initialization

In the case of exponential smoothing with trend, we have to initialize two parameters. Thus, a reasonable initialization requires at the least two demand observations to be in a position to judge whether demand is increasing rather than decreasing over time. Assuming that we choose to initialize the algorithm with this minimal information set, we can take the difference between two successive periods to estimate the demand trend[23]:

$$T_0 = Y_2 - Y_1.$$

As we have already discussed in the case of simple exponential smoothing, this procedure apparently violates the principle of nonanticipation, since T_0 incorporates demands from periods $t = 1$ and $t = 2$. In fact, we are just using a fit sample to initialize the smoothing algorithm. The same concept applies to all smoothing algorithms and initialization policies presented in the remainder of this chapter and we are not going to get back to it.

As to the initialization of the baseline demand at time 0, we shall properly exploit the demands of periods 1 and 2. Given the nonstationary demand process, demand Y_1 and Y_2 cannot be directly used to estimate the baseline level of demand at time 0. To use demand in $t = 1$ to estimate the baseline at time 0, we should remove the demand trend from it. To properly use demand observation from period $t = 2$, we subtract the trend twice:

$$B_0 = \frac{(Y_1 - 1 \cdot T_0) + (Y_2 - 2 \cdot T_0)}{2}.$$

These initial values depend on the specific demand realizations (draws) of the first two periods. Thus this approach can lead to significantly wrong estimates. For example, the initial demand trend T_0 might be negative even when demand tends to grow over time. Just like in the case of simple exponential

---

This is a more nervous statistic than the difference between the two most recent baseline demands $B_t - B_{t-1}$. Indeed, one can plug equation (3.16) into equation (3.18) and get

$$T_t = \beta[\alpha \cdot Y_t + (1 - \alpha) \cdot (B_{t-1} + T_{t-1}) - B_{t-1}] + (1 - \beta)(T_{t-1}), \qquad 0 \leq \beta \leq 1. \quad (3.20)$$

Hence

$$T_t = \beta \cdot \alpha[\cdot Y_t - B_{t-1}] + (1 - \beta \cdot \alpha) \cdot T_{t-1}, \qquad 0 \leq \beta \leq 1. \quad (3.21)$$

The previous equation shows that the weight of the previous trend T_t is very large in equation (3.18). For example, in case of $\alpha = \beta = 0.1$, 99% of the trend factor at time t is determined by the previous trend factor T_{t-1}. Basically, we apply both the α and the β smoothing factors. We first apply the smoothing factor α to estimate B_t, and then we further smooth this variable through β. Indeed, while the previous value of trend T_{t-1} has a weight of $1 - \alpha \cdot \beta$, the most recent demand observation has a weight $\alpha \cdot \beta$. In other words, the most recent demand observation has very little effect on trend, unless the smoothing parameter(s) is (are) very high.

This is neither good nor bad *per se*. One should simply account for that, when setting the smoothing parameter β; this might be larger in the case of the classic formulation (3.18).

[23]Notice that we assume to have n demand observations for $t = 1, 2......n$

smoothing, this sets a tradeoff between the quality of the initial estimates of the parameters through a fit sample, and the number of data points that can be used to judge the quality of the forecasting method in a test sample. In the interest of brevity we shall not repeat previous comments (see section 3.6.4).

Let us assume that we decide to use l periods to initialize the two parameters T_t and B_t. We can exploit these l data in various ways. Here we present the two major ones.

- First, we can use linear regression (see appendix A). In the linear regression, demand Y is the independent variable while time t is the independent one. Here we shall consider linear regression as a tool to interpolate demand data and identify a linear trend. Thus linear regression sets the parameters a and b of a straight line $y = a + b \cdot t$. These two parameters can be then used to initialize the baseline and the trend factors at time 0: $B_0 = a$ and $T_0 = b$.

- Second, we can use a simpler method that looks at the average demand levels and the average trend during the first l periods. We shall first estimate a trend factor to make demand observations in different periods comparable. During the first l periods, we observe $l - 1$ differences between successive demand periods, i.e., $l - 1$ observations of demand trend. Thus the initial trend level T_0 is given by the average of these $l - 1$ demand increases (or decreases):

$$
\begin{aligned}
T_0 &= \frac{\sum_{i=2}^{l} (Y_i - Y_{i-1})}{l - 1} \\
&= \frac{(Y_2 - Y_1) + (Y_3 - Y_2) + \cdots + (Y_{l-1} - Y_{l-2}) + (Y_l - Y_{l-1})}{l - 1} \\
&= \frac{Y_l - Y_1}{l - 1}.
\end{aligned}
$$

Notice how the average boils down to the difference of two single values: This method does not fully exploit all the information available, as we basically overlook all demand observations from period 2 to period $l-1$. This makes this approach simpler as well as less accurate than linear regression. One might think that, since we are basically using just two demand observations to initialize the trend factor T_0, we might just use the first two observations. Still, when we use $l > 2$ demand observations, the expected difference between Y_l and Y_1 is comparatively large and thus the estimate T_0 is less subject to noise. For example, if one tries to use this method to predict the weight of a newborn baby, the growth pattern can be estimated by comparing the weight in two consecutive days but such an estimate might be affected by various random events (for example, a stomachache can easily lead us to believe that the baby is losing weight, while it is actually and obviously growing). On the contrary, when you compare the weight over a two-week period, the estimate of the growth pattern is much more reliable.

Once we have generated an estimate of the trend factor, we can exploit it to make the l demand observations directly comparable and use them to initialize the baseline demand B_0.

We can tell the trend from the baseline demand and make all demand observations comparable so that we can use them to initialize the baseline B_0. To do so, we shall subtract from a generic demand observation Y_i the trend that we have observed during the i periods since time 0, $i \cdot T_t$. So we can initialize the baseline T_t as follows[24]:

$$B_0 = \frac{\sum_{i=1}^{l}(Y_i - i \cdot T_0)}{l}. \tag{3.22}$$

3.7.5   Drawbacks and limitations

Obviously, this forecasting method is effective to the extent that its assumptions hold (just like any quantitative method). On top of this, it has several drawbacks and limitations. As the forecasting horizon h grows, the model is more and more sensitive to any error in the estimate of the trend factor T_t. The model assumes that the trend we have observed in the past will last in the future. Actually, this is not a drawback *per se*, as all quantitative forecasting methods make some sort of assumption of stability of the demand pattern. However, we shall notice that this assumption can lead to poor performance at market "turning points." In such instances the forecasting method projects a growth (decrease) even when demand is starting to decrease (grow). This can open a wide gap between the company's expectations and actual demand. Also, the longer the forecasting horizon, the greater the problem since any error in the estimate of T_t is multiplied by h in formula (3.16). An example can clarify the concept. Let us assume that demand used to grow by g units per period, and that at time t_1 it takes a downturn and starts decreasing at a pace of s units per period. In period t_1 the forecasting method still projects a growth of g units and thus forecasts a demand $F_{t_1,h} = B_{t_1} + h \cdot g$ for period $t_1 + h$. So while the actual growth of demand stops in t_1, we project an increase up to period $t_1 + h$.[25] On the contrary, demand in period $t_1 + h$ is $h \cdot s$ units lower than it was in period t_1. Thus the forecasting error in period $t_1 + h$ is approximately equal to $h(s+g)$. Figures 3.20 and 3.21 show that the

---

[24] Notice that the initialization methods suggested in this chapter can also be used to forecast demand. These methods fall under the umbrella of decomposition methods. These methods basically derive estimates of the parameters of a demand pattern from a sample of demand observations and have no update whatsoever.

[25] This holds regardless of the smoothing factors. But in the case of small smoothing factors α and β, we continue to project a growth of demand for a much longer period of time, as the trend factor remains positive for a fair number of periods after t_1.

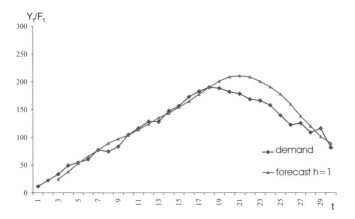

Fig. 3.20 Drawbacks of the trend model in case of demand downturn: $h = 1$.

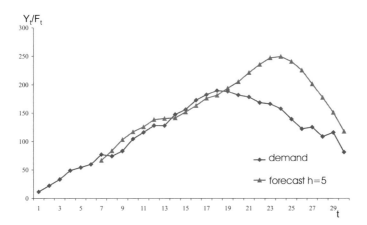

Fig. 3.21 Drawbacks of the trend model in case of demand downturn: $h = 5$.

forecasting error depends on the horizon h. Also, the error clearly depends on the responsiveness of the model and thus on its parameters, as we have already discussed in the case of simple exponential smoothing.

Also, using a linear demand model (either linear trend smoothing or linear regression) can lead us to unreasonable forecasts. For example, when the trend factor T_t is negative, this method might generate negative forecasts, especially in the case of long forecasting horizons and a markedly negative trend factor or low levels of the baseline B_t. Such unreasonable output shall be considered as a clear sign that the model is unfit for the demand at stake and thus a different forecasting method should be selected. For example, when the baseline demand B_t is 100 units and the trend factor T_t is –40 units, it might be reasonable to expect a decrease in demand for the next period, down to 60 units. For the following two periods, one might expect a 40% decrease rather than a 40-unit decrease. In other words, we might expect an exponential rather than linear demand model. In such cases, obviously, a multiplicative model that leads to an exponential demand pattern might be a more reasonable choice than the additive (linear) one we have described so far.

For the sake of completeness we show the formulas of the *multiplicative trend model*:

$$F_{t,h} = B_t(T_t)^h,$$
$$B_t = \alpha(Y_t) + (1 - \alpha)B_{t-1} \cdot T_{t-1},$$
$$T_t = \beta\frac{B_t}{B_{t-1}} + (1 - \beta)T_{t-1}.$$

3.8 EXPONENTIAL SMOOTHING WITH SEASONALITY

Seasonal fluctuations are a second source of nonstationarity. In several industries the average demand is neither increasing nor decreasing, but it still faces significant seasonal fluctuations, according to the time of the year, weather conditions, or day of the week. For example, think about the food industry that, in many developed countries, shows a negligible growth as neither the population nor the per capita consumption is increasing significantly. Still demand faces sharp fluctuations according to seasons and weather. Consider products whose consumption depends on weather conditions such as chocolate or ice creams and of event-related food such as turkey on Thanksgiving day in the USA and Christmas cakes in many other countries. On top of these yearly fluctuations we face weekly fluctuations as these products are mostly bought on Fridays and Saturdays (and on Sundays in countries where stores tend to be opened 7 days a week, such as the USA).

3.8.1 The demand model

We can model the effect of seasonality both as a percentage or an absolute variation (either increase or decrease) against average demand. For example, when we study demand fluctuations within the week, we can think that demand for grocery products on Saturdays is 80% above the average daily demand of 100.000€ per day. Also, we can think that demand on Saturdays exceeds the average demand by 80.000€. The two models behave differently in case of wide variation of demand during the year. They can behave rather differently both in peak- and low-demand weeks. The second model assumes that on Saturdays we sell 80.000€ more than on the average day of the week, during each of the 52 weeks of the year. On the contrary, the first model basically assumes that the demand increase on Saturdays is less than 80.000€ in low demand weeks (e.g., the beginning of the vacation period in large cities) and more than 80.000€ in high-demand periods (e.g., around Christmas).

In this section we discuss in detail the multiplicative model. We only provide the equations for the additive one. The logic, mechanics, and limitations of the additive model can be easily derived, *mutatis mutandis*, from the analysis of the multiplicative one.

Before we discuss the details of a seasonality model, we have to identify the periodicity of demand fluctuations we want to analyze. In other words, we want to identify the most relevant *season* for our forecasting purposes.

A first option is to study the periodicity of demand fluctuations within each year. In case the day is the time bucket, the season lasts 365 periods; in case the week is the time bucket, the season lasts 52 periods; and if the month is the time bucket, the season lasts 12 periods. A second option is to investigate the demand fluctuations within the week (in case of daily time buckets, the season lasts 7 days).

The choice of the appropriate season can be supported by the analysis of demand data to check whether they actually show the periodicity we have assumed. However, the choice shall not be totally data-driven. For example, in case demand data show a 4-day periodicity, we should look for a reasonable explanation of this demand behavior. If we cannot logically explain the 4-day periodicity, it might be simply a spurious statistical result. Thus, demand might behave differently in the future.

We denote by s the duration of the *season* we choose to analyze. For the sake of simplicity, in the remainder of this section we analyze the case of yearly seasonality with monthly time buckets to illustrate the underlying logic. However, the formulas do apply to different seasons and time buckets.

3.8.2 The algorithm

The factors of this forecasting method are the average level of demand B_t and the seasonality factors S_t. The average demand is a single parameter that is updated over time. On the contrary, to properly describe the demand fluctuations in a season, we need to have s different seasonality factors that are updated over time. In the case of 12 month seasons, we need to have a seasonality factor for each month of the year.

The demand model behind this forecasting tools is

$$F_{t,h} = B_t \cdot S_{t+h-s}, \qquad \text{for} \qquad h \leq s; \qquad (3.23)$$

more generally, when we consider a forecasting horizon that exceeds the single "season," we obtain

$$F_{t,h} = B_t \cdot S_{t+h-\lfloor \frac{h-1}{s}+1\rfloor \cdot s}. \qquad (3.24)$$

In other words, the forecast generated in period t for period $t+h$ considers (i) the most recent estimate of the average level of demand B_t and (ii) the most recent estimate of the appropriate seasonality factor. In case at the end of

December 2006 we want to forecast demand for January 2007, we consider the estimate of the average level of demand we created in December 2006. As to seasonality factors, one might be tempted to use the most recent estimate of seasonality generated in December 2006. However, this is very unreasonable as the seasonality of January might significantly differ from the seasonality of December (think of the Christmas effect for most product categories in most Western countries). Thus, we must use the most recent estimate of seasonality for the month of January, which was generated in January 2006.

The assumptions of this model lead us to generate the same forecast for the months of January 2007, 2008, 2009 and so on (if they are generated at the same point in time, that is with the same information set). These forecasts share the same average demand B_t and the same seasonality factor S_{t+h-s}. Indeed, this forecasting method assumes that the average demand does not change over time, though we face repetitive fluctuations. However, the forecasts of demand for January 2007, February 2007, March 2007 and so on, are (potentially) different, as we use different seasonality factors.

To make this forecasting method work, we shall now understand how to estimate the average demand B_t and the seasonality factors S_t. In this case, demand is a nonstationary process and thus we cannot estimate the average demand through the sum of demand observations from different periods. What would happen if we update the estimate of demand for ice cream in June, July, and August? This would lead us to overestimate the average demand we should expect in the average month of the year B_t and we would project a high demand for ice cream into fall and winter. We might be quite wrong, since the high demand for ice cream might simply depend on the season, and it might vanish in autumn and winter. To update the previous estimate of the average demand B_{t-1}, we have to understand whether the latest demand observation is higher or lower than we expected. So when we want to update the previous estimate of the average demand B_{t-1} with the June observation, we shall account for the seasonality of the specific month. Even if demand is higher than in the average month, we might not draw the conclusion that we shall increase B_t, as the relatively high demand for ice cream might somehow be expected.

In other words, we shall remove sources of nonstationarity from the latest demand observation Y_t so that it can be directly compared to the previous estimate of average demand B_{t-1}:

$$B_t = \alpha \frac{Y_t}{S_{t-s}} + (1-\alpha)B_{t-1}, \qquad 0 \le \alpha \le 1. \qquad (3.25)$$

Notice that if we assume seasonality to be additive rather than multiplicative we simply remove nonstationarity in a different manner. We subtract the seasonal increase (decrease) rather than divide by the seasonality factor. Moreover, we notice that we divide the latest demand observation Y_t by a seasonality factor S_{t-s}, which is one season old. At this point in time, S_{t-s} is the latest estimate of the relevant seasonality factor.

Once we have settled the estimate of average demand, we shall update the estimate of the seasonality factor. In our example, we shall estimate the seasonality factor for December 2006. The seasonality factor tries to capture whether the expected demand in a specific month (December) is above or below the average. If we expect the demand in December to be above the average monthly demand in the year, the seasonality factor is above 1; in case we expect the demand in December to be below the monthly average demand the seasonality factor is below 1. If the seasonality factor is 2, we expect the demand in December to be twice the average monthly demand. To update previous estimates we shall compare the last demand observation Y_t with the latest estimate of the average demand B_t. The ratio between these two variables tells us whether demand during the last month was above or below the monthly average. Then, we can use this ratio to update the last relevant seasonality factor; in our example we can update the seasonality factor for the month of December with the last estimate from December 2005. On the contrary, it does not make sense to use the demand observed in December 2006 to update the seasonality factor we have just estimated for November 2006 as these are totally incomparable figures. Thus, we can estimate the seasonality factor as follows:

$$S_t = \gamma \frac{Y_t}{B_t} + (1 - \gamma)S_{t-s}, \qquad 0 \leq \gamma \leq 1, \tag{3.26}$$

where γ is a smoothing parameter. In case the previous relevant seasonality factor was 1.5 and the demand observed in the last month was twice the average monthly demand B_t we shall increase our estimate of the seasonality factor and draw the conclusion that next December we might be expecting a demand that is between 1.5 and 2 times the average monthly demand, depending on the smoothing factor γ.

3.8.3 Setting the parameters

Like in previous cases, parameters α and γ determine the speed at which the parameters B_t and S_t are updated. Hence, these parameters influence the ability to filter noise and react promptly to changes in demand.

We notice that while the average demand B_t is updated each and every period, the seasonality factors are updated only once in a season (in our example, once every 12 months), that is, every s periods. So, if we want the seasonality factor to be as reactive as the average demand, we might want to choose a relatively large value of γ.

3.8.4 Initialization

This forecasting method too is recursive and thus must be initialized to be properly used. We have to initialize s seasonality factors S_t and one average demand B_0. This could lead us to believe that we need at least $s + 1$ demand observations. However, we only need s data points. The seasonality factors capture the difference between the demand in each specific month and the average month of the year. Thus, on the average they shall be equal to 1. This adds an extra constraint to our problem. Thus to initialize (and use) this method, at least s demand observations are required. With less than s demand observations (in our example 12 months) we cannot estimate the average demand in a season and thus we cannot estimate a single seasonality factor. Initializing demand is rather trivial when we only have s demand observations. B_0 is simply equal to the average of the s (12 in the example) demand observations, and the s seasonality factors are equal to the ratio between demand in the related period (month) and the average demand B_0:

$$B_0 = \frac{\sum_{i=1}^{s} Y_i}{s},$$

$$S_{j-s} = \frac{Y_j}{B_0} \qquad \text{for} \qquad j = 1, \ldots, s.$$

Just like in the previous cases, when we initialize with the very minimum set of data, errors might be considerable. In this specific case, each seasonality factor basically depends on a single demand draw that may be substantially different from its expected value (especially in the case of small time buckets and quite variable demand). Thus, in case more data are available, it is advisable to use more than s data points. We have already discussed the tradeoff we face when we choose the number l of periods that we use to initialize demand in section 3.6.4. If l is a multiple of s, we use "whole seasons." In this case, the simple average of the l observations is a good estimate for the average demand, as this metric is not influenced by seasonality since we take the average of demand over l/s seasons (years in the example). We can initialize the seasonality for a given month, say January, by comparing the average demand for all months of January in the fit sample to the initial average monthly demand B_0:

$$B_0 = \frac{\sum_{i=1}^{l} Y_i}{l},$$

$$S_{j-s} = \frac{\sum_{k=0}^{l/s-1} Y_{j+ks}}{B_0 \cdot l/s} \qquad \text{for} \qquad j = 1, \ldots, s.$$

In the slightly more complex case where we do not consider whole seasons and thus l is not a multiple of s, we still compare the average demand in January to the average demand B_0. The only minor issue is that in our fit sample we might have 3 months of January (say 2004, 2005, and 2006) and just 2 months of December (say 2004 and 2005). A simple average of all demand observations would not be a reasonable estimate for the initial monthly demand B_0, as it is influenced by the seasonality of January (which is overrepresented in the fit sample). Thus, we might want to compute first the average demand in each of the 12 months (average demand in January, February, March, etc.) and then take the average of these 12 (in general s)

Table 3.16   Demand data for a sport newspaper (data in thousands)

weekday	week 1	week 2	week 3	week 4	week 5
Tuesday	46	57	23	36	29
Wednesday	37	43	24	35	34
Thursday	19	35	34	43	38
Friday	50	50	60	50	52
Saturday	66	79	92	63	72
Sunday	95	81	81	110	91
Monday	121	114	123	116	113

figures. In this way, the estimate of average demand does not depend on the seasonality, as each month has an equivalent weight. The reader might want to try to translate the above concepts into formulas.

Example 3.15 Let us consider a large newsstand in Italy. Among other newspapers the newsstand sells sport newspapers. The dominant player in this business is the newspaper called Gazzetta. The newsvendor keeps track of demand (including any lost sales) and wants to forecast demand. The newsvendor places orders for copies of tomorrow's newspaper at the end of the working day. So, the forecasting horizon is one day. Now we are at the end of week 5 (Monday night) and he/she needs to plan orders for next Tuesday. So he/she need to generate a demand forecast. Table 3.16 shows data on the last 5 weeks of demand. Data show a clear seasonal pattern, as demand increases on Saturdays, Sundays, and Mondays, right before or after major sport events.

Also, given the nature of the product the time bucket is a single day since we need to plan inventories on each single day: Inventories leftover (unsold copies) on Tuesdays will not sell on Wednesdays. The newsvendor wants to have a distributional information about the future demand. Indeed, this distribution of demand is going to be used when setting the inventory levels (this is done later in example 5.10 on page 255). A point forecast is just not enough.

So the newsvendor wants to apply exponential smoothing with seasonality model to these data. We identify the 35 data points with $t = 1$ to $t = 35$ ($t = 1$ is Tuesday, week one; $t = 35$ is Monday, week five). The first decision is to set the fit sample and the test sample: In order to have a distributional information, we shall measure the forecasting error and thus should set aside a test sample. Let us assume that we want to have a test sample consisting of two weeks, thus we can use the first three weeks to fit the forecasting method to the data.

The first operation is the initialization of parameters. The initial estimate of baseline demand B_0 is the simple average of the first three weeks of demand (first 21 days). Since we take whole weeks (i.e., "whole seasons") the

Table 3.17 Initial seasonality factors

weekday	parameter	initial value
Tuesday	B_{-6}	0.6632
Wednesday	B_{-5}	0.5474
Thursday	B_{-4}	0.4632
Friday	B_{-3}	0.8421
Saturday	B_{-2}	1.2474
Sunday	B_{-1}	1.3526
Monday	B_0	1.8842

seasonality of demand has no impact on the baseline demand

$$B_0 = \sum_{t=1}^{21} \frac{Y_t}{21} = 63.33. \tag{3.27}$$

Notice that the initial estimate of the baseline demand refers to time 0. Once again one could be tempted to set $B_{21} = 63.33$ but this would mean that the forecast for period 22 is actually not based on any sort of exponential smoothing and thus should not be used to capture forecasting error of such a method, demand distribution, and uncertainty. With this figure we can now initialize the seasonality factors for the seven days of the week. Let us start with the initial seasonality factor for Tuesdays. We simply take the average (42) of demand in the three Tuesdays in out fit sample (46, 57, 23 units) and divide it by the baseline demand $B_0 = 63.33$. Thus the initial seasonality factor for Tuesdays is $42/63.33 = 0.6631$. The question then becomes: Which period does this seasonality factor refer to? The first Tuesday in our sample is period 1. Actually, initial factors precede the fit sample, and the first Tuesday before our fit sample was period $t = 1 - 7 = -6$. So the initial seasonality factor for Tuesdays is $B_{-6} = 0.6631$ [see equation (3.27), with j=1]. Similarly, we can derive initial seasonality factors for the seven days of the week, as table 3.17 shows.

Once we have initialized the parameters we can let the smoothing algorithm update them. Let us assume that $\alpha = 0.1$ and $\gamma = 0.2$. Let us walk you through the calculation for the first update. The updated baseline demand after we have observed period 1 is [see equation (3.25), where $t = 1$ and $s = 7$]

$$B_1 = 0.1 \cdot \frac{46}{0.6631} + (1 - 0.1) \cdot 63.33 = 63.93. \tag{3.28}$$

Similarly, we can update the seasonality factor for Tuesdays through equation (3.26), where $t = 1$ and $s = 7$:

$$S_1 = 0.2 \cdot \frac{46}{63.93} + (1 - 0.2) \cdot 0.6631 = 67.44. \tag{3.29}$$

Table 3.18 Baseline estimate B_t (data in thousands)

weekday	week 1	week 2	week 3	week 4	week 5
Tuesday	63.94	64.27	61.66	62.83	63.19
Wednesday	64.30	65.62	59.68	63.04	63.14
Thursday	61.97	67.16	61.27	65.87	64.32
Friday	61.71	66.43	62.47	65.18	64.14
Saturday	60.83	66.29	63.82	63.68	63.71
Sunday	61.77	65.49	63.40	65.50	63.79
Monday	62.02	64.95	63.64	65.12	63.48

Table 3.19 Estimates of the seasonality factors S_t

weekday	week 1	week 2	week 3	week 4	week 5
Tuesday	0.6744	0.7169	0.6481	0.6331	0.5983
Wednesday	0.5530	0.5734	0.5392	0.5424	0.5416
Thursday	0.4318	0.4497	0.4707	0.5072	0.5239
Friday	0.8357	0.8191	0.8474	0.8313	0.8272
Saturday	1.2149	1.2103	1.2565	1.2031	1.1885
Sunday	1.3897	1.3591	1.3428	1.4101	1.4134
Monday	1.8976	1.8691	1.8818	1.8617	1.8454

We can proceed with $t = 2, ..., 35$ to update the parameters B and S. Tables 3.18 and 3.19 show how the estimates are updated over time.

Had we been interested in a point forecast for period 36 (i.e., next Tuesday), we simply would have used the last estimate of demand baseline $B_{35} = 63.48$ and the relevant seasonality factor $S_{29} = 0.5983$. Using equation (3.23), we obtain the point forecast for the next Tuesday (period 36) as

$$F_{36} = F_{35,1} = 63.48 \cdot 0.5983 = 37.98. \qquad (3.30)$$

However, the newsvendor wants to have some distributional information about the demand on Tuesdays; thus, we have to investigate the expected forecast error. The smoothing algorithm suggests that we shall expect a demand for 37.98 units. However, so far we have no information about the confidence on that number. Actually, given we have set aside two weeks (week 4 and week 5) to test the performance of this forecasting method, we can investigate the forecasting error in these two weeks and reasonably assume that the expected error $[\mathrm{E}\,(Y_{36} - F_{36})]$ equals the average past error.[26] To capture the error in the test period, we shall generate the forecasts over the test period (F_t, $t =$

---

[26] Notice that, in this case, we assume that the expected error does not depend on the day of the week, even if different days have different demand expectations. In other words,

Table 3.20   Forecast $F_t = F_{t-1,1}$ (data in thousands)

weekday	week 4	week 5
Tuesday	41.25	41.23
Wednesday	33.25	34.08
Thursday	28.09	31.97
Friday	51.92	54.76
Saturday	78.49	78.42
Sunday	85.70	89.80
Monday	119.31	121.95

$22, ..., 35$). To generate the forecast for period 22, we shall use the most recent parameters. We use the baseline at time $t = 21$ and the seasonality factor of period $t = 15$, as equation (3.23) shows

$$F_{21,1} = F_{22} = B_{21} \cdot S_{15} = 63.64 \cdot 0.6481 = 41.25. \qquad (3.31)$$

Similarly, we compute the forecast for the remaining observations in the test sample (see table 3.20). Notice that while the parameters are estimated for the whole set of 35 observations, we only generate a forecast for the test sample, as using data from the fit sample to compare the forecast with the actual demand would not make sense.

Finally, we can compute the error we would have made in each of the 14 days in the test sample, had we adopted this algorithm in the past. For example, the error in period 22, is

$$e_{22} = Y_{22} - F_{22} = 41.25 - 36 = 5.25. \qquad (3.32)$$

By the same token, we can derive the errors for periods $t = 23, ..., 35$, as shown in table 3.21.

With these errors, we can compute our usual performance metrics. For example, the RMSE is 10.05. It really means that, if our assumption of a statistically stationary error holds, we shall expect a mean squared difference between our forecast for period 36 ($F_{36} = F_{35,1} = 37.98$) and demand Y_{36} to be 10.05^2 (bias is negligible and here we assume ME to be zero). In other words the demand in period 36 has an expectation of 37.98 thousand units and a standard deviation of 37.98 thousand units. In example 5.10 on page 255, we show how this distributional information can be used to make inventory decisions and how demand forecasting and inventory planning problems are strictly related.

---

we assume that a stable random noise overlaps the weekly fluctuations of demand. With a test sample longer than two weeks, we could test the assumption empirically. In this toy-example, data were generated according to this assumption, which is implicitly made.

Table 3.21 Errors $e_t = Y_t - F_t$ (data in thousands)

weekday	week 4	week 5
Tuesday	5.25	12.23
Wednesday	−1.75	0.08
Thursday	−14.91	−6.03
Friday	1.92	2.76
Saturday	15.49	6.42
Sunday	−24.30	−1.20
Monday	3.31	8.95

Also, our example can show the difference between demand variability and uncertainty. The standard deviation of the 35 demand observations is 34.48, whereas RMSE is just 10.05 units. While the standard deviation measures the variability of demand, RMSE captures our inability to forecast demand, that is, to predict demand fluctuations. In other words, in our example, demand is very variable, but some part of these fluctuations are predictable and due to weekly seasonality. Thus the forecasting error, that is uncertainty, is smaller than variability. □

3.8.5 Drawbacks and limitations

This forecasting method uses a wide range of factors and thus it requires a fairly large information set to operate properly. However, when we use a long past history to calibrate the model, we might end up using fairly old demand observations (e.g., 5 or 10 years old) that might have little to do with the current demand pattern.

Clearly, the larger the value of s, the more parameters we have to estimate and the less the information is available to estimate each single parameter. So the adoption of this method with a small time bucket and long seasons can be dangerous and lead to poor performance. Actually, some researchers found that the naïve approach ($F_{t,1} = Y_t$) might outperform the smoothing algorithm with seasonality for this reason (see [4]): A simple and very reactive method (with a short forecasting horizon) can adjust to changes in seasonality better than a complex one that might fail to estimate seasonality properly.

Example 3.16 Let us consider a company that uses a daily time bucket and wants to analyze demand fluctuations within a year. The company needs to estimate 365 different seasonality factors. Had the company adopted a monthly time bucket, we would have estimated only 12 parameters instead. Estimating daily seasonality factors might be counterproductive even if demand shows sharp seasonal fluctuations. In fact our estimate of seasonality might be so inaccurate that the seasonality factors increase rather than reduce the forecasting errors. Let us consider a company trying to forecast demand for ice cream with daily time buckets. In case April 6 was rainy the last two years, while next April 6 is going to be sunny, the forecast will turn out to be quite wrong as it will underestimate demand substantially. Indeed, we would estimate a low seasonality for April 6, and this would lead to a very low demand forecast for the next April 6. However, if next year April 6 is sunny (a more than reasonable chance in Italy), then error is going to be substantial.

Some correctives have been designed to improve the performance of the seasonal exponential smoothing. First, in many instances, demand is driven by weather conditions rather than the season *per se* (think about food products, apparel goods, and white goods). So one might want to investigate the relationship between weather (e.g., temperature and inches of rain) and demand (often through regression methods). This approach has a substantial advantage: We do not consider seasonality on April the 6 as totally different and independent from April 5, 4, 3 etc., and 7, 8, 9, etc. Hence, when we try to forecast demand for the next April 6, we exploit a much broader information set than in the case of traditional models (for an example of this approach to seasonality, see [2]).

Furthermore, one can feed the classic exponential model with seasonality with a smoothed demand, so that single events like rain on April 6 do not influence the seasonality factors as much. One option is to replace Y_t in formulas (3.25) and (3.26) with the average of periods from $t-3$ to $t+3$. In this case, the seasonality of April 6 does not depend on the specific weather conditions of April 6, but rather on the average condition of the week of April 6. This clearly improves the estimates of the seasonality factors and thus reduces forecasting errors.   ⊓

3.9 SMOOTHING WITH SEASONALITY AND TREND

3.9.1 The demand model

The last two sections have presented separate models to deal with demand trend and seasonality, respectively. In this section we present a forecasting model that combines these two features and thus can forecast a seasonal demand with trend. We assume that demand tends to grow (or decrease) in the long run, and we assume this trend to be linear. On top of this long-term trend, we observe seasonal fluctuations, that we assume to be multiplicative.

3.9.2 The algorithm

Given the above assumptions about demand, the forecasting model is:

$$F_{t+h} = (B_t + h \cdot T_t) \cdot S_{t+h-s \cdot \lfloor (h-1)/s+1 \rfloor}. \tag{3.33}$$

In other words, we take the baseline level of demand at time t (i.e., B_t) and the increase (decrease) we expect during the forecasting horizon h (i.e., hT_t). $B_t + hT_t$ is the level of demand we would expect in period $t + h$ if there was no seasonality or the seasonality factor of that period was 1. Thus, to generate an accurate forecast, we shall account for the seasonality of period $t + h$, which is a multiplicative factor that tells us whether we shall expect demand in period $t + h$ to lie above or below the general trend line (see figure 3.22). The term $S_{t+h-s \cdot \lfloor (h-1)/s+1 \rfloor}$ boils down to S_{t+h-s} if the forecast horizon h is not larger than s; the notation $\lfloor x \rfloor$ means that we round x down to the nearest integer number.

Like in previous cases, the second step is to design a procedure to update the $s+2$ parameters with the most recent demand observations. As to the baseline B_t, we shall simply combine the trend and the seasonal model properly. We shall de-seasonalize the last demand observation and add the last estimate of the trend factor T_{t-1} to the previous baseline demand B_{t-1}:

$$B_t = \alpha \frac{Y_t}{S_{t-s}} + (1 - \alpha)(B_{t-1} + T_{t-1}), \qquad 0 \le \alpha \le 1. \qquad (3.34)$$

As to the trend and seasonality factors we can adopt the equations we have designed in the last two sections:

$$
\begin{aligned}
T_t &= \beta(B_t - B_{t-1}) + (1 - \beta)(T_{t-1}), & 0 \le \beta \le 1; \\
S_t &= \gamma \frac{Y_t}{B_t} + (1 - \gamma)S_{t-s}, & 0 \le \gamma \le 1.
\end{aligned}
$$

3.9.3   Initialization

This is another recursive forecasting method and thus it need to be initialized. In this case we need at the very least $s+1$ periods. Indeed, we shall calculate $s + 2$ parameters, and we have one constraint on the s seasonality indexes: Their average must be 1. An alternative explanation is that to estimate the trend we must compare periods with the same seasonality. If in December demand for cakes is above the demand in January, we can hardly tell whether this is due to seasonality or trend. In our example of seasonality within the year and monthly time buckets, we need at the very least 13 demand observations to tell seasonality from trend.

If only this minimum information set is available, we can initialize the trend and seasonality factors as follows. We first estimate the trend factor

$$T_0 = \frac{Y_{s+1} - Y_1}{s} \qquad (3.35)$$

by taking the difference between the only two demand observations that are comparable, as far as seasonality is concerned.

This initialization, though often used, suffers from a significant problem: T_0 is affected by seasonality. Both demand observations Y_1 and Y_{s+1} depend on a (multiplicative) seasonality index[27] and thus their difference depends on the seasonality index as well. Understandably, in case the seasonality index of these two periods is close to 1, this is a minor issue and has no practical effects. On the contrary, in case the seasonality index is significantly above or below 1, it is a major concern and it is more appropriate to use at the least $2s$ demand observations, i.e., two whole seasons. In this way, we can estimate s differences between pairs of demands that share the same seasonality index (in our example, 12 pairs of months from successive years, e.g., January 2005 and January 2006, February 2005 and February 2006, and so on). In this case each single difference is affected by seasonality; however, the average of the s differences is actually affected by the average seasonality that is 1: By taking s differences, we cancel out the effect of seasonality. So in this case we initialize the trend factor as follows:

$$T_0 = \frac{1}{s} \sum_{i=1}^{s} \frac{Y_{s+i} - Y_i}{s} = \frac{1}{s^2} \sum_{i=1}^{s} (Y_{s+i} - Y_i) \qquad (3.36)$$

We can now use the above estimate of trend T_0 to tell the effect of trend from seasonality and thus estimate both the s seasonality factors and the baseline demand B_0.

To estimate the multiplicative seasonality factors, we shall compare the actual demand observations with the ones we would have expected, had there been no seasonality (i.e., with the ones we would have expected had there been a seasonality index equal to 1). To do so, in figure 3.22 we shall compare the actual demand observation Y_t with the corresponding point on the line $Y = B_0 + T_t \cdot t$. If points are above the line, the seasonality is greater than 1. If points are below the line, seasonality is lower than 1.

So we need to estimate B_0 to estimate the seasonality indexes. To do so, we need to make the demand observations in different time buckets comparable; therefore, we need to remove the trend from these observations by subtracting the expected growth $t \cdot T_0$ from demand observation Y_t.

If we initialize parameters with whole seasons (e.g., two seasons), $Y_t - t \cdot T_0$ is affected by seasonality; however, in the sample such seasonality cancels out as we consider whole seasons. Thus the estimate of the initial baseline demand is

$$B_0 = \frac{\sum_{i=1}^{l} (Y_i - iT_0)}{l}. \qquad (3.37)$$

If $s + 1$ demand observations are used to initialize the parameters, we should be a bit more careful. Indeed, periods 1 and $s + 1$ are overrepresented in the sample, and this might lead to a bias. For example, if we consider the January 2004 to January 2005 period, the month of January would be overrepresented in our sample (there are two months of January and just one February, one

---

[27]Note that this problem does not exist in the case of additive seasonality, as we take the difference between the two observations.

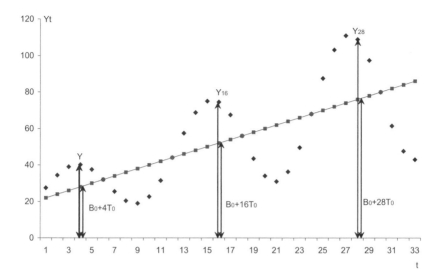

Fig. 3.22 The initialization process in case of exponential smoothing with seasonality and trend.

March, and so on). So if the product is in high demand during winter, we happen to overestimate demand, while if the product is mostly bought in summer, we tend to underestimate baseline demand.

To solve this problem, we first compute the average demand in each of the s (12) periods (months) in a season (year). Then we take the average demand across periods to get a baseline demand estimate with no trend and no seasonality:

$$B_0 = \frac{\left[(Y_{s+1} - (s+1)T_0) + (Y_1 - T_0)\right]/2 + \sum_{i=2}^{s}(Y_i - iT_0)}{s}. \qquad (3.38)$$

Finally, we shall compute the seasonality indexes. We compare the actual observation Y_t with the one we would have observed, had there been no seasonality (and no randomness in demand) $B_0 + t \cdot T_t$. For instance, if we use whole seasons (e.g., two seasons) to initialize the forecasting method, we have

$$S_{j,s} = \frac{\sum_{k=0}^{l/s-1} \dfrac{Y_{j+ks}}{(B_0 + (j+ks)T_0)}}{l/s - 1} \qquad 1, ..., s. \qquad (3.39)$$

Notice that we are not dealing with all of the possible options and thus our analysis is not exhaustive. The extension to other cases is left as an exercise for the reader.

Finally, this model shares the limits and drawbacks of the exponential smoothing with trend *or* seasonality. Also, the selection of the parameters shall follow the same process.

3.10   SIMPLE LINEAR REGRESSION

The forecasting models we have analyzed so far are widely adopted. However, in these models demand is just a function of time. On the contrary, in many real-life situations demand might depend on a variety of variables including advertising expenditures, weather, price, number of stores carrying the item, state of the economy, etc. A forecasting model that tries to capture these effects can be fairly complex for several reasons.

- First, demand might depend on many variables. For example, in grocery supermarkets, demand might depend on traffic (number of customers visiting the store), weather, the price of the item, promotions of the item at stake and/or substitute ones, religious events such as Easter or Christmas, and sport events such as the Soccer World Cup or the Olympic Games.

- Second, the relationships between the independent (i.e., explanatory) variables and demand can be complex and nonlinear. Let us assume that we have cut price by 50% and gained a 100-unit increase in demand. If we cut price by 100% and give the product away for free, we definitely should not expect a 200-unit lift.

Though the problem can be fairly complex in real life situations, in this section we address a relatively trivial situation where demand depends on a single independent variable. In other words, we illustrate the application of simple linear regression (see section A.10). The more general case of multivariate regression, i.e., situations where we explain and forecast demand through more than one independent variable, is dealt with in the web sections W.A.11 and W.3.11.

In section A.10 we describe in full detail the assumptions and properties of simple linear regression. This statistical method estimates, through empirical data, the linear relationship between a variable X we call independent and a variable Y we assume to depend on the first one. In our case, demand is the dependent variable.

As such, linear regression is just a tool to investigate the relationship between two variables. Thus it might be used to analyze the relationship between two variables, say demand for ice cream and temperature, demand for cars and Gross Domestic Product (GDP), and demand for fashion products and advertising. Once such a relationship is estimated, we can use it to forecast future demand, if we know the future values of the independent variable (or they can be predicted accurately). Indeed, a "perfect analysis" of the relationship between the demand for cars and the GDP does not help to forecast the demand for cars if we have no idea about the future of the economy. In the remainder of this section we assume the future values of X to be known with certainty (e.g., think about the prices the company sets). In the final part of the section we briefly discuss the consequences of uncertainty on the future value(s) of X (e.g., when estimates about the future GDP are available, though they are affected by some sort of error).

We assume that demand observations are drawn from a random linear process:

$$y_i = \alpha + \beta x_i + \epsilon_i; \tag{3.40}$$

where:

- i is the index that identifies the i-th observation of demand and of the variable that influences it;

- α and β are unknown parameters that influence the demand process; these parameters have to be estimated;

- ϵ_i is a normally distributed random variable with a expected value zero and standard deviation σ_ϵ (additional assumptions concerning statistical independence are pointed out in section A.10).

Also, we assume to have an estimate of the relationship between Y and x,

$$Y = a + bx. \tag{3.41}$$

Section A.10 shows how this relationship can be estimated, based on past observations of Y and x. The point forecast of Y (e.g., demand for cars) corresponding to $x = x_0$ is

$$\hat{Y}_0 = a + bx_0. \tag{3.42}$$

It is easy to show that $\hat{Y}_0$ is an unbiased prediction[28] of the future level of demand, since estimates a and b are unbiased. However, as discussed in the first section of this chapter, a point forecast is often meaningless, especially in the case of continuous variables such as Y_t (ϵ is continuous and thus Y is continuous as well).

So we should not only look at the expected level of future demand, we should also investigate the standard deviation of the estimate, that is, the

---

[28] Note that in section A.10 we are mainly concerned with the estimate of unknown *numbers* α and β. The demand Y_0, corresponding to a value x_0 of the independent variable is a *random variable* that we are trying to predict. This is conceptually different and, for instance, we should talk about *prediction* intervals rather than confidence intervals. In this chapter we are a bit sloppy at times. For the sake of simplicity, we use the term "standard error of estimate" when referring to $\text{See}(Y_0)$, which is conceptually not quite correct.

square root of the expected squared difference between the forecast $\hat{Y}_0$ and the actual demand Y_0:

$$Y_0 - \hat{Y}_0 = \alpha + \beta x_0 + \epsilon_0 - (a + bx_0) = (\alpha - a) + (\beta - b)\,x_0 + \epsilon_0. \qquad (3.43)$$

On the one hand, this can be interpreted as the forecasting error we shall expect. On the other hand, we can read the output of the forecasting process as a distribution of demand rather than a point forecast. The mean of the distribution is $\hat{Y}_0$ and the standard deviation is

$$\mathrm{See}(Y_0) \equiv \sqrt{\mathrm{E}\left[\left(Y_0 - \hat{Y}_0\right)^2\right]}.$$

In section A.10 we show that

$$\mathrm{See}_a \ \equiv \ \sqrt{\mathrm{E}\left[(\alpha - a)^2\right]} = \sigma_\epsilon \sqrt{\frac{\bar{x}}{\sum_{i=1}^{n}(x_i - \bar{x})^2} + \frac{1}{n}};$$

$$\mathrm{See}_b \ \equiv \ \sqrt{\mathrm{E}\left[(\beta - b)^2\right]} = \sigma_\epsilon \frac{1}{\sum_{i=1}^{n}(x_i - \bar{x})^2}.$$

Similarly, one can show that $\mathrm{See}(Y_0)$ is given by

$$\begin{aligned}
\mathrm{See}\,(Y_0) \ &= \ \sqrt{\sigma_\epsilon^2 + \frac{\sigma_\epsilon^2}{n} + \sigma_\epsilon^2 \frac{(x_0 - \bar{x})^2}{\sum_{i=1}^{n}(x_i - \bar{x})^2}} \\
&= \ \sigma_\epsilon \sqrt{1 + \frac{1}{n} + \frac{(x_0 - \bar{x})^2}{\sum_{i=1}^{n}(x_i - \bar{x})^2}}. \qquad (3.44)
\end{aligned}$$

We can read equation (3.44) and make sense of it. First, as n tends to infinity, the second and third terms under square root tend to zero (respectively, n and the number of terms in the summation grow), while the first one remains unchanged. Unlike the case of the estimates a and b of the parameters α and β, the prediction error does not go to zero, as n tends to infinity. Actually, as n tends to infinity the forecasting error tends to σ_ϵ.

Indeed, with an infinite number n of past observations we can perfectly estimate the relationship between Y and x, so we face no error in the estimates of α and β. However, this is just not enough to generate an error-free forecast. Indeed, a perfect estimate of the parameters leads to a perfect estimate of the expected level $(\alpha + \beta x_0)$ of the demand Y_0, that is, the nonstochastic part of the demand process. However, the random part of the process ϵ_0 still creates random fluctuations we cannot predict. Thus, it leads to forecasting errors, as figure 3.23 shows. By now the first term in equation (3.44) shall be clear, and we can devote our attention to the second and third one. They show the impact of errors of estimate of a and b. To clearly tell the contribution of these errors we shall assume that ϵ_0 is zero. In other words, we assume

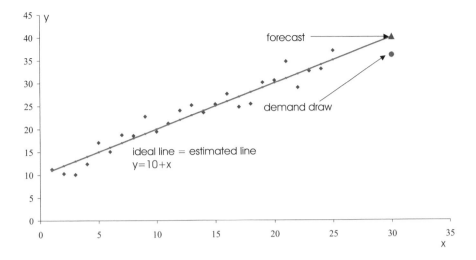

Fig. 3.23   The forecasting error due to the variability of the demand process.

that the random noise is zero and the new demand observation lies on the line $Y = \alpha + \beta x$. Therefore, any error is due to the wrong estimate of the parameters, rather than to the randomness of the demand process.

The second term σ_ϵ^2/n is just the variance of the n draws of ϵ we have studied to estimate the regression line $y = a + bx$. When the n observations tend to lie above the ideal line $Y = \alpha + \beta x$ (i.e., when the average of the n draws of ϵ is greater, or lower, than zero), the estimated regression line tends to lie above (below) the ideal one. Thus the estimate a tends to be larger than the actual parameter α. The error in the estimate of α leads to an error of estimate of Y_0 (see figure 3.24).[29]

Finally, the last term in equation (3.44) can be interpreted as the impact of errors of estimate of β on the accuracy of the demand forecast $\hat{Y}_0$. To isolate this effect, we set to zero the sources of errors we have discussed so far. More formally we assume that:

- $\sum_{i=1}^{n} \epsilon_i = 0$, i.e., we assume that the average of the n random draws is zero and thus the estimated line lies neither above nor below the ideal line;

---

[29]Notice that this is not the only source of error in the estimate of α. Indeed, even when the average noise $\sum_{i=1}^{n} \epsilon_t$ is zero, we might still face an error of estimate of α. Indeed, in this case the draws are on the average neither above nor below the ideal line $Y = \alpha + \beta x$. This means that the estimated line lies neither above nor below the ideal one. Still the estimate b of the slope might be wrong and (in case of $\bar{x} \neq 0$) this can lead to errors in the estimate of α (see section A.10). So one might more properly say that the first term shows the impact of errors in the vertical position of the estimated regression line, rather than errors of estimate of a *per se*.

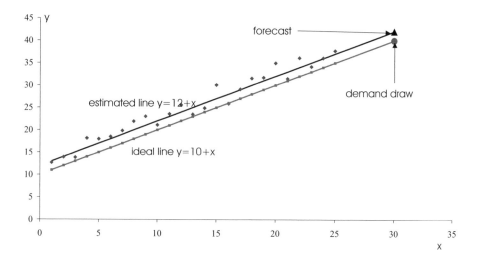

Fig. 3.24   Forecasting error due to a wrong estimate of α.

- $\epsilon_0 = 0$, i.e., we assume that demand Y_0 is exactly on the ideal line $Y = \alpha + \beta x$.

The third term in equation (3.44) can be interpreted as the error of estimate of β [see equation (3.44)] times $(x_0 - \bar{x})$. Why does the forecast accuracy depend on See$_b$ and on the distance between x_0 and $\bar{x}$? The definition of a shows that the estimated line goes through the barycenter of the demand observations $(x; Y)$. Also, since we assume that $\sum_{i=1}^{n} \epsilon_i = 0$, the ideal line $Y = \alpha + \beta x$ intersects the estimated one in the barycenter of demand observations. Thus the errors in the estimate of the slope (See$_b$) generate no effect on the inaccuracy of demand when $x_0 = \bar{x}$. On the contrary, the error in the estimate of the slope generates large errors when x_0 is far from the point $\bar{x}$ where the two lines intersect (see figure 3.25).

Concept 3.4 *The error of estimate is due to the randomness of demand process, and the errors in the estimate of the intercept (a vs. α) and slope (b vs. β) of the regression line.*

Equation (3.44) shows the standard forecast error and thus enables us to build confidence intervals of demand Y. The analysis above shows that the standard error of estimate reaches a minimum when $x_0 = \bar{x}$, since $\bar{x}$ is the barycenter of past observations and thus is the single point we have more information about. This relative abundance of information reduces the forecasting error.

Hence, the width of the prediction interval is affected by the distance between the barycenter of past observations $\bar{x}$ and the point x_0 for which we want to forecast demand. Figure 3.26 shows the confidence intervals (with

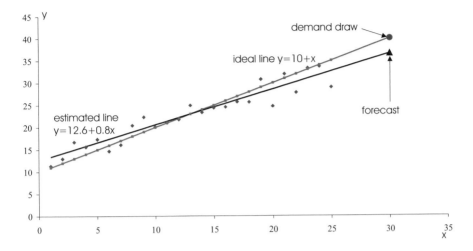

Fig. 3.25 Forecasting errors due to errors in the estimate of the slope of the regression line.

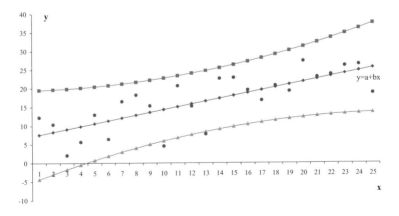

Fig. 3.26 Forecasting error as a function of x_0.

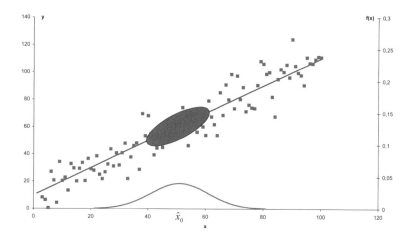

Fig. 3.27 Effects of a partial knowledge of x_0.

confidence level 90%) for various x_0. This figure shows that they are wider for x_0 significantly to the left or to the right of $\bar{x}$.

The above findings hold only under the tight assumptions of the regression model (see section A.10). When we study real-life data the assumptions are typically not fully met. For example, the relationship between x and Y might be linear only within a given range of x. Outside this range it might be nonlinear, and thus we should expect biased forecasts and far larger errors than the linear regression model predicts.

Finally, we investigate the effect of less-than-perfect information on X_0.[30] For example, consider a model where the demand for a given kind of food depends on the temperature. If we want to use this model for forecasting purposes, we should know the future temperature. However, the future temperature is uncertain and it is known, at best, in terms of probability distribution (or confidence intervals). Hence, when we use temperature to estimate future demand, we have an additional source of uncertainty. Geometrically, a partial information on X_0 means that we do not know exactly where, on the X axis, we shall read the relationship between Y and X. Hence, the confidence interval on Y is a sort of an area on the (X, Y) plane rather than a simple segment, as we face uncertainty on both X and Y. We do not know the right point on the X axis, and still for a given point on the x axis we only have distributional information on Y_0.

---

[30]Notice that in this case we use X_0 instead of x_0, as we do not know the future level of X and thus it can be interpreted as a random variable rather than a number.

3.10.1 Setting up data for regression

As we discuss in section A.10, linear regression relies on a rather wide range of assumptions on the demand process (more generally, a random process). Often, demand data hardly meet these assumptions and thus cannot be used for linear regression.

Example 3.17 When we want to investigate the relationship between temperature and demand of a food product or demand elasticity, i.e., the increase in demand due to promotions and/or price cuts, we might not be able to use straight demand data, as they might be affected by a significant seasonality that might lead to erroneous conclusions.

Let us consider a retail company that, during the weekends, cuts by 20% the price of a product that on the average sells 100 units/day. When we look at demand and price data we might be led to ascribe the whole increase in demand to price elasticity. On the contrary it is, at the least partially, due to weekly fluctuations that make demand increase on Saturdays and Sundays. So we might overestimate the price sensitivity of demand as we attribute both the seasonal fluctuation and the increase due to the price cut to the price elasticity of demand. For a more detailed discussion we refer to [7]. □

This is why we might want to "clean" the data before we apply linear regression to make them fit with the assumptions of the model and make sure the analysis is reliable. In the case of example 3.17 above, we should first remove the seasonality of demand from the dataset and then analyze price sensitivity to understand the relationship between price and demand. A second option is to use multiple regression that tries to estimate both effects at once.

W.3.11 FORECASTING MODELS BASED ON MULTIPLE REGRESSION

In the previous section, we have shown that sometimes linear regression models cannot fully capture the mechanism generating demand for a product.

Here we build on the theory of linear multiple regression, outlined in section W.A.11, by illustrating how this tool can be exploited to build more realistic forecasting models. Quite often, categorical variables are used to account for qualitative features, and this paves the way for quite sophisticated models. However tempting this may sound, we should always keep in mind that the adoption of overly sophisticated models may be counterproductive in practice. The more parameters we have to estimate, the more uncertainty we introduce. Hence, we should stick to a "principle of parsimony," and keep model complexity to a reasonable size.

3.12   FORECASTING DEMAND FOR NEW PRODUCTS

So far we have discussed forecasting methods that basically try to predict demand for a given product with a fairly long demand history either by extrapolating future demand from the time series or by reading the linear relationship with a variable (say price) that drives its behavior. Unfortunately, for new products we do not enjoy a long past history and thus we shall resort to other forecasting tools. We can resort to qualitative methods that leverage on the knowledge of experts; on the other hand, we can analyze other sources of relevant information on demand for new products and design forecasting techniques to properly exploit them.

3.12.1   The Delphi method and the committee process

The word Delphi refers to the sacred oracle in ancient Greece. Forecasts and advice from Gods were sought through intermediaries at this oracle. Leaders and generals in ancient Greece used to look for the advice and predictions of the Delphi oracle before any major war or political initiative.

Even today companies face very uncertain future events, and at times they do not enjoy any factual information or data, so they have to resort to some sort of oracle. Today's oracles are experts that over time have collected information about the future event we try to forecast and have developed an implicit interpretative model that puts them in a position to predict the future. Examples from today's world are fashion experts and designers trying to predict new fashion trends.

The Delphi method was originally designed for long-term technological forecasting; however, it can be used to forecast demand before the launch of a new product. In these circumstances no actual demand data are available and companies have to resort to a more qualitative process (see figure 3.3).

Probably, the most common qualitative forecasting process is the *committee process*, where a group of presumably expert panelists engage a discussion on the new product, its features, its positioning, pricing, etc. In this process the committee as a whole reaches an agreement on the future expected level of demand for a given product (more generally a given future event, in the remainder of the paragraph we use the specific case of a new product but statements apply, *mutatis mutandis*, to the more general case of a generic qualitative forecast).

This process clearly permits a high bandwidth communication among the experts and favors the exchange of information and the discussion on the implicit models of demand the various experts have. However, this process permits a social interaction that might have significant side effects.

In many social situations, individuals tend to be influenced by others and this might imply a loss of potentially relevant information. This is even more dangerous within the specific social system of a company. Indeed, within a company, conflicts among organizational units, incentive schemes, and hierarchy might prevent individuals from eliciting their opinion in a group. One example might make the point clear. One of the authors was simulating the committee process with a group of managers from various companies. They were trying to predict the demand for men's winter jackets. One member of the committee started the discussion with a bold statement: "Who's the idiot that can possibly wear that blue jacket?"[31] The personality of this "expert" was so strong that nobody dared to say a word and the committee finally "agreed" that the demand for the blue jacket was going to be very low. Quite interestingly, single experts were also asked to write what their personal expectation was. It turned out that a shy guy thought that the blue jacket was going to be the top seller, but he did not dare to defend his opinion in the public discussion. Actually, his opinion about the blue jacket was totally lost in the committee discussion. Similar problems can arise because of hierarchical relationships among experts rather than because of differences in personality.

Unfortunately, neither seniority nor a bold personality perfectly correlate with the ability to forecast demand for a new product. This really means that the committee process can lead to a loss of potentially relevant information.

Also, in a committee process we might face the "dictatorship of the majority," meaning that in many instances the majority of the group might disregard heterodox opinions and simply ignore them. This can be very dangerous in a very uncertain situation, where even odd scenarios might come true. As we have discussed, social interaction within the group can lead to the loss of potentially valuable information.

The Delphi method was actually designed to control interaction among experts. The original Delphi method is administered through questionnaires sent by mail. Clearly, nowadays it can be administered via e-mail. Experts can even sit in the same building or room.

Basically, the idea behind this method is that the interaction among panelists should be limited and formalized to avoid social interaction that might lead to loss of information. So in the basic Delphi method, panelist interact only through the administrator of the Delphi process. The steps of the process are the following:

- Building the panel of experts

- Development and test of the first round Delphi questionnaire

- Transmission of the first questionnaires to the panelists

- Analysis of the first round of responses

- Preparation of the second round questionnaires

---

[31]What made the situation really fun is that one of the authors was using the jacket during that winter!

- Transmission of the second round questionnaires to the panelists

- Analysis of the second round responses (Steps 5 to 7 are reiterated as long as desired or necessary to achieve stability in the results)

- Preparation of a report by the analysis team to present the conclusions of the exercise

Quite interestingly, the panelists are asked the same questions (e.g., how many units is the new product going to sell during the season?) multiple times. The logic behind this reiterated process is that iterations are the means of communication among panelists. After each round of the questionnaire, the responses are analyzed and the panelists are given a summary of the responses (e.g., the mean response, the standard deviation, and where he/she is in the distribution).

The idea behind this multiple interactions is to *tell noise from signal.* Let us consider a panelist that provided a response off the average. In other words, his/her opinion differs from the average one quite substantially. Once the first round is completed, the panelist is provided with the distribution of the responses and he/she figures out that other experts do not share his/her vision of the future. During the second round he/she basically has two options. The first is to stick to his/her opinion, while the second is to account for the other experts' opinions and revert to the mean. Actually, experts that do have a strong point on the demand for the new product tend to stick to their initial forecast. On the contrary, those that have no clear idea on the new product tend to revert to the average opinion of other experts. For these panelists, indeed, the average opinion of the other experts is a very relevant piece of additional information that lead them to significantly update their initial forecast. Such iterations of the process distinguish noise from signal as we retain grounded opinions off the average while we tend to discard those that are off by pure chance.

Concept 3.5 *The Delphi method relies on experts' knowledge to forecast, and it is designed to control social interaction in such a way that the signal about genuinely different opinions is kept, while random noise (i.e., lack of information and/or knowledge) is removed from the data.*

Another fairly important feature of the Delphi method is that it can capture the *disagreement among experts.* Indeed, while the committee process generates a single number, in the Delphi method each expert generates one number (actually a series of forecast, but the final outcome of the process is one forecast per expert). Thus in the latter case we can compute the disagreement among experts. This is actually a fairly important feature of the process, as several studies (e.g., see [6] and [8]) show that the disagreement among experts correlate with the degree of uncertainty. Such empirical studies basically confirm intuition. When a future event (think of a sport event) is basically certain, true experts can hardly have substantially different opinions, whereas when it is truly uncertain, each of them might have a different perspective. In other words, when experts tend to agree, they also tend to be right, whereas when they tend to disagree, their errors tend to increase. This is actually a very important feature of this process, as measures of uncertainty are needed to make several decisions, including sequencing of products in production planning (see [11]), sourcing decisions (see [9]), inventory planning (see chapter 4), etc.

Concept 3.6 *The disagreement among experts is a good proxy for demand uncertainty. When experts tend to agree, their error is relatively low and we face a low uncertainty. Vice versa, when they tend to disagree, very different demand scenarios might come true and we face a fairly uncertain demand.*

Basically, one can estimate the uncertainty the company is currently facing by looking at two information sets. First, we can use past predictions of the panel of experts to investigate the relationship between experts' disagreement and uncertainty as measured by the forecast error (difference between experts' forecast and actual demand). For example, in [6] and [11] authors suggest that the expected forecast error is twice the disagreement among experts. We can leverage on these relationships to gauge demand uncertainty for a product. We just need to measure the disagreement among experts and then read on the estimated relationship the expected error this entails.

Also, this method can provide us with relevant information to judge the real ability of the presumed experts to forecast demand. In the committee process the whole team generates a single number and thus it is hardly possible to judge to contribution of each member. On the contrary, in the Delphi method we can track the performance of single panelists and give different experts different weights or even remove some of them from the panel. Given the degree of uncertainty involved in this forecasting process, we shall not jump to the conclusion that a person is not good at forecasting demand for new products simply because one forecast was substantially wrong. Basically we are trying to estimate the average error one expert makes. This process is affected by substantial variability and we need a fairly large sample of forecasts to judge the true quality of the expert (i.e., the unknown parameter).

Drawbacks and limitations Just like all forecasting methods, Delphi has several drawbacks and limitations. First, the outcome of a Delphi forecasting process is nothing but the opinion of experts. Thus the results of this exercise are just as good as the experts and the information we provided them with (e.g., price, product description, advertising campaigns and budget, etc.) [14]. Thus the process needs high-quality intellectual capital to operate properly. Also, the process is rather long and time-consuming. This means that it does not fit emergency situations where a prompt answer is required. Moreover, it can be deployed only when the relevance of the decision at stake justifies the effort. For example, one might want to use a process like this for new

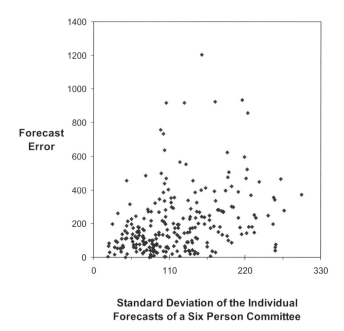

**Standard Deviation of the Individual
Forecasts of a Six Person Committee**

Fig. 3.28   Measuring uncertainty through disagreement among experts: source [6].

products, but it can hardly work for monthly forecasts of a wide range of existing ones.

Moreover, this process suffers from all weaknesses of qualitative methods. First it depends on clarity of the question(s) posed to the experts and of the objectives of the exercise (see [5]). For example, one should clarify the time frame that demand refers to (are we forecasting demand for 12 months or until current year's end?) and marketing levers (How many stores are going to carry the item? What price are we going to charge? And so on). A lack of clarity on the objectives and boundary conditions can lead to a very poor problem setting. This not only reduces the ability of experts to forecast demand but also reduces their commitment and interest in the process.

Experts, as well as managers of the whole process, might be biased by their objectives. For example, if one uses the forecast to set sales targets, the sales managers might be tempted to understate their expectations. On the other hand, if the forecast drives production and thus product availability, a sales manager might be tempted to overstate its forecasts. So one shall make very clear the purpose of the Delphi process and what data are going to be used for. Moreover, one might want to point out that experts are going to be judged on their forecasting accuracy to focus their attention on accuracy rather than on the potential effects of their opinions.

3.12.2   Lancaster model: forecasting new products through product features

The Delphi method relies solely on experts' opinion and it can be used to forecast demand for new products. However, even before the product is launched, one can use more structured models to predict sales.

A trivial possibility is to identify a "related" item and assume the new one is going to sell at a similar rate. It is rather interesting to discuss what the word "similar" really means. Often, this has two meanings. First, the existing one can be discontinued and a new one (actually a new version of the same product or a product in the same market segment) is going to take its place. A second meaning of the word "similar" is that the new product shares several features with the existing ones. For example, they might have a similar price, the same color, a similar design, etc.

Actually, one can try to generalize this rather trivial process and forecast the demand for a new product on the basis of its features.

The basic logic of this model is that customers are not interested in new products *per se* but rather in the features of the new product. For example, a customer might be looking for a red t-shirt, size large, rather than for a specific SKU. Clearly, this model of demand does not fit items that are somehow unique to the customers. For example, customers might want to buy the "Da Vinci code" book and not be interested in any other book on Da Vinci or the Opus Dei, or authored by Dan Brown. The average customer is just interested in that specific SKU. In other instances, customers might just look for a combination of features. For example, a customer might want to buy a package of spaghetti, Barilla brand, in a 1-kg package, and might be willing to switch to another brand or to another package size according to availability or price. Zara, a leading fashion retailer based in Spain, extensively analyzes the product features that happen to be most popular to design new products and adjust its assortment during the selling season. Basically, the retailer identifies the most popular color patterns, shapes, and accessories (e.g., buttons vs. zippers) and generates all possible combinations of the most successful product features to generate products with a very reasonable chance of success (see [9]).

So, to the extent that the model can be applied (i.e., products are not unique but rather can be described as set of features), one can try to predict the demand for a new product by looking at whether its features are popular in the current product assortment.

One way to estimate the popularity of these features is to use multiple linear regression. In this model the dependent variable are sales of the product and the independent variables are product features.

When one uses this model with statistics packages or even general-purpose software such as Excel, one should pay careful attention to the nature of the variables. These are often categorical variables. In other words, shape 1 and shape 2 are as different as shape 1 and shape 15. They are simply different options.

Other variables can be added to this model to account for seasonality and even experts' opinion simply by adding these variables as predictors in the linear regression.

3.12.3   The early sales model

Once the product is on the market, we can observe sales and judge the market potential for the product through its early sales. In the specific case of seasonal products (and more generally for products with a preset life cycle) such as fashion apparel, one could try to estimate demand up to season-end through the early sales. For example, one could try to forecast the season demand for a sandal based on sales in the month of March.

To do so, one has to estimate the relationship between the demand in the month of March and the demand for the total season. One can study the relationship between these two variables in the past to check whether actually demand in March is a good predictor of total season sales and, if this is the case, estimate the relationship. Notice that the underlying assumption is that the distribution of sales within the season next year will behave like it has behaved in past years. Thus, this process acknowledges that different products succeed to different degrees. Yet, it still somehow assumes that the future resembles the past. In particular it assumes that pattern of sales of any item within the season (in a product category) is the same year after year.[32]

Concept 3.7 *When season after season sales keep the same pattern over time, early sales are a good predictor of total season's sales.*

To judge the merits of this approach, one can draw the so-called "percentage done" curves. In other words, we can use demand data from past seasons and measure the percentage of total season sales accumulated by a given point in time in the season.

More formally, let $Y_{i,t}$ be the demand for item i at time t within the season, T the duration of the selling season, $C_{i,t}$ the cumulative demand for item i at time t, and $P_{i,t}$ the percentage of total season's sales of item i occurred by

---

[32]Notice that this concept is closely related to the seasonality model presented in section 3.8. However, there are two differences. First and foremost, in this case the product has less than one year or season of history, thus we cannot use the past demand to estimate the fluctuations of future demand. Therefore, we basically resort to related products to estimate how demand varies within the season. Also, in this case we do not look at sales in each time bucket (say one week), but rather at the sales up to a given point in time t.

Table 3.22   Demand data for a set of seasonal products

Time	Product 1	Product 2	Product 3	Product 4
1	206	127	488	192
2	291	193	674	289
3	565	296	925	334
4	841	482	1017	394
5	826	469	861	370
6	590	273	607	238
7	514	242	541	224
8	514	269	610	214
9	482	194	497	156
10	452	194	427	157
11	529	186	461	160
12	483	190	472	173
13	465	187	482	166
14	461	214	532	188
15	466	228	528	192
16	568	210	526	206
17	427	191	464	141
18	394	154	408	166
19	400	189	444	163
20	347	134	357	131
21	314	122	354	121
22	294	109	328	101
23	304	112	347	119
24	286	97	327	103
25	233	85	287	91
26	199	101	262	84
27	216	105	235	84
28	229	112	263	108
29	221	102	259	91
30	176	96	200	93
31	143	73	202	86
32	146	64	212	70
33	127	73	203	65
34	137	81	216	76
35	117	76	212	72
Demand in the season	12963	6030	15228	5618

Table 3.23   Percentage done of season demand curves

Time	Product 1	Product 2	Product 3	Product 4
1	2%	2%	3%	3%
2	4%	5%	8%	9%
3	8%	10%	14%	15%
4	15%	18%	20%	22%
5	21%	26%	26%	28%
6	26%	31%	30%	32%
7	30%	35%	34%	36%
8	34%	39%	38%	40%
9	37%	42%	41%	43%
10	41%	45%	44%	46%
11	45%	49%	47%	49%
12	49%	52%	50%	52%
13	52%	55%	53%	55%
14	56%	58%	56%	58%
15	59%	62%	60%	61%
16	64%	66%	63%	65%
17	67%	69%	66%	68%
18	70%	71%	69%	70%
19	73%	74%	72%	73%
20	76%	77%	74%	76%
21	78%	79%	77%	78%
22	80%	80%	79%	80%
23	83%	82%	81%	82%
24	85%	84%	83%	84%
25	87%	85%	85%	85%
26	88%	87%	87%	87%
27	90%	89%	88%	88%
28	92%	91%	90%	90%
29	93%	92%	92%	92%
30	95%	94%	93%	93%
31	96%	95%	94%	95%
32	97%	96%	96%	96%
33	98%	97%	97%	97%
34	99%	99%	99%	99%
35	100%	100%	100%	100%

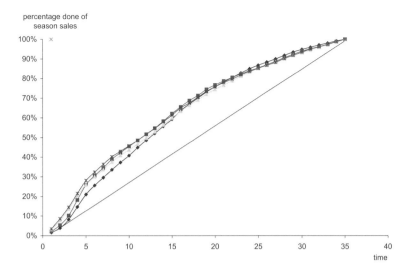

Fig. 3.29 Predicting season's sales with the early sales: percentage done curves.

time t:

$$C_{i,t} = \sum_{\tau=1}^{t} Y_{i,\tau};$$ (3.45)

$$P_{i,t} = \frac{C_{i,t}}{C_{i,T}}.$$ (3.46)

In this case, time t is the time elapsed since the beginning of the selling season. So the second week of the selling season of year '05 and '06 both refer to $t = 2$.

If the graph of percentage done shows that products in the past had a rather similar behavior as in figure 3.29, then we know how demand behaves within the season. Actually, we are not interested in sales on a specific day or in a specific week. We just notice that by time t we tend to sell a given percentage of the total season sales.[33]

In this case, a more trivial approach is to estimate the average distribution of sales as the average of the percentage done curves of all N products sold in past season(s).

$$\overline{P_t} = \frac{1}{N} \sum_{i=1}^{N} P_{i,t} = \frac{1}{N} \sum_{i=1}^{N} \frac{C_{i,t}}{C_{i,T}}.$$ (3.47)

---

[33] Notice that the products we analyze shall be related to the products we want to forecast during next season. For example, they should belong to the same product category in order to share the same demand pattern during the selling season.

We can then use this percentage done curve to estimate total season sales for the new product j as

$$\hat{C}_{j,T} = \frac{C_{j,t}}{\overline{P}_t}. \tag{3.48}$$

This procedure is often adopted though it is rather crude, as we assume that $\mathrm{E}(W) = \mathrm{E}(W/X) \cdot \mathrm{E}(X)$, which in general is not true and holds only if the two variables are independent. If this process is followed, we tend to measure the accuracy of this prediction as the standard deviation of $P_{i,t}$ (i.e., the percentage of total season sales occurred by time t for various products), which again is an oversimplification as the real uncertainty depends on the standard deviation of $1/P_{i,t}$.

A more appropriate description of the process is to investigate the relationship between the two variables through a regression. We can use past seasons' data to estimate the relationship between the sales up to time t, $C_{i,t}$, and total season sales, $C_{i,T}$. In particular, if we assume that the relationship is proportional, we can investigate the following relationship:

$$C_{i,T} = \alpha + \beta \cdot C_{i,t} + \epsilon.$$

Once we have estimated the parameters α and β through a and b, we can use them to estimate total season sales for product j as follows:

$$\hat{C}_{j,T} = a + b \cdot C_{j,t}$$

Linear regression provides us with more information on the distribution of errors and enables us to estimate errors and thus uncertainty [see equation (3.44)].

If the curves of various products show rather different behaviors over time, we might want to investigate the drivers of such differences. First, products might have inherently different demand patterns. For example, sandals and shoes that are part of a spring–summer collection might show different behaviors and we might want to tell one from the other by drawing two separate percentage done curves for the two clusters of products. Also, the actual selling pattern of products might be influenced by actions and decisions of the company. For example, demand might be influenced by the number of stores carrying the item, the current price and the availability of the product variants. All these variables might distort sales and thus open a gap between the natural demand pattern and the actual sales pattern we observe. For example, two products might have a relatively similar demand pattern over time, but one might take off at a later stage simply because it is delivered to stores at a later stage. Also, one product can take off at a given point in time simply because its price was significantly reduced.[34] Finally, sales might dip

---

[34] This is a very relevant issue in those countries where retailers can freely reduce price at any point in time. In other more regulated countries, such as Italy, the retailers can reduce prices only during the off-price season (e.g., January 10–February 15).

at a given point in time simply because stores are running out of the product. This means that either the product actually stocks-out or that inventories are so low that sales are reduced.

Example 3.18 Indeed, for some products the inventory level drives sales. This theme is widely investigated for fast moving goods such as grocery. Interestingly, similar findings hold in the case of slow moving goods. In the case of shoes, a US company has estimated that when less than ten pairs of a given shoe (style-color combination) are available, sales start declining. Indeed, with less than ten units the size distribution is broken and store managers start pulling back the product and even salespersons might not suggest the product to a consumer simply because he/she does not know whether the right size is available. The salesperson might prefer to suggest another product not to disappoint the consumer and embarrass him/herself.

3.13   THE BASS MODEL

The Bass model is a classic tool for the analysis of new product introduction from the marketing field [3]. The model is designed to forecast the adoption pattern of new durable products. It is a so called *diffusion model*. In other words, the model tries to forecast the adoption pattern of a new product. In particular, the model is aimed at *durable* products: For such products, multiple purchases of the same item are unlikely, so the model assumes that each potential adopter buys only one unit. In other words, given the number m of potential adopters the number of units sold over the lifetime of the product is by definition m. So rather than forecasting the size of a market, the model analyses how demand varies over time. The demand for the new product at time t is Y_t, and it corresponds to the growth in the number of actual adopters during period t. In other words, the demand Y_t is the difference between the adopters at time t, N_t, and the adopters at time $t-1$, N_{t-1}:

$$Y_t = N_t - N_{t-1}. \tag{3.49}$$

Furthermore, the model assumes that there are two basic adoption processes. On the one hand, some potential customers, called *innovators* adopt the new product at a given rate p simply because they come to appreciate its features. On the other hand, other potential users, called *imitators*, simply imitate current users. This second demand generation process actually depends on the number of current adopters that can be imitated, as well as the number of current nonadopters that can imitate them. This is the so-called "word of mouth" or "contagion" effect, and a parameter denoted by q accounts for it.

Hence, the probability that a potential adopter actually adopts the product in current period t is

$$p_t = p + q \cdot \frac{N_{t-1}}{m}. \tag{3.50}$$

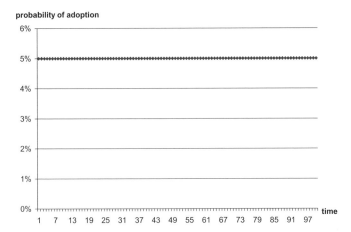

Fig. 3.30 Bass model: probability of adoption p_t with no imitation ($q = 0$). Case of $p = 0.05$.

Hence, the probability that a potential user adopts the product at time t depends on the rate of innovative adoption plus the rate of imitation times the percentage of current adopters. Basically, there are two probabilities of adoption and the second one is fully deployed only when the number of actual adopters reaches its maximum level m: $p+q$ is the probability of adoption of the "last customer."

This probability of adoption is multiplied by the number of potential new customers that have not adopted the product, so far:

$$Y(t) = p_t \cdot (m - N_{t-1}) = \left(p + q \cdot \frac{N_{t-1}}{m} \right) \cdot (m - N_{t-1}). \qquad (3.51)$$

In this model the shape of demand over time depends on the two parameters p and q. While parameter p tells the initial degree of adoption, the parameter q tells whether the product all of a sudden becomes very popular because of imitation and so demand reaches a peak.

The combination of these two parameters fits very different patterns of adoption. In the case of $q = 0$ there is no imitation and thus the percentage of customers that adopt the product is steady (see figure 3.30). This means that the adoption pattern follows a logarithmic curve (see figure 3.31) and the demand is decreasing as we have a constant probability of adoption but a decreasing number of potential adopters.

When we change the imitation parameter to $q = 0.1$, adoption pattern and demand change substantially. In this case, the probability of adoption increases with time (see figure 3.33) and the demand shows a peak in period 6

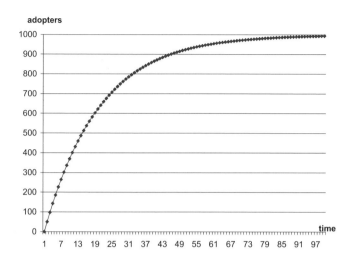

Fig. 3.31   Bass model: adoption pattern with no imitation ($q = 0$). Case of $p = 0.05$.

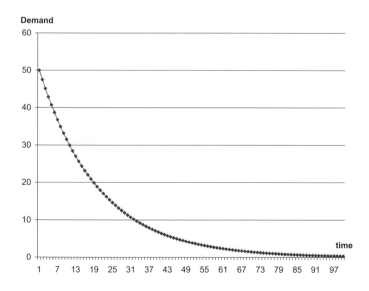

Fig. 3.32   Bass model: demand pattern with no imitation ($q = 0$). Case of $p = 0.05$

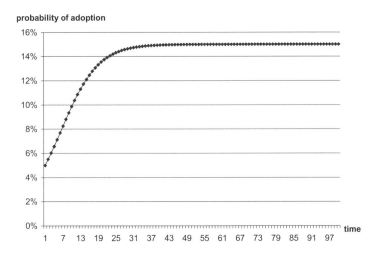

Fig. 3.33 Bass model: probability of adoption. Case of $p = 0.005$ and $q = 0.1$.

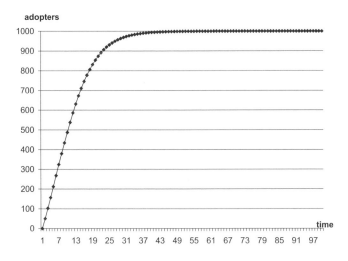

Fig. 3.34 Bass model: adoption pattern. Case of $p = 0.005$ and $q = 0.1$.

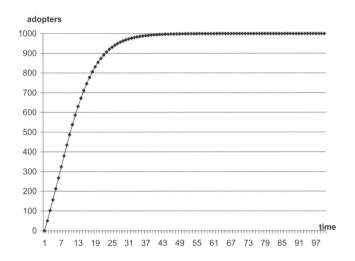

Fig. 3.35 Bass model: demand pattern. Case of $p = 0.005$ and $q = 0.1$.

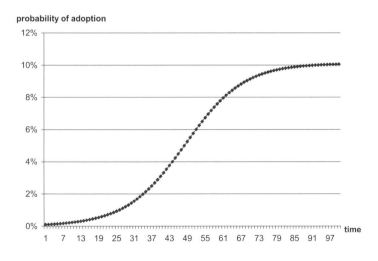

Fig. 3.36 Bass model: probability of adoption. Case of $p = 0.001$ and $q = 0.1$.

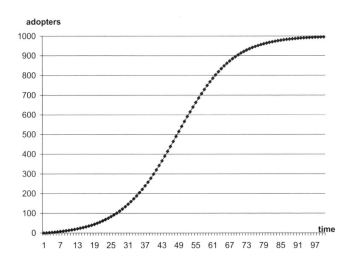

Fig. 3.37   Bass model: adoption pattern. Case of $p = 0.001$ and $q = 0.1$.

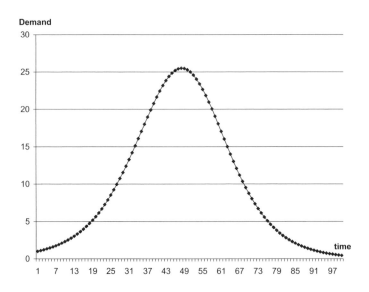

Fig. 3.38   Bass model: demand pattern. Case of $p = 0.001$ and $q = 0.1$.

(see figure 3.35). Previous periods have a lower probability of adoption while following ones have a relatively low number of potential adopters left. The number of adopters shows a similar pattern (see figure 3.34).

However, when we reduce p to 0.001 things change substantially. Like in the previous case, the probability of adoption changes over time. Interestingly, the pattern of adoption is very different from the previous case. In this case the product fever due to imitation starts at a later stage. Indeed, the imitation effect plays a significant role only once the innovators have built a significant base of users that current nonadopters can imitate. The buildup of this minimum number of users depends on the speed of the adoption by innovators, p. In case it is slow (i.e., there's a low probability of adoption p), it takes more time to build the critical mass of adopters that can be imitated by nonadopters (see figure 3.36). This change in p_t has an effect on the number of current adopters that shows a classic S shape (see figure 3.34). Demand starts very slowly as only innovators adopt the product in the early stages. As a fair base of adopters is established, the imitation effect starts playing a role and demand increases. Finally, demand decreases as the number of adopters reaches the maximum level m and we run out of new potential adopters (see figure 3.38).

SInce the demand patterns can be so different according to the parameters p and q, we should figure out how to estimate the parameters properly. A wrong estimation of the parameters might lead us to believe that future demand will look like figure 3.32 while it might resemble figure 3.35.

The Bass model is a centerpiece in the literature on new products and several procedures to estimate parameters have been suggested, but they are beyond the scope of this book. Here we report the estimation process suggested by Bass in his 1969 article.

We first restate the model as

$$Y(t) = \left(p + q \cdot \frac{N_{t-1}}{m} \right) \cdot (m - N_{t-1})$$
$$= pm + (q - p)\, N_{t-1} - \frac{q}{m} \cdot N_{t-1}^2.$$

By setting

$$a = p \cdot m, \tag{3.52}$$
$$b = q - p, \tag{3.53}$$
$$c = -\frac{q}{m}, \tag{3.54}$$

we can then restate the problem as follows:

$$Y(t) = a + b \cdot N_{t-1} + c \cdot N_{t-1}^2. \tag{3.55}$$

Equation (3.55) can be used with any statistical package to estimate parameters a, b, and c through multiple regression.[35]

Once these estimates are available, we can derive estimates of m, p, and q through equations (3.52) to (3.54). We know that $p = \frac{a}{m}$. Also, $q = -c \cdot m$. Substituting in (3.53) we have

$$b = -c \cdot m - \frac{a}{m}, \qquad (3.56)$$

that can be restated as a quadratic equation

$$a + bm + cm^2 = 0. \qquad (3.57)$$

Hence,

$$
\begin{aligned}
m &= \frac{-b + \sqrt{b^2 - 4ac}}{2c}, \\
q &= -c \cdot m, \\
p &= \frac{a}{m}.
\end{aligned}
$$

Once we have investigated how the model can be estimated, we should discuss how these estimates can be used. Actually, this really depends on where the product is in its life cycle. Before the product (or new technology) is launched, these estimates cannot be possibly derived from product's demand data. What one can do in these cases is to build estimates for the various parameters, based on various sources of information:

- The potential market size m can be estimated through analysis of product performance, price, etc., to estimate the number of customers that might potentially buy the product (for further discussion see [10] and [16]).

- The innovation and diffusion parameters p and q can be derived from past introduction patterns of comparable products. The assumption behind this approach is that the pattern of adoption of a new product depends (mainly) on the products category it belongs to (see 3.12.2). There might be significant differences among products, but still we lack the data to judge the current product and thus resort to the most reliable piece of information we have, that is, the category it belongs to.

---

[35] Notice that we are using a, b, and c as both estimates and parameters. Indeed, the model does not provide any formulation of demand as a stochastic process. Also, here we estimate a, b, and c, and then we obtain estimates of m, p, and q. However, the best estimate of a function of parameters a, b, and c might not be the function of the best estimate of those parameters. In other words, this is a robust empirical solution that happens to work, but it is not grounded in solid statistics.

At a later stage, as the product starts selling and more demand data on the specific product is available, we can estimate parameters for the specific product as shown in equation (3.55). Notice that the usage of product-specific estimates rather than category-related ones depends on the availability of data. Once we have three data points, we are in a position to estimate the three parameters of the Bass model m, p, and q. However, they might be so poor that it might still be worth using the past estimates based on the category.

3.13.1   Limitations and drawbacks

The Bass model is one of the best-known models for new product demand forecasting. A wide literature has applied, discussed, and improved the model.

A major stream of research has highlighted that the diffusion depends on the marketing levers of the company. For instance, the adoption pattern might depend on the pricing policies. Actually, if the price is reduced at a given point in time, the demand is likely to increase. Also, competitive variables might be a second source of variations in the adoption pattern. For instance, the adoption of the model can be influenced by the launch of competitor products or their price over time.

REFERENCES

1. J.S. Armstrong. Forecasting by Extrapolation: Conclusions from 25 Years of Research. *Interfaces*, 14:52–66, 1984.

2. E. Bartezzaghi, R. Verganti, and G. Zotteri. The Impact of Aggregation Level on Forecasting Performance. *International Journal of Production Economics*, 93-94:479–491, 2005.

3. F.M. Bass. A new Product Growth for Model Consumer Curables. *Management Science*, 15(5):215–227, 1969.

4. M. Dekker, K. van Donselaar, and P. Ouwehand. How to Use Aggregation and Combined Forecasting to Improve Seasonal Demand Forecasts. *International Journal of Production Economics*, 90:151–167, 2004.

5. A.L. Delbecq, A.H. Van de Ven, and Gustafson D.H. *Group Techniques for Program Planning: A Guide to Nominal Group and Delphi Processes*. Scott, Foresman, Glenview, IL, 1975.

6. M.L. Fisher and A. Raman. Reducing the Cost of Demand Uncertainty through Accurate Response to Early Sales. *Production and Operations Management*, 44(1):87–99, 1996.

7. F.X. Frei and D. Campbell. *Pilgrim Bank (A): Customer profittability, case 9-602-104.* Harvard Business School Publishing, Boston, MA, 2005.

8. V. Gaur, S. Kesavan, A. Raman, and M.L. Fisher. Estimating Demand Uncertainty Using Judgmental Forecasts. Unpublished, 2005.

9. P. Ghemawat and J.L. Nueno. *Zara: Fast Fashion, case 9-703-497.* Harvard Business School Publishing, Boston, MA, 2003.

10. D.B. Godes and E. Ofek. *XM Satellite Radio (A) case 9-594-009.* Harvard Business School Publishing, Boston, MA, 2004.

11. J. Hammond and A. Raman. *Sport Obermeyer Ltd., case 9-695-022.* Harvard Business School Publishing, Boston, MA, 1996.

12. S.H. Hum and H.-H. Sim. Time-Based Competition: Literature Review and Implications for Modelling. *International Journal of Operations & Production Management*, 16(1):76–90, 1996.

13. S.G. Makridakis, S.C. Wheelwright, and R.J. Hyndman. *Forecasting: Methods and Applications.* Wiley, New York, 1997.

14. J. Martino. The Precision of Delphi Estimates. *Technological Forecasting and Social Change*, 1(3):293–299, 1970.

15. S. Nahmias. Demand Estimation in Lost Sales Inventory Systems. *Naval Research Logistics*, 41:739–757, 1994.

16. E. Ofek. *Forecasting the Adoption of a New Product, Note 9-505-062.* Harvard Business School Publishing, Boston, MA, 2005.

17. V.K. Rangan and M. Bell. *Dell On-Line.* Harvard Business School Publishing, Boston, MA, 1998.

18. A. Shah. *CVS: The Web Strategy.* Harvard Business School Publishing, Boston, MA, 1999.

19. J.G. Wacker and L.G. Sprague. Forecasting Accuracy: Comparing the Relative Effectiveness of Practices between Seven Developed Countries. *Journal of Operations Management*, 16(2):271–290, 1998.

20. G. Zotteri and A. Raman. Estimating Retail Demand and Lost Sales. Unpublished.

4

Inventory Management with Deterministic Demand

4.1 INTRODUCTION

Managing distribution logistics effectively requires coordination of both *information* and *material* flows in the supply chain to gain *efficiency*, that is minimize costs, and *efficacy*, that is meet demand.

In chapter 1 we have discussed the various functions of inventories and showed that they can be deployed at various stages of the supply chain from raw materials, to components, to finished goods both at central distribution centers, at local warehouses, and at single retail stores. This short overview shows that inventory management is a rather broad and complex topic. Thus, we need a framework to identify single problems, tell the differences among them and design specific solutions. The first step is to identify the variables along which the various inventory problems differ, i.e., dimensions of the problems' space. Such variables are going to be used to classify both problems and solutions presented in the following:

- nature of inventories and of the supply chain;

- nature of demand;

- available information set;

- set of objectives the solution tries to achieve (e.g., which metric is used to measure inventory efficiency? And how do we measure customer service?).

Nature of inventories and of the supply chain.   To properly set the inventory management problem, we need to identify the *supply chain* we refer to. For example, inventory management policies designed for a single warehouse often perform poorly in a multiechelon supply chain where the purchase quantity might depend on inventory levels and demand both upstream and downstream. Also, as we have seen in chapter 1, multiechelon supply chains can be linear, converging, or diverging; each structure may have its peculiarities.

Supply chains with deterministic *lead times* (LT) differ from supply chains with stochastic ones. In the latter case, indeed, there is no tight relationship between purchase/production plans and deliveries to the warehouse, and this makes planning harder. Moreover, we must choose the *set of products* we intend to manage. Indeed, a *single item* supply chain is relatively easy to model and manage. Understandably, in supply chains where interactions among products are weak (e.g., they are neither substitute nor complements, they do not share production equipments, transportation means, or warehouse space) we can pretend that the various inventory problems are independent. On the contrary, modeling and managing *multi-item* supply chains where products are complements, share limited production or transportation capacity, etc., is more complex.

Another relevant feature of the *product* is the ratio between the *product life cycle* and the purchasing *LT*. Such a ratio tells whether the planning problem is *static* (i.e., decisions are taken at one point in time) rather than *dynamic* (decisions are taken at multiple points in time). For example, in the case of products with a very short life cycle and a relatively long production lead time, such as newspapers, we can decide how much inventory to carry only before we start selling. Such problems are *static*, as only one decision is made and

- the decision is not going to be updated at a later stage;

- the current decision has no effect whatsoever on the future ones as products expire.

Example 4.1 In the case of newspapers, the number of copies is set the night before the product starts selling; moreover, any units left unsold at the end of the day expire, as they become yesterday's news. Furthermore, any units left unsold today do not reduce the requirement of copies of tomorrow's newspaper. Thus, we make a single decision on the number of copies we want to print at one single point in time. Hence, the newsvendor problem is *static*.
□

A related, though different classification variable catches whether we are making just one purchase decision rather than multiple decisions. In other words, we can face both *single-period* and *multiple-period* ones. To tell the difference between these two variables let us consider the following examples (see also chapter 1).

Example 4.2 Let us consider the case of a European retailing chain that sells fashion apparel products. In this business the assortment changes completely each and every season. Also, let us assume that the retailer purchases products from Asia so that LT are too long to readjust the assortment and the quantities based on early season sales (see section 3.12.3). So, say that the retailer places orders for the fall/winter collection only once a year. This makes the problem *static*, meaning that all decisions are taken at one point in time with one information set (i.e., the problem is *single-stage*). Nevertheless, the retailer might decide to receive goods at various points in time. For example, the retailer might decide to receive some goods in early August and some additional quantities in early October, in order to reduce the inventory investment in August and September. This makes the problem multiperiod. Indeed, the decisions are taken for various periods of time. The inventories received in August are designed to meet demand during the first portion of the season, while goods received in October are designed to meet demand during the last portion of the season (see section 5.2.1). So this example shows a case of static and multiperiod problem that contrasts with the case of the newsvendor that is a static and single-period problem. The next example finally shows a case of dynamic and multiperiod problem. □

Example 4.3 Some products have such a long life cycle that it can be considered to be infinite. For example, some standard packaged goods such as dry pasta or frozen beans often have a rather long *shelf life* and thus goods left over at the end of one day are carried to the next day. Also, these products remain unchanged over time and thus we can have multiple deliveries over time. This makes the inventory planning problem a multiperiod one. Also, the delivery quantities are not the same forever. For example, a planner might decide production or purchase quantities for dry pasta once a week. These decisions are updated as more information becomes available. So, we plan inventories for multiple periods rather than for a single one. Also, we make decisions at various points in time with different information sets. This makes the problem *dynamic* as well. □

In the next two chapters we only consider single-period problems and those multiperiod problems where the life cycle (and shelf life) of the product is so long that we can ignore the end-of-life-cycle (and end-of-shelf-life) issues and costs (we only briefly discuss a two-period problem in section 5.2.1). The following example illustrates a kind of multiperiod problem where end-of-life/end-of-shelf-life issues must be dealt with.

Example 4.4 The case of fresh food in supermarkets is actually an intermediate problem where we face both the complexity of end of life, that is typical of the single-period problems, and the complexity of multiple periods. So, it is an intermediate problem that sums the complexities and hurdles of both extreme situations. These products often have roughly one month of shelf-life. For example, yogurt expires in roughly three weeks after the production is completed. So, during the product shelf life we can replenish the store multiple times. On the other hand, the product shelf life is quite limited and we cannot rule out the risk that the product expires on the shelf (or that the shelf life left is so short that customers do not buy it anymore). These intermediate problems are much more complex to model since, like in the case of multiperiod problems, we need to model the impact of decisions at time t on the initial condition at time $t + 1$, but we also have to model the age of products in the supply chain and the probability that they expire.   ☐

Nature of demand.   The nature of demand is a crucial issue, as meeting demand is the problem the firm tries to address through an appropriate inventory management policy.

Two important features of demand are the degree of *certainty/uncertainty* and the degree of *variability*. These two basic concepts are often erroneously considered synonyms (see [10]). Nevertheless, we shall separate them and clearly tell the difference between them. As the following examples show, demand can be very variable and certain, or stable and uncertain.

Example 4.5 The production of machineries industry faces sharp fluctuations of demand that depend on the economic outlook, as it might or might not lead firms to make investments in additional capacity or replace old machineries. In this industry, plus or minus 50% year-to-year variations are rather common and make demand very variable. However, such products are *Engineered To Order* (ETO). In other words, each single machine is designed or, at least, partially customized according to industrial customers' needs. Thus, production cannot be possibly completed before the customer order is received and the product is fully designed and engineered. Purchasing departments in this industry face a variable demand over time; however, when they place an order for a component, they often know customer demand. As we just said, demand variability is objective, while uncertainty is subjective and refers to an information set available to the forecaster: The purchasing planner in the machinery industry typically knows demand very well when he/she places purchase orders; on the contrary, when the *management* of the same company prepares next year's budget, typically it faces a significant uncertainty about next year's orders and turnover. Hence, the same demand is certain for the purchasing planner while the management perceives a great deal of uncertainty as it plans over a much longer horizon (at the least one year) and does not enjoy a very relevant piece of information, i.e., customer orders (usually the order portfolio covers just a few months).   ☐

Example 4.6 In the car industry, most Original Equipment Manufacturers (OEMs) choose one (*single sourcing*) or two suppliers (*parallel sourcing*) for a specific component (e.g., a speedometer) for a specific model (e.g., FIAT Punto). OEMs do their best to keep the utilization rate of the very expensive production lines steady; thus, also the demand for components is relatively flat (at least for those that are common to all product variants). However, for suppliers, demand might be very uncertain for several reasons. First, while designing a car with an expected demand of 100 units per day, several suppliers bid for the supply of speedometer. Thus, each of them faces a significant uncertainty: Demand could be either 0 (if a competitor is selected) or around 100 units per day (if the supplier itself is chosen). Once a supplier wins the bid, it still faces some additional uncertainty as the new car might be more or less successful than we initially expected. Thus, according to the level of sales, the OEM might use one, two, or even more production lines to assemble the car. As we can see, the demand for car components is likely to be fairly flat over time, but it is very uncertain before the new model is launched and, even more so, before the supplier is selected.   ☐

The *variability* of demand (and, more in general, of a process) is easy to capture through statistics such as standard deviation or the coefficient of variation when, as it is often the case, an adimensional metric is more appropriate. These statistics capture the variation of demand over time as they compare single observations (draws) of demand to their average.

On the contrary, the concept of *certainty/uncertainty* is more complex and subtle, as it refers to an information set available to the forecaster that tries to predict demand with a given forecasting horizon. Demand certainty/uncertainty depends on the extent to which the forecaster can predict the future level of demand with a given horizon and thus is subjective, meaning that it depends on the subject that is forecasting. For example, the future level of price of the shares of a given company might be very uncertain for many analysts before quarterly results are released but, at the same time, it might be relatively easy to predict for the CEO of the company that might already know them (at least roughly).

Demand can be either *continuous or discrete*. In some instances demand is basically a continuous process where small (infinitesimal) orders are continuously collected. For example, think about the sales pattern of Coke cans in any large chain of grocery stores. In theory the demand for cans of Coke is discrete, as one cannot buy 0.7853 units. However, aggregate demand is so large that this discreteness is somehow irrelevant for our planning problems. In other instances, demand is discrete as single customer orders are large as compared to the average demand. In this case the number of orders per unit of time is very small and thus orders are not received continuously. For example, consider the case of large machineries where each order is a separate project and several salespersons might follow a single customer.

The first case is easier to model than the latter, as we can model the demand pattern rather than each single order. The latter demand pattern is also called lumpy demand, as orders occur only sporadically in large lumps.

In the rest of this chapter and in the bulk of chapter 5 we assume demand to be a continuous process for the sake of simplicity and because this type of demand covers a large share of real-life problems. In chapter 6, we discuss how lot sizing policies can turn a continuous final consumer demand in a rather discrete demand upstream in the supply chain. We refer the reader to a more specific literature on discrete lumpy demand (e.g., see [9], [11], [8]).

Finally, to capture the nature of demand we shall look at *Delivery Lead Times* (DLT) customers expect or desire, i.e., the speed that customers require. If customers expect a zero DLT, the company knows customer orders only when they are delivered. Brick and mortar grocery stores are a typical example of this situation: Customers literally walk into the store and take what they need on the spot. In other instances, customers place their orders before their desired delivery date. For example, in the car industry customers (in Europe and only partially in the USA) place their order for a custom built car a few weeks in advance of the required delivery date. Also, the same happens with e-grocery stores, where customers might place orders a few days in advance of the requested delivery date. If DLT larger than zero, companies can take advantage of this advance notice of customers' need and plan their inventories accordingly.

Example 4.7 The comparison between various cases from the *retail industry* can shed some light on the meaning and effects of customers' DLT. As consumers, we expect zero DLT at fast food restaurants, while we are willing to wait for our fresh fish to be cooked for 40 minutes or so in a top-end restaurant. Thus, in the first case the retailer carries finished goods (e.g., burgers) and keeps a relatively limited assortment, whereas top-end restaurants only carry raw materials and produce a very wide range of finished products (dishes) to customers' order. Also, customers' expectations might depend on the product category. In most furniture stores (with the remarkable exception of IKEA) most customers place their orders a few weeks before the requested delivery date. This enables retailers to plan their inventories accordingly and make sure goods are delivered to the warehouse just a few days before they are shipped to consumers.   ⬚

In case of strictly positive Delivery Lead Time, the order portfolio can give us a very relevant piece of information that can be used to make proper plans. However, for the sake of simplicity in this book we assume DLT to be zero.

Set of information available   As we already mentioned, inventory problems differ in the degree of demand uncertainty. This is just a specific case of a broader concept: More generally, the applicability of various inventory management methods depends on the availability of various pieces of information.

Example 4.8 Some grocery supermarkets adopt automatic replenishment for some of their products, whereas others rely on employees that walk the store, look at current inventory levels, and place orders. The choice between one solution and the other depends on several issues. One of them is the ability of information systems to reliably capture current inventory levels on hand in the store (just think about what happens when a customer buys one vanilla and one banana yogurt and, given the identical prices, the vanilla one is scanned twice; or when products are damaged or stolen).

To fully appreciate the impact of this problems, let us assume that the optimal inventory level is 100 units. If the information system does not capture the current inventory level accurately, it cannot place an optimal replenishment order that raises inventories to 100 units. Several grocery retail chains think that an automated and centralized replenishment system might reduce labor costs, cut inventory investment, and improve product availability and customer service. However, they still use a manual process, as they know that employees can have a much clearer picture on current inventory levels than an automatic and centralized information system can. Other companies have decided not to surrender to inventory data inaccuracy (errors in inventory records). These companies spend time, money, and efforts to reduce such errors rather than live with them and adopt manual ordering. For example such companies audit their inventories more frequently or adopt more fancy techniques such as the so-called zero balance walk. In other words, these companies noticed that it is very easy to count zero units on the shelf and tell the system the products that are currently out of stock. This practice enables them to capture actual stockouts; also, they can check the inventory level of many products (on the average, 8% of the items in a grocery store are out of stock; see [2]) with a very limited effort.

Even these companies, though, adopt automatic replenishment only for long shelf life products. For fresh products, the central information system might know the quantity available in the store, but it cannot possibly capture other relevant pieces of information. For example, the system does not know the expiration dates of the 10 units of yogurt left on the shelf; also, the system ignores whether fruits and vegetables on the shelves are good-looking or are getting rotten.   □

In classic inventory models we assume to know the current inventory level. Only recently the issue of inventory data accuracy has been investigated empirically (see [4], [5], [6]) and theoretically (see [3]). Given the scope of this book, we assume that the decision maker knows current inventory levels perfectly.

Among other pieces of information, in inventory theory we study whether the information on inventory levels is available continuously (*continuous review*) rather than sporadically (*periodic review*). In other words, we tell the cases where we can monitor inventories continuously from the cases where we can know inventory levels only with a given periodicity. Indeed, in periodic review systems we shall acknowledge that in the time between reviews, inventories are out of control and fluctuate freely according to demand.

In the past, periodic review systems were often used because continuous ones were way too expensive. Today, the cost of such systems has gone down substantially. Thus, an information system that can track inventory level in real time is no longer an issue in most developed countries. However, the availability of such an system is just one of the requirements for the adoption of continuous review planning systems. For example, in many companies, purchase orders proposed by the computerized system must be approved (and at times modified) by planners and time and resource constraints might force the planners to look at purchase plans only once a month. In this case, the Information Technology infrastructure supports continuous review but the company is forced to use a periodic review system, as any piece of information released between two purchase plans would not be used to make better decisions and thus would be totally ineffective: The company works as if the real-time information was not available.[1]

Objectives Finally, to properly set the inventory problem, we should identify the objectives the company is trying to achieve. Such objectives are heterogeneous. However, we often try to translate them into costs to make the objective function scalar rather than vectorial. If and when this is a viable option, the objective function can be optimized.[2] In this chapter, we fully explore the flipsides of this approach. Also, cases where several objectives cannot be turned into a scalar objective function are investigated.

Finally, to fully identify an inventory management system, we have to specify the objectives the company (or organization) is trying to achieve. As we just discussed, such objectives can be fairly heterogeneous. However, we often try to turn them into a single cost function so that we can deal with a scalar rather than a vector and optimize it. We do so by associating each single objective with a unit cost. For example, we turn the service objective into a cost by stating that a single stockout is worth a given monetary amount. In the next chapter, we discuss the issues this approach raises and we investigate cases where we cannot turn several objectives into a single cost function.

The most common cost categories are:

- *Purchase costs*, i.e., the amount of money required to buy the goods.

- *Ordering costs*, i.e., the costs associated to an order or a lot. Such costs can be setup costs in a production environment where the warehouse is supplied by the company's production plant, but they can also be fixed transportation costs the customer pays for (or the supplier charges for); also, they can be administrative costs of order processing, receipt, and inspection of inbound materials.

---

[1]Notice that in current information systems we have so-called alerts. These tools call for the attention of the planner under "critical" conditions. This really means that the decision maker can plan the product once each month, but on the other hand the system can call for his/her attention when inventories run high/low, demand has any odd behavior, and so on. This really means that the decision makers have a blend of the two processes. They review decisions periodically, but at the same time the computerized system controls inventories and demand continuously and calls for decision-makers' attention when it is needed.

[2]See section B.7 on multiobjective optimization.

- *Inventory costs*, i.e., the cost for warehouse space occupancy, financial cost of holding money in inventories, loss of value of goods carried both because they might perish (think of fresh goods that lose weight and can rot) or they might lose value because of technological innovation (think of the value of older PC when a newer one is launched) or fashion (think about the value of a garment at the beginning of the selling season and after the end of it).

- *Costs of lack of service* to customers; one of the functions of inventories is to enable the company to meet demand with short Delivery Lead Times. Poor inventory management can lead to fairly low service levels. Obviously, customer service can be defined in several ways that entail very different cost functions. For example, in the case of drugs for spring allergies the lack of a product might not be too serious; on the contrary, the lack of B– blood in a hospital might be a very dangerous situation.

In this chapter, we start with the simplest problem:

- single warehouse (single echelon);

- infinite life cycle and thus multiperiod problem;

- stationary and continuous demand (later extended to variable but perfectly predictable demand);

- known (deterministic) lead times, demand, and inventory levels;

- the objective is to minimize the sum of inventories and ordering costs.

The above system is deterministic and thus we do not need to tell the continuous review case from the periodic review one. Once we know the status of the system at any point in time t_0, we can derive the status of the system at any other point in time t.

Section 4.2 deals with the simple case of a single product with zero Lead Time, and section 4.3 discusses the robustness of the model. The following sections extend the model in several directions. Section 4.4 describes the case of deterministic but nonzero lead times; section 4.5 deals with the so-called finite production rate case, that is, when the products in a lot are not delivered all at once, but are rather delivered progressively. Section 4.6 discusses the multi-item case, and nonlinear costs are investigated in section 4.7. Finally, in section 4.8 we illustrate a few examples of how variable but known demand can be managed by deterministic optimization models.

4.2 ECONOMIC ORDER QUANTITY

If demand is constant, the model that supports decisions is the so called Economic Order Quantity (EOQ). In the simplistic conditions of the model (deterministic, constant, continuous and known demand and zero lead times) we can completely fulfill demand and therefore the cost related to lack of service is zero. Indeed, in this case we only need to order any amount of goods when we run out of the product. The goods are replenished immediately and thus all customers are served.

Moreover, the basic model described in this section assumes no quantity discounts. Under these assumptions, the purchase cost is not a relevant performance measure, since in the long run the quantity purchased depends on the demand rate rather than on the purchasing policy. The purchasing cost per unit time is just equal to the demand rate times the unit purchase cost.

Under these circumstances the only relevant costs are:

- Ordering costs C_{or}; in the basic model we assume that ordering cost are a fixed cost the company incurs each single time an order is placed; thus the total ordering cost is equal to the number of orders placed times the fixed cost of each single order. In other words, we assume that the ordering cost is a linear function of the number of orders placed.

- Inventory costs C_{in}; inventory costs too are a linear function of the average inventory level; we assume that other variables such as the maximum inventory level are irrelevant, while in real-life applications they might matter (think of the size of the warehouse that depends on the maximum inventory level rather than on the average one).

Given the above setting, we clearly face a tradeoff: A purchasing policy that entails frequent purchases of a few units at a time leads to relatively low average inventory level, but incurs the fixed purchasing cost very frequently. On the contrary, buying large quantities infrequently causes an increase in the average inventory level but reduces the number of orders and saves on ordering costs.

Let us introduce some notations to describe this tradeoff more precisely:

- d is the demand rate, i.e., the number of units or quantity of demand per unit of time (e.g., units/month or kg/year);

- A is the fixed ordering cost in units of value per lot (e.g., € or \$ per lot);

- h is the inventory holding cost in units of value per unit of product held in inventory, per unit of time (e.g., € per unit per month). Sometimes, this cost is stated as a percentage of the unit purchasing cost u; hence, we have $h = h\% \cdot u$, where

  - u is the unit purchasing cost;
  - $h\%$ is the percentage holding cost of a unit of value for a unit of time; we may note that this quantity plays the role of an interest rate for the money tied up in inventories;

- Q is the order quantity, i.e., the amount that is purchased when inventory drops to zero;

- T is the time between two consecutive orders, i.e., the periodicity of the ordering process, that we also call the *order cycle*.

We note that in this simplistic case the decision maker can only play with one decision variable. Once the purchase quantity (lot) Q has been set, the purchase frequency T is fixed as well. Vice versa, once the frequency T is set, quantity Q is fixed as well.

Example 4.9 Consider a company with a demand for 100 units/month. This company can order once a month. In this case, the company orders 100 units at a time. Also, the company can order 600 units twice a year. Finally, the company can order 1200 units once a year. This example shows that the choice of the order quantity implies the order frequency and vice versa.   ⬜

Given a demand rate d, the order quantity Q is consumed in Q/d periods (e.g., months) and therefore an order is placed each $T = Q/d$ periods.

Finding the economic order quantity calls for the definition of an objective function expressing the total cost, over a given time period, as a function of the decision variable Q. Before we try to write the objective function, we should understand the dynamics of the inventory system. Any stockout is avoided by placing orders exactly when inventories are depleted. Also, it would not make sense to place orders before the inventory level reaches zero, as we would just increase inventory holding costs and at the same time would be ordering sooner rather than later (that is, we would anticipate the ordering cost). Right after the order is placed, the purchase quantity Q is delivered to the warehouse (as LT is zero) and the inventory level I immediately reaches Q. Such inventory buildup is progressively consumed at the constant demand rate d. The inventory level displays a typical *saw-tooth pattern* (figure 4.1) and fluctuates between the minimum level zero and the maximum level Q.

Once the dynamics of the system is clear, we can write the total cost function, which consists of two terms:

$$C_{tot} = C_{or} + C_{in}$$

The ordering cost depends on the fixed ordering cost (fixed cost for a lot) A and on the number of orders placed in a period (unit of time). The latter variable is equal to the demand in a period divided by the lot size Q. Indeed, in the long run, demand equals the purchased quantity that in turn equals the number of lots times the order size Q.

Example 4.10 For example, let us assume that a company sells 1000 units per month. In the long run, a company that sells 1000 units per month must buy 1000 units per month. We do not mean that the company buys exactly 1000 each and every month. In some months, the ordering policy might lead

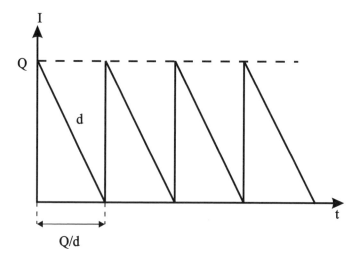

Fig. 4.1 Inventory dynamics in the EOQ system for LT $= 0$.

the company to buy slightly more than 1000 units while in other months the company might buy slightly less than 1000 units. But on the average a company that sells 1000 units per month buys 1000 units per month. Purchases in excess of 1000 units per month would lead to a progressive increase in inventories. On the contrary, if the company purchases less than 1000 units per month, sooner or later it is going to run out of inventories and a stockout is going to occur. A company that purchases 1000 units can purchase 10 lots of 100 units per month, 2.5 lots (i.e., some months 2 orders, other months 3 orders) of 400 units per month, and so on. Thus, in general the number of lots (in a period) is equal to demand (in a period) divided by the lot size Q.

Hence,

$$C_{or} = A \cdot \frac{d}{Q} = A \cdot \frac{1}{T}.$$

As to inventory holding costs, we can refer to figure 4.1 to understand that the average inventory level is $Q/2$, as inventories fluctuate linearly between 0 and Q. Thus, the inventory holding cost term is

$$C_{in} = h \cdot \frac{Q}{2}$$

and total cost is

$$C_{tot} = A \cdot \frac{d}{Q} + h \cdot \frac{Q}{2}. \tag{4.1}$$

Figure 4.2 shows the total cost as a function of the order quantity Q. Taking

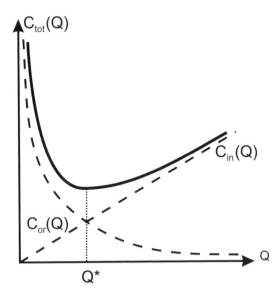

Fig. 4.2   Total cost per unit time as a function of the order quantity Q.

the first-order derivative with respect to Q, we can prove that the optimal value of Q is

$$Q^* = \sqrt{\frac{2Ad}{h}} \tag{4.2}$$

and thus the optimal purchasing frequency is

$$T^* = \sqrt{\frac{2A}{dh}}. \tag{4.3}$$

We can read the mathematical results economically. Equation (4.2) shows that:

- As the fixed ordering cost A increases, we shall increase the lot size Q to incur the high cost A less frequently.

- As the demand rate d increases, we shall buy larger lots Q; indeed, with a higher demand we use up Q items more quickly and thus order more frequently, with a corresponding increase in the total ordering cost; this makes ordering costs more crucial and a larger Q helps keeping them under control.

- As the holding cost h increases, the company is less and less willing to carry inventories and thus cuts the lot size Q to reduce the average inventory level $Q/2$.

We can evaluate the minimum total cost by substituting Q^* for Q in equation (4.1):

$$
\begin{aligned}
C_{\text{tot}}^* &= C_{\text{tot}}(Q^*) = A \cdot \frac{d}{Q^*} + h \cdot \frac{Q^*}{2} \\[2ex]
&= A \frac{d}{\sqrt{\dfrac{2Ad}{h}}} + h \cdot \frac{\sqrt{\dfrac{2Ad}{h}}}{2} \\[2ex]
&= \sqrt{\frac{Adh}{2}} + \sqrt{\frac{Adh}{2}} \\[2ex]
&= \sqrt{2Adh}.
\end{aligned}
\tag{4.4}
$$

The above equation show that minimum cost is reached where the ordering cost equals the inventory holding cost (see also figure 4.2).

To apply the above EOQ model, we must learn how to quantify the parameters of the model. Estimating the demand per unit time d is a fairly easy task, provided we have adequate information about past demand data.[3] On the contrary, measuring A and h is not that easy.

Ordering costs   The ordering cost A includes all costs that linearly depend on the number of lots the company manufactures or orders. As we will see in our further discussion, some costs might depend on the number of lots in some contexts while in others they might not. We are not in a position to tell what shall be included in a generic situation. Here we just list the major costs that *might* be included and provide the reader with guiding *principles* that can help him/her to figure out whether in his/her specific context they shall or shall not be included.

- *Administrative costs.* Administrative costs might depend on the number of lots purchased. As the number of orders increases, the number of invoices, proofs of delivery, etc., can increase and thus the number of employees and related costs might increase as well. Having said that, the estimate of A is often biased as *all* administrative costs are allocated to A, while *some* of them should not be considered as they do not depend on the number of orders placed (a very common practice is to divide the total administrative cost by the number of orders). For example, the cost of administrative software is not proportional to the number of orders; it is actually very likely to remain the same, no matter whether we place 100 or 150 orders. Moreover, some costs depend on the number

---

[3]We should always keep in mind that information systems typically record *sales*, and not demand. If stockouts do not occur too frequently, we may consider recorded sales as a proxy for demand.

of orders but not linearly (think of the cost of labor). Using a term from Economics, we call them *semivariable costs*, since they are not variable costs *strictu sensu*, as they are flat within a given rage (of orders in this case), but they increase outside this range. For example, consider a company with two administrative persons that manage 100 orders per month; also let us assume that the company needs to raise the number of orders to 101. It is very likely not to incur any additional labor cost; also, the company definitely does not pay for 2.02 employees. On the contrary, if the company needs to manage 150 orders, an additional employee is hired rising the number of administrative persons to 3. The additional cost is the gross cost of the employee plus any related cost such as a PC, software licenses, etc. Finally, does the cost of administrative employees drop if the number of orders drops from 100 to 50? Well, it depends. In case the "spare" employee can be fired or used productively elsewhere in the company, the reduction in the number of orders can turn into a reduction of administrative costs. Otherwise, cutting the number of orders has no real economic benefit (think of highly regulated labor markets such as Italy and France). More generally, costs are subject to *hysteresis*; that is, an increase in output (number of orders) followed by an equal decrease in output might not take the costs back to initial stage, as it is often much easier to increase costs (headcounts in this case) than cut them.

However, the current trend toward outsourcing of administrative activities and/or more flexible contracts to absorb workload peaks make administrative costs more and more variable. Thus, administrative costs may depend on the number of lots, but such relationship is often nonlinear. This makes the EOQ model a linear approximation of a nonlinear cost.

- *Transportation costs.* This is a second cost component that can depend on the number of lots. Indeed, moving goods from production (or an upstream warehouse) frequently and in small batches might increase transportation costs. However, these costs too might not depend linearly on the number of orders. A linear relationship between number of orders and transportation costs implies that the cost of a single delivery does not depend on the quantity transported. Such an assumption clearly does not hold, at least for large variations in the quantity transported. In the case of a private fleet of trucks, if the quantity increases above the capacity of a given truck, the company can either use a larger (and thus more expensive) truck or increase the number of trips. Both options increase transportation costs. Also, in the case of a private fleet, costs like insurance and depreciation of trucks are fixed, at least in the short run, and therefore they should not be included in A.

In the case of a third-party fleet the problem is even more complex. In the case of "point-to-point" transportation (i.e., a direct trip from source to destination), the cost might not depend on the quantity transported; in such a case, if the truck capacity is not saturated, one more order may cost practically nothing (to a first approximation). On the contrary, in the case where goods from several sources and moving toward several destinations share the same truck (think of FedEx, DHL, national postal services, etc.) the cost typically depends on the volume or weight of goods transported. Indeed, the weight and volume of the goods can set some relevant constraints to the routing problem (see chapter 8) and this makes planning the trips harder and potentially less efficient.[4]

Thus, in the EOQ model we consider transportation costs to be a linear function of the number of orders, but we shall be aware that this is an approximation that can be crude at times.

- *Receiving, inspection, and handling costs.* When an order is received at a warehouse, personnel needs to perform a series of time consuming activities. Documentation shall be checked and recorded in the information system, quality shall be controlled, and finally goods should be placed on shelves. The duration of some of these activities does not depend on the number of units delivered but rather on the number of orders. For example, handling a pallet with 10 or 12 units takes exactly the same amount of time (provided that 12 units fit on a single pallet). However, if the number of units increases above the maximum capacity of a pallet, the workload and thus the cost increases as well.

  What has been said about administrative costs applies to these costs as well: They are often semivariable and subject to hysteresis. Moreover, many companies outsource warehousing activities, and this turns these costs into truly variable ones.

- *Setup cost.* In case a warehouse is supplied by a production process, setup costs of production machines contribute to the ordering cost A. Also in this case, the specific condition of the company should be carefully considered when one tries to measure setup costs. Setting up a machine creates direct costs that clearly depend on the number of setups (think of the need to use washing material, or the scrapped production at the beginning and at the end of each lot in continuous production processes such as food). These direct costs shall always be accounted for.

  Other costs might/might not be included in the ordering cost A. Cost of workers setting up the machines and downtime (i.e., lack of production

---

[4]The problem is getting more and more complex, as companies are trying to match supply and demand at times with online auctions. For example, transportation might be much more expensive from China to Italy than vice versa. Such pricing policies known as *revenue management* at times can lead to odd pricing strategies. For example, consider air-fares: Just to give one example, at times a round trip can be less expensive than a one-way ticket.

while setting up the machines) are often considered to be part of setup cost. As to personnel we shall use the same *caveat* we have discussed in the case of administrative costs: Can we really save the money? Can we really use the worker effectively in case we do not use him/her to set up the machine? As to downtime, its cost really depends on operational condition of the firm:

- if production capacity is saturated (fully utilized) a single setup prevents the company from producing other goods and serving other customers; thus in this case a setup creates opportunity costs;

- in the case of spare capacity, on the contrary, the setup does not have any significant flipside and might be basically cost-free. Also, setup costs shall only be considered when the plant supplying the warehouse produces on a lot-for-lot logic. That is when the delivery lot equals the production lot and thus *one additional delivery entails an additional setup.*

Inventory holding cost.   The EOQ model assumes an infinite product life cycle. No product has a infinite life cycle, so *strictu sensu* we cannot apply the EOQ model to any real-life problem. However, the model can be applied when the end of the product life cycle is so far into the future that costs of goods leftover at the end of the life cycle are irrelevant for our problem. Under this assumption, inventory costs are just the cost of holding inventories in the warehouses. Still, measuring h appropriately is not trivial, as several variables contribute to it.

- *Financial costs.* The investment in inventories increases the working capital and the need for financial resources to run the company. Thus, inventories imply not only more need for capital, but also a larger financial cost of capital.[5] Still, what is the right measure of the financial cost of capital for a company? Is it the average cost of debt? Is it the cost of equity? If so, what is the cost of equity capital? For a thorough discussion we refer to any textbook on accounting systems and finance (e.g., [7]), while here we just provide the basic ideas. The cost of capital really depends on the current financial conditions of the specific company. Consider a company with spare cash invested in short-term bonds. Any increase in the inventory investment reduces the spare cash, the investment in bonds, and their interests. In this case the percentage cost of capital $h\%$ is equal to the interest one gains on short-term bonds. Note that these are opportunity costs (the investment prevents the company from making money, rather than creating an actual cost) but this is just

---

[5]The larger the working capital the larger the need for capital and the higher the risk of insolvency. This means that money lenders ask for a higher return on their capital as a reward for the higher risk.

irrelevant for a cold-blooded, rational decision maker. However, the cost can be much higher. Let us consider a company with financial problems that has very large debts. Under these circumstances, debt might be fairly expensive. An extra investment in inventories might require extra cash that in turn requires an additional (i.e., marginal) debt with some additional (marginal) costs. In this case, the cost of inventories can be higher than in the former case as borrowing rates are usually higher than lending rates.

Clearly, in real contexts an extra screw might not increase the cost of debt at all. More generally, very small changes in the inventory investment for a specific SKU might not require any additional debt (or might not reduce the investment in short term bonds). However, we just aim at figuring out the average financial cost of an extra (i.e., marginal) investment of one Euro in inventories.

Also, one might wonder what happens if the contract with suppliers sets three or four month terms of payment. Does that change the holding cost, as the company can hold goods for four months without any real financial exposure? The answer is no, it does not matter. The key idea is that no matter what the payment conditions are, any increase in the average inventory level increases the financial exposure of the company and thus the need for working capital. Terms of payment can really make a difference for the overall financial exposure of the company. For example, a company that used to pay suppliers one month after goods were received and then moves to three months might improve the financial exposure substantially and might cut $h\%$ (see note 4.2 on page 203). However, given the terms of payment, if inventories increase by 1€, the working capital increases by 1€ and its cost goes up by $h\%$ €, no matter what the terms of payment are. Thus, a larger purchase quantity Q implies an increase in inventories and thus an incremental holding cost. Indeed, a change in terms of payment basically changes the in-transit stock (see chapter 1), i.e., basically changes the point in time (and in the supply chain) from which the company financially holds inventories. As we have learned in chapter 1, companies hold both in-transit stock and cycle stock. The former is influenced by terms of payment and is not a function of order quantity Q (1.8). If LT is shorter than terms of payment the net in-transit stock can be negative. On the contrary, cycle stock depends on the order quantity Q, it can only be positive, and it is not a function of terms of payment.

Example 4.11 Let us consider a European company that imports goods from East Asia. It takes roughly one month to transport goods from East Asia to Europe. Let us assume that bills are paid two months after goods are shipped. Also, the demand for product A (let us assume this is the only one product for the sake of simplicity) is 1200 units/year.

The cost is 1€/unit and the holding cost $h\%$ is 10%. The company is considering two policies:

- order 100 units once a month;
- order 400 units every four months.

In the first case, 100 units are shipped on January 1, delivered on February 1, and paid on March 1. Goods delivered on February 1 are sold during the month of February. Thus, the financial exposure of the company fluctuates between 0 on February 1 and -100 on February 28. This pattern is the repeated for February shipments, March shipments and so on (see figures 4.3 and 4.4).

In the second case, 400 units are shipped on January 1, delivered on February 1 and paid on March 1. Goods delivered on February 1 are sold in February, March, April and May. Thus, the financial exposure of the company changes as follows. It starts at 0 on February 1, when goods start being sold before they are paid. It reaches -100 on February 28, then on March 1 400€ are paid and it reaches 300. Then it progressively decreases to 0 on May 31 (see figures 4.5 and 4.6).

The two patterns above can actually be interpreted as the overlap between a positive cycle inventory and negative in-transit inventories. In the first case, cycle inventories vary between 0 and 100; when we add a negative working capital of -100 units as goods are paid one month after they are delivered, we get the pattern described in figure 4.4. In the latter case we can interpret figure 4.6 as the overlap between cycle inventories varying from 0 to 400 with a negative in-transit inventory level of 100 units (see figure 4.5).   □

- *Warehousing costs.* The investment in inventories requires not only capital but also warehouses where goods can be stored. When the company owns its warehouses, these costs are semivariable and subject to hysteresis: Until spare space is available in the current warehouse(s), the cost does not vary significantly.[6] However, if we run out of space in the warehouse, we have to either build (or buy) a new one or just rent additional space. Variable costs such as insurance premia or the energy cost for refrigerated goods should be added on top of these variable costs. As discussed for ordering costs, things change substantially if the company outsources warehousing since costs might be "more linear," i.e., cost might vary linearly with the average inventory level. For example, a contract with the third-party logistic provider might set a cost per pallet per month.

---

[6]In some very specific instances such as refrigerated warehouses, the cost of running the warehouse might depend on the amount of goods carried, as the amount of energy required to keep temperature constant might depend on the mass kept in the warehouse.

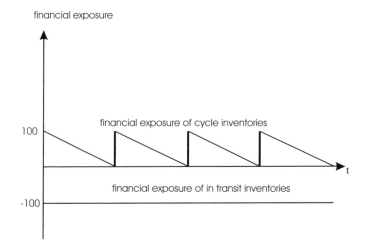

Fig. 4.3   Components of financial exposure of policy 1 over time.

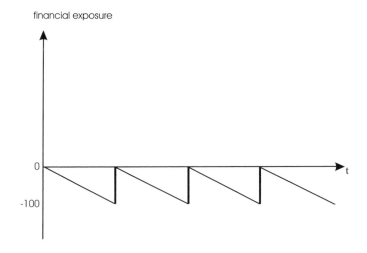

Fig. 4.4   Overall financial exposure of policy 1 over time.

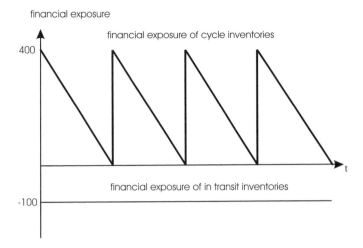

Fig. 4.5 Components of financial exposure of policy 2 over time.

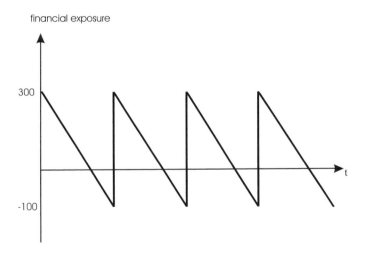

Fig. 4.6 Overall financial exposure of policy 2 over time.

- *Loss of value of inventories.* Holding inventories in a warehouse can reduce the value of the goods carried. For example, in the case of furniture, handling goods in the warehouse might damage them and reduce their value substantially. In other instances, such as fresh fruit and vegetables, products might lose weight or rot.

4.3 ROBUSTNESS OF EOQ MODEL

In the previous section we showed that measuring parameters of the EOQ is not a trivial task. Forecasting d and estimating A and h is all but trivial, hence we should check whether the model is *robust.* In other words, we have to understand whether the decisions the model suggests are reasonable even when the input data are not 100% exact, but they are affected by some error. By "reasonable" in this context we mean that decisions lead to a performance (cost in our case) that is nearly optimal. This is the reason why we measure the robustness of the model by the deviation from optimal cost.

Just to show an example, we try to understand what happens if the ordering cost A is not estimated properly. Let us assume that the error in the estimate of A is Δ. The company would believe the cost A to be $A + \Delta$ and the EOQ model would suggest to choose a quantity $Q_1 = \sqrt{2\,(A + \Delta)\,d/h}$, rather than optimal quantity $Q^* = \sqrt{2Ad/h}$. One could be tempted to measure the robustness of the EOQ model as the difference between Q_1 and Q^*. Tempting though this might sound, we are not really interested in this metric. Indeed, it would capture whether our decision Q_1 is close to the optimal decision Q^*, whereas we are more interested in whether our decision leads to nearly optimal costs. Thus, we shall measure the difference in total cost.

The suboptimal quantity Q_1 leads to a higher cost than the optimal one $C^*_{tot} = C_{tot}(Q^*) = \sqrt{2Adh}$. The cost we get by setting $Q = Q_1$ is

$$
\begin{aligned}
C_{tot}(Q_1) &= A \cdot \frac{d}{Q_1} + h \cdot \frac{Q_1}{2} \\
&= A \frac{d}{\sqrt{\frac{2 \cdot (A + \Delta) \cdot d}{h}}} + h \cdot \frac{\sqrt{\frac{2 \cdot (A + \Delta) \cdot d}{h}}}{2} \\
&= \sqrt{\frac{dh}{2}} \cdot \left(\frac{A}{\sqrt{A + \Delta}} + \sqrt{A + \Delta} \right).
\end{aligned}
$$

If we compare the above cost with the optimal one C^*_{tot}, we can measure the percentage increase in cost due to the error Δ in the estimate of A:

$$
\frac{C_{tot}(Q_1)}{C^*_{tot}} = \frac{\sqrt{\frac{dh}{2}} \cdot \left(\frac{A}{\sqrt{A + \Delta}} + \sqrt{A + \Delta} \right)}{\sqrt{2Adh}} = \frac{\sqrt{\frac{A}{A + \Delta}} + \sqrt{\frac{A + \Delta}{A}}}{2}. \tag{4.5}
$$

As equation (4.5) shows, a 10% error in the estimate of A increases the cost only by 0.1%. This finding sheds a new light on the estimation problems discussed in section 4.2. Parameters of the EOQ model might be hard to estimate. But even sizable errors lead to small increases in cost.

Similar analyses can be performed on the effects of errors in the estimate of h and d, and similar results can be obtained.

A related but distinct case is the one where not all values of $Q \in \mathbb{R}+$ are a viable option as there are some constraints. For example, many products are distributed to retailers in casepacks (apparel products, packaged goods, etc.) that cannot be broken and are called *minimum order size* or *minimum lot*. In other instances, minimum lots are set for marketing purposes rather than for logistic ones.

Moreover, it is often convenient to set the ordering frequency in such a way that warehouse operations can be easily managed. For example, it is very easy to receive goods from a given supplier once a day or once a week. On the contrary, receiving each 1.7 days is much more complex as the time of delivery would keep on changing. The EOQ model does not consider any constraint on the solution Q. Hence, the solution Q^* is very likely not to be viable. We can deal with this problem in two ways. First we can re-define the problem. The second and more convenient solution is to choose the feasible solution Q_f which is closest to the optimal one Q^*.

Concept 4.1 *When inputs to the model are uncertain or not precise, we should perform some sort of sensitivity analysis. In sensitivity analysis we want to capture the quality of the solution in terms of extra cost (loss of potential profit) we might face. In other words, we are interested in performance. We are not really interested in whether the solution we suggest is actually close to the optimal one.*

The cost of this solution $C_{\text{tot}}(Q_f)$ is higher than the unconstrained minimum $C_{\text{tot}}(Q^*)$, that could be reached if Q^* was a viable option, i.e., if there was no constraint on the solution. However, the increase in cost $C_{\text{tot}}(Q_f) - C_{\text{tot}}(Q^*)$ is often fairly small. We can compute the percentage increase in costs by comparing the cost of Q_f with C_{tot}^*:

$$
\begin{aligned}
\frac{C_{\text{tot}}(Q_f)}{C_{\text{tot}}(Q^*)} &= \frac{\dfrac{hQ_f}{2} + \dfrac{Ad}{Q_f}}{\sqrt{2Adh}} \\
&= \frac{Q_f}{2}\sqrt{\frac{h^2}{2Adh}} + \frac{1}{Q_f}\sqrt{\frac{A^2 \cdot d^2}{2Adh}} \\
&= \frac{Q_f}{2}\sqrt{\frac{h}{2Ad}} + \frac{1}{2Q_f}\sqrt{\frac{2Ad}{h}} \\
&= \frac{1}{2}\left(\frac{Q_f}{Q^*} + \frac{Q^*}{Q_f}\right).
\end{aligned}
\tag{4.6}
$$

Equation (4.6) shows that rounding the optimal solution Q^* to the closest feasible solution increases the total cost marginally. For example, in the case the nearest feasible solution is 20% larger than the optimal solution Q^*, the cost increases only by 1.67%. Therefore, we can take the solution of the EOQ and round it to the nearest[7] feasible solution with a very limited increase in costs.

Example 4.12 Company B produces canned soups and sells them in 12-units casepacks. One of the customers is a large hypermarket that gets direct deliveries from the supplier. The ordering cost for the hypermarket is 10€ which covers administrative costs, handling costs, and quality checks. Demand is 100 units/week. A single can of soup costs 1€ to the hypermarket. Cost of inventories is 0.4% per week and covers financial cost and variable warehousing costs. Given the above inputs, the optimal solution is $Q^* = 707.1$ unit. However, we cannot buy 0.1 units, and even purchasing 707 pieces at a time is not a viable option. The two options we have are 696 and 708 units. Given these two options, we clearly choose to purchase lots of 708 units. The delivery of case packs of 12 units increases the cost by less than 0.01%. This increase is very likely to be offset by savings in handling costs at the warehouse. ⧠

The EOQ model can be extended in several ways by changing the most unrealistic assumptions of the basic EOQ model. Such extensions are the subject of sections 4.4–4.7.

4.4   CASE OF $LT > 0$: THE (Q, R) MODEL

The first easy extension of the EOQ model is the case of nonzero and deterministic LT. In such a case, we cannot wait until we run out of inventories to order, as the order quantity Q is not readily available. On the contrary, we should order LT units of time before we run out of stock. Given the demand rate d, we shall place an order when the inventory level reaches the so-called *reorder point*, also known as *reorder level* $R = LT \cdot d$. In other words, we order when we have just enough inventory to meet demand during the LT. The system works this way: Each time inventory reaches the level R, an order of Q units is placed. Q units are then delivered LT periods after the order is placed.

In such a system we have two decision levers, i.e., variables we can control: Q and R. However, the pattern of inventories over time behaves just like in

---

[7]As equation (4.6) shows, the right metric for distance is $Q_f/Q^* + Q^*/Q_f$, which is a geometric distance. In fact, two quite different decisions, such as increasing the optimal quantity by a factor of 2 (plus 100%) or decreasing it by a factor of 2 (-50%), result in the same increase in cost of 25%. On the contrary, increasing the optimal quantity by 50% raises the total cost by only 8.3%.

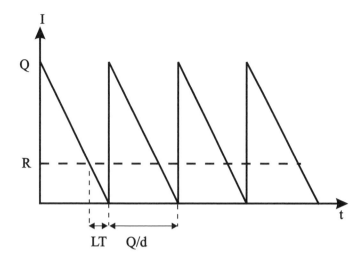

Fig. 4.7 Inventory pattern in the EOQ model, case of LT > 0.

the case of the EOQ model (see figure 4.7). Hence, nonzero lead times have no influence on the choice of Q. Indeed, just like in the basic EOQ model, Q units are delivered exactly when inventory reaches the zero level, and an order is placed every Q/d time units. Thus, the optimal quantity is the EOQ in this case as well. Notice that the two control levers are set independently: Q^* sets the optimal purchase quantity to minimize the sum of inventory and ordering costs, whereas R picks the right timing for ordering so that Q^* is delivered exactly when it is needed.

This rather simple case forces us to introduce a new variable: the *inventory position*. If a warehouse manager checks the inventories a few seconds after an order is placed, he/she would be tempted to place a second order, as inventories on hand might look low since they are below the reorder point R. This would be very dangerous, since we might keep on ordering Q units several times. Such a series of orders might then be delivered over a short period of time, leading to a skyrocketing increase in inventories. But how can we avoid such problems?

In inventory management we should not only look at current inventory level physically in the warehouse, called *inventory on hand*. We must consider physical inventory in the warehouse plus the outstanding orders, i.e., all orders that have been placed but have not been delivered yet.[8] We call this new

---

[8]In the next chapters we will investigate the stochastic case. In this more complex setting, we might experience a stockout and thus customer orders might be backlogged. As we will see, in such a case the inventory position is equal to inventory on hand plus outstanding orders sent to the supplier(s) minus unmet orders from the customer(s). The case of DLT >

variable *inventory position* (IP). This is the variable we should be looking at when we ponder whether we should or should not be placing an order.

When we look at the above example, we can see that the inventory position IP can easily solve the problem. Orders are placed when and only when IP reaches the reorder level R. Therefore, right after the order for Q units is placed, a warehouse manager is not even tempted to place a second purchase order. Indeed, the inventory position increases to $R + Q$ immediately after the order for Q units is placed (even if these units have not been delivered yet).

Concept 4.2 *In inventory management and planning the current on-hand inventory might be a misleading figure. We would rather carefully consider the inventory position, that accounts for inventory on hand as well as for incoming orders and customer backorders (if any), in order to give us a more dynamic picture of our current inventory level.*

Just like in the basic EOQ model, inventory on hand fluctuates between 0 and Q units (with an average of $Q/2$ units), while the inventory position varies between R and $R + Q$ units, with an average of $R + Q/2$. In the case of relatively short LT (LT $< Q/d$) the inventory position differs from inventory on hand when the order is placed, as it jumps from R to $R+Q$, while inventory on hand remains unchanged at R. Only when the order quantity Q is delivered, the inventory position equals inventory on hand as there are no more outstanding orders, i.e., we no longer wait for the supplier to deliver an order (see figure 4.8). On the contrary, in the case of relatively long LT (LT $> Q/d$), the warehouse is always waiting for at least one order to be delivered, and thus the inventory position is always greater than the inventory on hand, by definition.

4.5   CASE OF FINITE REPLENISHMENT RATE

So far we have discussed cases where the quantity ordered is delivered in a lump of Q units. This is generally true when the warehouse receives goods from an upstream warehouse. On the contrary, when the warehouse is served by a production plant, goods might be progressively delivered as they are produced. This happens when each single unit is delivered to the warehouse as production is completed. This changes the dynamics of inventories in the warehouse, so the cost function and the optimal quantity Q^* change as well.

The finite replenishment rate r is the number of units delivered per unit of time. Obviously, r shall be greater than d. When d is greater than (or equal to) r, the production rate is not sufficient (or barely sufficient) to meet

---

0 sets similar problems, as there is a series of orders placed by the customers that have not been met yet.

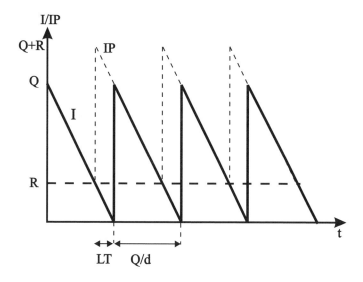

Fig. 4.8   On-hand inventory and inventory position, case of LT < Q/d.

demand, thus we keep on producing continuously. In this case, the order quantity problem is not properly set. In the remainder of this section we assume r to be greater than d.

When the inventory level reaches zero, it does not immediately increase by Q units. It increases progressively at a rate $r - d$. The $r - d$ growth rate is the result of an inflow of r units of product per unit of time and an outflow of d units of product per unit of time. This process goes on until all Q units of the production lot are delivered. It takes Q/r periods to complete the production lot Q with a production rate r. Thus, when the lot is completed, the inventory level has reached $(r - d)\,Q/r$ units. This quantity can be rewritten as $Q - d \cdot Q/r$; in other words, this means that the maximum inventory level is equal to the production lot Q minus the demand (d) that has occurred while the lot was being delivered (over a period of time Q/r). Once the production is over, inventory starts decreasing at a rate d. The inventory level reaches its maximum when the production lot Q is completed. Therefore, inventory increases linearly between 0 and $(r - d)\,Q/r$ at a $r - d$ rate and then it decreases linearly between $(r - d)\,Q/r$ and 0 at a d rate, as figure 4.9 shows.

Hence, the average inventory level is $\dfrac{(r - d)\,Q}{2r}$ and the total cost function is

$$C_{\text{tot}} = A \cdot \frac{d}{Q} + h \cdot \frac{(r - d) \cdot Q}{2r}; \tag{4.7}$$

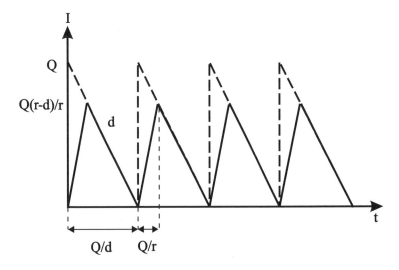

Fig. 4.9 Inventories pattern in case of finite rate of delivery.

by taking the derivative with respect to Q we can show that

$$Q^* = \sqrt{\frac{2Ad}{h} \cdot \frac{r}{r-d}}. \tag{4.8}$$

We shall now read the formula to understand it. The finite production rate EOQ suggests that when production rate r equals demand rate d the production lot Q is infinite; that is, we shall continuously produce to meet demand. Also, when r is barely greater than d, the production lot tends to be very large, as inventories build up very slowly in the warehouse (at a $r - d$ rate). Finally, as $r \to +\infty$, equation (4.8) tends to look like the EOQ formula, which can be considered just a specific case of this more general one.

4.6 MULTI-ITEM EOQ

So far we have investigated single-item problems. However, very often we can find several items in a warehouse. This makes modeling the problem harder.

One might want to model the multi-item problem as a series of independent single-item problems. This is an easy way out, but it comes at a cost. When we use this approach, we basically treat a single warehouse with several products as a series of independent warehouses, each with a single product. Companies actually build a single warehouse to leverage on joint economies of scale; that is, to take advantage of the savings one can gain by managing several products under one roof. Hence, looking at a multi-item problem as a series of single-item ones might be a simplistic rather than a simple solution to a complex problem. Indeed, this might miss some crucial feature of the problem and lead to a poor or unfeasible solution.

Example 4.13 Let us consider a warehouse with two products α and β sourced from a single supplier. Demand for product α is 100 units/month while demand for product β is 300 units/month. Placing an order costs 800€, no matter whether one or two products are ordered. The cost is fairly high as the supplier is located in China and transportation costs are significant. Holding one unit of α for a month costs $1 €$, while holding one unit of β for a month costs $16/27 €$.

If we optimize order quantities for the two products separately, we get the following optimal quantities: $Q_\alpha^* = 400$ and $Q_\beta^* = 900$. Once order quantities are defined, we can try to overlap orders to gain some savings. The best option is the following: order α on 4th, 8th, and 12th month, and order β on the 3rd, 6th, 9th, and 12th month. This order plan implies 6 orders per year, that is 0.5 orders per month.

Thus, the monthly cost is

$$
\begin{aligned}
C_{\text{tot}} &= 0.5 \frac{\text{orders}}{\text{month}} \cdot 800 \frac{€}{\text{orders}} + 1 \frac{€}{\text{unit} \cdot \text{month}} \cdot 200 \, \text{units} \\
&\quad + 16/27 \frac{€}{\text{unit} \cdot \text{month}} \cdot 450 \, \text{units} \\
&= 866.6 \, €/\text{month}.
\end{aligned}
$$

Intuition suggests that this is very likely not to be the best option, as potential synergies between the two products are not fully exploited since often we do not take advantage of shared ordering costs. For example, if we order product β once every 4 months, the total cost is

$$
\begin{aligned}
C_{\text{tot}} &= 0.25 \frac{\text{orders}}{\text{month}} \cdot 800 \frac{€}{\text{order}} + 1 \frac{€}{\text{unit} \cdot \text{month}} \cdot 200 \, \text{units} \\
&\quad + 16/27 \frac{€}{\text{unit} \cdot \text{month}} \cdot 600 \, \text{units} \\
&= 755.55 \, €/\text{month}.
\end{aligned}
$$

This solution is clearly crude, and it is likely not to be optimal; yet, it proves that optimizing the order quantity for single items might not be a good idea. The next subsections show how we can find optimal solutions to multi-item problems. ▯

4.6.1 The case of shared ordering costs

Example 4.13 above shows that setting EOQ for each single item independently might be sub-optimal as it might not fully exploit the potential advantages of a joint and coordinated policy. We first investigate the case of shared ordering costs, where the cost of an order A depends neither on the number of product types ordered nor on their quantities, just like in example 4.13.

To apply the results of single-item EOQ to the multi-item case, we shall first identify the mix of products. In the deterministic case, such a mix is easy to set, since deliveries to the warehouse and inventories at the warehouse have exactly the same mix as demand. Thus, we define a *bundle* of products with the right mix of items. For example, if demand for Coke is twice the demand for Sprite, when planning joint deliveries we can refer to a "virtual" composite product consisting of 2 Cokes and 1 Sprite. Note that for our purposes, a bundle (i.e., virtual composite product) of 4 Cokes and 2 Sprites would work just as well.[9]

Example 4.14 In example 4.13, the ratio between the demands of product α and β is 1:3. Therefore, we can define a bundle consisting of 1 unit of α and 3 units of β. Demand for this bundle of products is 100 units per month. Holding one bundle for 1 month means holding 1 unit of α and 3 units of β; hence the holding cost of one bundle is $1 + 3 \cdot 16/27 = 25/9$. The ordering cost is 800 €, as it is a fixed cost that does not depend on quantity.

Thus, we can apply the EOQ model to the bundle of products and find that the optimal purchase quantity for the bundle is $Q^* = 240$ units, which really means setting the purchase quantities of the two products to $Q_\alpha^* = 240$ and $Q_\beta^* = 720$. This choice implies a cost of 666.6 €, which is well below the two costs found in example 4.13.   □

It is interesting to notice that the joint optimal quantity for two (or more) products is lower than the optimal quantity for single products. Let us compare two different scenarios.

1. In the first scenario, products share a common cost A, as the company places one single order with multiple products and receives just one delivery. Therefore, reducing the order frequency by one order per month, will only save A Euro, but it will increase the inventory levels and thus holding costs of both products.

2. In the second scenario, each single product requires a separate order and a separate delivery and thus products do not share a common fixed cost A. Here, reducing the ordering frequency of, say, item α saves ordering cost A, but it increases inventories and holding cost of item α only.

We see that in the first case we have a stronger incentive to increase order frequency and thus order smaller quantities. An extra order makes us save inventory investment and holding costs on several products, rather than a

---

[9]Note that this holds true when we have no constraint on the units or case-packs, that is when demand and deliveries have no lower bound on actual units and can be considered as continuous variables for all practical purposes. For example, demand for fast moving goods like Coke can actually be considered continuous even if it is produced and distributed in units. On the contrary, retail demand for jewels can hardly be modeled as a continuous variable.

single one. Another way to see this problem is that when two or more products are delivered together and a bundle is defined, the holding costs add up while the shared ordering cost remains unchanged. Thus, pulling together various products (e.g., buying them from the same supplier or receiving them from the same warehouse) leaves the numerator of the EOQ formula unchanged (in particular A) while it increases the denominator (in particular h), thus reducing optimal purchase quantities.

This result takes us back to a basic concept discussed in section 2.1.2. Joint deliveries of multiple products reduce inventory investments as the ordering frequency can be increased. That is why many grocery stores and retail stores of large chains (such as Wal-Mart) receive the bulk of their deliveries from a central distribution center rather than from suppliers. These chains rather hold goods in a central warehouse and then deliver to stores very frequently with trucks filled with goods from several suppliers rather than get direct (and sparse) deliveries from suppliers with a relatively limited assortment. The savings on inventory costs are well worth the cost of a distribution center.

4.6.2   The multi-item case with a constraint on ordering capacity

In section 4.2 we showed that the ordering cost often depends on the degree of utilization of administrative resources and warehouse personnel. For example, if the utilization of employees involved in ordering and receiving materials is low and they can neither be fired nor be utilized in any other way, the cost of an order A can be very close to zero.

In many instances, to choose the appropriate fixed cost A, the company needs to figure out an ordering policy to exploit limited resources. In the example just discussed, the EOQ problem is one where the limited ordering capacity shall be allocated to all items in the company's assortment. Similarly, the company might have a limited ability to receive goods in the warehouse or a limited transportation capacity, as the number of trucks in the company's fleet is fixed. In our further analysis we assume that products are ordered from separate suppliers and are delivered separately. Thus we assume that they do not share ordering costs.

If this is the case, the company aims at minimizing the inventory holding cost, subject to a constraint on the total number of orders placed with a proper allocation of the limited capacity to each single product. In other words, the company tries to keep inventories under control subject to a constraint on the overall number of orders.

This is a nonlinear optimization problem that can be written as follows:

$$\min \quad \frac{1}{2}\sum_{i=1}^{N} h_i Q_i,$$

$$s.t. \quad \sum_{i=1}^{N} \frac{d_i}{Q_i} \leq F; \tag{4.9}$$

where:

- i is the item index,

- N is the total number of item types,

- d_i is the demand per unit of time for product i,

- h_i is the cost of holding one unit of product i for one unit of time,

- Q_i is the purchase quantity for item i,

- F is the ordering capacity per unit of time, i.e., the maximum number of orders that can be placed in a unit of time.

We notice that as the total number of placed orders increases, the purchase quantity of each single item i decreases and so does the overall inventory holding cost. Therefore, we know that the ordering capacity constraint is active in the optimal solution. Hence, we can replace the inequality constraint with an equality one in equation (4.9).

To solve this problem, we can resort to the method of Lagrangian multipliers . First, we write the Lagrangian function:[10]

$$\mathcal{L}\left(Q_1, \ldots Q_i, \ldots Q_N, \lambda\right) = \frac{1}{2} \sum_{i=1}^{N} h_i \cdot Q_i + \lambda \left(\sum_{i=1}^{N} \frac{d_i}{Q_i} - F \right). \qquad (4.10)$$

In principle, we should add a non-negativity constraint on decision variables Q_i. We assume a so-called *interior optimum*, which means that no non-negativity constraint is active and the optimal solution is such that $Q_i^* > 0$. Also, this condition is required for the constraint (4.9) to make sense. Then, we enforce first-order optimality conditions by computing derivatives with respect to the N variables Q_i and λ.

$$\frac{\partial \mathcal{L}\left(Q_1, \ldots Q_i, \ldots Q_N, \lambda\right)}{\partial Q_i} = \frac{h_i}{2} - \frac{\lambda \cdot d_i}{Q_i^2} = 0 \qquad \forall i,$$

$$\frac{\partial \mathcal{L}\left(Q_1, \ldots Q_i, \ldots Q_N, \lambda\right)}{\partial \lambda} = \sum_{i=1}^{N} \frac{d_i}{Q_i} - F = 0,$$

thus

$$Q_i^* = \sqrt{\frac{2\lambda^* d_i}{h_i}} \qquad \forall i, \qquad (4.11)$$

$$\sum_{i=1}^{N} \frac{d_i}{Q_i^*} = F. \qquad (4.12)$$

---

[10] See section B.4. In case of equality constraint, how the constraint is added to the objective function and the sign of the Lagrangian multiplier are irrelevant. However, the sign we have adopted helps us to read the results economically and would work in case of an inequality constraint as well.

When we look at these optimality conditions, we notice that equation (4.11) resembles the EOQ formula. The only minor difference is that the ordering cost A is substituted by the optimal Lagrangian multiplier. To understand the formula we shall resort to the economic meaning of Lagrangian multipliers discussed in section B.4.2 (shadow prices).

From the discussion in section B.4.2, we know that the multiplier tells us the extent to which a (marginal) increase in the ordering frequency can decrease the inventory holding cost of a product. The shadow price interpretation of the Lagrangian multipliers guides us in the design of a procedure to identify the optimal solution. Indeed, optimality conditions (4.11) and (4.12) are a system of nonlinear equations. In general, such a system can only be solved numerically, even though in specific cases like this one, a closed form solution is easy to find.[11] Even if this is an easy case, we prefer to suggest an iterative algorithm to find the optimal value of the multiplier. One reason is that the procedure can be applied in a more general setting, when an analytical solution cannot be find (see the multi-item newsvendor problem in section 5.2.1). Another reason is that the approach lends itself to a nice economic interpretation.

From equation (4.11), we see that the multiplier plays the role of an ordering cost, which is consistent with its shadow price interpretation. If the value of λ is too small (i.e., smaller than the optimal one), we order small quantities Q_i too frequently, and the total number of orders exceeds capacity. If λ is too large, on the contrary, capacity is not fully utilized and inventories are excessive, leading to suboptimal solution. We should look for the right ordering cost, which results in the full utilization of our ordering capacity.

1. Choose an initial value for λ.

2. Use this value of λ to calculate the N quantities Q_i^*.

3. If $\sum_{i=1}^{N} \frac{d_i}{Q_i^*} > F$, increase λ.

4. If $\sum_{i=1}^{N} \frac{d_i}{Q_i^*} < F$, decrease λ.

---

[11] By plugging expression (4.11) for Q_i^* into (4.12), we see that

$$\sum_{i=1}^{N} \frac{d_i}{\sqrt{\frac{2\lambda d_i}{h_i}}} = F \quad \Rightarrow \quad \sum_{i=1}^{N} \sqrt{\frac{d_i \cdot h_i}{2\lambda}} = F \quad \Rightarrow \quad \frac{1}{F} \sum_{i=1}^{N} \sqrt{\frac{d_i \cdot h_i}{2}} = \sqrt{\lambda}.$$

Hence,

$$\lambda = \frac{1}{F^2} \left(\sum_{i=1}^{N} \sqrt{\frac{d_i \cdot h_i}{2}} \right)^2,$$

which can be plugged back into equation (4.11) to find the optimal order quantities.

5. If the current solution satisfies the constraint, at least within a given tolerance, stop; otherwise go back to step 2.[12]

This search procedure can be interpreted as demand/offer mechanism by which we aim at finding the right ordering cost: If the resource is too cheap, it is over-utilized, and we must increase its cost; if it is too expensive, we should lower its cost in order to achieve full utilization.

A common finding is that ordering (or setup) costs are often overstated, leading to an unnecessarily large amount of stock. Indeed, a fixed ordering cost A is used in the EOQ formula, even when from an economical standpoint the orders imply no marginal cost, as resources are fixed. The EOQ approach may be justified as a simplification of the above procedure, when we apply a fixed cost λ to provide an incentive to "properly" use the limited ordering capacity. Finally, even in this case we might face estimation problems. Above all, we should keep in mind that in the long run F is a decision variable as well, and we might wonder what is the appropriate ordering capacity for our company.

4.7 CASE OF NONLINEAR COSTS

The basic EOQ model assumes that cost parameters $h = h\% \cdot u$ and A do not depend on quantity Q. Thus purchasing costs are assumed to be proportional to the total quantity purchased (which in the long run equals demand). Also, the EOQ model assumes that ordering costs are proportional to the number of orders placed (i.e., the number of lots). In many real contexts these assumptions hardly hold, as cost parameters depend on the purchase quantity Q. Very often, suppliers are willing to offer discounts to customers that place large orders (*quantity discounts*). Also, some of the costs included in A might depend on Q. Transportation costs might depend on Q, as this cost can be semivariable; when the quantity Q exceeds the capacity of a small truck, a larger and more expensive truck is required.

Let us consider a company that places orders for a single product and pays for direct deliveries (i.e., there is no interaction with other products and the product purchased from the supplier can fully use the capacity of means of transportation). Also, we assume that the company can choose among three means of transportation, say 1, 2, and 3, with capacity constraints

---

[12] A simple but effective search algorithm is bisection. It is often used to solve scalar nonlinear equations with a single unknown variable. We identify two values of λ, λ^- and λ^+, that lead to overutilization and underutilization of the ordering capacity F, respectively. We know that the optimal value of λ lies in the (λ^-, λ^+) range. We consider the midpoint of the interval, $\lambda_m = (\lambda^- + \lambda^+)/2$, and we check whether it leads to over- or underutilization of capacity. We continue our search accordingly, by setting $\lambda^- = \lambda^m$ or $\lambda_+ = \lambda_m$, respectively. We see that the interval bracketing the solution is always bisected. The process is repeated until the range (λ^-, λ^+) is "small enough."

Q_1, $Q_2 > Q_1$, and infinite, respectively. Using the three different means of transportation implies three different ordering costs $A_1 < A_2 < A_3$.

In this case, we cannot choose of the best order quantity by enforcing a first-order optimality condition like in the basic EOQ model, as the cost function need not be differentiable (nor continuous) for all $Q \in \mathbb{R}$. Nevertheless, solving the problem is rather simple. The overall cost function consists of pieces of convex functions (see figure 4.10), and we can take advantage of this property while searching for the optimal solution. To start our analysis, we can draw the cost functions for the three means of transportation (see figure 4.10). Figure 4.10 shows that as A increases, the cost function is shifted upward and the optimal quantity increases. Economically, this means that as the fixed cost increases, the overall cost and the economic order quantity increase as well.

We might be tempted to choose the first means of transportation, as it keeps the ordering costs to a minimum. This view neglects the capacity constraints. Although the cost function 1 is lower than others, we might not be able to purchase the optimal quantity Q_1^*, as it might be greater than the maximum capacity Q_1. In other words, the optimal quantity Q_1^* might not fit on the relatively small means of transportation 1 that cuts ordering costs down to A_1.

If this is the case, we can leverage on the convexity of the cost function 1 to identify the best viable solution. The cost function 1 is decreasing up to Q_1^*, thus the optimal viable solution is Q_1, i.e., the maximum quantity that can be transported on the first means of transportation.

As to the second means of transportation, we assume that the optimal quantity Q_2^* is lower than the maximum capacity Q_2. Notice that under these assumptions the third means of transportation is basically not even an option. Indeed, we know that the cost function of the second means of transportation is lower than the cost function of the third one. Also, the second means of transportation can operate at the optimal level Q_2^*. Thus, no matter what quantity Q_3 we want to carry, the cost of the third means of transportation is higher than the cost of the second means of transportation.

Hence, the solutions we shall consider are:

- means of transportation 1, quantity Q_1;
- means of transportation 2, quantity Q_2^*;

within this set of options we shall pick the one with the lowest cost. In the example of figures 4.10 and 4.11, the best option is to use the second means of transportation and order Q_2^* units at a time.

More generally, we exploit the convexity of constituent cost functions to solve these problems.[13] We can identify the optimal quantity for each interval and then compare the various local optima to tell the global optimum.

---

[13]To be more precise, the convexity of each single piece of the overall cost is just exploited to guarantee that the optimum of each piece is either the point of stationarity, or an extreme

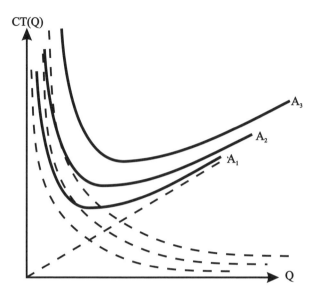

Fig. 4.10   Cost functions for various fixed ordering costs A.

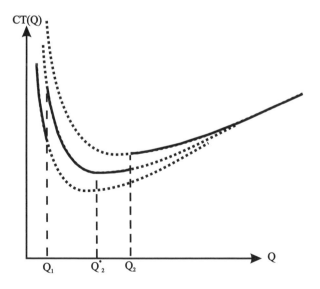

Fig. 4.11   Overall cost function.

A similar procedure can be used in the case of *quantity discounts*, i.e, when the unit cost of the product u depends on the purchase quantity Q. In this case, a third cost should be added to the cost function (on top of ordering and inventories costs): the purchasing cost. Indeed, larger lots might reduce the total purchase cost, as quantity discounts might cut the average unit cost.

We can distinguish two different kinds of discount[14]:

- *all-unit discount*: the discount applies to the whole quantity purchased;

- *marginal unit discount*: the discount applies only to the marginal quantity, i.e., on the quantity that exceeds a given minimum threshold.

For example, some suppliers might offer a 20% discount when customers purchase more than 100 units. Other suppliers might offer a 20% discount on units in excess of 100.[15]

In the former case the cost function might be discontinuous, as reaching the minimum quantity required to gain a given discount (say Q_1) might actually reduce the overall purchase cost and, as a consequence, holding cost. In other words, oddly, the last unit required to qualify for the discount might have a negative marginal cost: Purchasing Q_1 units costs less than ordering $Q_1 - 1$ units.

Example 4.15 Alpha is a retail company that sells product a. This product is purchased from company Beta. Demand for a is 1,000 units a week. Company Beta has a rather complex pricing policy. For orders below 10,000 units, it charges 4€ per unit; for orders larger than or equal to 10,000 and less than 50,000 units it charges 3.75€ per unit; and finally it charges 3.5€ per unit for orders of at least 50,000 units. Placing an order costs 500€. The holding cost for one week is 1% of the unit cost.[16] Company Alpha wants to properly set the purchase quantity from Beta. The price discount is tempting, but managers are wondering whether they should be buying very large lots of 50,000 units.

---

point of the pertaining interval; then we select the minimum over all of the sub-intervals. If we had to minimize a piecewise-concave function (of a single variable), we would use much the same strategy; the only difference would be that the minimum of a concave function would always be one of the two extreme points of each interval.

[14] Notice that in this section we only investigate cases where the unit cost u depends on the purchase quantity Q. Some of the reasons why a company might give such discounts are going to be investigated in chapter 7. Furthermore, in that chapter other kinds of discount are going to be investigated. For example, a company might reward customers for the overall amount of revenue they generate, rather than for the size of each single order. In our current framework these pricing policies would be basically irrelevant, as we consider d to be an exogenous variable that logistic managers just try to forecast.

[15] This is basically the logic of taxation on personal income in most Western countries.

[16] Notice that assuming a holding cost given by a fixed percentage of the purchase costs implicitly means that we assume that most holding cost are financial in nature. However, other costs such as warehousing do not depend on the unit cost of the item, but they depend on its physical features such as volume instead.

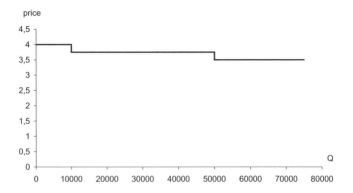

Fig. 4.12   Overall cost function.

If the unit cost is 4€ per unit, the EOQ formula suggests buying

$$\sqrt{2 \cdot 500 \cdot 1,000/ (4 \cdot 1\%)} = 5,000 \text{ units.}$$

This quantity implies monthly holding and ordering costs for

$$\sqrt{2Adh} = \sqrt{2 \cdot 500 \cdot 1,000 \cdot (4 \cdot 1\%)} = 200€/\text{month.}$$

In this case, these are not the only relevant costs. The purchasing cost (cost of goods) is not fixed as the unit cost u depends on Q (see figure 4.12). Thus the relevant total cost consists on ordering, inventory, and purchasing costs and it is $200€/\text{month} + 1,000 \text{ unit/month} \cdot 4€/\text{unit} = 4,200€/\text{month}$.

In case we want to pay just 3.75€/unit, the EOQ formula suggests to buy $\sqrt{2 \cdot 500 \cdot 1,000/ (3.75 \cdot 1\%)} = 5,164$ units. Unfortunately, this is not consistent with the pricing policy of the supplier Beta. To enjoy the price reduction, we must order at least 10,000 units. This would imply an overall ordering cost of

$$\frac{500€/\text{order} \cdot 1,000 \text{ units/month}}{10,000 \text{ units/order}} = 50€/\text{month,}$$

and an overall holding cost of

$$\frac{1}{2} \cdot 3.75€/\text{unit} \cdot 1\%/\text{month} \cdot 10,000 \text{ units} = 187.5€/\text{month.}$$

The sum of ordering and holding costs is in this case (237.5€), higher than in the former one (200€), since the purchased quantity is far larger than the unconstrained optimal one. However, purchasing costs make up for the increase in holding costs. They drop by 250€, down to 3750€. Notice that if company Alpha buys 9,999 units it pays 3,996€, whereas the cost of 10,000 units is 3,750€. Basically, the $10,000^{\text{th}}$ unit has a negative price of $-246€$, so buying one additional unit makes saves money. The overall cost drops to 3,987.5€/month, making this alternative more attractive than the first one.

This initial result might lead us to believe that buying large quantities makes us save money. Let us check whether we can save any more money by cutting the price down to 3.5€/unit. In this case the EOQ formula would suggest to purchase 5,345 units, well below the minimum quantity required to enjoy such a low price. Thus, we need to purchase 50,000 units; given the convexity of the cost function and the requirement to get the discount, this is the best viable option we have. In this case, the ordering cost drops to

$$\frac{500\,€/\text{order} \cdot 1,000\,\text{units}/\text{month}}{50,000\,\text{units}/\text{order}} = 10\,€/\text{month}.$$

Also, purchasing cost drops to 3,500€/month. However, these savings are more than offset by the increase in holding cost that reaches

$$\frac{1}{2} \cdot 3.5\,€/\text{unit} \cdot 1\%/\text{month} \cdot 50,000\,\text{units} = 875\,€/\text{month},$$

leading to a total cost of 4,385€/month, making this low unit-cost alternative the least attractive for company Alpha.   ⬜

Concept 4.3 *A lower purchase cost often implies a higher fixed ordering cost (e.g., because we import from low cost countries). We must carefully manage the trade off between the variable purchase cost and the fixed ordering one.*

The case of discounts on marginal quantities is solved by a similar approach, which takes into account a few differences:

- The overall purchase cost, as a function of the order quantity, is a continuous piecewise-linear function, with kinky points corresponding to price breaks; because of discounts, the slope of the linear pieces is decreasing with respect to order quantity.

- The unit holding cost should be evaluated by taking the *average* unit purchase cost into account.

- The total cost function is continuous as well, even though not differentiable for all $Q \in \mathbb{R}$.

4.8   THE CASE OF VARIABLE DEMAND WITH KNOWN VARIABILITY

So far, we have studied cases of deterministic and constant demand. In the next chapter we discuss the case of uncertain demand, that is unpredictable variability. In this section we illustrate some examples of variable though pre-dictable demand. The objective is to show how modeling by mixed-integer linear programming, introduced in section B.6.2, can be used in complex con-texts. All of these models assume that the planning horizon can be split in periods, also known as time buckets. Time buckets are quanta of time: For example, if the time bucket is the single week, we never look at single days or single hours.

The simplest problem is basically a generalization of the EOQ problem to the case of deterministically variable demand. We have a product whose demand in time bucket t is d_t, for $t = 1, \ldots, T$. The objective is to meet demand at the minimum cost. We denote by h the unit holding cost and by A the fixed ordering cost, just like in the case of the EOQ model:

$$\min \quad \sum_{t=1}^{T} (hI_t + A\delta_t)$$
$$\text{s.t.} \quad I_t = I_{t-1} + x_t - d_t \quad \forall t,$$
$$x_t \le M_t \delta_t \quad \forall t,$$
$$I_t, x_t \ge 0, \ \delta_t \in \{0, 1\},$$

where x_t is the quantity ordered (and immediately delivered, assuming a zero lead time) during time bucket t, I_t is the inventory level at the end of the time bucket t, and δ_t is a binary (i.e., Boolean or 0/1) variable that is set to 1, if we order during t. When we order in time bucket t, we also pay a fixed cost A.

This model is a simplified version of the lot sizing model proposed in section B.6.2, since in this case we have no capacity constraints that create some sort of interaction among products. The constant M_t is the typical big-M that we use to link continuous variables to binary ones to model fixed costs. We know we do not buy more goods than we need to meet current and future demands, so we can set:

$$M_t = \sum_{\tau=t}^{T} d_\tau.$$

This is a mixed-integer linear programming (MILP) model, and apparently branch and bound methods are required to solve it. Actually, it can be solved very efficiently by exploiting the properties of the optimal solution. This analysis, though, is beyond the scope of this book.[17]

---

[17]We can show that the optimal solution is such that $I_{t-1}^* \cdot x_t^* = 0$, that is we order only when the warehouse is empty. Thus the optimal solution is such that $x_t^* = \sum_{k=t}^{t+\tau} d_k$. The optimal lot size satisfies the needs of the current period and of a given number τ of future periods. The issue is actually finding this number of periods τ. The problem can be boiled down to finding the shortest-path on a graph. For example, see [1] for a brief but readable description.

In more complex situations, we actually have to resort to branch and bound methods. The first case we show deals with suppliers' selection.

Example 4.16 Let us consider the case of N products, indexed by $i = 1, \ldots, N$. The demand over the next T time buckets is known to be d_{it}. Products are bought from various suppliers. The suppliers' index is $j = 1, \ldots, J$. Each supplier can sell a subset $\mathcal{I}_j \subset \{1, 2, \ldots, N\}$ of products. For each product i we have a set of alternative suppliers $\mathcal{J}_i \subset \{1, 2, \ldots, J\}$. Suppliers charge prices c_{ij} that might be different for different products i and different suppliers j. We sell product i at a price p_i. However, the least expensive supplier for each product might not be the optimal solution. We assume that, like in example 4.13, there is a fixed cost component A_j we incur each time we send an order to a supplier j, no matter what and how much we order. We can think of A_j as a fixed cost for transportation, which depends on the geographical distance from the supplier. If an Italian firm sources from a Chinese supplier, probably the unit cost is low, but the distance increases the fixed cost A_j so that only large batches make economic sense. Also, we consider another fixed cost a_{ij}, which we have to pay when ordering product i from supplier j. This fixed cost is smaller than A_j, and it can model costs for lot inspection to control both quality and quantity of the product. These costs might depend on the supplier: For example, a certified supplier might reduce these costs, since we might not need to check each and every lot.[18] Products can be stored in the warehouse for a holding cost h_i.[19]

Let us assume that the initial inventory level I_{i0} is given, and that we want to reach a target inventory level H_i at the end of the planning horizon consisting of T time buckets (otherwise the model shows a "border" effect and leaves all inventories empty at the end of the planning horizon). The objective is to maximize profit, assuming that customers are not willing to wait, and thus demand cannot be backlogged.). First, we define decision variables:

- $x_{ijt} \geq 0$ is the quantity of product i we purchase from supplier j during t, assuming immediate delivery and defining the variables only for $i \in \mathcal{I}_j$;

- $\delta_{jt} \in \{0, 1\}$ is a binary variable that tells us whether during t we place an order to supplier j;

- $\gamma_{ijt} \in \{0, 1\}$ is a binary variable that tells whether during t we order a product i from supplier j;

---

[18] In manufacturing, we face similar issues when we have product families. In this case we might incur a relatively large cost when we switch between product families, and a minor setup cost/time when we switch among products within the product family.
[19] Actually, the holding cost might depend on the supplier. For the sake of simplicity we neglect the issue, as it is sometimes done in accounting practice. In case we want to model this appropriately, we shall introduce different inventory variables for products sourced from different suppliers.

- $I_{it} \geq 0$ is the inventory level of product i at the end of time bucket t;

- $z_{it} \geq 0$ is the quantity of i sold in period t.

To link binary variables and continuous ones, we need a "big-M" for each product in each single period, which gives us the maximum quantity it makes sense to purchase.[20] It is the sum of present plus future demands for the item plus the desired ending inventory:

$$M_{it} = \sum_{\tau=t}^{T} d_{i\tau} + H_i.$$

The resulting model is

$$\max \quad \sum_{i=1}^{N}\sum_{t=1}^{T} p_i z_{it} - \sum_{i=1}^{N}\sum_{t=1}^{T} h_i I_{it}$$

$$- \sum_{i=1}^{N}\sum_{j\in\mathcal{J}_i}\sum_{t=1}^{T} c_{ij}x_{ijt} - \sum_{i=1}^{N}\sum_{j\in\mathcal{J}_i}\sum_{t=1}^{T} a_{ij}\gamma_{ijt} - \sum_{j=1}^{J}\sum_{t=1}^{T} A_j\delta_{jt}$$

s.t. $\quad I_{it} = I_{i,t-1} + \sum_{j\in\mathcal{J}_i} x_{ijt} - z_{it} \qquad \forall i,t,$

$\qquad I_{iT} = H_i \qquad \forall i,$

$\qquad x_{ijt} \leq M_{it}\gamma_{ijt} \qquad \forall t,i,j \in \mathcal{J}_i,$

$\qquad \gamma_{ijt} \leq \delta_{jt} \qquad \forall t,j,i \in \mathcal{I}_j, \qquad\qquad\qquad\qquad (4.13)$

$\qquad z_{it} \leq d_{it} \qquad \forall i,t,$

$\qquad x_{ijt}, I_{it}, z_{it} \geq 0, \quad \delta_{jt}, \gamma_{ijt} \in \{0,1\}. \qquad\qquad\qquad (4.14)$

Just like in the example B.13 from page 573, formulation (4.13) is computationally more efficient than the equivalent one:

$$\sum_{i\in\mathcal{I}_j} \gamma_{ijt} \leq |\mathcal{I}_j|\cdot\delta_{jt} \qquad \forall t,j$$

where $|\mathcal{I}_j|$ is the cardinality of the set, that is, the number of different products the supplier j can supply.

The model we just discussed might look complex. Actually, through appropriate reformulation and state-of-the-art software solvers, near-optimal solutions can be found. Indeed, in real applications the key hurdles are the uncertainty on demand data d_{it}, ill-defined customer priorities, etc., rather than CPU time. However, the strength of this modeling approach is flexibility. Let us consider a slight variation of the problem.

---

[20]See example B.12 on page 571.

Example 4.17 In example 4.16 we assumed that the company tries to maximize profit, but it need not meet all demand necessarily. Now, let us assume that the company wants to fully meet demand. In this case, the objective is to minimize costs since revenues are a constant. Also, we can remove decision variable z_{it} that models the number of units we want to sell.

Let us assume that the fixed cost A_j depends on the transportation cost from the supplier j. Also, let us assume that we can use different trucks to transport goods from supplier j, just like in section 4.7. For the sake of simplicity, let us assume that we only have two kinds of truck, a small one and a large one. The volume and weight capacities of the small truck are denoted by C_V and C_W, respectively (measured, say, in cubic meters and tons). We call v_i and w_i the unit volume and weight of product i. Let us assume that the large truck is large enough to transport any amount of goods we might reasonably wish. The cost depends on the kind of truck and on the distance traveled. Thus we have fixed cost $A_j^{(1)}$ for the small truck and $A_j^{(2)}$ for the large truck. For this model we can use the same notations we have introduced for the previous one, but we must separate decision variables in two groups corresponding to the two kinds of trucks. We introduce two groups of binary variables $\delta_{jt}^{(1)}$ and $\delta_{jt}^{(2)}$ for the small and large trucks respectively. Similarly, we have two sets of variables $x_{ijt}^{(1)}$ e $x_{ijt}^{(2)}$. Using the same notation as above, we can write the model below:[21]

$$\min \quad \sum_{i=1}^{N}\sum_{t=1}^{T}h_i I_{it} + \sum_{i=1}^{N}\sum_{j\in\mathcal{J}_i}\sum_{t=1}^{T}c_{ij}x_{ijt}$$

$$+ \sum_{i=1}^{N}\sum_{j\in\mathcal{J}_i}\sum_{t=1}^{T}a_{ij}\gamma_{ijt} + \sum_{j=1}^{J}\sum_{t=1}^{T}A_j^{(1)}\delta_{jt}^{(1)} + \sum_{j=1}^{J}\sum_{t=1}^{T}A_j^{(1)}\delta_{jt}^{(2)}$$

s.t. $\quad I_{it} = I_{i,t-1} + \sum_{j\in\mathcal{J}_i}x_{ijt} - z_{it} \qquad \forall i,t,$

$\quad I_{iT} = H_i \qquad \forall i,$

$\quad x_{ijt} \leq M_{it}\gamma_{ijt} \qquad \forall t,i,j \in J_i,$

$\quad x_{ijt} = x_{ijt}^{(1)} + x_{ijt}^{(2)} \qquad \forall t,i,j \in J_i,$

$\quad \gamma_{ijt} \leq \delta_{jt}^{(1)} + \delta_{jt}^{(2)} \qquad \forall t,j,i \in \mathcal{I}_j,$

$\quad \sum_{i\in\mathcal{I}_j}v_i x_{ijt}^{(1)} \leq C_V\delta_{jt}^{(1)} \qquad \forall t,j,$

$\quad \sum_{i\in\mathcal{I}_j}w_i x_{ijt}^{(1)} \leq C_W\delta_{jt}^{(1)} \qquad \forall t,j,$

---

[21] Notice that the constraint $\delta_{jt}^{(1)} + \delta_{jt}^{(2)} \leq 1$ is actually redundant. Indeed, in the case the large truck is used, there is no reason whatsoever to use the small truck as well. It just adds extra fixed costs and thus the model self selects a solution where $\delta_{jt}^{(1)} + \delta_{jt}^{(2)} \leq 1$.

$$I_{it}, x_{ijt}, x_{ijt}^{(1)}, x_{ijt}^{(2)} \geq 0, \ \gamma_{ijt}, \delta_{jt}^{(1)}, \delta_{jt}^{(2)} \in \{0, 1\}.$$

Notice that there is no guarantee that there is at least a feasible solution to this model. In the case of tight constraints on capacity, we might not be in a position to fully meet demand. In this case we would need a model with slack variables like in example B.2 from page 546. ⬜

We conclude this chapter by reminding the reader that we can use linear approximations of nonlinear concave functions to model quantity discounts, as we have shown in section 2.3.

REFERENCES

1. P. Brandimarte and A. Villa. *Advanced Models for Manufacturing Systems Management*. CRC Press, Boca Raton, FL, 1995.

2. D. Corsten and T. Gruen. Desperately Seeking Shelf Availability: An Examination of the Extent, the Causes, and the Efforts to Address Retail Out-of-Stocks. *International Journal of Retail & Distribution Management*, 31:605–617, 2003.

3. N. DeHoratius, A.J. Mersereau, and L. Schrage. Retail Inventory Management when Records Are Inaccurate. Available from web page http://faculty.chicagogsb.edu/adam.mersereau/research/, filename dehoratius_mersereau_schrage_061023.pdf, 2006.

4. A. Raman, N. DeHoratius, and Z. Ton. The Achilles' Heel of Supply Chain Management. *Harvard Business Review*, May:2–3, 2001.

5. A. Raman, N. DeHoratius, and Z. Ton. Execution: The Missing Link in Retail Operations. *California Management Review*, 43(3):136–152, 2001.

6. A. Raman and Z. Ton. *Operational Execution at Arrow Electronics, case 9-603-127*. Harvard Business School Publishing, Boston, MA, 2003.

7. S.A. Ross, R.W. Westerfield, and B.D. Jordan. *Fundamentals of Corporate Finance (6th Ed.)*. McGraw-Hill, New York, 2003.

8. R. Verganti. Order Overplanning with Uncertain Lumpy Demand: A Simplified Theory. *International Journal of Production Research*, 35:3229–3248, 1997.

9. J.B. Ward. Determining Reorder Points when Demand is Lumpy. *Management Science*, 24:623–632, 1978.

10. N. Watson. *Paper and More, case 9-604-093*. Harvard Business School Publishing, Boston, MA, 2006.

11. G. Zotteri and R. Verganti. Multi-Level Approaches to Demand Management in Complex Environments: An Analytical Model. *International Journal of Production Economics*, 71:221–233, 2001.

5

Inventory Control: The Stochastic Case

5.1 INTRODUCTION

The vast majority of distribution systems do not enjoy the benefits of certainty. They face several sources of uncertainty:

- *Demand* uncertainty: In many supply chains, at least a subset of decisions have to be made before customers place their orders, i.e., with a partial knowledge of future demand.

- Uncertainty on *delivery quantities*: Suppliers might not deliver the quantity we have ordered, either because they face production problems (e.g., strikes or machine breakdowns) or because a portion of the delivery quantity does not fully meet minimum quality standards.

- Uncertainty on suppliers' *delivery lead time*: Suppliers can be late because of production problems, transportation problems, or simply because their capacity is overbooked.

- Uncertainty on current *inventory level* in a warehouse: this is due to wrong tracking of inflows and outflows (e.g., think of shrinkage in a supermarket).

Though all sources of uncertainty could be modeled in our analysis, we focus on demand uncertainty, which is the most classic and often the main source of uncertainty in supply chains. So, unless we specifically mention other sources of uncertainty, in the remainder of this chapter we focus on *just one* source of uncertainty. As to uncertain lead times, we refer to [5] and [10]. As to uncertain delivery quantities, see [9]. As to uncertainty on current inventory levels, we refer to [3], [6], [7], and [8].

Different kinds of uncertainty   To better understand the concept of uncertainty, we should appreciate the various shades of this rather broad concept.

1. *Uncertainty on **draws** of the random variable.* We can assume we know the probability distribution from which the single demand observations are drawn. For example, the demand can follow a normal distribution with known mean and standard distribution (e.g., a weekly demand with a normal distribution, a mean of 100 units, and standard deviation of 10 units). Under these conditions, the future level of demand is uncertain but it is known at least in terms of probability distribution. With this knowledge we can compute the probability that demand will fall in a given range or will be below a given threshold. In this case, managing inventories is like betting at the casino. We know the odds in advance. A good inventory manager, like the bank in a casino, knows that in the long run he/she is going to win, though the outcome of each single decision might be very uncertain.

2. *Uncertainty on the **parameters** of the uncertain variable.* Under more critical conditions we might know the shape of the demand distribution, but we might have an imperfect knowledge of its parameters. In this case we do not even know the probability distribution of demand from which we draw demand observations. This case is more complex than the previous one, as we cannot attribute a probability (or probability density) to each of the possible levels of demand. This actually resembles bets on sport events. Soccer matches have only three possible outcomes. Thus the probability distribution is trinomial. However, the probability that F.C. Inter is going to beat A.C. Milan is not known in advance of the event.

   Nevertheless, when we know the probability distribution of the unknown parameters, the problem boils down to the previous case. We can resort to the envelopment of the demand distribution. In other words, we attribute a probability (density in the case of continuous variables) $f(x|\mathbf{p})$ conditional on the vector of parameters $\mathbf{p}$. Also, we assume to have a probability distribution $g(\mathbf{p})$ for all possible vectors $\mathbf{p}$ (see [1] for an example of how this problem can be tackled).

   In this case, the probability density of x can be estimated as the integral over all possible values of $\mathbf{p}$ of the conditional probability $f(x|\mathbf{p})$

$$f(x) = \int_{-\infty}^{+\infty} f(x|\mathbf{p}) \cdot g(\mathbf{p}) \cdot d\mathbf{p}. \tag{5.1}$$

Hence, even when the parameters of the distributions are unknown, if we know their distribution we can derive the probability distribution of demand and the problem boils down to the previous, simpler case.

Example 5.1 Let us consider a situation where we have two dices. One dice has 6 faces, while the other one has 10 faces. We randomly generate a draw by the following process. The random variable of interest is the dice-roll. First we randomly select one of the two dices and then roll it. When we start the process, we really do not know the probability that we will draw number 2. If the first dice is selected the probability of that event is $1/6$, if the second dice is selected the probability of that event is $1/10$. So, we know the probability that we draw a 2 conditioned on the number of faces of the dice. Also, we know the probability that we will select a dice with 6 faces $(1/2)$ rather than the dice with 10 faces $(1/2)$. Using the theorem of total probabilities (see section A.2), we can calculate the unconditional probability that we will draw a 2, based on the two conditional probabilities. These are $1/6$ in the 50% of the cases where the first dice is selected, and $1/10$ in the 50% of the cases where the second dice is selected. This means that the (*a priori*) unconditional probability is $1/2 \cdot 1/6 + 1/2 \cdot 1/10 = 2/15$. The same result obviously holds for 1, 3, 4, 5, and 6. On the contrary, things are different in the case of 7, 8, 9, and 10. In these cases, the probability is zero when the first dice is selected and thus the unconditional probability is $1/20$. Notice that obviously the sum of all probabilities is still 1.   ⬚

3. *Uncertainty on the* **shape** *of the demand distribution.* In this case we cannot attribute any probability distribution to demand (e.g., we do not know whether it is normal, lognormal, uniform, gamma, etc.). Thus we do not even know the parameters that control the distribution (e.g., μ and σ for the normal distribution a and b for the gamma distribution) and, in case, what their distribution is.

Concept 5.1 *In a logistic system there are different* sources *of uncertainty (demand, delivery quantities and lead times, current inventory levels) as well as different* kinds *of uncertainty, ranging from uncertainty on the value of each single demand draw, to the uncertainty on the parameters of the demand distribution and, finally, on the shape of the demand distribution.*

In this book we analyze only the first case; that is, we assume to have a demand distribution and face uncertainty on each single future draw of demand, but we assume to know its parameters [though we include the cases where we know the probability distribution of the parameters, see equation (5.1)].

Finally, we shall remind the reader that probability can be interpreted both as a *frequency of occurrence* or as a *subjective* estimate. The second is more relevant when no objective data are available to estimate probabilities. This is very typical of brand new products with no past history whatsoever (see section 3.12).

Effects of uncertainty.   Uncertainty, has several consequences for supply chains. First, an uncertain demand can lead to stockouts that in deterministic conditions can be easily avoided (though actually limited capacity can lead to stockouts).

Example 5.2 Let us consider the demand for fresh bread in a bakery shop and let us assume that it is normally distributed with an average of 100 kg and a standard deviation of 10 kg. In this case, manufacturing 100 kg of bread (i.e., the expected demand) entails a 50% chance of stockout. The demand distribution is symmetric and thus the probability that demand is above its expected value is 50% (in symmetric distributions the mean is also the median). The baker can reduce the probability of stockout with an extra investment in inventories. However, this increases the cost of inventories. Also, under the normal distribution assumption, this probability is never zero (though from a practical standpoint it might be virtually zero for all relevant managerial purposes).   ☐

Thus, in order to model and manage inventories under uncertainty, we should understand *what happens when we experience a stockout*, i.e., what happens when we run out of inventories. In real contexts, stockouts can have various and heterogeneous effects that often are hard to capture and measure.

Example 5.3 In grocery stores a customer can substitute a stocked-out product with a substitute, postpone the purchase, leave the store to visit another one, and maybe never come back, if the new store meets his/her needs. Also, some recent papers (e.g., see [2]) have highlighted that the reaction to a stockout really depends on the frequency of stockouts. Somehow customers seem to forgive sporadic errors. However, they interpret frequent stockout as an advance notice of future stockouts and adapt their buying behavior accordingly.   ☐

However, inventory management models typically do not investigate this rich array of possible situations and just analyze the two extreme scenarios.

- *Lost sales assumption.* In the "lost sales" case we assume that unmet demand is completely lost and thus the customer goes to another supplier and does not accept a delayed delivery.

- *Backorder assumption.* In the "backorder" case, unmet demand is backordered to the next period; so in this case we assume that customers are willing to wait. We call *backorder* or *backlog* the demand not met in previous periods we still have to fulfill.

These alternative assumptions basically tell us the *mechanics* of a stockout, that is, they tell us what happens to the inventory system when we run out of goods. In the first case, customer just walk away from us. In the second case, customers are patient and basically enter a waiting line.

In most models we assume demand to be stationary and we assume it does not depend on the service level we provide. However, in real-life situations, low service levels (frequent stockouts) can reduce demand, at least in the long run.

The cost of a stockout In uncertain conditions we can face a stockout, and this makes the cost of a stockout a relevant cost.

The cost of a stockout depends on customers' reactions. In example 5.3, if the customer is willing to substitute the stocked-out product with a substitute with an equal (or even higher) margin, the cost of a stockout might be small or even negative. On the contrary, if the customer faced with a stockout leaves the store and never comes back, the cost is substantial (we call this cost "customer lifetime value").

Once again, inventory models simplify this wide range of possible situations and classify them in two categories:

- *Cost depends on the occurrence of a stockout.* We can model the cost of a stockout in several different ways. A first option is to assume that the cost of a stockout depends on the number of stockouts, and thus assume that the occurrence of the stockout is the cost driver. In other words, we can assume that the lack of service is a problem *per se*, small or large that it might be.

- *Cost depends on the size of the stockout.* A second, more frequent assumption is that the cost of a stockout depends on its size. In other words, the stockout of Barilla pasta number 5 in the 500-g package can be both a minor or a major issue. In case we stocked out at 8 p.m. on a Monday night and we failed to meet demand for 10 more units, then the stockout is a minor problem. In case we stocked out at 9 a.m. on a Saturday morning and we had to turn away demand for hundreds of units, the cost of the stockout is substantial. Notice that in the former case (cost depends on occurrence of stockout) we would have considered these two fairly different situations to be equal. Indeed, in both cases a stockout has occurred.

While the previous classification tells the mechanics of the inventory system, this classification tell us the *economics* of the inventory system. In other words this classification tells us what is the economic effect of a stock out. Does it depend on the occurrence of a stock out or on its size?

At first sight, the first option might look odd. However, in many situations it can depict real life quite effectively. Both the former and the latter cost functions are simplified and stylized models. While in some contexts the latter fits better, in other contexts the former performs better.

Example 5.4 In continuous production systems such as steel plants, starting the production process involves significant setups. Thus, in these cases, we do not care about the duration of the stockout of raw materials or energy.[1] Rather, the occurrence of a stockout is *per se* a problem that generates additional costs. In recent years Italy, Central Europe and the USA have experienced significant blackouts. The interruption of electrical energy was the cause of significant costs to restart power plants, the distribution network, and other industrial plants. In this case, a lack of supply caused a problem, no matter what its duration was. On the contrary, in the case of grocery stores the duration of the stockout, the number of upset customers, and the size of the product shortage are very relevant pieces of information to tell the cost of the stockout.   □

Parameters of the cost function.   Once we have chosen appropriate *stockout cost function*, we shall set its *parameters*. Thus in both cases we shall answer one of the basic, but still among the hardest questions in the field of operations management:

What is the cost of a stockout?

All researchers agree that this is a tough question to answer. In this book we do not even try to generate a complete set of rules or variables to gauge the cost of a stockout. We simply provide a list of variables one might want to carefully consider when he/she tries to capture the cost of a stockout.

- *Cost depends on the occurrence of a stockout.* In this case we write the cost function as the (expected) number of stockouts, times the cost of a single stockout p, that is, the *penalty cost* of a single stockout.[2] We shall capture the effect of a stockout on customer goodwill and any additional contractual penalty cost.

---

[1]This is actually a first, rough cut approximation, as the duration of the stockout might matter as well, but still this simplification captures the problem effectively as the occurrence of the stockout matters more than its duration.

[2]Notice that we are implicitly assuming that the decision maker is risk neutral, as we are simply looking at the expected number of stockout and neglect the variance (or standard deviation) in the number of stockouts. We are going to make such an assumption in all stochastic models presented in this book. This is a very reasonable assumption, as these are operational level decisions that are going to be repeated multiple times over a fairly large number of products. Under these circumstances, risk-averse decision makers might care about the uncertainty on overall performance of the logistic system (e.g., aggregate profit from multiple products). However, they are very likely to be risk neutral at the item level, since the large number of decisions taken on single items in each single unit of time (say week) guarantees that even risky decisions at the single item level are not risky from the standpoint of the overall collection of items.

- *Cost depends on the size of the stockout.* In case the cost depends on the size of the stockout, we shall capture the demand that we were not able to fulfill immediately. Also, we shall estimate the cost of not fulfilling one unit of demand for the company: We call this parameter p_u, that is, the penalty cost for each unit of demand not fulfilled immediately (in the backorder case, demand is fulfilled at a later stage, whereas in the case of lost sales demand will never be fulfilled and is lost).

In both cases, various costs are relevant:

- *Loss of customer goodwill or loss of image.* Stockouts have obvious short-term effects; however, very often the long-term ones can be even more important, though harder to capture. For example, frequent stockouts can lead current customers to switch to a different supplier. These costs are crucial but hard to measure; indeed, it is very hard to figure out the reason why a customer has defected our company. Once again the grocery retail chain is telling. A customer that has decided to stop shopping at a given supermarket because he/she cannot find the product he/she was looking simply leaves the store. He/she does not tell the managers what he/she was looking for and what led him/her to the decision to switch to another retailer. The estimate of the *customer lifetime value* is an attempt to capture how much the *customer goodwill* is worth. This concept captures how much a customer is worth. Nevertheless, it still leaves two questions unanswered. First, what is the probability that a stockout turns into a lost customer? Second, does the upset customer influence the behavior of other potential users? In other words, can a stockout experienced by customer A influence the buying behavior of customer B? Marketing research suggests that in some industries where customers have rather infrequent purchases, this "word of mouth" effect can be substantial. Retail companies often can hardly estimate such cost but try to retain customers in two ways:

  - On the one hand they use marketing levers such as promotions or advertising, to increase *store loyalty*, that is, they try to induce customers to first select the store and only at a second stage, once they are in the store, choose the product within the assortment offered by the store.[3]

---

[3]While retailers try to increase store loyalty, manufacturers try to increase *brand loyalty*. In other words, they try to push consumers in the opposite direction: they push consumers to first select the products and then select a retailer that carries it. Both actions try to reduce the side effect of a stockout. Indeed, the retailer tries to convince the customer that in case a product is stocked out or no longer available he/she should select another one from the store assortment rather than walk away from the store. Manufacturers try to convince the consumer that the product is very unique and he/she should move to another store rather than switch to another brand.

- Also, retailers often identify sets of "must-have products," i.e., products that often customers consider to be absolutely necessary. For these products, companies offer a very high service level since a stockout might cause large damages and lead to the loss of several customers. Coke, Nutella, and personal hygiene products are examples of three items with very high brand loyalty. For these products a stockout is very likely to turn into a lost sale or, even worse, a lost customer. In other cases, the consumer might not care about the single product, but it might care a lot about the availability of a product category. In other words, the customer might be very willing to switch among brands in the product category to the extent that at least one option is viable. For example, a customer might be very willing to switch to a different brand of low fat milk. However, the customer might leave the store to move to another retailer in case no low fat milk is available or, even worse, no milk at all is available. The customer and his/her family want to have milk for breakfast and might be willing to visit another store to get it.

- *Penalties.* Final consumers tend to sanction bad vendors that experience a stockout with a reduction in purchases. Most industrial customers use contractual penalties, as well. In other words, industrial customers tend to let the vendors pay for the cost their lack of service creates, even when such costs are incurred at the customer's site. The contract makes the estimate of the cost of a stockout (p or p_u) rather simple, though it might overlook substantial issues such as customer goodwill (a customer might get paid for the lack of service but might still be upset).

Other variables are relevant only when the cost of the stockout depends on its size.

- *Lost sales.* If customers (or a portion of them) decide not to buy the item when it is not available (lost sales), a stockout causes, at the very least, a loss of margin.[4] In other words, had stock been available, it would have increased revenue by the unit price of the item. However, this additional unit would have added extra variable costs. Thus the net effect is the additional margin, that is, the difference between unit price and variable costs.

  This is what economists call an opportunity cost. In other words it is not a actual cost with a negative (outgoing) cash flow. It is actually a lost opportunity to make money and have a positive (incoming) cash

---

[4]By "margin" we mean the difference between the marginal price, (which in linear, i.e. standard, contracts equals the unit price) and the marginal cost, that is, the variable unit cost.

flow. These are not actual costs for the company and generate no negative cash flow. This is why these costs are often overlooked, though for a rational decision maker there is no factual difference between opportunity costs and actual ones. Indeed, for a rational decision maker there is no difference between an actual cost and the missed opportunity to generate revenues and margins. Also, we shall keep in mind that the cost of lost sales should account for the margin gained through a surrogate product, in case customers (or a subset of them) are willing to substitute.

- *Complementary products.* Often we tend to measure the cost of a stockout through the margins of the stocked-out product. In some instances, however, a customer that cannot find one item might take the whole shopping basket to an alternative supplier. In this case the cost of the stockout can be substantial as it might include the loss of margin on a rather large set of products.

Example 5.5 Let us consider a consumer that wants to prepare a barbecue for the weekend. In case charcoal is stocked-out in a supermarket, the customer is very likely to leave the store right away. Indeed, it would not make sense to purchase meat for the grill, which is useless without the charcoal. However, estimating these complex effects in a real-life context is all but trivial. One should estimate demand of meat as a function of the charcoal inventories. This is not impossible to do *per se*. What makes it very hard are several other variables that can influence demand for meat, such as season, weather, average price, news on food heath (e.g., foot and mouth disease), and so on. Thus, telling the net effect of the availability of charcoal from others is not an easy task. What makes the problem even more complex is the fact that the impact of the availability of one item might depend on the availability of other items. Let us assume that a supermarket carries two kinds of charcoal (A and B). The average consumer is very likely to substitute charcoal A with B, in case A is stocked out. If B is available, the cost of the stockout for A is actually negligible. On the contrary, if B is stocked out as well, the cost of the stockout is substantial, as we lose margins on charcoal and complementary products such as meat.   ☐

Often companies cannot perform such a complex analysis and just study so-called *shopping baskets*, that is, they investigate which products tend to be sold together, both to estimate the cost of a stockout and to design an appropriate store layout.[5]

---

[5] A classic example of shopping basket analysis shows that in-depth data analysis can provide counterintuitive and interesting insights in customer behavior (as well as supply chain

Metrics for service level. In chapter 1, we mentioned that there are several kinds of services including tracking information, dependability of deliveries, and so on. In this chapter we focus on product availability that is a relevant metric for service and is the one most directly impacted by forecasting and inventory management policies.

So we shall define metrics to measure the availability of products to capture the service we are getting from our suppliers and are delivering to our customers.

We start from the single product/single warehouse case in a static environment (stationary demand distribution, and stable inventories) to clearly show the logic behind the service level metrics. The concepts we discuss in this rather simple case still hold in more complex situations, though they shall be properly adapted. In this rather simplistic context we can design two metrics of service level that mirror the two possible costs of the stockout.

- *Type I Service Level.* When the cost driver is the occurrence of a stock-out, rather than its size, the frequency of a stockout in a given time frame (e.g., frequency of a stockout within a week) might be a relevant metric of service level. For example one might want to control how often a production line is stopped because the plant runs out of a given component. This is a purely *ex post* metric of service. To gauge service level *ex ante* we calculate the probability of a stockout within a given time-frame (say a week). For example, let us assume that demand for a newspaper at a newsstand follows a probability distribution $f(x)$. If the newsvendor purchases N units of the newspaper, he can offer his customers a type I service level (SL_I) equal to

$$SL_I = \sum_{x=0}^{N} f(x) \qquad (5.2)$$

  in case of discrete distributions, and

$$SL_I = \int_0^N f(x)\,dx \qquad (5.3)$$

  in case of continuous distributions whose support is $\mathbb{R}^+$.

  Type I service level captures the probability of a stockout in a given time frame. Thus, comparing the type I service levels of different companies might not make sense as it might be measured over different time frames. For example, an publisher of newspapers might have a 95% probability of completely meeting demand (i.e., avoid any stockout) in a day. A

---

behavior). These analyses showed that customers who buy diapers also tend to buy beers. Indeed, parents of newborn kids tend not to go out at night and tend to drink beer at their place rather than at trendy pubs.

second publisher might have a 90% probability of meeting the demand for weekly magazines. This does not mean that the former publisher is better than the latter at managing the supply chain. To properly compare service levels, we should refer service levels to the same given time frame. For example, if we assume that the first publisher provides the market with the same 95% service level throughout the week, the probability that he/she does not incur a single stockout during the week is $(0.95)^7 = 69.8\%$. Thus in any given week we are more likely to observe a stockout of newspapers than of magazines.

- *Type II service level.* When the size of the stockout is a relevant matter, we shall design a second metric for service levels that compares the demand actually met with the overall potential demand. So to express such a metric we need to be able to measure the demand that was not immediately met. In some contexts, such as industrial customers or catalogues, it is relatively easy to do so as customers tell the company what they want. In other industries, such as brick-and-mortar retailing, the customer who cannot find the item he/she is looking for leaves the store and is very likely not to leave any information about the item(s) he/she would have purchased had it (they) been available (no information is left in the IT systems and very often no information is given to the sales-personnel in self-service environments such as most supermarkets). In this case too, we have to estimate the service level *ex ante* rather than just measure it *ex post*.

The expected type II service level is:

$$SL_{II} = \frac{\sum_{x=0}^{N} x \cdot f(x) + \sum_{x=N+1}^{+\infty} N \cdot f(x)}{\sum_{x=0}^{N} x \cdot f(x)} = \frac{\mathrm{E}[x] - \sum_{N+1}^{+\infty} (x - N) \cdot f(x)}{\mathrm{E}[x]},$$

(5.4)

in the case of discrete demand distribution;

$$SL_{II} = \frac{\int_{0}^{N} x f(x)\, dx + \int_{N}^{+\infty} N f(x)\, dx}{\int_{0}^{+\infty} x f(x)\, dx} = \frac{\mathrm{E}[x] - \int_{N}^{+\infty} (x - N) f(x)\, dx}{\mathrm{E}[x]},$$

(5.5)

in the case of a continuous demand distribution with support $\mathbb{R}^+$.

Clearly, these definitions show that these two metrics are very different and mean very different things. Type I service level measures a probability (of not stocking out), whereas type II service level is a ratio between the demand we expect to serve and the demand we expect to face. So while type I service

Table 5.1   Differences between type I e type II service levels

Period	1	2	3	4	5	6	7	8	9	10
Demand	1	1	1	1	1	11	1	1	1	1
stockout	0	0	0	0	0	1	0	0	0	0
Demand met	1	1	1	1	1	2	1	1	1	1

level measures the *percentage of periods* during which we expect no stockouts, the second is a *percentage of demand* that we expect to meet.

Concept 5.2 *There are several different metrics for service level and we shall carefully investigate the definition of any service level metric to understand what it means and whether it can be compared across companies, business units, regions, or products. In particular the type I service level measures the probability of a stockout over a given period of time, while type II service level measures the percentage of demand met from stock.*

Also, these definitions gave us a chance to show how we can "translate" equations from the continuous to the discrete case and vice versa. In most of the remainder of this book we use continuous variables. The reader can derive the equations for the case of discrete distributions easily. In the few instances where we use discrete distributions, we explicitly warn the reader.

Example 5.6 To show that the two metrics are very different, we consider a newsstand that faces the demand distribution for a newspaper in table 5.1. The newsvendor has decided to carry two copies of the newspaper.

The newsvendor stocks-out only one day out of ten. So the type I service level is 90%. However, the type II service level, that is the percentage of demand met, is significantly lower. The newsvendor was able to meet demand for 11 units out of 20. Thus the type II service level is just 55%.   ⬜

This simple example shows that the two definitions are very different and thus jumping to the conclusion that a company that offers a 70% service level is worse than a company that offers a 90% service level might be misleading. We must fully understand the metrics used and their meaning to properly compare them.

- *The multiperiod case with backorders.* In the multiperiod case with back-order, customers are willing to wait for products that are temporarily stocked-out. In this case, one might want to add metrics for late deliveries. We can leverage on tools and concepts from queueing theory. Also, backordered customers might change the distribution of inventory consumption in the next period and complicate the estimate of type II service level.

- *The multiproduct multilocation case.* Many chains consists of several stores and warehouses that carry thousands of items. One might want to have aggregate measures of service level. For example, in a supermarket chain one might want to know the service level of the yogurt sections. In these cases one can use weighted averages of service levels for the item/location combinations. Obviously one can design various averages. For example, one might consider the stockout of any item equally important. This means that the stockout of cans of Coke is just as important as the stockout of spicy soy sauce (which, in the average supermarket, hardly sells a fraction of what Coke can sell). On the contrary, if the average accounts for number of units sold, turnover, or margins, the picture looks fairly different. In the case of industrial firms (i.e., firms that sell products and services to other firms), one has to decide whether the service level is measured for each order rather than for each line.[6] In the former case, one might consider the order completed only when all units of all lines are delivered. So, in general, measuring service level for orders rather than for lines is a more conservative measure of service level. So the multiproduct and/or multilocation problems adds new complexity to the concept of service level. In these cases we shall be even more careful when we want to compare across companies, business units, etc., as numbers might be hardly comparable.

The remainder of the chapter discusses inventory management problems and techniques. We start from the simple case (section 5.2) involving a single product and single period, and then we move to dynamic problems that are introduced in section 5.3. We discuss (Q, R) policies in section 5.4; section 5.5 introduces periodic review policies, which are discussed in sections 5.6 and 5.7.

5.2   THE NEWSVENDOR PROBLEM

With respect to the classification of inventory problems introduced in section 4.1, the newsvendor problem is:

- single-product, single-period and thus static, single-echelon;

- demand is uncertain, though we know its distribution;

- the objective is to minimize the (expected) cost of inventories and service; in each period we have to place one order, as inventories left over

---

[6]Industrial firms place formal orders that consist of various lines. The order typically comes form a single customer and requires a single delivery at a single point in time. The customer might want several different items to be delivered at once. Each single line refers to a single item and states how many units (of the item) the customer wants to receive.

in the previous period cannot be used any longer (think of newspapers left over yesterday); this makes the ordering costs constant and thus irrelevant.

This problem is called the newsvendor problem, as it resembles the problem a newsvendor faces each and every morning, when he/she needs to decide how many copies of a newspaper he/she wants to purchase from the publisher or distributor. In this case, ordering costs do not depend on the purchase quantity and thus are an irrelevant variable in the model. The vendor shall balance two contrasting objectives. He/she wants to fully meet customers' demand, take all the opportunities to sell the product, and minimize cost of service. On the other hand, buying too many units of the newspaper might leave some units left over (i.e., unsold) at the end of the day.

Formally, we can define two costs:

- *Cost of the stockout.* In the remainder of this section we consider the unit cost of the stock out to basically consist of the loss margin (thus we ignore for the sake of simplicity other issues such as customer goodwill or the possibility that a customer switches to a substitute product; our results, though, hold in the more general case). So in the remainder of this section the unit stockout penalty p_u is identified with the margin m that the newsvendor can gain by selling one unit of the newspaper, sold at a price p to the consumer and bought at a cost u from the distributor or the publisher.

- *Cost of excess inventories* that is the units left over at the end of the day. The cost is c equal to the difference between the purchase cost u and the salvage value v of the newspaper at night (i.e., the residual value of the product at night, that is, the amount of money we can get back from the publisher, the distributor, or the value of scrap paper).[7] Notice that, in this case, we do not consider the stock holding cost (that is, the cost of inventories in the EOQ model), since goods are held for a very short period of time and the holding cost is negligible, as compared to the cost of stockouts and of excess inventories.

---

[7]Oddly, in most countries the publishers and distributors give the newsvendors full credit for the units left unsold at the end of the day. In other words, $u = v$. As we will see later in this section, this situation seems to suggest that overstocking creates no additional costs. So the newsvendor might be tempted to order very large quantities as excess inventories are not expensive to them. Nevertheless experience tells us the this is not the case, since often newsvendor run out of newspapers by day-end. So there is an apparent gap between what the newsvendor model seems to suggest and the actual behavior of real newsvendors. Are newsvendors stupid or is the model not working properly? Actually, neither hypothesis is true. Indeed, there are other costs associated to very large purchase orders. First, newsvendors have limited space in the newsstand and thus excess inventories might mess up the operations of the store (see section 5.2.1). Also, there are some administrative costs associated to receiving and returning the copies left unsold. For example, at the end of the day the newsvendor needs to count the units left unsold and packs them, so that the distributor's truck can pick them up the next morning.

Table 5.2 Probability distribution

Units(x)	1	2	3	4	5	6	7	8	9
Probability p(x)	0.1	0.1	0.1	0.1	0.2	0.1	0.1	0.1	0.1

When one faces demand uncertainty, he/she might be tempted to treat the problem "as if" it were deterministic; that is, he/she might be tempted to neglect demand randomness. A simple example can show that this is actually not a good idea.

Example 5.7 A newsvendor faces an uncertain and discrete demand that follows the distribution in table 5.2. He/she sells the newspaper for 1€, buys it at 80 cents, and can give it back to the publisher for 75 cents, in case some units are left over at the end of the day.

A first option is to buy 5 units, as the expected demand is 5 units. Also, 5 units is the single most likely demand scenario (mode of the distribution). The newsvendor does not know his/her profit $\pi(Q)$ in advance, since it depends on the random demand. However, he/she can compute the expected profit $E[\pi(5)]$ of this policy:

$$E[\pi(5)] = \sum_{x=5}^{9} 5 \cdot 1 \cdot p(x) + \sum_{x=1}^{4} [x \cdot 1 + (5 - x) \cdot 0.75] \cdot p(x) - 5 \cdot 0.8 = 0.75 \, €.$$

When demand is equal to or larger than inventories (5 units), we sell 5 units and the revenue is 5 units·1€/$unit = 5€$. If the demand is lower than inventories (5 units), sales equal demand. The cost of this policy is the purchase cost of the 5 units, that is, 4 €. This apparently sensible policy turns out to be all but optimal.

To begin with, we can investigate the marginal profit of the 6th unit, that is the additional amount of money the newsvendor would make if he/she purchased 6 units instead of 5. The 6th unit has a certain cost of 0.8€ and an uncertain revenue. It is 1€ if it is sold, while it is 0.75€ when it is left over at the end of the day. The 6th unit is going to be sold if demand is at the least 6 units. Thus the probability selling it to the final consumer at 1€ is 40%, that is $P\{X \geq 6\}$. Hence, the probability that it will be sent back to the publisher for 0.75€ is 60%.

Hence, revenue is uncertain, but we can compute the expected marginal revenue of the 6th unit that is 1€·0.4+0.75€·0.6 = 0.85€. In other words, if the vendor decides to carry the 6th unit on top of the first 5 units, revenue is expected to increase by 0.85€. Given the marginal cost of 0.8€, the marginal profit is then 0.05€. The vendor can increase the profit by 0.05€ simply by buying an extra unit. This means that the vendor shall expect a profit of

0.8€ if he/she buys 6 units:

$$E\left(\pi\left(6\right)\right) = \sum_{x=6}^{9} 6 \cdot 1 \cdot p(x) + \sum_{x=1}^{5} \left[x \cdot 1 + (6-x) \cdot 0.75\right] \cdot p(x) - 6 \cdot 0.8 = 0.8\text{€}.$$

☐

More generally we can easily show that the objective function is concave, as the expected marginal return is decreasing with respect to the stocking quantity Q (as the quantity increases, the probability that demand will be high enough to sell the Qth unit decreases, and thus marginal revenues decrease) while the marginal cost is constant. Hence, also units 1–5 have a positive marginal profit. Thus the apparently reasonable decision to buy 5 units is suboptimal. One could be tempted to draw the conclusion that when demand is uncertain, we shall always carry extra inventories, i.e., carry more inventories than we expect to sell. This is actually a wrong conclusion. The solution depends on the economic parameters of the problem. In some cases we over-stock, while in other cases we understock.

Example 5.8 Let us assume that the publisher in example 5.7 decides that for some reason he does not collect unsold copies for 0.75€ and, thus, the salvage value of unsold copies drops to zero. In this case, the marginal revenues from the 6th unit would drop to 1€ · 40% + 0€ · 60% = 0.4€, making the marginal profit negative (−0.4€) and the 6th unit unprofitable.

Also, the marginal profit of the 5th unit would be negative as well (−0.2€). So we would buy less than 5 units, and would increase the risk of stockouts (we would experience a stockout when demand equals or exceeds 5 units).

☐

Concept 5.3 *Uncertainty changes the fundamentals of the decision-making problem. Thus, choices that are optimal under deterministic conditions turn out to be fairly inappropriate for uncertain ones.*

This example challenges one of the so-called "golden rules" of logistics: We shall have a 100% service level. This very simple example shows that a 100% service level can be counterproductive for some companies. The costs of such an high service level might exceed the potential benefits.

As we have seen in the newsvendor example, the optimal inventory and the optimal service level depend on the *economics* of the company, that is, costs and margins. Increasing the service level without changing the economics is an error that can have consequences.

Concept 5.4 *A 100% service level is not always a recipe for success. The choice of the optimal service and inventory level depends on the economics of the company.*

Once these basic concepts are set, we can move on to the next stage and create a more rigorous model to answer the question, How many copies of the newspaper should the newsvendor buy from the publisher? We can deal with this question in two different ways, though obviously they lead to the same answer. A first approach uses economic theory and investigates marginal revenue and marginal cost. A second approach leverages on mathematics and optimization, and it stems from the profit function and its differentiability.

An economic approach to newsvendor problem.   We can interpret the newsvendor problem as a simple economic problem where the newsvendor is just an economic agent who tries to maximize its profits. To do so, the newsvendor invests in inventories when they are expected to generate profit, whereas he/she stops investing when the reward for such an investment is negative. In the previous section we have shown that the marginal return of the investment in inventories is decreasing. In other words, the expected profit for the first unit is higher than the expected profit from the second unit, which in turn is higher that the expected profit from the third unit and so on. Indeed, the probability of selling unit $Q + 1$ decreases as Q increases. On the contrary, the purchase cost is a linear function of Q and the marginal cost of unsold inventories increases (the probability that the item is unsold at the end of the day increases). Thus, the profit function is concave.

This implies that in the case of continuous demand distributions, a local maximum of profit function is a global maximum as well. In the case of discrete demand distributions we just need to solve the problem in the continuum and then choose between the two nearest integer solutions (floor and roof of the continuous solution). In the remainder of this section we only deal with the continuous distribution problem, as we have shown that discrete problems can be easily solved through a continuous relaxation.

To find the optimal solution in the continuous case, we shall just equal marginal returns and marginal costs. In the newsvendor model this means that the newsvendor keeps on buying additional copies of the newspaper to the point where the marginal return he expects from the additional copy is larger than (or equal to) the additional cost he expects to face. Notice that in the model we assume the decision maker to be *risk neutral.* At a first sight this might look like a fairly odd assumption, as most decision makers are *risk adverse.* However, if the decision maker repeats the decision each and every morning and maybe he/she makes the decision on several newspapers (more generally when the decision is repeated multiple times and the impact of each decision is very limited[8]), even a risk-adverse decision maker tends to behave like a risk-neutral one.

---

[8]This means the we shall not include decisions that, in the case of bad luck, can lead to immediate bankruptcy or, in general, irreversible problems.

Now we just need to turn these concepts into math. We can write the marginal expected profit from the Qth unit as:

$$\frac{\partial \mathrm{E}\left(\pi\left(Q\right)\right)}{\partial Q} = m \cdot P\left\{X \geq Q\right\} - c \cdot P\left\{X < Q\right\}; \qquad (5.6)$$

where:

- $\pi'(Q)$ is the expected marginal profit from the Qth unit;

- $P\left\{X \geq Q\right\}$ is the probability that demand X is larger than or equal to inventory level Q.

We do not know the marginal (i.e., additional) profit we gain through the Qth unit, but we know that we have a probability $P\left\{X \geq Q\right\}$ of selling it and if we sell it we gain a margin m. On the contrary, if we cannot sell the Qth unit we face a cost c and this scenario has a probability $P\left\{X < Q\right\}$, since in all scenarios where demand is lower than Q, the Qth unit is left unsold.

Now we can restate equation (5.6) as

$$\frac{\partial \mathrm{E}\left(\pi\left(Q\right)\right)}{\partial Q} = m \cdot \left(1 - F\left(Q\right)\right) - c \cdot F\left(Q\right), \qquad (5.7)$$

where $F(Q) \equiv P\left\{X \leq Q\right\}$ is the cumulative distribution function of random demand; also note that, in the continuous case, $P\left\{X \leq Q\right\} = P\left\{X < Q\right\}$. Then we set the marginal profit to zero to find the maximum; it is easy to show that for continuous demand distributions we have

$$F\left(Q^*\right) = \frac{m}{m+c}. \qquad (5.8)$$

A mathematical approach to the Newsvendor Problem   We can also write the expected profit function and calculate its derivatives to find the maximum. The expected profit function is

$$\mathrm{E}\left[\pi\left(Q\right)\right] = m \cdot \left(\int_0^Q x f(x)\,dx + \int_Q^{+\infty} Q f(x)\,dx\right) - c \cdot \int_0^Q (Q - x) f(x)\,dx.$$

In other words, the expected profit is equal to the margin we expect to enjoy minus the costs of inventories we expected to face. Expected margins are equal to the unit margin times the expected number of units sold, i.e., demand x, when this is lower than the quantity on hand Q, and Q in all other cases. Cost of unsold inventories is related to the difference between inventories Q and sales: We only have $(Q - x)$ units left over when demand (x) is lower than inventories (Q).

Now we have to take the first-order derivative of the profit function with respect to Q to maximize it. To do so, we use Leibniz's rule, which is used to differentiate integrals.

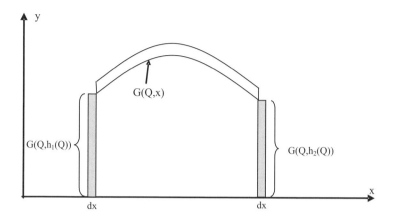

Fig. 5.1 Graphical illustration of Leibiniz's rule.

Technical note (Leibniz's rule). Consider a function $g(Q, x)$ of two variables, and let us define a function of Q through the integral:

$$G(Q) = \int_{h_1(Q)}^{h_2(Q)} g(Q, x)\, dx.$$

Notice that $G(\cdot)$ does not depend on x, which is used as an integration variable. In many applications we need the derivative of $G(Q)$ and Leibniz's rule suggests how to find it. Under suitable technical assumptions, Leibniz's rule says that

$$\frac{\partial \int_{h_1(Q)}^{h_2(Q)} g(Q, x)\, dx}{\partial Q}$$
$$= \int_{h_1(Q)}^{h_2(Q)} \frac{\partial g(Q, x)}{\partial Q} dx + g(Q, h_2(Q)) \cdot h_2'(Q) - g(Q, h_1(Q)) \cdot h_1'(Q).$$

The intuition behind Leibniz's rule is that the integral can vary either because the upper endpoint of the integration interval is shifted, or because the lower endpoint is shifted (there is a minus in the formula because if the lower endpoint increases, the overall value of the integral moves in the opposite direction if g is positive) or, finally, as the function $g(Q, x)$ is changed. This point is illustrated in figure 5.1. ◻

We can use Leibniz's rule to compute the derivative of the expected profit function with respect to Q:

$$\frac{\partial E(\pi(Q))}{\partial Q}$$

$$= m \cdot \left(Q \cdot f(Q) + \int_Q^{+\infty} f(x)\, dx + Q \cdot (-1) \cdot f(Q) \right) - c \cdot \int_0^Q f(x)\, dx$$

$$= m \cdot \int_Q^{+\infty} f(x)\, dx - c \cdot \int_0^Q f(x)\, dx$$

$$= m\,(1 - F(Q)) - c \cdot F(Q) = 0.$$

We can use the above equation to obtain (5.8).

Also, we can show that this a maximum by checking the second-order derivative:

$$\frac{\partial^2 \left(\mathrm{E}\left(\pi\left(Q \right) \right) \right)}{\partial^2 Q} = -m \cdot f(Q) - c \cdot f(Q) < 0 \qquad \forall Q.$$

Thus, no matter what approach we follow, we still find the same conditions for optimality and thus the optimal quantity Q^*.

Once we have derived the optimal condition (5.8), we shall reflect upon the process we have followed.

- We have made no explicit assumption on the demand distribution to derive equation (5.8); hence, (ruling out pathological cases) the results are essentially distribution-free.

- The solution thus suggests that the optimal quantity Q^* depends on two basic ingredients:

  - First, the optimal solution depends on the *economics* of the item/firm we are planning for, that is the margin m and the cost of excess inventories c

  - Second, the demand cumulative distribution $F(x)$.[9] This fairly simple model clearly shows that there is no such thing as one "right" level of inventories that fits all companies. Often companies and managers tend to ask point blank: We have k units in stock (often inventories are measured as number of months of supply); is this the right level of inventories? Do we have too much inventories? Or maybe too little? Is it more or less than the competitors? The rather simple newsvendor model is telling us that we cannot possibly answer that question without a proper analysis of the economics of the company.

---

[9]When demand is discrete (in our case, it is expressed by integer numbers), there might not be any value x that satisfies equation (5.9). However, given the convexity of the cost function (concavity of the profit one), we shall only consider the solutions x_1 and $x_2 = x_1 + 1$ that are the two solutions where $F(x)$ is just lower than and just greater than $\frac{m}{m+c}$. More formally, $x_1 = \max x$, s.t. $F(x) \leq \frac{m}{m+c}$, and $x_2 = \min x$, s.t. $F(x) \geq \frac{m}{m+c}$.

- Also, we notice that equation (5.8) suggests the service level that the newsvendor should provide. Indeed, $F(Q)$ is the probability that demand is met completely; thus, it is the type I service level we shall expect if we carry Q units. So the newsvendor model suggests an optimal type I service level as a function of the economics of the company and of the item.

The next step is to identify the inventory level Q^* that guarantees the required type I service level. Obviously, Q^* depends on the probability distribution of demand. The quantity Q that leads to a 90% service level in a newsstand in a small town is definitely insufficient for a newsstand at a New York City railway station.

We need to identify the demand distribution (and its cumulative function $F(\cdot)$) to properly set the optimal value Q^* as follows

$$Q^* = F^{-1}\left(\frac{m}{m+c}\right). \tag{5.9}$$

Example 5.9 Let us consider a seller of milk fresh from the cow. This seller buys fresh milk from the cattlemen in the Alps at 0.5€/liter and sells it for 1.1€/liter in the city. The product perishes quickly, as it is does not go through any thermic or chemical process. Also, the seller wants to sell only very fresh milk to support the high-price policy. Milk left over at the end of the day is sold to a pig farmer at 0.2€/liter. Demand for milk is a random and stationary process that follows a normal distribution with an expected value of 100 units/day and standard deviation of 20 units/day.

The seller has to decide how many liters he wants to buy from the cattlemen. As the reader can easily see, this situation closely resembles the newsvendor problem: At the end of each single day, all the units are sold either to the final consumers at a margin m or to the pig farmer at a loss c. This makes the problem static since the decisions taken at time t have no impact whatsoever on successive decisions.

Thus the milk seller just wants to apply the findings of the newsvendor problem.

In this case, if the product is sold at full price to the end-consumer he gains a margin $m = 1.1€ - 0.5€ = 0.6€$, whereas if the product is sold to the pig-farmer the seller loses $c = 0.5€ - 0.2€ = 0.3€$.

Hence the seller shall seek a service level

$$LS_I(Q^*) = F(Q^*) = \frac{0.6}{0.6 + 0.3} = 66.6\%.$$

Up to this stage we have not used any information on the demand distribution. Now, to move forward, we need this information to identify the quantity Q^* that leads to the optimal service level (66.6%).

In the case of the normal distribution, we can use the standardized normal distribution. In the appropriate tables we can find z so that $F(Q^*) = 0.666$.

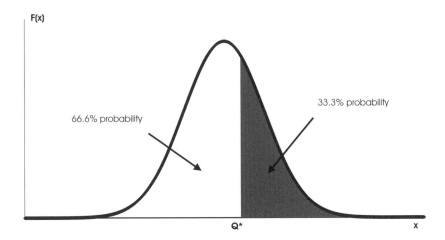

Fig. 5.2   Solution of the newsvendor problem: $F(Q^*) = 66.6\%$ and normal distribution.

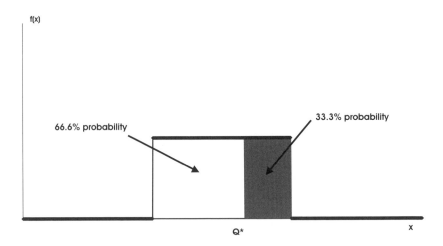

Fig. 5.3   Solution of the newsvendor problem: $F(Q^*) = 66.6\%$ and uniform distribution.

Tables suggest that we select $z = 0.429$. Hence, the seller maximizes expected profit by purchasing $Q^* = 100 + 0.429 \cdot 20 = 108.58$ liters of milk per day.

Clearly, in the case of different demand distributions (either different shapes or different parameters) we would have reached different conclusions (compare figure 5.2 with figure 5.3). ▯

Example 5.10 Finally, we want to show how demand forecasting and inventory planning can be integrated. We refer back to example 3.15 on page 149. We consider the inventory planning the newsvendor faces for day 36, that is, next Tuesday. He/she needs to make a decision on how many copies of the newspaper he/she wants to orders. We know that he/she expects a demand for 37.98 units with a standard deviation of 10.05. Notice that when planning inventories we care about the demand uncertainty rather than demand variability. When the newsvendor will be faced with the decision on the number of units to buy on Sunday, the expected demand will be substantially higher and thus the quantity will be substantially higher. Nevertheless, when planning inventories for Tuesday we really care about the expected demand and the uncertainty in demand for Tuesday and are not interested in the variability of demand within the week. For the sake of simplicity, let us assume we can model the demand distribution as a normal distribution (demand is discrete, but given the large number of units involved the approximation by a continuous variable is rather reasonable). To find the best decision we have to investigate the economics of the newsvendor. The newsvendor sells the newspaper for 1€ and buys it for 0.8€ from the editor. At the end of the day the newsvendor gets back 0.7€ for each unit left unsold. In this case, the margin for the newsvendor is 0.2€, while the cost of inventories is 0.1€. We can use these economic parameters in equation (5.8) to derive the optimal service level

$$F(Q^*) = \frac{0.2}{0.2 + 0.1} = 66.66\%. \tag{5.10}$$

Finally, we need to use the standard tables for the Normal distribution to find the optimal quantity Q^* that satisfies the above equation. The standard tables for the normal distribution suggest that the relevant z for a 66.66 service level is $z = 0.429$. Thus the optimal quantity is $Q = 37.98 + 0.43 \cdot 10.05 \approx 42.30$. ▯

Concept 5.5 *Demand forecasting provides a key input to inventory planning. Both the expected demand and the demand uncertainty are provided by the forecasting process.*

Once the Newsvendor problem is solved, we can try to read the result economically. Equation (5.8) suggests that the service level one shall provide increases as the margin increases. As m increases we have a greater and greater incentive to buy more units and accept the risk that they might be left unsold. On the contrary, if the cost of inventories left unsold grows, we shall reduce the purchase quantity to limit the probability that this fairly expensive scenario comes true.

Also, equation (5.8) suggests that when the margin m is larger than the cost of inventories c, we shall achieve a greater than 50% type I service level. In the opposite case, it is advisable to reach a less than 50% type I service level.

This simple result shows that it is very unreasonable to assume that all companies should seek the same service level. In particular, for many companies a 100% service level can be a very silly strategy.

In the case of symmetric demand distributions, to gain a service level above 50%, we shall select an inventory level above the expected demand (a so-called overproduce, overbuy, or overstock policy), whereas when the target service level is lower than 50% we shall deliberately choose an inventory level below the expected demand (a so-called underproduce, underbuy, or understock policy).

The newsvendor problem is not just a powerful prescriptive tool, but also an interesting interpretative tool can help us read economic behaviors that at first sight might look odd.

Example 5.11 A first application of the newsvendor problem is apparel products with a high fashion content. We know very well that retailers in this industry have a fair amount of products left over at the end of the season. In the Western world the winter selling season peaks at Christmas. Most retailers at the beginning of each year "discover" that units bought for the previous Christmas season were excessive and there are left over goods that shall be liquidated through a sale. What is really surprising is that this happens year after year and retailers seem not to be able to adjust their purchase quantities to reduce the amount of goods sold at a discount.

A first simplistic reading of the phenomenon is that apparel retailers are just optimistic by nature and tend to overestimate demand for almost all products they sell.

Actually, this has more to do with economic incentives to overstock than with forecasting. Apparel goods are often manufactured in the Far East and thus have rather long lead times because of both relatively limited responsiveness and long transportation lead time to major Western markets (3 to 4 weeks to transport goods via ship to Europe). These products are sold over a relatively short period of time (one season at maximum). So many companies only place one order for the whole season to Asian suppliers (though there is just one advance order we might have multiple deliveries – see chapter 4).[10] Therefore, the purchase planning for fashion products resembles the

---

[10]Notice that to judge whether the company is in a position to place more than one order and thus adjust purchases to meet demand, we shall compare the time it takes to read early demand – i.e., the amount of time to collect a statistic on demand that significantly improves accuracy over forecasts generated before the beginning of the season – and supplier's lead

newsvendor problem. Before we start selling, the inventory quantity is set. Later we try to sell the product at full price and finally we get rid of left over inventories (excess inventories) just like in the case of the newsvendor. So the structure of the problem is exactly the same, though products are different, lead times are different, product life cycle is different and the means to get rid of excess products are different (returns vs. end-of-season sale). Companies in the fashion apparel industry enjoy fairly large margins. Many small stores have a 100% markup, i.e., when they buy an item for 100€, they sell it for 200€. Cost of inventories is rather limited: Though end-of-season discounts are significant the cost of inventories is significantly below the margin of 100€. Thus retailers offer a very high service level. For example, consider a product purchased for 100€. The retailer adds transportation and handling costs for additional 20€. The full price is 200€ while the sale price (so-called salvage value, i.e., the value of the product after the end of the selling season) is 110€. In this case the optimal service level is

$$\frac{200 - 100 - 20}{(200 - 100 - 20) + (100 + 20 - 110)} = 88.9\%.$$

This means that, for any product with these economics, it is advisable to have a 88.9% probability of not stocking out, i.e., a 88.9% probability that some units will be left over at the end of the season and only a small probability 11.1% of selling-out the item before the end of the full-price season (with continuous distributions the probability that demand perfectly matches supply is zero). Thus the Newsvendor problem provides a very clear reading of an apparently odd behavior. ⬜

Example 5.12 The above example might lead us to believe that in real-life situations, companies tend to over-stock. A counterexample helps us understand that this is not actually the case. Italians like to eat fresh bread and thus typically buy fresh bread daily from small bakery shops that bake their own bread. Bakers typically bake bread once a day (the production process is very long and they start preparing bread as early as 2 a.m.). So before the sun rises, the baker makes the production decision a few hours before the store opens and he/she can start selling. This is exactly what happens to the newsvendor. All Italian consumers know very well that at a few minutes before the bakery store closes the vast majority of bakers run out of most kinds of bread.

A first reading of this behavior is that bakers are pessimistic by nature and tend to underestimate potential demand. On the contrary, these retailers know their business very well. They know that a piece of bread sold for 50 cents has a relatively high cost; let us assume a variable cost of 40 cents for

---

time with the product life cycle. In the case of apparel goods imported from Asia, we often have a limited ability to react and thus the problem is a static one, though in the case of multiple deliveries and significant holding costs, it can be a multiperiod problem.

raw materials and energy. Also, bread unsold at the end of the day has a very low value. It can either be turned into dry grated bread, frozen at a significant cost, or sold as food for animals. Let us assume that the salvage value of a piece of bread is 10 cents. Under our simple assumptions, the baker shall seek a type I service level of

$$\frac{30 - 25}{(30 - 25) + (25 - 10)} = 25\%.$$

Thus, the baker shall run out of bread three days out of four. Again, the newsvendor problem helps us make sense of an apparently odd behavior that is actually economically sound. Notice that here we do not argue that mom and pop bakers in Italy do their math an optimize the Newsvendor problem. Actually most of them make no formal calculation to determine the production quantity. However, they face the same decision each and every day. So, they experience both stockouts and excess inventories and they immediately understand that the cost of a stockout is smaller than the cost of excess inventories. Therefore, given that they repeat the decision over and over again, they can adjust their stocking policy over time and get the right balance empirically.[11] Unfortunately, not all businesses enjoy this opportunity. Thus, it would be advisable to understand the newsvendor model and get it right the first time.

$\Box$

Example 5.13 In our examples we have compared two different product categories. Now we want to compare the service level that different companies offer on a given product category. In most large cities, one can hardly find fresh fish in a fish shop late in the afternoon. On the contrary, if one enters the best fish restaurant in town, say at 10 p.m. one is very likely to find exactly the kind of fish he/she wants. So one might wonder why one cannot find fresh fish in the afternoon in a shop while in the same town one can find fish a few hours later in a restaurant.

Again we can find the right solution to this puzzle by looking at margins. The margins of a good restaurant are usually far higher than those of a shop. So the restaurant is willing to overbuy so that it can meet the demand from an occasional customer late at night. On the contrary, the fish shop tends not to overbuy, as the amount of money it makes does not justify the risk of having fishes left over at the end of the day.

So, quite interestingly, two stores with the same upstream supply chain (they are very likely to buy the fish from the same distributors or fisherman) have very different inventory policies. $\Box$

---

[11] Interestingly, some bakers started to offer discounts between 7 p.m. and closing time to get rid of the excess inventories and increase the salvage value. This enables these bakers to offer a higher service level.

5.2.1   Extensions of the newsvendor problem

The newsvendor problems can be extended in several different ways. In this book we consider two extensions:

- Multi-item newsvendor problem

- Two-period newsvendor problem

Multi-item newsvendor problem.   The basic newsvendor problem considers only one product. Notice that a single-item problem can be applied to companies with many items that are independent (e.g., no complementarity or substitution on the demand side, no supply or budget constraint on the resource side). In this case a multi-item problem is just the collection of single-item problems.

On the contrary, when the various items somehow interact, the problem becomes more complex. Here we only investigate the case of items that share common resources. These common resources constrain the optimal quantity of single items. Such constraints can stem from various issues ranging from space available on the shelves in stores, to limited production capacity, limited budget for a product category, etc. No matter what is causing the constraint, we assume we can write it as an upper bound on the total quantity for the I products:

$$\sum_{i=1}^{I} Q_i \cdot r_i \le R,$$

where r_i is the amount of the limited common resource consumed by one unit of item i, R is the available amount of the common resource, and Q_i is the quantity of item i we decide to stock. We want to maximize the profit of the I products in our assortment. Thus we can restate the profit function as follows:

$$g\left(Q_1, ..., Q_i, ..., Q_I\right) = \mathrm{E}\left(\sum_{i=1}^{I} \pi_i\left(Q_i\right)\right) = \sum_{i=1}^{I} \mathrm{E}\left(\pi_i\left(Q_i\right)\right).$$

We are assuming that the profit functions of the I items are independent. In this case "independent" has nothing to do with *statistical* independence; we may write the function in this separable and additive form, since there is no interaction either on the demand side (e.g., no substitution or complementarity) or on the cost side (no joint economies of scale).

In this case, we can use Lagrangian multipliers to find the optimal solution. We define the Lagrangian function

$$h\left(Q_1, \ldots, Q_i, \ldots, Q_I, \lambda\right) = g\left(Q_1, \ldots, Q_i, \ldots, Q_I\right) - \lambda\left(\sum_{i=1}^{I} Q_i \cdot r_i - R\right).$$

Note that the sign in the Lagrangian function depends on the fact that we are *maximizing* the objective function subject to a budget constraint (see section B.4.1).

Notice that we have an inequality and that we cannot simply assume that the constraint is active (i.e., the equality is verified) for the optimal solution (contrast this situation with the multi-item EOQ problem). Thus, we shall refer to Kuhn–Tucker conditions (see theorem B.6 on page 558). The multiplier is bound to be positive and the complementary slackness condition must hold as well. This condition simply states that there are two possible cases:

1. The constraint is not active, that is, the common resource is actually not fully utilized as it is actually abundant; in this case the multiplier is zero.

2. The multiplier is strictly greater than zero and the constraint is active and thus it can be treated as an equality constraint.

In the former case the constraint is not active and thus we shall just set $\lambda = 0$ in the condition below. In this case the optimal solution is just the solution of the single-item unconstrained newsvendor problem, since the only constraint is basically irrelevant.

In the latter case the optimal conditions are

$$
\begin{aligned}
\frac{\partial h \left(Q_1, ..., Q_i, ..., Q_I, \lambda\right)}{\partial Q_i} &= \frac{\partial \mathrm{E}\left(\pi_i\left(Q_i\right)\right)}{\partial Q_i} - \lambda \cdot r_i \\
&= m_i \cdot \left(1 - F_i\left(Q_i\right)\right) - c_i \cdot F_i\left(Q_i\right) - \lambda \cdot r_i = 0 \qquad \forall i; \\
\frac{\partial h \left(Q_1, .., Q_i, ..Q_I, \lambda\right)}{\partial \lambda} &= \sum_{i=1}^{I} Q_i \cdot r_i - R = 0;
\end{aligned}
$$

hence

$$
F_i\left(Q_i^*\right) = \frac{m_i - \lambda \cdot r_i}{m_i + c_i}; \tag{5.11}
$$

$$
\sum_{i=1}^{I} Q_i \cdot r_i = R. \tag{5.12}
$$

Once we have found the solution, it is interesting to read the economic message the mathematical solution is sending us. Optimality conditions suggest that if the budget constraint is binding, the resource may have a positive shadow price λ. Given this shadow price, the margin (profitability) of each product should account for the opportunity cost of the shared resource. For example, we should consider the cost of the shelf space used by each unit of the product, or the cost of the limited production capacity consumed, etc.

Example 5.14 Let us consider a product with a price of 100€, a purchase cost of 60€, and a salvage value of 40€. Let us assume that the production of one unit consumes 1 unit of production capacity that could be otherwise be employed for 10€. A first option to solve the problem is to use the findings of the basic single-item newsvendor problem. This suggests that we should reach a $\frac{40}{40+20}$ = 66.6% service level. However, this approach might lead us to overuse the limited production capacity that could be effectively used otherwise as we overlook the value of the limited capacity. So, we can be a little bit more sophisticated and account for the consumption of the valuable capacity. If we use one unit of capacity to manufacture one unit of the product, we might gain a 40€ margin. However, to do so we give up a 10€ margin we could have gained by using the capacity differently. Hence, if we account for the capacity consumed, the margin we can hope to gain with one additional unit is 30€.

Cost of inventories, too, shall account for the opportunity cost of the valuable capacity. The cost of inventories is the difference between the purchase cost (60€) and the salvage cost (40€). Also, we should consider the opportunity cost of capacity. In other words, this product has not only a 60€ direct cost but also the cost of capacity that could have generated a profit of 10€, had it been employed differently. So manufacturing a unit that is then left unsold at the end of the period implies direct costs for 20€ plus 10€ we could have gained had we used the capacity to manufacture other products. So the total cost of inventories left unsold is 30€.

So if we restate the solution of the newsvendor problem to account for the cost of capacity rather than using just the direct margin m_i and direct cost of inventories c_i, we end up writing equation (5.11).

In other words we shall provide a $\frac{30}{30+30}$ = 50% type I service level. This is exactly what equation (5.11) suggests to do: $\frac{40-10\cdot1}{40+20}$, where $\lambda = 10$ is the unit cost of capacity and $r_i = 1$ is the consumption of capacity to manufacture one unit of the product.   ☐

The Lagrangian multipliers' method uses exactly this logic. However, there is a significant difference: The Lagrangian multipliers find the optimal value for λ, that is, the opportunity cost of the shared resource. So Lagrangian multipliers do not require the value of the shared resource as an input. Indeed, the shadow price of the resource (i.e., the opportunity cost) is endogenous to the multi-item problem, as it depends on the stocking quantity of the whole set of products, their consumption of the shared resource, and their margins. For example, if products are produced in limited quantities, the capacity might be abundant and basically have a zero value. On the contrary, if products are in high demand and have large margins, the opportunity cost of capacity is relatively large.

So far we have just derived optimality conditions. In other words, had we found the optimal solution, it would satisfy these conditions. Now we want to design an algorithm to find an optimal solution:

1. $\lambda = 0$, if the solution satisfies the constraint, then stop; otherwise proceed.

2. If the current solution violates the constraint, increase the opportunity cost of capacity λ; otherwise, reduce it (while still keeping it positive).

3. Calculate optimal quantities Q_i with equation (5.11).

4. If the constraint is satisfied with a given tolerance, stop; otherwise goto 2.

This solution process identifies an optimal solution simply because it checks first whether the unconstrained optimal solution is feasible and we can immediately stop. If this is not the case, we try to identify a price λ for the shared resource in such a way that it can be fully utilized (100% utilization rate) and efficiently allocated among the I products. A too-low opportunity cost λ leads to an excessive utilization of the shared resource and thus we shall increase it. On the contrary, a too-high opportunity cost λ leads us to underutilize capacity. Thus it is convenient to use capacity utilization to capture whether λ is too high or too low.

Finally, it is interesting to discuss the special case of products that share the same margins m_i, cost of inventories c_i, and consumption of the common resource r_i. In this case, the optimal solution is to give all products exactly the same service level. Once again math suggests a solution with a clear economic message: If we have two identical products and a limited capacity to manufacture or purchase them, why should we provide a higher service level for one rather than the other? A higher type I service level means that we have a lower probability of selling the last (marginal) unit and a higher probability (just equal to the service level) of not selling the last unit. A simple re-allocation of capacity from the higher service level product to the lower service level one increases the chances of selling the last unit manufactured. Clearly, this increases the expected profit.

Also, notice that the same reasoning applies to the case where margins m_i, inventory costs c_i and consumption of the shared resource r_i, though not equal across products, are proportional. Indeed, if product 1 has 50% of the margins and of the inventory costs of product 2 ($m_1 = 0.5 \cdot m_2$ and $c_1 = 0.5 \cdot c_2$) and it also consumes 50% less of the shared resource ($r_1 = 0.5 \cdot r_2$), then basically one unit of item 1 is equivalent to 0.5 units of item 2. So, also in this case we should provide the same type I service level for all products.[12]

---

[12]Notice that in the case of discrete demand we might not be in a position to reach exactly the same service level on all products. One possible way out is to allocate the limited capacity to the product that has the largest expected profit. A description of a greedy heuristic might be

- $C = R$
- $Q_i = 0 \qquad \forall i$
- $O_i = 1 \qquad \forall i$
- Do until $\sum_{i=0}^{I} O_i = 0$

Table 5.3 Data for multi-item newsvendor problem

Product	A	B	C	D	E
Selling price	100	130	170	80	80
Purchase cost	60	70	80	50	50
Salvage value	40	50	60	45	45
Expected demand	1000	500	500	1500	2000
Standard deviation	250	300	350	100	350

Table 5.4 Unconstrained solution to the multi-item newsvendor problem

Product	A	B	C	D	E
Optimal service level	0.667	0.75	0.818	0.857	0.857
Optimal quantity	1108	702	818	1607	2374
Optimal purchase cost	66480	49140	65440	80350	118700

Example 5.15 Let us consider a buyer in the fashion business. The company has long lead times, so the buyer only buys once in the season. The buyer manages a section of the company's assortment, say women's parkas. In the assortment we have five products. Table 5.3 shows selling price, purchase cost, and salvage value for the five products in the assortment. Table 5.3 also shows expected demand for the season and its standard deviation. Let us assume that the buyer has a limited budget and can only purchase products for 330K€.

With the data in table 5.3 we can derive the optimal and unconstrained purchase plan that is described in table 5.4.

- Find
$$j = \text{argmax}_i \left\{ \frac{[(1 - F(Q_i + 1)) \cdot m_i - F(Q_i + 1) \cdot c_i]}{r_i}, \text{s.t.} \, O_i = 1 \right\};$$
- If $(1 - F(Q_j + 1)) \cdot m_j - F(Q_j + 1) \cdot c_j < 0$ then stop
- If $r_j > C$ then $O_i = 0$ else
- $Q_j = Q_j + 1$
- Next

In this process we basically allocate the limited capacity R to all products. O_i are Boolean variables that capture whether we are still allocating the capacity to product i or we stopped either because no additional unit fits in the limited capacity available or because adding an extra unit is no longer profitable.

Table 5.5   Optimal constrained solution to the multi-item newsvendor problem

Product	A	B	C	D	E
Service level	0.470	0.578	0.675	0.576	0.576
Quantity	981	559	659	1519	2067
Purchase cost	58860	39130	52720	75950	103350
Constr.Optimum / Unconstr.Optimum	0.89	0.80	0.81	0.94	0.87

Table 5.4 provides us with several insights. Interestingly, different products have different target inventories. First, products D and E have the highest service level since they have good margins (30€) and a very low risk of obsolescence (5€). Though these two products share the same economics and thus the same service level, the stocking quantity of product E is higher than the stocking quantity for D since E has both a higher uncertainty (350 vs. 100) and a higher demand expectation (2000 vs. 1500). At the other extreme of the spectrum, product A has a relatively high cost of inventories (20€ = 60€ – 40€) as compared to the margin (40€ = 100€ –60€). Understandably, product A has the lowest service level and we produce just a few units more than we expect to sell (108 vs. 304 for E). So we can see that different products have different stocking quantities according to their economics and demand distributions. So these results are very consistent with the findings of the basic newsvendor problem. Unfortunately, this optimal plan is infeasible because the total purchase cost is 380110€, which is well above our budget of 330K€. This means that we have an opportunity cost for the budget and should look for a shadow price to give the planners of the five items an incentive to reduce the stocking quantity. In our problem, the purchase costs of the items represent the consumption of scarce resources r_i. Thus our solution should tend to reduce the production quantity, *ceteris paribus*, of those items with a large purchase cost. The optimal solution is displayed in table 5.5. We can find the optimal solution below with $\lambda^* = 0.197$. What does this mean? Basically, we would be willing to invest up to 0.197€ for one additional euro of budget. So the optimal value $\lambda^* = 0.197$ is not only useful to find the optimal allocation of the limited budget, it is also a valuable mean to judge how much we are willing to spend to move (marginally) our constraint. When we apply this optimum level of λ, we can find the optimal solution displayed in table 5.5.

Now let us try to make sense of this constrained solution. In particular the analysis of the last line of the table is telling. It is the ratio between the constrained solution and the unconstrained one. This ratio basically tells us the extent to which the constrained solution differs from the unconstrained one. The closer the ratio is to 1, the more the two solutions are similar, the closer to 0 the more the two solutions differ. The first relevant observation is that we decrease the stocking quantity of all five products to meet the budget. However, we do not reduce the stocking quantities of all items proportionally. While we reduce the stocking quantity of item D by roughly 6%, we cut the inventories of item B by roughly 20%. Why do we do that? Also, the service level on item A is below 50%, while the service level for item C is still 67.5% (notice that while in the previous case D and E had a higher service level than C, in the constrained solution it is just the opposite).

We actually tend to decrease the stocking quantity of products that are relatively uncertain (compare product E with product D), as we are not willing to give up a production that is relatively safe; that is, we are quite sure we are going to sell (the reader might want to reduce the uncertainty, that is standard deviation, of item D to zero and repeat the above exercise: In this case the stocking quantity of D is not affected by the budget allocation). Also, we tend to reduce more significantly the stocking quantity of expensive items (see the ratio for products B and C), since reducing the stocking quantity for these two expensive items is a very effective mean to cut the total purchase cost. If we reduce the stocking quantity by one unit of B and one unit of C we save as much money (150€) as we save when we reduce the stocking quantity of D (or E) by three units. A reduction in the stocking quantity of B and C is 50% more effective than the reduction in the stocking quantity of D or E. As we can see, the multi-item problem allocates capacity in a rather brilliant way according to the features of the products.   ☐

Two-period newsvendor problem.   Up to this stage, we have investigated the classic Newsvendor problem in a single-period context. In the next sections we are going to discuss dynamic problems where product life cycle is so long that we can neglect end-of-life in our planning problem. The two-period newsvendor problem is an intermediate situation between these two extremes.

Let us consider a product with uncertain demand and a rather short life cycle that makes stock holding costs relatively irrelevant, like in the case of the newsvendor problem. However, unlike in the newsvendor problem, we assume that we can replenish the product during its life cycle. We assume we can observe the initial pattern of demand (see section 3.12.3) update and improve our forecast and then place a second order that is going to be delivered before the end of the season, as figure 5.4 shows.

So according to the classification of inventory problems proposed in section 4.1 the two-period newsvendor problem is

- single-product, single-echelon, and multistage, that is dynamic;

- demand is uncertain, though we know its distribution;

- the objective is to minimize the cost of inventories and service; for the sake of simplicity we assume we always place two orders, so we neglect

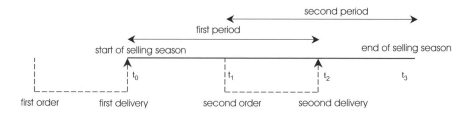

Fig. 5.4   Timeline for the two-stage Newsvendor problem.

the possibility of making just one order at the beginning of the season to save fixed ordering costs (note that this is a fairly attractive strategy for products with a relatively low uncertainty whereas it is definitely not a good option for very uncertain products).

Example 5.16 Some fashion companies can replenish their products during the season as they can enjoy short enough lead times. For example, in the case of products for the spring–summer collection, retailers can place two orders: One to be delivered at the beginning of the selling season. The second is placed once some sales are observed, but early enough for products to be delivered by (say) early May (any delivery beyond this deadline would be too late and the chances of selling the units would be too low).   ▢

In these situations we can make two decisions:

- First, we place the *initial order* that is delivered at the beginning of the product life cycle.

- Second, we observe initial demand and place a *second order* that is delivered before the end of the product life cycle. Hence, the problem is two-stage, not only two-period.

Given this situation, we call the time interval between the beginning of the product life cycle and the second delivery *first period*. We call *second period* the time interval between the placement of the second order and the end of the product life cycle. Figure 5.4 shows that these two periods overlap.

The second order. The decision on the second order resembles the classic newsvendor problem.[13] The optimal quantity shall balance (i) the cost of stockouts in case demand exceeds inventories and (ii) the cost of inventories in case goods are left over at the end of the season. The only one difference

---

[13] Notice that this holds when the LT is very short or when demand is *backordered*, in case of stockouts. In case of long LT and lost sales, we should account for any sales we might lose before the second delivery.

between this case and the basic newsvendor problem is that when we make this decision we might have some inventories on hand (in case we have not completely sold out the first lot Q_I). This is actually a marginal difference. We shall simply subtract the quantity on hand (i.e, the quantity from the first order not sold as yet) from the optimal quantity. When inventories on hand exceed the optimal quantity, we simply place no additional order.

Example 5.17 Let us consider a retailer that sells Christmas cakes and places two orders in a season. The decision on the last delivery is made on December 1st. On December 1st, 700 cakes are on hand. Given the current demand trend, we expect a demand for 1000 units by season end. Demand is normally distributed and has a standard deviation of 200 units. Christmas cakes cost 1€ and are sold at a full price of 5€. After the end of the season they are sold for 0.8€.

Given these data, we shall reach a $\frac{4}{4+0.2} = 95.24\%$ service level, as the newsvendor problem suggests. To reach that target we need $1000 + z(95.24\%) \cdot 200 = 1000 + 1.67 \cdot 200 = 1334$ units; 700 of these 1334 unit are already on hand. Thus, we shall place an order for 634 (1334 – 700) additional units. Had we had more than 1334 units, we would have chosen to order zero units or sell some units to another retailer, in case this is a viable option. ☐

The initial order.[14] The first order might resemble the newsvendor problem, but actually we have a second chance to purchase products and this might lead us to be slightly more conservative, since we can have a second chance to increase inventories in case demand happens to be higher than we initially expected. When we set the initial purchase quantity, we face two risks:

- The risk that inventories are insufficient and a stockout occurs before the second order is delivered. Thus when we estimate the cost of the stockout we shall compare the stocking quantity with the demand over the first period. Indeed, demand after the first period can be met by the second purchase order delivered in t_2.

- In this scenario the cost of holding inventories is relatively negligible and thus we shall focus on the cost of excess inventories at the end on the season. Thus to capture this cost we shall compare initial inventories to the demand during the whole life cycle of the product $(t_0 - t_3)$.

We identify the variables that refer to the initial period $(t_0 - t_1)$ with subscript I, while we identify the variables that refer to the whole product

---

[14] We emphasize that the approach we describe here is actually a reasonable solution heuristic. Among other things which could complicate the problem, we are not considering issues related to correlation between the demand in the two periods; we refer the reader to [4] for a full treatment.

life cycle $(t_0 - t_3)$ with subscript T. Thus the cost of inventories and cost of stockout are:

$$C_{in} = c \cdot \int_0^{Q_I} (Q_I - x) \, f_{d_T}(x) dx,$$

$$C_{so} = m \cdot \int_{Q_I}^{+\infty} (x - Q_I) \, f_{d_I}(x) dx.$$

Notice that the cost of inventories depends on the demand for the whole selling period, since products lose value only at the end of the "season." In our example, we have no fixed costs or minimum order size (these are common issues in real-life problems) and we have no stock holding costs (which in real-life problems are actually negligible). This makes the purchase quantity bought in the initial order as expensive as the quantity purchased in the second order. However, postponing part of the purchases enables us to observe early demand. This reduces uncertainty and thus reduces mismatches between supply and demand.

So what distinguishes this problem from the basic newsvendor problem is that the two probability distributions refer to different time frames. However, both can be differentiated with respect to the purchase quantity Q_I. Using Leibniz's rule we can show that

$$
\begin{aligned}
C_{tot} &= c \cdot \int_0^{Q_I} (Q_I - x) f_{d_T}(x) dx + m \cdot \int_{Q_I}^{+\infty} (x - Q_I) f_{d_I}(x) dx \\
\frac{\partial C_{tot}}{\partial Q_I} &= c \cdot \int_0^{Q_I} f_{d_T}(x) dx - m \cdot \int_{Q_I}^{+\infty} f_{d_I}(x) dx \\
&= c \cdot F_{d_T}(Q_I) - m \cdot [1 - F_{d_I}(Q_I)] = 0.
\end{aligned}
\tag{5.13}
$$

These are conditions for optimality; before we try to use them to find a solution, we want to make sense of them and try to understand what they are saying. Equation (5.13) suggests that an increase in the initial purchase quantity Q_I increases the cost of inventories in case the last unit Q_I is left unsold at the end of the season (this scenario has a probability $F_{d_T}(Q_I)$ and a cost c). An increase in Q_I also reduces the cost of the stockout in the first period of the season, when demand exceeds inventories (this scenario has a probability $[1 - F_{d_I}(Q_I)]$ and a cost m), we keep on increasing the stocking quantity to the point where the expected savings on the cost of lost sales in the first part of the season are greater or equal to the additional expected costs for excess inventories at the end of the selling season.

Concept 5.6 *When planning short life cycle products with more than one deliveries the first delivery shall trade-off two risks. On the one hand, we want to order enough units to meet demand up to the next delivery (even in scenarios where demand is higher than we expected) . On the other hand, we want to make sure that the first delivery is not too high, not to leave some goods left over at the end of the season (even in scenarios where demand is lower than we had expected).*

We cannot find the optimal quantity Q_I^* in closed form unless we specify the demand distribution for the demand in the first period and in the whole season. Once the two demand distributions are identified, we can find the optimal solution through a search procedure to find the value of Q_I that satisfies equation (5.13). The search is rather simple. As Q_I increases, the cost of excess inventories $c \cdot F_{d_T}(Q_I)$ increases, while the cost of lost sales $m \cdot (1 - F_{d_I}(Q_I))$ decreases. Thus the left-hand side of equation (5.13) decreases in Q_I.[15]

5.3 MULTI-PERIOD PROBLEMS

In the previous sections we have discussed the newsvendor problem (and some extensions of it). The newsvendor problem is a static (single stage) and single period problem, as decisions at time t have no influence whatsoever on successive periods. Indeed, inventories left over at the end of period t are just sold at salvage value. In this section we discuss the more common case of dynamic problems (actually, in the last subsection we have discussed a first dynamic problem) with multiple periods. In these problems, decisions at time t have an impact on decisions and performance at time $t + 1$. All planning problems where the life cycle exceeds the replenishment lead time (plus the time required to read demand trends) are dynamic. However, in the following sections we only investigate those where the life cycle is so long that it can be considered to be infinite. In these cases, we can neglect the end-of-life-cycle costs. So we do not investigate situations where either technological innovation or fashion changes make the current product obsolete. Also, we do not consider situations where the product has a limited shelf life.[16]

In sections 5.4–5.7 we discuss problems that are:

- single product, single level, dynamic and multiperiod;

- with uncertain demand, with a known and stationary demand distribution, deterministic LT;

- where the objective is to minimize stock holding costs, ordering costs and costs of service.

---

[15] Notice that in this case the optimization problem has been written as a cost minimization problem, while in the basic newsvendor problem we have maximized profit. Obviously, both problems can be written either one way or the other and the end-result does not change at all. We have decided to write the two models with different procedures to show different approaches to modeling. We suggest the reader to try to rewrite the classic newsvendor problem as a cost minimization problem and rewrite the two-stage newsvendor problem as a profit maximization problem and check that results do not change, as basic logic suggests.

[16] Note that all products sooner or later expire, but the issue here is whether the expiration date entails a relevant cost for our planning problem. For example, canned food can have a 5-year shelf life; in this case the constraint is basically irrelevant from any practical standpoint, as companies basically never carry it for 5 years.

Before we discuss the details of these planning problems we need to under-stand the effects of uncertainty on the planning process. In chapter 4 we have shown that, in the case of deterministic demand, setting the purchase quan-tity (order size Q) or the ordering period (time T elapsing between orders) is basically the same. Once we have set the order quantity Q, the frequency is implicitly set (it is equal to the order size Q divided by the demand rate d) and vice versa. A first significant difference between certain and uncertain con-ditions is that, under uncertain conditions, purchase quantity and frequency are no longer deterministically linked. As the following example shows, such a relationship still exists; however, it is stochastic rather than deterministic. Indeed, the frequency is still the ratio between the purchase quantity Q and the demand, but demand is a random variable.

Example 5.18 Let us consider a company with a zero LT. Such a company could order Q units when inventory level reaches zero, even under uncertain conditions, but let us assume that the order is issued when inventory reaches a reorder point R (see figure 5.5, possibly setting $R = 0$). The frequency of such orders is not fixed but rather depends on demand. If demand is very high (i.e., higher than its expected value), quantity Q is sold out in a short period of time (see the second time between orders in figure 5.5). On the contrary, if demand is very low (i.e., significantly lower than its expected value), the time between two successive orders is relatively long (see the third time between orders in figure 5.5). Going back to example 4.9, if the company orders once a month, some months it orders more than 100 units (when demand was higher than expected) while in other months it orders less than 100 units (when demand was lower than expected). Similarly, if the company places orders for 100 units at a time, sometimes we can wait for more than a month to place the next order (when demand is lower than expected) while we can have two orders in a month in case demand exceeds expectations.   ⬚

Concept 5.7 *While facing an uncertain demand, we have no deterministic relationship between the order quantity and the order frequency. Thus, we design planning policies that fix one parameter (e.g., the order size) and let the other fluctuate (e.g., order frequency) according to demand fluctuations.*

For example, a company can set the order size to $Q = 100$ and place an order once 100 units have been sold. A second option is to order once a month the quantity that was sold. In other words, one parameter is fixed and the second one fluctuates according to demand. This is why the inventory planning methods are called *fixed quantity* and *periodic* or fixed period.

5.4 FIXED QUANTITY: THE (Q, R) MODEL

A first option in managing inventories is to set a quantity Q that is ordered each time inventories reach the reorder point R. Obviously, this planning

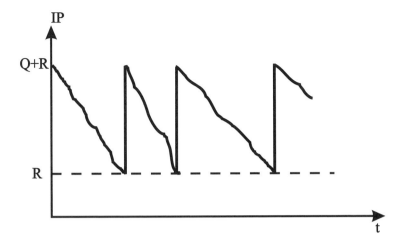

Fig. 5.5 Time between orders as a function of demand in the case of fixed order size Q.

method can be adopted if and only if one can continuously control inventories (and demand is a continuous process with no batches) to make sure we can place an order exactly when inventories reach the reorder point R. This scenario is called *continuous review*. In the past, this was a rather critical assumption, while today it is relatively easy to control inventories of several SKUs in several warehouses by the minute through appropriate technology (as mentioned in section 4.1, technology is just one portion of the equation, as a company needs to design business processes appropriately and execute them accurately, in order to make sure that inventory data are collected accurately and in a timely fashion).

A company shall record all transactions properly, taking into account defective products, stolen products, errors in deliveries from suppliers, etc., to have reliable information on current inventory levels. While we shall keep that in mind and acknowledge it is hard job, in this book we assume, like in the bulk of literature on inventories, that the company can perfectly know the current inventory level.

Also, before we discuss the details of this inventory model we shall go back to chapter 4 and recall the relationship between physical inventories and inventory position. Our decisions (e.g., should I place an order? How many units should I buy?) are based upon inventory position. On the contrary, stockouts are generated by a lack of physical inventories.

If we overlook backlog for a second, the two variables are equal when we have no open order, i.e., we are not waiting for any delivery from the supplier. If we assume that we can have at maximum one open order, when it is delivered the physical inventories and inventory position coincide (see figure

5.6). Our assumption holds when LT is lower than the time between orders. In other cases we can have more than one open order.

Though our results are derived under this simple assumption to build intuition, they hold under more general conditions. So they apply even when LT is longer than the time between orders and thus we can have more than one open order at a time.[17] For the sake of simplicity, we normally refer to the case of $LT < f$, where there is one outstanding order at the most.

The dynamics of the inventory system. We shall describe and intuitively understand the dynamics of the inventory system before we get into the details of parameters optimization.

First, when facing an uncertain demand it might be inappropriate to set the reorder point R to $LT \cdot E(d)$, that is, the expected demand over the replenishment LT. Indeed, demand uncertainty makes draws that differ from the expected value rather likely (actually in the case of a continuous distribution, the probability that demand during the LT equals its expected value is zero).

The physical inventories in the warehouse one second before the order is delivered is equal to the reorder point minus the demand over the LT. Given that demand is a random variable, also the physical inventories just before the delivery of the order are a random variable (it is a parameter R minus a random variable). So the physical inventory level is as uncertain as the demand over the lead time is and the two random variables follow the same demand distribution (in terms of shape and variability, though understandably they have a different expectation in general).

After the order is placed, the inventories can follow various patterns over time. In figure 5.6 the top trajectory represents a case where demand over the lead time was lower than we expected, the intermediate trajectory represents a case where demand was equal to its expected value, and finally the bottom line represents a case where demand was higher than we expected. Figure 5.6 shows that when the demand distribution is symmetric, the probability distribution of physical inventories a few moments before the order is delivered is symmetric as well. So in 50% of the cases a reorder point $R = LT \cdot E(d)$ would be insufficient to meet demand over the lead time and we would experience a stockout.[18]

The newsvendor problem has proven that in general it is not advisable to have a 50% probability of stockout, as this policy is optimal under very specific conditions. On the contrary, for many companies it is appropriate to reduce the probability of stockouts, while for others it might be appropriate to set a service level target below 50%.

---

[17]Notice that these findings hold in the even more general case of stochastic LT, if orders do not cross.

[18]In the case of discrete distributions we should also account for the probability that demand is exactly equal to the inventory level so the probability of a stockout might be slightly below 50%.

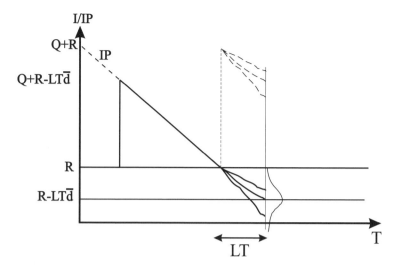

Fig. 5.6   Various patterns of inventories during the LT.

In general, the reorder point can differ from $LT \cdot \mathrm{E}(d)$. In most cases, we set a reorder point $R > LT \cdot \mathrm{E}(d)$ to increase the service level (type I) above 50%. The quantity $R - LT \cdot \mathrm{E}(d)$ is called *safety stock* (SS). These inventories are used only when demand exceeds expectations to meet unexpected demand. So, in general, the parameters that are optimal in deterministic conditions are not optimal under uncertain ones.

Concept 5.8 *Under uncertain conditions, the reorder point R differs from the expected demand over the LT. Just like in the case of the newsvendor problem, the economics of the business suggest us whether we shall carry more inventories than we expect to sell over the LT $(R > LT \cdot E(d))$ or vice versa.*

Concept 5.9 *We call the difference between the reorder level and the expected demand safety stock, that is inventories we plan to use only when demand exceeds its expectation $SS = R - LT \cdot E(d)$.*

Notice that in our analysis we have neglected the various scenarios of demand *before* we reach the reorder point R. Actually, demand can show very different patterns before the reorder point as well as after the reorder point is reached. The demand pattern before inventory position reaches the reorder point is not a crucial issue, as it only determines *when* we reach R and thus *when* we place an order. On the contrary, the demand pattern *after* we reach R is crucial as we wait for the order quantity Q to be delivered and we hope not to experience a stockout in the meantime. In other words, we hope that the inventory level R is going to be large enough to fully meet demand.

The cost function   Once the dynamics of the inventory system are clear, we can try to write the cost function and derive optimal levels for the control parameters Q and R. In this situation there are three relevant costs:

- *Cost of inventories.* The cost of inventories in this case is the holding cost. Indeed, the infinite life cycle of the product makes other costs (disposal, loss of value, markdown) irrelevant for our purposes. Given the logic of the (Q, R) system, the inventory position fluctuates between the minimum level R and the maximum level $R+Q$ and thus the average inventory position is $R + Q/2$.[19]

  While IP is the variable we watch to make decisions, the physical inventories in our warehouse determine our costs. On the one hand, physical inventories (also known as on-hand inventories) are the inventories we actually carry in the warehouse and thus we pay for their holding cost. On the other hand, we incur the cost of the stockout when we run out of physical inventories. When a customer that expects a zero DLT finds an empty shelf he/she is upset regardless of incoming orders over the next few days that make $IP > 0$.

  So we need to investigate the level of physical inventories, which we call I, to calculate costs. Physical inventories I reach their minimum just before the delivery of Q units, and they reach their maximum level just after the delivery. These inventory levels are actually random variables, as they depend on the demand over the LT. So we cannot tell, *ex ante* what their future level will be. However, we can study the distribution of inventories a few seconds before the delivery of Q units and consider their expected level R minus the expected demand over the LT.[20]

---

[19]Notice that in general this information about the minimum and maximum level of inventories is not enough to draw the conclusion that the average quantity is the average between the maximum and the minimum. To draw that conclusion, we have to prove that the inventory level over time is, in a sense, uniformly distributed between the maximum and the minimum. By "uniformly distributed" we do not really mean that if we observe inventory level at a specific time instant, we see a uniform distribution; if we take our observation just after issuing an order, this is certainly not the case. Rather, we mean that *over time*, i.e., along a sample path, we see a uniform distribution. To make the idea rigorous, one can resort to observers arriving at a random time, according to a Poisson process, but we prefer leaving such complications aside. In our case the proof is rather trivial, as the demand process is stationary and thus inventories are consumed equally when the warehouse is almost full (i.e., when the inventory level is high) and when the warehouse is almost empty (i.e., when the inventory level is low). This means that there is no reason whatsoever to presume that the inventory level stays at a very high level or at a very low level for a long period of time. Thus all levels of inventories between the maximum and the minimum are basically equally likely.

[20]Notice that what we suggest in the main text is actually an approximation, as any model actually is. Indeed, physical inventories are bound to be non-negative. So, we should somehow ignore all cases where demand exceeds inventory level. In these cases, inventories drop to zero. In other words, the expected inventory level before the quantity Q is delivered is $R - (\mathrm{E}(d_{LT}|d_{LT} < R) \cdot F_{d_{LT}}(R) + R \cdot (1 - F_{d_{LT}}(R)))$. In other words, the availability

Thus, the expected inventories just before the delivery of Q units ordered at time t_0 (i.e., at the end of the planning cycle) can be reasonably approximated by

$$\mathrm{E}\left(I(t_0 + LT^-)\right) = R - \mathrm{E}\left(d_{LT}\right) = R - \mathrm{E}(d) \cdot LT.$$

Now it is easy to show that the expected level of physical inventories at the beginning of the next cycle (so-called maximum expected level of inventories), right after the delivery of Q units, is

$$\mathrm{E}\left(I(t_0 + LT^+)\right) = R + Q - \mathrm{E}\left(d_{LT}\right) = R + Q - \mathrm{E}(d) \cdot LT,$$

and the expected level of average inventories $\mathrm{E}(I)$ is

$$\mathrm{E}\left(I\right) = R + Q/2 - \mathrm{E}(d) \cdot LT,$$

as inventories fluctuate between the maximum level and the minimum level and are uniformly distributed between these two values.

---

of inventories censores the demand distribution. Also, the distribution of inventories a few seconds after the quantity Q is delivered really depends on the customers' willingness to wait. If customers do not wait, i.e. under the lost sales assumption, the inventory level after the delivery is just equal to $R - (\mathrm{E}(d_{LT} | d_{LT} < R) \cdot F_{d_{LT}}(R) + R \cdot (1 - F_{d_{LT}}(R))) + Q$. Indeed, in this case we have no list of customers waiting for the delivery and thus the quantity Q is just added to the inventory available before the delivery. On the contrary, in the backorder case, the inventory position just after the delivery is equal to $R + Q - \mathrm{E}(d)$, as in this case any demand in excess of the inventory level R is just backordered and thus once the quantity Q is delivered, the whole demand is fulfilled and inventories are consumed (we deliberately ignore situations where we still have a list of customers right after the delivery of a lot Q). Thus as we can see, the approximation we have made in this book works better for the backorder than for the lost sales case. For example, let us consider the case of a uniform demand distribution U(100,200) over the LT. Also, let us assume that the reorder point R is 180 and the order quantity Q is 500. Under the lost sales assumption, inventories before the delivery are going to be 0 with a 20% probability and the remaining 80% of probability is uniformly distributed between 0 and 80 units. Also, the inventory level just after the delivery is 500 with a 20% probability and the remaining 80% of probability is uniformly distributed between 500 and 580. Thus the expected inventory level at the beginning of the next cycle is 532 rather than 530. This means that not only our model simplifies the dynamic of inventories at the end of any planning cycle, but this also has an effect on inventories in the next cycle. Let us now consider how things change under the backorder assumption. Under the backorder assumption the inventory distribution before the delivery remains unchanged. So, even in the backorder case we are cutting some corners. Indeed, we are overestimating the inventory level and thus the inventory holding cost (e.g., we might run out of inventories before the end of the cycle). However, the physical inventory level at the beginning of the next cycle is $R + Q - \mathrm{E}(d_{LT}) = 180 + 500 - 150 = 530$. In this case any demand unsatisfied at the end of the previous cycle is delivered as the lot Q is delivered. Finally, we shall notice that this approximation is very crude in case of low service levels. At one extreme, when R is lower than the expected demand, we might draw the conclusion that the expected level of physical inventories at the end of any planning cycle is negative! On the contrary when the service level target is high, unmet demand is negligible and thus our simplification is reasonable, especially in the case of backordered demand. Luckily, most products with a long life cycle have a fairly high service level, as the cost of inventories is the cost of holding inventories rather than the loss of value like for high-tech or fashion products and thus our simplifying assumptions turn out to be reasonable.

Thus the expected cost of inventories for this policy is

$$C_{in} = h \cdot (R + Q/2 - \mathrm{E}(d) \cdot LT).$$

- *The ordering cost.* While the inventory level depends on both control parameters R and Q, the ordering cost only depends on the lost size Q only. The reorder point R tells when to order, while the parameter Q tells how much to order and thus how often we order. The number of orders in a period is actually a random variable that depends on the level of demand. However, we can compute the expected number of orders that is equal to $\mathrm{E}(d)/Q$. Thus the expected ordering cost is

$$C_{or} = A \cdot \frac{\mathrm{E}(d)}{Q}. \tag{5.14}$$

- *Cost of lack of service.* To describe the cost of lack of service we shall choose one of the two scenarios according to the driver of the cost function:

  - *cost due to the presence of a stockout;*
  - *cost due to the size of the stockout.*

The cost of the stockout is different under the two scenarios (difference in economic consequences of a stock out) and thus the optimal policy can change accordingly. So the two cases are going to be treated separately.

Cost depends on the occurrence of a stockout. In this case, the cost of the stockout is equal to the cost of a single stockout, i.e. the cost a company incurs each time it runs out of the product, times the number of stockouts we expect to face. In turn, the number of stockouts then depends on two factors:

- the number of chances to experience a stockout, that is the number of planning cycles, when we wait for the delivery of the Q units of the product and inventories might fall short of demand;

- the probability that in each planning cycle the demand actually exceeds the stocking quantity R and we face a stockout.

Thus, under these assumptions the expected stockout cost is

$$\begin{aligned}
C_{so} &= p \cdot \frac{\mathrm{E}(d)}{Q} \cdot \int_{R}^{+\infty} f_{d_{LT}}(x)dx \\
&= p \cdot \frac{\mathrm{E}(d)}{Q} \cdot (1 - F_{d_{LT}}(R)),
\end{aligned}$$

where $F_{d_{LT}}(x)$ and $f_{d_{LT}}(x)$ are the cumulative and the density function of demand over the LT.

Under these assumptions the total cost is

$$
\begin{aligned}
C_{tot} &= C_{or} + C_{in} + C_{so} \\
&= A \cdot \frac{\mathrm{E}(d)}{Q} + h \cdot (R + Q/2 - \mathrm{E}(d) \cdot LT) \qquad (5.15) \\
&\quad + p \cdot \frac{\mathrm{E}(d)}{Q} \cdot (1 - F_{d_{LT}}(R)). \qquad (5.16)
\end{aligned}
$$

Cost depends on the size of the stockout. Under this assumption the cost of the stockout depends on the number of units of demand (in the remainder of the chapter we call them customers, assuming that a single customer buys a fixed and small quantity –say one unit– of the product) unmet, because of a stockout. In this case, to quantify the cost of a stockout, we should measure the cost p_u of not meeting one unit of demand and the number of units not immediately delivered to the customers from stock. Once again this latter variable is equal to the number of planning cycles times the number $n(R)$ of units of demand we expect not to deliver from stock in each planning cycle. Clearly, this depends on R: the higher R, the lower the unmet demand. Under this assumption, we can write the expected stockout cost as

$$
C_{so} = p_u \cdot \frac{\mathrm{E}(d)}{Q} \cdot n(R) = p_u \cdot \frac{\mathrm{E}(d)}{Q} \cdot \int_R^{+\infty} (x - R) f_{d_{LT}}(x) dx.
$$

Thus, under this assumption the total cost function is

$$
\begin{aligned}
C_{tot} &= C_{or} + C_{in} + C_{so} \\
&= A\frac{\mathrm{E}(d)}{Q} + h\left(R + \frac{Q}{2} - \mathrm{E}(d) \cdot LT\right) + p_u \frac{\mathrm{E}(d)}{Q} n(R). \quad (5.17)
\end{aligned}
$$

Solution process Given the total cost function we can try to find a solution in two ways. The first option is clearly to minimize the total cost and find an optimal level of the control parameters Q and R. This approach is clearly the first best option but requires a reliable estimate of the costs of a stockout (p and p_u). As we have discussed in this chapter, finding a reasonable estimate for such cost parameters is actually a real challenge. So many companies rather prefer to set minimum targets for service level and then try to minimize ordering cost and stock-holding costs (the idea is that the minimum requirement on service level keeps the cost of the stockout under control). This approach might look simplistic but it is actually fairly reasonable when the measures of the cost of a stockout are very unreliable. Indeed, in this case the manager provides an indirect estimate of the cost of the stockout by requiring a minimum service level. While setting the minimum service level the manager implicitly balances the cost of the stockout with the holding cost (and to a minor extent the ordering cost). The higher the service level requirement, the higher the implicit estimate of the stockout cost and vice versa. So it is just matter of finding the most appropriate and accurate way

	Optimization	Constraint on service level
Cost depends on the size of the stock-out	SEC 5.4.1	SEC 5.4.2
Cost depends on the occurence of a stock-out	SEC S.5.8	SEC 5.4.3

Fig. 5.7 Various scenarios for the (Q,R) problem: schema of the section.

to estimate the cost of the stockout. It can be either a direct measure of the consequences of a stockout (p or p_u) or an indirect measure that is a minimum service level requirement.

As figure 5.7 shows, there are four possible combinations of cost structure (occurrence vs. size of the stockout) and solution process (optimization vs. minimum service level requirements). In the next three sections we investigate three of these four cases, while we address the fourth and more complex case in supplement S.5.8 at the end of the chapter; the supplement, which is technically more involved than the rest of the chapter, may be safely skipped.

5.4.1 Optimization of the (Q,R) model in case the stockout cost depends on the size of the stockout

When we have a reliable estimate of the stockout cost p_u, we can compute the derivative of the total cost function (5.17) with respect to the control parameters Q and R.

$$\frac{\partial C_{tot}}{\partial Q} = -\frac{A \cdot \mathrm{E}(d)}{Q^2} + \frac{h}{2} - \frac{p_u \cdot \mathrm{E}(d) \cdot n(R)}{Q^2} = 0,$$

$$\frac{\partial C_{tot}}{\partial R} = h + \frac{p_u \cdot \mathrm{E}(d)}{Q} \cdot n'(R) = 0.$$

To proceed with our problem, we should evaluate $n'(R)$; that is, we have to understand how the number of customers (units of demand) unsatisfied changes as a function of the reorder point R. Once again, on the one hand, we can use rigorous math; on the other hand, we can use our intuition to support math.

Increasing the reorder level R by an infinitesimal quantity dR has no effect whatsoever on the number of unhappy customers (i.e., unmet demand) in all demand scenarios where the reorder point was already high enough to fully meet demand over the LT.

On the contrary, increasing the reorder level up to $R + dR$ pays off in all scenarios where R is not enough to fully meet demand and thus we experience a stockout. In these cases, dR additional units decrease the number of unsatisfied customers (i.e., the unmet demand) by dR units. The last step is to remember that the probability that this second scenario occurs (i.e., the probability of a stockout) is $1 - F_{d_{LT}}(R)$. Thus the dR increase in reorder point leads to a dR reduction in unmet demand with a probability $1 - F_{d_{LT}}(R)$, that is, the probability that the additional dR units actually turn out to be really useful. Thus

$$n'(R) \quad = \quad \frac{d\,(n(R))}{dR} = -\frac{dR \cdot (1 - F_{d_{LT}}(R))}{dR} = -\left(1 - F_{d_{LT}}(R)\right).$$

More formally we use Leibniz's rule to differentiate $n(R)$ (obviously this process leads to exactly the same results)

$$n'(R) \quad = \quad \frac{\partial \int_{R}^{+\infty} (x - R) \cdot f_{d_{LT}}(x) \cdot dx}{\partial R}$$

$$= \quad -\int_{R}^{+\infty} f_{d_{LT}}(x)\,dx + 0 - 0 \cdot f(R_{d_{LT}})$$

$$= \quad -\left(1 - F_{d_{LT}}(x)\right).$$

Thus we can re-write the optimality conditions as

$$-\frac{A \cdot \mathrm{E}(d)}{Q^2} + \frac{h}{2} - \frac{p_u \cdot \mathrm{E}(d) \cdot n(R)}{Q^2} = 0,$$

$$h - p_u \cdot \frac{\mathrm{E}(d)}{Q} \cdot (1 - F_{d_{LT}}(R)) = 0; \qquad (5.18)$$

hence

$$Q^* = \sqrt{\frac{2\mathrm{E}(d) \cdot (A + p_u \cdot n(R^*))}{h}}, \qquad (5.19)$$

$$F_{d_{LT}}(R^*) = 1 - \frac{h \cdot Q^*}{p_u \cdot \mathrm{E}(d)}. \qquad (5.20)$$

Before we use these optimality conditions to look for the optimal solution (Q^* and R^*), we shall read the economic message they are sending. Equation (5.19) resembles the Economic Order Quantity. The one difference is that under uncertain conditions the fixed ordering cost A is replaced by $A + p_u \cdot$

$n(R)$. In fact, an order cycle implies not only the ordering cost when the inventory position reaches R, but also the risk of stocking out before the quantity Q is delivered. While the first is a certain cost, the second one is an uncertain one. However, we know its expected value is $p_u \cdot n(R)$. In other words, when facing uncertainty (in a (Q, R) system) a company might want to increase the order size both (i) to avoid very frequent orders (and the related cost A) and (ii) to reduce the number of orders since each time an order is placed, the company faces the risk that the demand during the LT exceeds the reorder point R, a stockout occurs and some customers cannot find the product they were looking for. Clearly, this further incentive leads companies to increase Q above the EOQ quantity. Such incentive is stronger and stronger as the cost of a stockout p_u increases and the reorder point R decreases, leading to an increase in the probability of a stockout. Equation (5.20) is derived from (5.18) and marginally analyzes the effects of an (infinitesimal) unit increase in the reorder quantity. This equation basically compares the cost and the benefits of such an increase in R. Increasing R leads to an increase in the stocking quantity and thus in the holding cost h [in fact, the derivative of the holding cost component C_{in} with respect to R is $\partial C_{\text{in}}/\partial R = h \cdot \mathrm{E}(I)/R = h$; see equation (5.14)].[21] On the benefits side, increasing R reduces the cost of a stockout by p_u with a probability $1 - F_{d_{LT}}(R)$ during each of the $\mathrm{E}(d)/Q$ planning cycles. Thus the second equation basically compares the marginal stock-holding cost with the marginal reduction in the cost of stockout, suggesting to increase R until the savings on the stockout cost exceed the additional holding costs.[22]

Also, this equation suggests that R shall be reduced, as Q increases (this is the flipside of equation (5.19)). Indeed, if the purchase quantity Q is relatively large the planning cycle is relatively long and thus it might not make sense to carry an extra unit of safety stock (that is R) for a very long planning cycle to slightly reduce the probability and/or size of a stockout at the end of it.

---

[21] Notice that this is a result of our simplifying assumption on the minimum level of inventories (see footnote 20 on page 274). Actually, the additional unit can be sold before the end of the planning cycle and thus might be held for less that a whole planning cycle. In particular, if R guarantees a β type II service level, we expect the product to be available on the average during a fraction β of the cycle. Thus in the more complete model the term h is multiplied by β. Finally, β is obviously a function of the control parameters Q and R as $1 - \beta = n(R)/Q$.

[22] Notice that while the marginal cost of inventories is constant (h), the savings on the stockout cost are decreasing in R, as the probability of a stockout decreases and thus the additional investment in inventories is more and more likely to be completely ineffective. Also, notice that the holding cost is equal to h because we assume that the additional unit of inventories is always carried. As discussed in footnote 20 on page 274, we are cutting some corners and deliberately ignore the fact that we might run out of inventories and thus we might not always carry the marginal unit. Again this assumption, however, is fairly reasonable, given that for most companies and most products it is advisable to reach a fairly high service level. Thus we seldom sold out, that is we seldom sell the last unit carried in inventories and almost always carry it until very late in the planning cycle.

Concept 5.10 *Under uncertain conditions the fixed cost of an order cycle includes both the fixed ordering cost and the expected cost of a stockout. Thus the order lots tend to be larger to reduce the number of cycles and thus the opportunities to stockout.*

Once the economic message equations (5.19) and (5.20) are sending is clear, we use them to look for the optimal solution (Q^*, R^*).

We shall notice that so far we have made no assumption whatsoever on the demand distribution, we just used the generic functions $F_{d_{LT}}$ and $n(R)$. Obviously, we need to specify the demand distribution to find the optimal solution. So we just need the basics of statistics to deploy the solution for various demand distributions, that is, according to the various situations we might face.

Using equations (5.19) and (5.20) to find the optimal solution is not trivial. Indeed, in equation (5.19) the optimal order quantity Q^* depends on R^*, while in equation (5.19) the optimal reorder point R^* depends on Q^*. One way out of this loop is to iteratively find the solution. For example, we can start with a reasonable order quantity $Q_0 = EOQ$ to find a first tentative reorder level R_0 through equation (5.20). This first estimate of the reorder point R_0 can then be used to find a better order quantity Q_1 and so on. The iterative process can be stopped when $Q_i \approx Q_{i-1}$, where the tolerance depends on the specific decision at stake.

Example 5.19 Let us consider a product whose cost is 10€, with a LT of 6 months, a holding cost of 20%/year, and a cost of a stockout of 25€ per unit, as we expect the customer to leave the company in case he/she routinely cannot find the product he/she looks for.

Also, let us assume that the ordering cost is 50€ and the demand over the LT is normally distributed with a mean of 500 units and a standard deviation of 100 units.

In this situation we can start from the EOQ quantity as a first, rough-cut estimate Q_0 of the optimal order quantity:

$$Q_0 = \sqrt{\frac{2A \cdot \mathrm{E}(d)}{h}} = \sqrt{\frac{2 \cdot 50€ \cdot 200\,\text{units/y}}{20\%/\text{y} \cdot 10€/\text{units}}} = 100\,\text{units.}$$

We can then use Q_0 to derive a first, rough-cut estimate of the reorder level R_0:

$$F(R) = 1 - \frac{2\,\dfrac{€}{\text{units} \cdot \text{y}} \cdot 100\,\text{units}}{25\,\dfrac{€}{\text{units}} \cdot 200\,\dfrac{\text{units}}{\text{y}}} = 96\%.$$

Now the distribution of demand comes into the picture and starts playing a role. For some distributions such as the Uniform or the Exponential we can find R through integrals, while for others such as the normal distribution we cannot solve the integral in close form and thus we resort to standard tables. To reach a 96% type I service level we shall choose a level z equal to 1.75. In other words, the first rough cut decision is to carry 1.75 standard deviations more than we expect to sell during the LT. Thus we only need to find the right parameters and refer them to the demand over the LT:

$$R_0 = 100 + 1.75 \cdot 25 = 144.$$

Once we have obtained R_0, we want to use it to improve the decision on lot size through equation (5.19). To do so, we shall derive the quantity $n(R)$. Once again this function depends on the demand distribution. For some distributions such as the uniform or the exponential, we can get the function in close form. For other distributions such as the Normal, we resort to standard tables.

In the case of normal distribution the estimate of $n(R)$ is based on the standardized *loss function*

$$L(z) = \int_z^\infty (t - z)\phi(t)dt,$$

where $\phi(t)$ is the density of the standard normal distribution. We can show that:

$$n(R) = \sigma_{LT} \cdot L\left(\frac{R - \mathrm{E}(d_{LT})}{\sigma_{LT}}\right) = \sigma \cdot L(z),$$

where $z = (R - \mathrm{E}(d_{LT}))/\sigma$. The loss function can be found on statistical tables in most reference books in statistics. We can use such tables to calculate a new (and improved) order size Q_1. In the case of the Normal distribution we can proceed as follows:

$$n(R_0) = \sigma \cdot L(z) = 25 \cdot L(1.75) = 25 \cdot 0.0162 = 0.405,$$

and thus the "new optimal order size" is

$$Q_1 = \sqrt{\frac{2 \cdot 200 \frac{\text{units}}{\text{y}} \cdot \left(50€ + 25 \frac{€}{\text{units}} \cdot 0.405 \text{ units}\right)}{2 \frac{€}{\text{units} \cdot \text{y}}}} \simeq 110 \text{ units}.$$

We can find the optimal solution $Q^* = 111$ e $R^* = 143$ with a type I service level $F(R) = 95.73\%$ by simply iterating the above process.

It is rather interesting to notice that the initial solution $Q_0 = $EOQ suggests an order size below the optimal level Q^*, as the EOQ model overlooks the cost of stockouts in equation (5.19). This is why the initial reorder level is higher than the optimal one and the initial type I service level (96%) is higher than the optimal one: The initial order size is relatively small and thus we face a relatively large number of planning cycles and therefore reduce the risk of stocking out in each order cycle.   ▯

We shall notice that in the above example the text provides data on expected demand and standard deviation over the LT. What shall we do when data refer to a different time horizon?

We can refer to example A.10 of page 464. The example shows that we simply need to scale demand expectation linearly. For instance, in case we want to derive the expected demand for a week given the total demand for the year, we must simply divide the demand expectation for the year by the number of weeks in a year (52).[23] As to the standard deviation, the example investigates the case of non-autocorrelated demand, that is, demand at time t is independent from demand in previous an successive periods. Under such an assumption, the standard deviation can be scaled back with the *squared root* rule. For example, the demand for the week is $\sqrt{1/52}$ times the standard deviation of demand for the year.

Example 5.20 Let us assume that the total year demand for a product is Normally distributed with an expected value 1000 and standard deviation 250. LT is two months. The optimal reorder point shall cover demand over the LT. But what is the distribution of demand over the LT? Actually, we do not know the shape of the distribution, but for the sake of simplicity we can assume that the distribution at the month level is Normal. We still need to find the expectation and the standard deviation of demand over a two-month period. We assume that the year consists of 12 identical months (of roughly 30 days) and assume the demand in each month is independent. Thus we can show that

$$\mathrm{E}\left[d_{LT}\right] = 1000/6 \approx 166.67; \qquad \sigma_{d_{LT}} = 250/\sqrt{6} \approx 102.06.$$

We shall notice that the model of demand shows some weaknesses. Indeed, while the probability of negative values of demand was negligible for the demand distribution for the year, it is sizeable for the demand for 2 months (the ratio $E(d)$ to σ is 4 for the year and 1.63 for the two-month period). This suggests that while the normal assumption for the demand for the year might work fairly well, this model shows some strains for the bimonthly demand. But we shall also notice that this might not be a crucial issue for inventory models. We typically consider fairly high service level targets. This means that what really matters is whether the demand distribution we adopt properly describes the right tail of the actual demand distribution. If the demand

---

[23] We assume a stationary demand and thus assume that all weeks share the same expected demand. So this finding does not apply directly to the case of significant seasonality.

distribution we choose fails to properly describe the left tail of the actual distribution, this tends not to be a problem for our purposes and we still can consider the model of demand fair enough, as it leads to nearly optimal decisions.   ∎

5.4.2   (Q,R) system: case of constraint on the type II service level

Often companies can hardly estimate the cost of a stockout and thus cannot write and minimize the total cost function. To keep under control the cost of stockout (that are still relevant even if we fail to accurately measure them), companies often set a minimum service level target. In an odd way this is an indirect estimate of the cost of a stockout, as high service level requirements imply a "gut feeling" that stockouts are expensive and vice versa. In such a situation we can resort to two approaches: One is simple and straight, while the other one is more complex, though better performing and more elegant.

Disjoint choice of Q and R   The first option is to split the total cost function into two disjoint parts.

$$
\begin{aligned}
C_{tot} &= C_{or} + C_{in} + C_{so} \\
&= A \cdot \frac{\mathrm{E}(d)}{Q} + h\left(R + Q/2 - \mathrm{E}(d) \cdot LT\right) + P_u \frac{\mathrm{E}(d)}{Q} \int_R^{+\infty} (x - R)\, f_{d_{LT}}(x)\, dx \\
&= \underbrace{A \cdot \frac{\mathrm{E}(d)}{Q} + h \cdot Q/2}_{\text{Economic Order Quantity}} \\
&\quad \underbrace{+\, h \cdot (R - \mathrm{E}(d) \cdot LT) + P_u \cdot \frac{\mathrm{E}(d)}{Q} \cdot \int_R^{+\infty} (x - R)\, f_{d_{LT}}(x)\, dx}_{\text{Cost constrained by the service level requirement}}
\end{aligned}
$$

Hence, a solution to this problem is to set $Q = EOQ$ to minimize the first part of the objective function, while the service level requirement keeps the second part of the cost function under control. In this case, we set a constraint on the type II service level as the size of the stockout rather than its occurrence matters. This approach basically makes the choice of Q independent from R and vice versa. So Q is set as if there were no uncertainty. R is the only control lever in charge of managing uncertainty and is set to reach the minimum service level target.

Example 5.21 Going back to example 5.19, we can assume that the company cannot properly estimate the cost of the stockout p_u and thus simply requires a 95% type II service level. The order size is set according to the EOQ model $Q = EOQ = 100$ units.

We now select the reorder point R in such a way that we meet 95% of demand from stock. In a planning cycle, the average demand is Q and the expected unsatisfied demand is $n(R)$. Thus the percentage of unsatisfied demand in each planning cycle is $n(R)/Q$. Since we want at the least a 95% type II service level, we set this to 5%. More generally, when we want to achieve a β type II service level, we want to set this ratio to $1 - \beta$:

$$1 - \beta = \frac{n(R)}{Q}; \tag{5.21}$$

hence

$$
\begin{aligned}
n(R) &= (1 - \beta) \cdot Q \\
&= (1 - 0.95) \cdot 100 = 5 \, \text{units},
\end{aligned}
$$

and thus we can leverage on the properties of the normal distribution and resort to the standardized normal distribution:

$$n(R) = 5 = \sigma \cdot L(z) = 25 \cdot L(z). \tag{5.22}$$

Thus $L(z) = 0.2$, $z = 0.49$ and $R = 100 + 0.49 \cdot 25 = 112.25$.   ▯

Though simple and straight, this approach is not optimal. Indeed, it fixes Q first and then uses only R to reach the service level target, while equation (5.21) shows that to reach a given service level target one can act on Q as well as R. Actually, it is easy to show that this approach leads to a suboptimal Q because it overlooks the effect of larger orders on the number of planning cycles and thus the number of potential stockouts and the unsatisfied demand.

Cost of the stockout implicitly estimated through a minimum service level requirement. A second and slightly more sophisticated approach is to optimize the parameters Q and R jointly to meet the service level target. This approach tries to elicit the cost of the stockout from the minimum service level requirement. When the service level required is relatively high, managers implicitly believe the stockout to be very expensive; while when they require a low service level they implicitly believe that the stockout is not as expensive. So we can use equation (5.20) to measure the cost of the stockout implicit in the managers' requirements:

$$\hat{p}_u = \frac{h \cdot Q}{\mathrm{E}(d) \, (1 - F_{d_{LT}}(R))};$$

this estimate can be used to derive the optimal quantity Q by substituting p_u with $\hat{p}_u$ in equation (5.19):

$$Q = \sqrt{\frac{2\mathrm{E}(d) \cdot \left[A + \dfrac{h \cdot Q \cdot n(R)}{\mathrm{E}(d) \cdot (1 - F_{d_{LT}}(R))} \right]}{h}},$$

that is, a quadratic equation in Q:

$$Q^2 = \frac{2 \cdot \mathrm{E}(d) \cdot \left[A + \frac{h \cdot Q \cdot n(R)}{\mathrm{E}(d) \cdot (1 - F_{d_{LT}}(R))} \right]}{h} = \frac{2 \cdot A \cdot \mathrm{E}(d)}{h} + \frac{2 \cdot n(R)}{1 - F_{d_{LT}}(R)} \cdot Q.$$

We can derive the optimal solution (we ignore the negative solution for obvious reasons):

$$Q^* = \frac{n(R^*)}{1 - F_{d_{LT}}(R^*)} + \sqrt{\left(\frac{n(R^*)}{1 - F_{d_{LT}}(R^*)} \right)^2 + \frac{2 \cdot A \cdot \mathrm{E}(d)}{h}}. \qquad (5.23)$$

We now need to simultaneously solve equation (5.23) and the condition on service level:

$$n(R^*) = L(z) \cdot \sigma = (1 - \beta) \cdot Q^*. \qquad (5.24)$$

Before we use the above equations to find an optimal solution, we shall notice that the order size is larger than the EOQ, since terms $\frac{n(R^*)}{1 - F_{d_{LT}}(R^*)}$ account for the cost of stockouts at the end of the planning cycle.

Equations (5.23) and (5.24) are not independent and thus we shall solve them through iterative methods. In this case, too, we can start with a reasonable order size $Q_0 = EOQ$ to then derive from equation (5.24) a first rough-cut reorder quantity R_0. We can then derive from equation (5.23) an improved order quantity Q_1. Finally, we iterate this process until the solution converges.

Example 5.22 Going back to example 5.21, we can start from $Q_0 = EOQ = 100$. We can use this order size to derive $R_0 = 100 + 0.49 \cdot 25 = 112.25$, as we have already shown. This solution can be used to calculate an improved order size Q_1

$$\begin{aligned} Q_1 &= \frac{5}{1 - F(0.49)} + \sqrt{\left(\frac{5}{1 - F(0.49)} \right)^2 + \frac{2 \cdot 200 \cdot 50}{2}} \\ &= \frac{5}{0.31} + \sqrt{\left(\frac{5}{0.31} \right)^2 + \frac{2 \cdot 200 \cdot 50}{2}} = 116.21; \end{aligned}$$

thus

$$n(R) = 5\% \cdot 116.21 = \sigma \cdot L(z) = 25 \cdot L(z),$$

and $L(z_1) = 0.232$, $z_1 = 0.395$, and $R_1 = 100 + 0.395 \cdot 25 = 109.875$. With one more iteration we get $Q_2 = 116.10$; $z_2 = 0.39$ $R_2 = 100 + 0.39 \cdot 25 = 109.75$, which is basically the steady state solution.

It is rather interesting to compare this solution with example 5.21. In this case we have a larger order size Q and a smaller reorder point R as both parameters are used to reach the target service level. As we have said multiple times, a similar solution process can be used for other demand distributions, *mutatis mutandis.* □

5.4.3 (Q, R) system: case of constraint on type I service level

Section 5.4.2 shows that, at times, penalty costs are hard to estimate and we resort to constraints on service level (see section 5.4.2).

$$
\begin{aligned}
C_{tot} &= C_{or} + C_{in} + C_{so} \\
&= A \cdot \frac{\mathrm{E}(d)}{Q} + h\left(R + Q/2 - \mathrm{E}(d) \cdot LT\right) + p\frac{\mathrm{E}(d)}{Q} \int_{R}^{+\infty} f_{d_{LT}}(x)\, dx \\
&= \underbrace{A \cdot \frac{\mathrm{E}(d)}{Q} + h \cdot Q/2}_{\text{Economic Order Quantity}}
\end{aligned}
$$

$$
\underbrace{+\, h \cdot (R - \mathrm{E}(d) \cdot LT) + p \cdot \frac{\mathrm{E}(d)}{Q} \cdot \int_{R}^{+\infty} f_{d_{LT}}(x) \cdot dx}_{\text{Cost constrained by the service level requirement}}
$$

In this case, too, we can split the cost function in two. The first part is basically the EOQ problem, while the second part depends on R and includes costs that can be kept under control through a constraint on service level. Thus we can set Q according to the EOQ model and set R in such a way that the minimum service requirement is reached.

Example 5.23 Going back to example 5.19, we can assume that the company has no reliable estimate of the penalty cost of a stockout p. So the company requires a 95% type I service level. Just like in previous examples, $Q = EOQ = 100$ units. Now we need to set R in such a way that the risk of a stockout in any planning cycle is 5%, so $F(R) = 0.95$. To solve this equation, we must know the demand distribution, and in this example we assume a Normal demand distribution. In the table of the standard Normal distribution, we can find z that guarantees a 95% type I service level: $z = 1.64$. Hence $R = \mathrm{E}(d_{LT}) + z \cdot \sigma = \mathrm{E}(d) \cdot LT + z \cdot \sigma = 100 + 1.64 \cdot 25 = 141$ units. We can compare this solution with example 5.21. A 95% service level leads to different reorder points R depending on whether the requirement is on the type I rather than type II service level. This example shows that a type I service level requires more inventories than a type II service level (at the least for a normal distribution and most "well-behaved" demand distributions).[24]

☐

---

[24] The type I service level considers a planning cycle where demand exceeds inventories (the reorder point R) by one unit basically a failure, as it counts as a stockout, full stop. On the contrary, the type II service level considers it as a fairly good result. Indeed, the type II service metric would consider the service level in such a planning cycle to be $1 - 1/Q$. So, in general, the type I service level sets tighter requirements and thus requires more inventories. The only exception to this general rule are demand distributions with a very long right tail. For example, when demand is very low (say 1) in most cycles (say 99%) and is very large (say 901) in others (1%) a very limited amount of inventories can be enough

This procedure is used very often in practice and in many basic software tools. However, we shall underline that it is suffers from structural problems. First the service level is measured as the probability of not stocking out in each planning cycle. For example, a 95% service level implies that we experience a stockout in only 5% of the planning cycles. This means that the (expected) number of stockouts in any given time frame (e.g., a year) depends on the number of planning cycles, that is, the number of chances to experience a stockout.

Example 5.24 Let us consider a company with a 95% service level. If on the average the company only places one order per year, then we have only a 5% probability of stocking out in a year and we expect to experience a stockout once each 20 years. On the contrary, if on the average we place an order a month, the probability of zero stockouts in a year drops to $0.95^{12} = 54\%$ and we basically face a stockout each 2 years.   ▯

5.5 PERIODIC REVIEW: S AND (s, S) POLICIES

Continuous review policies have a significant drawback: Different products reach the reorder level at different points in time. As section 4.6 suggests, coordinating orders among products might be very appropriate. Placing orders of various products at the same time can reduce some fixed cost (such as ordering or transportation) that the various products can share (joint economies of scale).

This is the reason why periodic review systems are often used. In periodic systems we place an order each τ periods. We assume we can observe the inventory levels periodically or, at least, we use the information on inventory levels to make planning decisions periodically.

To understand the mechanics of periodic review systems, we must first determine the relevant planning horizon to set inventory levels. A first, erroneous intuition might suggest that the planning horizon is the LT. So intuition might suggest that we set inventories according to the demand over the LT, like in the case of continuous review policies.

Example 5.25 A simple counterexample shows that setting inventories according to demand over the LT might be just not enough. Let us consider a company that orders some products once a year and receives them with a LT of one week. Setting inventories according to the weekly demand is definitely insufficient, as the company would run out of the products long before the end of the year and the next delivery, one year down the road.   ▯

to cover most planning cycles, but still it serves a small fraction of expected demand. In our example, $R = 2$ meets demand in 99% of the cases (type I service level = 99%) but meets a very small fraction of overall demand (type II service level = 10.1%). However, we shall notice that these are rather infrequent situations.

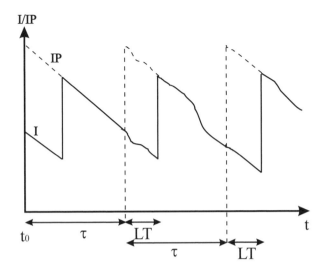

Fig. 5.8 The out-of-control period in periodic systems.

Example 5.26 Let us consider a company that orders stationery products once each 3 months to a supplier that delivers in a month. Let us focus our attention on the order the company places in early January. This order is delivered in early February. The next order will be placed in early April and will be delivered in early May. Thus the order placed in January (together with any existing inventories) shall cover demand until the next delivery, that is, the beginning of May. □

More in general, in periodic review systems, inventories shall cover demand over the LT plus the time τ between two consecutive orders (that is, also the time between two consecutive deliveries, given the deterministic lead times). Indeed, the order placed at time t_0 is delivered at time $t_0 + LT$, and the successive order is placed at time $t_0 + \tau$ and is delivered at time $t_0 + \tau + LT$ (see figure 5.8). Hence, the order issued at time t_0 shall cover demand over $LT + \tau$ periods. Such a period $LT + \tau$ is the so-called *Out Of Control period* (OOC period) since, once the order in $t = t_0$ is placed, inventories are out of our control and they depend only on demand fluctuations for $LT + \tau$ periods, until the quantity ordered in $t_0 + \tau$ is delivered in $t_0 + \tau + LT$ (notice that in the supply chain perspective demand is an exogenous variable we just want to meet, though we shall acknowledge that for the company at large it is a variable we can at least try to influence, e.g., through marketing efforts).

Concept 5.11 *In the case of periodic review systems, inventories shall cover the out-of-control period. In other words, when we place an order we shall cover demand up to the delivery of the next order. When orders are placed with a periodicity τ, the order shall increase the inventory position to a point where it is enough to cover demand over a period $\tau + LT$.*

Once we have introduced the core concept, we can define the basic periodic inventory policies:

- A first policy is S. Under this policy, each τ periods we place an order and the inventory position increases up to S. This is the so-called *base stock policy* or *order-up-to policy*.

- The main drawback of the S policy is that it can lead us to place very small orders in case the inventory position is barely below the quantity S. To avoid this problem and make sure orders are "large enough", we do not place orders if the inventory position is above s. So, under this policy called (s, S), each τ periods we check inventories. If inventory position is equal to or lower than s, we place an order and take the inventory position back to S, if the inventory position is greater than s we do not place any order. This way we make sure that the order quantity is at the very least $S - s$.

5.6   THE S POLICY

As usual, before we start modeling the inventory policy and try to set the parameters, we shall understand the basic dynamics of the policy. The inventory dynamics is described in figure 5.9. Let us consider time t_0, when we place an order. At time t_0 we place an order and immediately increase the inventory position up to S. Obviously, physical inventories do not increase on the spot. The quantity ordered, $S - IP_{t_0^-}$, is delivered at time $t_0 + LT$ (in this instant, inventory position equals physical inventories when $LT < \tau$).

When we are delivered, the physical inventory is a random variable that depends on demand over the LT. Thus we cannot tell the exact value of inventories at this point in time. We can just compute its expected value at time $t_0 + LT^+$: $\mathrm{E}\left(I(t_0 + LT^+)\right) = S - LT \cdot \mathrm{E}(d)$.

This is the highest inventory level we expect to have in our inventory (actually the maximum possibly conceivable level of physical inventories we might have is just S, when demand during the LT is zero.). Physical inventories start decreasing according to the demand, and they keep on going down up to next delivery at time $t_0 + LT + \tau$. Obviously, even the inventory level at this time is a random variable, so we cannot tell its exact value. We can just compute the distribution and the expected value $S - (LT + \tau) \cdot \mathrm{E}(d)$ (the maximum conceivable value is still S, when demand is zero over $LT + \tau$)[25]. This inventory level is called safety stock. Indeed, these are inventories one uses just when demand exceeds its expectation, that is they are designed to

---

[25] In this case as well we deliberately ignore the fact that inventories are a non-negative variable, thus the probability distribution and the expected level of inventories is slightly different. We refer the reader to footnote 20 on page 274.

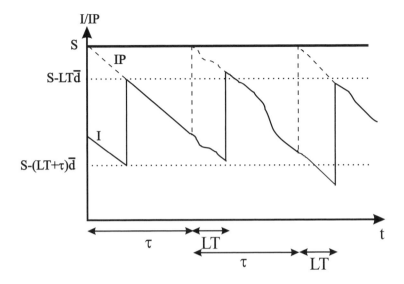

Fig. 5.9 Dynamics of inventories in the case of S policy.

manage demand uncertainty. So physical inventories are expected to fluctuate between $S - LT \cdot E(d)$ and $S - (LT + \tau) \cdot E(d)$ in each cycle. So the quantity ordered on the average is $\tau \cdot E(d)$ and average physical inventory is $S - (LT + \tau/2) \cdot E(d)$.

Once we have understood the dynamics of the system, we can try to set the control parameters τ and S. A first difference between the S and the (Q, R) policy is that once we have set τ, the ordering cost are deterministically fixed, while in the previous case Q had an influence on the expected ordering cost that still was a random variable that depended on demand.

Like in previous sections we can try to write the total cost function:

- The *ordering cost* is equal to the fixed ordering cost A divided by the ordering period τ that determines how often we incur the fixed cost A:

$$C_{or} = \frac{A}{\tau};$$

- The *inventory holding cost* is equal to the unit holding cost h times the average inventory level $S - (LT + \tau/2) \cdot E(d)$:

$$C_{in} = h \cdot (S - (LT + \tau/2) \cdot E(d)).$$

- Finally, as we have already seen in previous sections, we shall compute the cost of a stockout. When the cost of the stockout depends on the size of the stockout, the stockout cost is

$$C_{so} = \frac{p_u}{\tau} \cdot \int_S^{+\infty} (x - S) f_{d_{LT+\tau}}(x) \, dx; \qquad (5.25)$$

while when the cost of the stockout depends on the occurrence of the stockout, the expected cost in a period is

$$C_{so} = \frac{p}{\tau} \cdot \int_S^{+\infty} f_{d_{LT+\tau}}(x)\, dx. \tag{5.26}$$

Interestingly, both in equation (5.25) and in equation (5.26) the relevant demand distribution refers to demand in the out-of-control period $\tau + LT$.

From the previous equations we can derive the total cost function. In this section we only investigate the case where the cost of the stockout depends on the size of the stockout. The concepts and solutions we suggest for this case apply to the other case as well. We leave the derivation of the second case to the reader.

$$C_{tot} = \frac{A}{\tau} + h \cdot [S - (LT + \tau/2) \cdot \mathrm{E}(d)] + \frac{p_u}{\tau} \cdot \int_S^{+\infty} (x - S) f_{d_{LT+\tau}}(x)\, dx \tag{5.27}$$

Optimization problems in the S policy.   Now we can take the derivatives of the total cost function with respect to τ and S to identify the optimal control parameters. Optimal values of control parameters τ and S satisfy the equations below:

$$
\begin{aligned}
\frac{\partial C_{tot}}{\partial \tau} &= \frac{A}{\tau^2} - \frac{h \cdot \mathrm{E}(d)}{2} - \frac{p_u}{\tau^2} \cdot \int_S^{+\infty} (x - S) f_{d_{LT+\tau}}(x)\, dx \\
&\quad + \frac{p_u}{\tau} \cdot \frac{\partial \int_S^{+\infty} (x - S) f_{d_{LT+\tau}}(x)\, dx}{\partial \tau} \\
&= \frac{A}{\tau^2} - \frac{h\mathrm{E}(d)}{2} - \frac{p_u}{\tau^2} \int_S^{+\infty} (x - S) f_{d_{LT+\tau}}(x)\, dx \\
&\quad + \frac{p_u}{\tau} \int_S^{+\infty} (x - S) \frac{\partial f_{d_{LT+\tau}}(x)}{\partial \tau}\, dx = 0; \tag{5.28}
\end{aligned}
$$

$$
\begin{aligned}
\frac{\partial C_{tot}}{\partial S} &= h + \frac{p_u}{\tau} \frac{\partial \int_S^{+\infty} (x - S) f_{d_{LT+\tau}}(x)\, dx}{\partial S} \\
&= h + \frac{p_u}{\tau} \int_S^{+\infty} [-f_{d_{LT+\tau}}(x)]\, dx \\
&= h - \frac{p_u}{\tau} \cdot [1 - F_{d_{LT+\tau}}(S)] = 0. \tag{5.29}
\end{aligned}
$$

While equation (5.29) does not pose significant problems, the last term in equation (5.28) can be problematic. The first two terms in equation (5.28) resemble the EOQ problem, as they capture the effect of an increase in τ

on the ordering cost and on the inventory holding cost. The last two terms capture the impact of a (marginal) increase in τ on the stockout cost. In particular, the term $-\frac{p_u}{\tau^2} \cdot \int_S^{+\infty} (x - S) f_{d_{LT+\tau}}(x) \, dx$ shows that as the time between orders increases, the number of stockouts decreases simply because the number of cycles and thus the number chances to experience a stockout decreases. This term does not set specific challenges when we want to find an optimal solution. The term $+\frac{p_u}{\tau} \cdot \int_S^{+\infty} (x - S) \frac{\partial f_{d_{LT+\tau}}(x)}{\partial \tau} \, dx$ tells the effect of τ on the demand distribution during the out of control period. This is all but trivial, as both the expected demand and the standard deviation of it change as τ changes. Also, as τ changes, the shape of the demand distribution might change (though we shall remind you that the sum of normal demand distributions is still normal and thus in case of normal identically distributed distribution we can obtain this function) and thus we cannot really write the function in the general case. Then taking the integral of this derivative might not be trivial.

One reasonable heuristic. For the above computational problems we often resort to heuristics. In particular, we hardly can find the optimal value for τ. So we set it to minimize the first two terms in equation (5.28). Then the ordering frequency is $\tau = \sqrt{\frac{2A}{h \cdot \mathrm{E}(d)}}$, that is, exactly the ordering frequency that we achieve when we try to order, on the average, the EOQ (when several products share the same fixed order A, we can use the process described in section (4.6.2) and then set the order frequency τ as the ratio between the optimal quantity Q^* and the expected demand $\mathrm{E}(d)$ for the bundle). [26] In practice we then round τ so that it can be managed in a real context: It is very hard to place an order each 1.7313 weeks. Instead, one orders each week or once each 2 weeks.

Once we have set τ, S is the only control lever left. The optimal level of S can be derived from equation (5.29):

$$F_{d_{LT+\tau}}(S^*) = \frac{p_u - h \cdot \tau}{p_u}. \tag{5.30}$$

This result might look odd at first sight: Why is the holding cost multiplied by τ? Is not $\tau + LT$ the relevant time frame to set inventories?

A more careful reading of the equation sheds some light on its meaning. The manager of the warehouse keeps an additional (marginal) unit of inventories in the warehouse for the whole cycle τ between two successive deliveries just in case it is needed at the end of the cycle to avoid a stockout or at least

---

[26] Notice that in this case the mix of products in each single order can change according to demand. Indeed, if we keep the mix of purchases fixed, any random fluctuation in the mix of demand turns into fluctuations in the mix of inventories. So while we set the control parameter τ according to bundles of products we do not actually buy in bundles with a preset mix.

reduce its size. So the holding cost over a whole cycle $h \cdot \tau$ counterbalances the reduction in the cost of a stockout p_u.[27] Also, as usual we should notice that our findings required no assumption on the demand distribution so far. Now we need to make assumptions on the demand distribution to find the value of S that satisfies equation (5.30). To solve this equation, we need to know the cumulative demand d distribution $F_{d_{LT+\tau}}(\cdot)$

$$S = F_{d_{LT+\tau}}^{-1} \left(\frac{p_u - h \cdot \tau}{p_u} \right).$$

A similar reasoning applies to the case of cost of the occurrence of a stockout. One we can still set the review period τ in such a way that we tend to order the EOQ quantity. But in this case the optimal S is

$$S = f_{d_{LT+\tau}}^{-1} \left(\frac{h \cdot \tau}{p} \right).$$

S policy with a constraint on service level.   Also in the case of the S inventory policy, we might have a hard time estimating the cost of a stockout (p or p_u) and thus we might implicitly estimate it through constraints on service level.

When the constraint is on the type I service level, we call γ the minimum service level required. In this case, the solution is rather trivial. We should simply set γ in such a way that the probability that demand is lower than S is γ (i.e., the probability of a stockout in a cycle is $1 - \gamma$):

$$S = F_{d_{LT+\tau}}^{-1} (\gamma). \tag{5.31}$$

Example 5.27 Let us consider a company that places orders to international suppliers once a month. The suppliers deliver in 3 months. Let us assume that monthly demand follows a normal distribution with mean 200 and standard deviation 40 and is not autocorrelated over time. Also, let us assume we have no reliable estimate of the cost of a stockout and thus require a minimum type I service level of 98% to limit the number of stockouts. In this case, S shall be high enough to make sure that inventories are higher than demand

---

[27] Notice that again we assume that the inventories are held during the whole cycle, whereas they might be sold before the end of the cycle and, in this case, they are going to be held for less than a cycle. So our model makes a simplifying assumption (see footnote 20). Also, it is interesting to read the relationship between the above equation and the newsvendor model. In this case the profit we gain when we sell the marginal unit is p_u. However, when that happens, we still hold the marginal unit for τ. So the actual net marginal profit is $m = p_u - h \cdot \tau$. On the contrary, in case we cannot sell the marginal unit in the cycle τ it generates a cost $c = h \cdot \tau$. So when we apply the newsvendor formula in the correct way (i.e., net of holding costs we ignore in the newsvendor problem), we reach the same results we have derived here.

over the out-of-control period in 98% of the cases. Thus S is simply equal to
$S = 4 \cdot 200 + z(0.98)\sqrt{4} \cdot 40 = 800 + 2.05 \cdot 80 = 964$.

Thus S is $S = 4 \cdot 200 + z(0.98)\sqrt{4} \cdot 40 = 800 + 2.05 \cdot 80 = 964$. $\quad\square$

If the constraint is on the type II service level, we call the minimum service level δ. In this case the situation is slightly more complex. Given the constraint, only a $1 - \delta$ portion of the demand in a cycle shall be not met from stock. If the order is placed each τ periods, the average demand in a cycle is $\tau \cdot E(d)$. Thus demand we expect not to meet in a cycle is $\tau \cdot E(d) \cdot (1 - \delta) = n(S)$ at maximum. If demand happens to be normally distributed, we have $n(S) = L(z) \cdot \sigma_{\tau+LT}$. As the above equation shows (see the term $\sigma_{\tau+LT}$), the inventory level S depends on the demand over the whole out-of-control period $LT + \tau$, though we compare the demand not met to the total demand in a cycle τ. Also, in this case, a simple example can help us make sense of this apparently odd concept.

Example 5.28 Let us go back to example 5.27. We just slightly change our assumptions: We assume that the 98% minimum service level requirement refers to a type II service level. In other words, 98% of customers shall find in stock the product they want. This means that only 4 customers a month (or less) should not find the product they want. Thus we know that the expected demand unmet before the next delivery (delivery frequency is equal to the review period of one month, given the deterministic lead times) shall be equal to 4 units.

Having said this, we shall now find the right level of S to reach that minimum requirement. For example, when we place an order in early January, it is going to be delivered in early April. The delivery quantity shall cover demand up to the beginning of May. So the inventory position S shall be large enough to cover demand up to the beginning of May; that is, it shall cover the out of control period of four months.[28]. Given our assumptions, the mean of the demand over the $LT = 4$ months is 800 units, while the standard deviation is 80. Also, we know that demand is normal and the demand we expect not to meet

---

[28]Notice that the backorder assumption is crucial. Indeed, if unmet demand gets lost, the distribution of sales is not exactly equal to the demand over the out-of-control period. Let us a make an odd, though telling, example. Let us assume that for some reason we are out of stock and we forgot to place any order over the last three months. So basically in early January we place an order that is going to be delivered in early April. Up to that time, the whole demand is going to get lost. So it does not make sense to increase the inventory position to meet demand over the whole out-of-control period as basically demand is going to get lost over the first 3 months. In this case we simply set S (and thus the order quantity) to meet demand during the month of April. More in general, in the lost sales case we set the ordering quantity to meet demand in a period τ. The demand over the LT still plays a role, as it influences the inventory level when the order is delivered and thus the total units available to meet demand in cycle. Finally, notice that the lost sales and the backorder case are very different in the case of very low service levels, like in the case we have just discussed in this note. However, they behave rather similarly in the more common case of a very high service level.

in any cycle is $n(S) = L(z) \cdot \sigma_{LT+\tau} = 4$. So we shall select $z \mid L(z) = \frac{4}{80} = 0.05$. In the tables for the normal distribution we can select $z = 1.255$. Thus, $S = E(d_{LT+\tau}) + z \cdot \sigma_{LT+\tau} = E(d) \cdot (LT+\tau) + z \cdot \sigma_{LT+\tau} = 800 + 1.255 \cdot 80 = 900.4$.

$\square$

Example 5.29 We go back to example 3.14 from page 126 to see how we can integrate the forecasting process with inventory planning. In example 3.14 we have selected the moving average with step 5 ($k = 5$) as our forecasting algorithm. Also, while selecting the method we estimated that it yields a RMSE of 11.90. We basically consider this to be the uncertainty in the demand estimation. Let us now assume we are in period 24 and want to plan the deliveries for period 26. Our point forecast (i.e., our predicted demand) is the average of the last 5 demand observations.

$$E(d) = F_{24,h} = (103.58 + 87.95 + 110.83 + 103.87 + 115.57) / 5 = 104.36.$$

As discussed in example 3.14, the LT is 2 days and the delivery frequency is daily. So we face an out-of-control period of 3 days. Thus we shall set the S parameter to meet a demand over three days. The expected demand over the out-of-control period is $3 \cdot 104.36 = 313.08$. As to the standard deviation of demand, as usual we shall make some assumptions on the correlation among demand observations over the out of control period. If we assume demand fluctuations to be independent, we can use the square root rule. In this case, the standard deviation of demand in the out-of-control period is $\sqrt{3} \cdot 11.9 = 20.61$.[29]

Now let us assume that we want to achieve a 95% type II service level. This really means that in each cycle only 5% of demand can be lost. The order cycle is daily, so we can loose up to 5% of daily demand. Thus $n(S) = 5\% \cdot 104.36 = 5.22$ So we shall select $z \mid L(z) = \frac{5.22}{313.08} = 0.0166$. In tables for the normal distribution we can select $z = 1.74$. Thus, $S = 313.08 + 1.74 \cdot 20.61 = 348.94$.

$\square$

5.7 THE (s, S) POLICY

The order up to policy S is often used, but it can lead to quite irrational replenishment decisions, as we might order a very small quantity when inventories are just slightly below the order up to level S. In these cases, one might prefer not to order to save the fixed ordering cost A and slightly reduce the service level he/she offers.

---

[29] Actually if one wants to be precise he/she shall measure the error we make with a forecasting horizon of one, with a forecasting horizon of two and with a forecasting horizon of three and then check whether the three errors are correlated. Here we implicitly assume that the error we make with a 2 days horizon is equal to the error we make with a one day and a three days horizon.

This is the reason why the (s, S) policy was designed. Under this policy, each τ periods we check the inventory position. If the inventory position is above s, we do not place any order. Otherwise, we place an order and take the inventory position up to S. Though this policy can be very effective (the S policy is just a special case of the (s, S), so the latter is by definition more flexible and potentially can perform better), it is actually quite hard to model for several reasons:

- The number of orders placed in a given time frame is a random variable that depends on the number of times the inventory position goes below s at the points in time $(t_0, t_0 + \tau, t_0 + 2\tau, t_0 + 3\tau \ldots)$ when we check it.

- The expected service level offered in each planning cycle is actually a random variable; while in some planning cycles we take the inventory position up to S with an order, in others the starting inventory position is below S (though above s) and thus the service level (no matter whether type I or II) decreases.

- Even calculating the average inventory level is actually all but trivial; we know that the inventory position reaches S and goes below s. However, it is hard to estimate the distribution of the inventory position. Indeed, we do not know whether it depends on demand over $LT + \tau$ (in case the inventory position goes below s during the first order cycle, that is the inventory position is below s in $t_0 + \tau$), or on demand over $LT + 2\tau$, $LT + 3\tau$, etc. (in case this happens in two, three, or more cycles).

For these reasons we often resort to heuristics. The most frequently recommended is to refer to the (Q, R) model as follows:

- The quantity ordered is the fixed quantity Q in the (Q, R) policy. In the (s,S) model, on the contrary, the quantity ordered is variable. So we try to set the parameters in such a way that on the average we order Q units. One approximation we can make is $Q = S - s$. Such an approximation basically assumes that when we place an order the inventory position is exactly s. When τ is relatively small and thus we check the inventory position rather frequently, this is a minor issue. In those cases where the inventory position is actually below s, the gap between our simplifying assumptions and reality is negligible. When τ is substantial and thus we check the inventory position once in a while, the inventory position might be significantly below s and thus the average order can be significantly larger than $S - s$.

Example 5.30 Let us assume

- $\tau = 1$ month;
- monthly demand is normally distributed with mean 100 and standard deviation 10;

- demand shows no autocorrelation;

- $S = 250$;

- $s = 100$.

Finally we assume, for the sake of simplicity, that an order is placed at time $t = 0$ and the inventory position reaches 250. At time $t = 1$, after one month, the inventory position is a random variable, that is, Normally distributed with mean 150 (250–100) and standard deviation 10. This really means that the probability that at time $t = 1$ inventories are below $s = 100$ is actually negligible (the expected value of inventories is 5σ above the threshold s). The next month, at time $t = 2$ the probability distribution of the inventory position is a normally distributed random variable with mean 50 (250−2·100) and standard deviation 14.3 (10·$\sqrt{2}$) and we have an almost 100% probability of placing an order. So the order is very likely to be placed at time $t = 2$ and though its size is actually a random variable, the expected value is actually $S = 250$ minus the expected level of the inventory position at time 2 (50 units). So, though $S − s = 150$ units, the expected order size is 200 = 250 − 50. This large difference is due to the relatively low frequency of inventory control. As the frequency of control increases and the time between inventory controls τ reaches 0, the periodic review system is more and more similar to the continuous review ones and thus the approximation to the (Q, R) policy is more and more effective. ⬚

- The threshold s plays a role that resembles the reorder point R in the (Q, R) policy, so we can set s at the same level we would have selected for R in a continuous review system: The key idea is to make the (s, S) mimic the (Q,R) system. In this case too, the question is how good the approximation is. For this control parameter as well, the issue is whether the order is placed exactly when the inventory position is s or is significantly below it. In this case, too, the larger the value of τ, the more the approximation is crude.

To solve the problems of this simple heuristic we can, (i) increase the frequency of periodic controls (reduce τ), (ii) perform more sophisticated statistical analysis to capture the actual inventory position when we place an order, (iii) resort to simulation to check the performance that various levels of the control parameters (τ, s, and S) can generate, or finally, resort to dynamic programming. All options lay outside the scope of this introductory book.

S.5.8 OPTIMIZATION OF THE (Q,R) MODEL WHEN THE COST OF A STOCKOUT DEPENDS ON THE OCCURRENCE OF A STOCKOUT

We can try to jointly optimize the two parameters Q and R also when the cost of a stockout depends on the occurrence of a stockout. We can follow a process that resembles 5.4.1. We write the total cost function (5.16) and compute its derivatives with respect to Q and R:

$$\frac{\partial C_{tot}}{\partial Q} = -\frac{A \cdot \mathrm{E}(d)}{Q^2} + \frac{h}{2} - \frac{p \cdot \mathrm{E}(d)}{Q^2} \int_R^{+\infty} f_{d_{LT}}(x)\, dx$$

$$= \frac{h}{2} - \frac{\mathrm{E}(d)}{Q^2} \cdot (A + p \cdot (1 - F_{d_{LT}}(x))) = 0,$$

$$\frac{\partial C_{tot}}{\partial R} = h - \frac{p \cdot \mathrm{E}(d)}{Q} f_{d_{LT}}(R) = 0$$

and derive conditions for optimality:

$$Q^* = \sqrt{\frac{2\mathrm{E}(d) \cdot [A + p \cdot (1 - F_{d_{LT}}(R^*))]}{h}}, \tag{5.32}$$

$$f_{d_{LT}}(R^*) = \frac{Q^* \cdot h}{p \cdot \mathrm{E}(d)}. \tag{5.33}$$

In this case, too, the optimal lot size Q suggests that the cost of a cycle is greater than the ordering cost A, since in each cycle we run the risk of stocking-out. The one difference is that in this case the cost of the stockout in a cycle depends on a probability $(1 - F_{d_{LT}}(x))$ rather than the expected level of unfulfilled demand $n(R)$. Also, the condition for optimality resembles equation (5.20) since the marginal cost of inventories h is compared to the reduction in the stockout cost $p \cdot f_{d_{LT}}(R)$ in each of the $\mathrm{E}(d)/Q$ planning cycles. While in the case of (5.20), $F(R)$ is a growing function of R, $f_{d_{LT}}(R)$ is not a monotonous function of R (at the least not for all density functions). On the contrary, for all symmetric demand distributions, if there is one solution to equation (5.33) there must be at the least another one.

Example 5.31 For example let us consider a normal demand distribution with mean of 100 units and a standard deviation of 20 units. We can reach a $f_{d_{LT}}(R) = 0.015$ both for $R = 85$ and $R = 115$ (see figure 5.10).   ☐

Intuitively, while one of the solutions is a maximum, the other is a minimum of the cost function. Indeed, in a symmetric demand distribution, the marginal savings from a marginal increase in inventories dR is $p \cdot f_{d_{LT}}(R) \cdot dR$ for each planning cycle. For R below the expected level of demand, we face increasing returns for our investment in inventories since in this range $f_{d_{LT}}(R)$ is a growing function of R. This rules out all points below the mean as potential

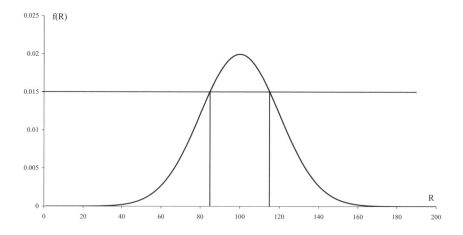

Fig. 5.10 Condition for optimality on the reorder point R.

candidates for optimality (more in general this rules out all points between a minimum and a maximum and below the first maximum). Indeed, the second derivative is

$$\frac{\partial^2 C_{tot}}{\partial^2 R} = -\frac{p \cdot E(d)}{Q} f'(R).$$

Thus where the density function is growing ($f'(R) > 0$) we have stationarity points that are either maximums or saddle points of the cost function. Once again, economic intuition and math perfectly match.

Example 5.32 Let us go back to example 5.31. Let us assume the 85th unit generates a reduction in the cost of stockouts that justifies the extra holding cost h. Then all units 86 to 135 are worth the inventory investment. Indeed, they share the same cost of inventories h but generate an even greater reduction in the cost of the stockout (since the probability density is greater than in the case of the 85th unit).   ⧠

This cost structure suggests that in the case of symmetric demand distributions (and more in general in the case of demand distributions with a mode greater than zero) the cost function has a local minimum in $R = 0$ that shall always be considered as a potential candidate. It shall then be compared with those points to the right of the mode(s) (in the case of symmetric distributions with one mode, it is also the mean) that satisfy equation (5.33). When $R = 0$ is the optimal solution, the company deliberately decides to experience a stockout in each and every planning cycle. This raises the cost of a planning cycle to $A + p$ and leads the company to increase the order quantity Q up to

$$\left[Q^* = \sqrt{\frac{2E(d) \cdot (A+p)}{h}} \right].$$

Also, there is no guarantee that the density function is such that at least one point satisfies equation $\left[Q^* = \sqrt{\frac{2\mathrm{E}_{(d)} \cdot (A+p)}{h}} \right]$. Indeed, if demand is very uncertain (and thus the demand distribution is very flat) the marginal gain from a marginal increase in inventories might be very low and it might not justify the investment in inventories.

Even from the standpoint of the solution process, the situation is all but trivial. As discussed in section 5.4.1, equations (5.32)–(5.33) can only be used through iterative methods, as they are not independent. However, in this case we might want to start our search procedure from the maximum of the density functions. In other words, we set R_0 equal to the mode of the demand distribution. We start our search from the single point with the largest return on the inventory investment. This first rough-cut estimate of R is then used in equation (5.32) to get a first rough-cut estimate of Q^*, Q_0. We can then use this estimate Q_0 to get a better estimate R_1 of R^*, and so on.

REFERENCES

1. K.S. Azoury. Bayes Solutions to Dynamic Inventory Under Unknown Demand Distributions. *Management Science*, 31(9):1150–1160, 1985.

2. D. Corsten and T. Gruen. Desperately Seeking Shelf Availability: An Examination of the Extent, the Causes, and the Efforts to Address Retail Out-of-Stocks. *International Journal of Retail & Distribution Management*, 31:605–617, 2003.

3. N. DeHoratius, A.J. Mersereau, and L. Schrage. Retail Inventory Management when Records Are Inaccurate. Available from web page `http://faculty.chicagogsb.edu/adam.mersereau/research/`, filename `dehoratius_mersereau_schrage_061023.pdf`, 2006.

4. M.L. Fisher and A. Raman. Reducing the Cost of Demand Uncertainty through Accurate Response to Early Sales. *Production and Operations Management*, 44(1):87–99, 1996.

5. C. Nevison and M. Burstein. The Dynamic Lot-Size Model with Stochastic Lead Times. *Management Science*, 30(1):100–109, 1984.

6. A. Raman, N. DeHoratius, and Z. Ton. The Achilles' Heel of Supply Chain Management. *Harvard Business Review*, 5, 2001.

7. A. Raman and Z. Ton. *Borders Group Inc., case 9-601-037*. Harvard Business School Publishing, Boston, MA, 2003.

8. A. Raman and Z. Ton. *Operational Execution at Arrow Electronics, case 9-603-127*. Harvard Business School Publishing, Boston, MA, 2005.

9. Y. Wang and Y. Gerchak. Periodic Review Production Models with Variable cost, Random Yields and Uncertain Demand. *Management Science*, 42(1):130–137, 1996.

10. D.C. Whybark and J.G. Williams. Material Requirement Planning under Uncertainty. *Decision Sciences*, 7(4), 1976.

6

Managing Inventories in Multiechelon Supply Chains

6.1 INTRODUCTION

In the two previous chapters we have studied inventory management for a single warehouse. However, most distribution (as well as manufacturing) systems consist of more than one echelon. This makes the problem more complex and makes modeling harder. Indeed, in a multiechelon system the optimal inventory policy depends on both inventories and inventory policies of all other warehouses in the system, both upstream and downstream.

Example 6.1 In the food industry a typical supply chain has several echelons. Upstream, manufacturers tend to centralize production in a relatively small number of plants to enjoy economies of scale that in this industry (like most process industries) are quite sizeable. In Europe, a single plant can serve the whole continent (for high-value products such as yogurt and fresh pasta) or a whole country (one exception is that of low value per kg products such as drinking water whose market tends to be fairly local). These large plants feed warehouses for finished products within the plant. These warehouses feed distribution warehouses where products coming from various plants of the same manufacturer are held. These distribution warehouses feed the central distribution centers (CDC) of the retail chains. The manufacturer's distribution warehouses guarantee frequent and quick deliveries to the retailers that cannot be achieved from the central, and thus on the average far, production plants. Also, local distribution centers decouple production lots from distribution quantities. Production needs fairly large lots of a relatively limited assortment (for any given plant) to run smoothly while

303

customers want frequent deliveries of a variety of products. Local warehouses make these apparently noncompatible ends meet.

The retailers' CDC finally deliver to the single stores (at times through a network of independent agents or distributors). Finally, in a broader perspective the supply chain ends with the inventories in the refrigerators and cupboards of the final consumers. This might look like an irrelevant issue since the amount of inventories in each house is very small. However, the number of these small warehouses called houses and apartments is enormous. For example, to understand the demand pattern for some basic and price-sensitive fast-moving consumer goods the time between promotions is a key variable. For example, in the now classic case of diapers, the final consumer's consumption is actually very stable and predictable. Many parents willing to keep the cost of these fairly expensive items tend to buy in large quantities during promotions. So, when one wants to understand whether the next promotion is going to be successful, one should consider not only the discount offered, but also the time since the last promotion (for the same item and/or for the product category in case customers are willing to switch among brands). This variable catches the inventory level at the consumers' place and thus their willingness to buy a large quantity. Indeed, in this case the price reduction does not increase the aggregate consumption (families with no children hardly buy diapers because they are cheap) but rather pushes consumers to concentrate purchases over time and at a single retailer.   ☐

Moreover, multiechelon distribution system might involve more than one organization. For example, in a distribution chain we might have a producer of raw materials, a manufacturer of the finished product, wholesalers, and retailers. Each of these players has his/her own economic objectives that might not fully overlap with those of other members of the supply chain. So when we study multiechelon supply chains, we shall not only look for the optimal plan but we should also wonder what are the objectives each of the players is trying to reach and ask ourselves whether they are compatible or conflicting.

In other words, to fully understand and optimize a supply chain, we shall merge the perspectives of an economist and the perspective of an engineer.[1].

Basically a classic engineer tries to find the best possible algorithm to find optimal solutions to nontrivial problems. The perspective of the engineer is that managers of a supply chain are not bright enough to run their supply chain so they need some support from algorithms and computers to design better plans. So in the engineers' mind, men and women are very willing to implement optimal solutions, if somebody suggests such optimal solution to them.

---

[1] Professor A. Raman of the Harvard Graduate School of Business originally developed this telling example.

The classic perspective of an economist is actually quite the opposite. Economists believe that men and women are extremely brilliant "economic beasts." For example, economists assume that people immediately change their willingness to spend or borrow money according to the interest rates. Unfortunately, economists acknowledge that these economic beasts are rather selfish and are only interested in their welfare rather than in the performance of the whole supply chain. This is the reason why for an economist an optimal plan might be quite likely not to be ever implemented for a very simple reason: It might not be good for one (or more) of the players in the supply chain that might have the power to call the plan off.

This really means that to manage a multiechelon supply chain with several decision makers and organizations, we shall definitely design optimal plans (i.e., be good "engineers") but at the same time we have to design a network of contracts, incentives, and ways to share the benefits of the plan that makes sure the plan actually is implemented and can improve actual performance (i.e., be a good "economist").

Concept 6.1 *In multiechelon supply chains we shall (i) design rules and algorithms to identify good solutions to rather complex problems, and (ii) design contracts or incentives in such a way that all relevant players are willing to implement these solutions and push in the same direction.*

The next chapter analyzes the relevance of conflicting objectives and incentives in a distribution chain. In this chapter we assume that managers of the warehouses in the system belong to one single organization or at the very least that the various organizations have agreed on a system of incentives to share the benefits of an optimal solution that makes them all very willing to minimize the total cost of the supply chain.[2]

Example 6.2 The case of large grocery retail chains can be insightful. When a product stocks out, the average customer is very likely to switch to a substitute product. For example, when the consumer was looking for a specific kind of chips, he/she might switch to a different brand, to the private label, or to another snack. That is why the stockout might create a limited damage to the retailer (at least in the short run) that is still very likely to sell some sort of snack. On the contrary, in the manufacturer's perspective this is a loss of margin and turnover. What makes things worse, is that this stimulates consumers

---

[2]Notice that we deliberately use the term organization rather than firm. Indeed, often working for the same firm is just not enough for people to share the same objectives. For example, in a large multinational manufacturer of white goods the managers of some European subsidiaries are not willing to share demand and inventory data with the managers of the central logistic center for spare parts, as they fear that these pieces of information might be used against them. Such lack of information definitely worsens the overall performance of the company, but some European subsidiaries still think this situation is in their best interest. These subsidiaries behave like independent organizational units.

to try substitute products that they might like. This creates a misalignment between the objectives of the manufacturers and retailers. While producers have all kinds of incentives to reduce the frequency of stock-outs, retailers pay for holding the inventories in the stores and in their warehouses. Quite interestingly, some recent studies on stock-outs in grocery retailing show that, on the average, 8% of the products in a supermarket are stocked out. Retailers seem to be OK with such an apparently bad performance (given the high traffic and relatively high volumes of these stores) while most manufacturers fell off the chair and believe this to be an unacceptable performance.   ⧠

Multiechelon systems can be very diverse and thus can set a wide variety of problems and issues.

A first variable we can use to classify them is the *structure of the distribution system*. In chapter 1 we have shown that a multiechelon system can be:

- *linear*, if each warehouse receives goods from a single supplier and ships goods to at most a single warehouse;

- *distributive*, *divergent* or *arborescent*, if each warehouse receives goods from at most a single warehouse but can ship to more than a single warehouse;;

- *assembly* or **convergent**, if each warehouse delivers to at most a single warehouse but can receive goods from various warehouses.;

Obviously, these are the basic structures, while in the more general case each single warehouse can both ship to and receive from many other warehouses.

In the case of linear, arborescent, or convergent systems we can define the number of echelons in the system. In the remainder of this chapter we number the echelons from downstream. So the first echelon of the supply chain serves the final customer. Warehouses that serve the first echelon are in the second echelon, and so on (for an example of this numbering of the echelons see figure 6.1). In this chapter we only investigate supply chains with 2 echelons since the complexity of the model increases significantly as the number of echelons grows. However, the analysis of this relatively simple problems gives us a chance to shed some light on some concepts that apply to the more general case of supply chains with 2 or more echelons.

Example 6.3 Going back to example 6.1, the distribution chain for a given grocery product is arborescent, since a single production plant serves several local warehouses; each of them serves several retailers' central distribution centers that, in turn, serve several stores. If we consider the consumers to lay outside of the boundaries of the supply chain (to some extent the definition of what lays outside and what lays inside the supply chain is arbitrary), we

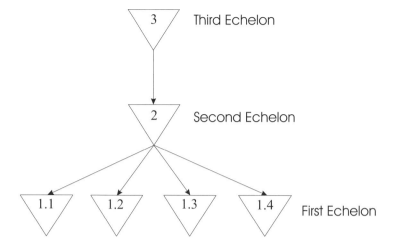

Fig. 6.1   Example of a three echelon supply chain.

can identify 4 echelons in the chain. In this case, stores are the first echelons, and warehouses at the production plants are the fourth echelon.   ☐

Our definition of the echelons depends on the flow of materials among warehouses. Often, though, echelons are also different in nature. For example, the first echelon often consists of stores that differ from other warehouses in several ways. First, consumers enter these warehouses while they do not visit other warehouses. This also means that safety measures are tighter in this echelon than in others. Also, this specific kind of warehouses looks really different, as they are designed to raise the interest of the consumer. In the second level we have a warehouse that differs from other echelons because inventories held come from a variety of brands and manufacturers. Also, inventories held there are (typically) on the retailers' balance sheet.

When we define the echelons of a supply chain through the characteristics of the warehouses (say stores vs. non-stores) we can define *horizontal transshipments*; that is, shipments among warehouses that belong to the same echelon.[3]

Example 6.4 In retailing, in luxury business, and in the apparel/footwear business in particular, horizontal shipments among stores in a single city or

---

[3]Indeed, if we define echelons through the flows of goods and two stores exchange goods we must simply draw the conclusion that the structure is neither linear, nor arborescent, nor convergent and we cannot define the echelons. If store A ships to store B, B should be in the first echelon and A in the second, though this contrasts with the fact that A also sells to the final consumer. Things get even worse if A also receives goods from B. Is A a client or a supplier to store B? Neither A nor B can be said to belong to echelon 1 or 2. They are an odd mix and our numbering of echelons simply fails.

area is a common practice. Retailers use this mean to meet customers' requests for products/colors/sizes momentarily out of stock.   ⬚

However, in this chapter we only consider models that do not allow horizontal shipments for several reasons.

- First, in several cases, horizontal shipments are not used for organizational issues related to the incentives of the store managers and salespersons. For example, the store manager might not want to send some of his/her units of the hottest product to the manager of another store that might be competing with him/her for bonuses. Also, some retail chains (e.g., Zara) discourage this practice to commit the manager of each store to selling the products he/she carries and to carefully select from the company's assortment the products that fit the local market.

- Second, for many products with a low value per unit of volume or weight, moving goods from one store to another might not make economic sense since the cost of point to point transportation[4] might significantly exceed its benefits. This is why in grocery retailing we basically have no shipment among stores.

- Finally, horizontal shipments make modeling harder both because it is hard to tell what is the demand for a given node of the network (demand might come both from downstream and from other warehouses in the same echelon) and because LT becomes stochastic and bimodal, as the goods might be delivered from upstream warehouses (typically with a longer LT) or from warehouses in the same echelon (typically with a shorter LT).

In this chapter, section 6.2 shows that we can manage a distribution network both with (i) local and detailed information (so called *installation-stock*) and (ii) with a global and aggregate information (so-called *echelon stock*). Section 6.2 also discusses pros and cons of these two options. Section 6.3 introduces the rather broad theme of coordination in a supply chain showing the main causes of lack of coordination, their root causes, and some possible remedies. Section 6.4 shows a first inventory problem with a two-echelon linear system with certain demand. Section 6.5 shows how to plan inventories when facing demand uncertainty in a two-echelon supply chain consisting of I warehouses in the first tier and one transit point that feeds them. This model suggests a heuristic and discusses the role and functions of transit points; that is, warehouses where inventories merely transit for a few hours. Finally, section 6.6 analyzes the more complex case where warehouses in the first echelon are

---

[4]As we will discuss in chapter 8, point-to-point transportation (in our case, store-to-store) contrasts with hub-and-spoke transportation where one of the nodes of the network works as a connecting point, just like in the case of air transport, to consolidate traffic and gain economies of scale.

supplied by a warehouse that carries products. This section also shows that, in multiechelon systems, demand uncertainty generates some uncertainty on lead times.

6.2 MANAGING MULTIECHELON CHAINS: INSTALLATION VS. ECHELON STOCK

In a multiechelon distribution system the optimal inventory level in a warehouse might in general depend on inventories and demand in all other nodes of the network. However, accounting for these pieces of information requires a lot of real-time information on the current inventory level and demand in each warehouse.[5] Often such information is not complete and is not available quickly enough, and reliably enough. So we try to design inventory policies that can lead to good global performance though they make decisions based on local information. Often decision makers at one warehouse only have information on their own warehouse and simply have no information on the inventory levels upstream and downstream. *Installation Stock* measures the inventory position of a given warehouse through local-only information. Under the Installation Stock logic, the inventory position of a given warehouse (or installation) is just inventories on hand plus incoming orders minus customer backorders. As the reader can immediately realize, this logic neglects information on inventories and orders in other nodes of the distribution network.

This approach contrasts with the more complex *Echelon Stock* logic. This logic measures the Echelon Inventory Position as the sum of the inventory positions in the warehouse plus all the inventory positions in the downstream warehouses. So this logic requires global information to work properly. The Echelon Inventory Position is greater than (or equal to) the Installation Stock Inventor Position. Also, the Echelon Stock gives a broader perspective on the current inventory level in the distribution chain. For example, this second logic might show that a second-tier warehouse, with no inventories on hand and no incoming orders, might still not need to place any order to suppliers, simply because the stores this warehouse delivers to are overstocked. So in the near future this warehouse might simply not need any inventories.

Example 6.5 Let us consider the distribution chain in figure 6.1. Let us assume that the Installation Stock Inventory Position for each warehouse is that shown in table 6.1. As we can see, the Echelon Inventory Positions are greater than (or equal to) the Installation ones. The Inventory Position of the two logics is obviously the same for warehouses in the first echelon, since there is no downstream warehouse.

---

[5]This section was inspired by [1].

Table 6.1   Inventory Position, Installation Stock, and Echelon Stock

Warehouse	1.1	1.2	1.3	1.4	2	3
Installation Stock Inventory Position	20	20	20	20	20	0
Echelon Stock Inventory Position	20	20	20	20	100	100
LT	1	1	1	1	1	1
Demand	5	5	5	5	–	–

Also, the Echelon Stock Inventory Position for the warehouse in the second echelon is equal to its Installation Stock Inventory Position plus the sum of the inventory positions of all warehouses in the first echelon. Finally, the Echelon Stock Inventory Position of the warehouse in the third echelon (warehouse 3) is equal to the Echelon Stock Inventory Position of warehouse 2 since the Installation Stock of warehouse 3 is zero.

A planner in warehouse 3 needs to decide how many units he/she wants to order. If he/she looks at the problem with an Installation Stock logic, he/she would be tempted to place an order since the inventory position is zero (the size of the order depends on the specific policy the manager adopts, purchasing LT, distribution LT, and demand). On the contrary, if we look at the problem with the Echelon Stock logic, the distribution system looks fairly well stocked (Echelon Stock Inventory Position is 100 units). At the very least with an Echelon Stock logic, we wonder whether we should be ordering at all. This decision still depends on the specific inventory policy the manager adopts, purchasing LT, distribution LT, and demand. But still while the Installation Stock logic seems to suggest that we obviously shall place an order, the Echelon Stock logic might not suggest to place an order. For example, let us assume that demand at each store is deterministic and equals 5 units per period, we adopt a continuous review period, and all LTs are 1 period (purchasing LT to warehouse 3, distribution LT from warehouse 3 to warehouse 2, and distribution LT from warehouse 2 to the stores in the first echelon). In this case, in the distribution system we have enough inventories for 5 periods ($100/(4 \cdot 5)$), while the out-of-control period (which is the sum of all lead times) is just 3 periods. Thus the inventory level in the system is more than enough and we decide not to place any order.   ⬚

Example 6.5 shows that the two logics are basically different eyeglasses that give us a very different reading of the current situation in the supply chain. This really means that the two logics can lead to very different decisions. Example 6.5 also shows the fundamental advantage of the Echelon Stock policy: Decisions are based on a global perspective on the current status of the whole supply chain rather than on local information on the local warehouse.

Table 6.2 Inventory Position, Installation, and Echelon Stock, in example 6.6

Warehouse	1.1	1.2	1.3	1.4	2	3
Installation Stock Inventory Position	80	0	0	0	20	0
Echelon Stock Inventory Position	80	0	0	0	100	100

Concept 6.2 *The Echelon Stock logic looks at the inventory level in the downstream supply chain, and thus decisions can be based on a global information.*

Example 6.5 might lead us to believe that Echelon Stock can always outperform Installation Stock. Example 6.6 complements example 6.5 and shows the fundamental weakness of this logic.

Example 6.6 Let us consider the supply chain structure in example 6.5. Let us now assume that inventory levels are those displayed in table 6.2. Just like in the previous case the Echelon Stock logic seems to suggest that there is no need to place a purchase order for warehouse 3. However, a more detailed analysis shows that stores 1.2, 1.3, and 1.4 immediately need 5 units per period to fulfill their demand and thus (in case we cannot transfer products from store 1.1) we shall deliver 15 units per period from warehouse 2. So inventories in warehouse 2 are just enough for one period. So clearly warehouse 3 needs to place an order so that inventories required to meet demand in stores 1.2/1.4 can enter the distribution system as soon as possible. ⬜

Example 6.6 shows rather apparently the fundamental weakness of the Echelon Stock logic. It sums the Inventory Positions of all warehouses that lay downstream of the warehouse we are planning for. So, for the Echelon Stock logic, excess inventories in one location can counterbalance a lack of inventories in another location. In the long run, excess inventories in one warehouse reduce the need for inventories, so all goods entering the supply chain can be devoted to the warehouses currently lacking inventories. In the long run, any unbalance in the distribution of inventories can be smoothed and inventory levels can be rebalanced. However, in the short run we can rebalance inventories only if we can move some goods from the overstocked warehouse(s) to the under-stocked warehouse(s) through horizontal shipments. Otherwise, the warehouse(s) lacking inventories generates a requirement for inventories that is not counterbalanced by excess inventories in other installations. This major issue can only be tackled if the excess inventories are actually moved horizontally to the store that needs them or if the planning method only considers the minimum between actual inventory level and the optimal one. In this second case, we basically ignore any excess inventories in one part of

Fig. 6.2   Linear distribution chain with N echelons.

the supply chain so that it cannot counterbalance any lack of inventories in other parts of the supply chain.

Concept 6.3 *The Echelon Stock logic looks at the aggregate inventory level in the downstream supply chain and thus does not properly capture any unbalance in the distribution of inventories within the supply chain.*

6.2.1   Features of Installation and Echelon Stock logics

We investigate a linear distribution chain with N echelons to further understand the features of the Installation and Echelon Stock logics. Also we assume that the inventory policy that each of these warehouses adopts is a (Q, R) continuous review system (see figure 6.2). We investigate how the (Q, R) policy works under the Installation Stock and the Echelon Stock logics.

We introduce some notations to model and study the Echelon Stock and the Installation Stock logic.

- Q_n is the lot size for the nth stage of the supply chain. We assume that the two logics share the same order size, as there is no reason why we would order a larger quantity under one of the two logics.

- $IP^i_{n,t}$ is the Inventory Position in the nth tier of the supply chain with the Installation Stock logic.

- $IP^e_{n,t}$ is the Inventory Position in the nth tier of the supply chain with the Echelon Stock logic.

- R^i_n is the reorder level in the nth tier of the supply chain with the Installation Stock logic.

- R^e_n is the reorder level in the nth tier of the supply chain with the Echelon Stock logic.

Also, we make the following assumptions:

- We assume the lot size at stage n is a integer multiple of lot size at stage $n - 1$, that is,

$$Q_n = j \cdot Q_{n-1}, \qquad j \in \mathbb{Z}_+ = \{1, 2, 3, \ldots\}.$$

  This assumption is actually very reasonable since the upstream warehouse n receives orders of minimum size Q_{n-1} and thus it seems reasonable that warehouse n places orders that are multiples of "quantums of demand" Q_{n-1}.

- Also, let us assume that initial conditions are such that

$$R^i_n < IP^i_{n,0} \leq R^i_n + Q_n,$$
$$R^e_n < IP^e_{n,0} \leq R^e_n + Q_n.$$

  In other words, we assume that the initial inventory position lays in the long-run min–max range. In other words, these assumptions make sure we have no initial transient state. Notice that if these assumptions do not hold, we simply have to wait until we place the first order for each warehouse in the chain to make sure the transient state is over and our assumptions hold;

- Finally let us assume that when a customer places an order, the supplier immediately receives it (LT for the information flow is zero, so the LT only consists of time required to handle and transport goods). Under these assumptions, when the customer in stage n places an order, his/her inventory position (both Installation and Echelon) grows immediately, but at the same time the Installation Inventory Position of the supplier decreases by the same amount.

Property 1: Installation Stock $(\boldsymbol{Q}, \boldsymbol{R})$ policy is nested. An inventory policy is *nested* if, when a warehouse in echelon n places an order, also warehouses at echelons 1 to $n - 1$ served (directly or indirectly) by the warehouse in echelon n are placing an order as well.[6]

---

[6]Actually in case of arborescent supply chains at the least one warehouse for each downstream echelon orders at the same point in time.

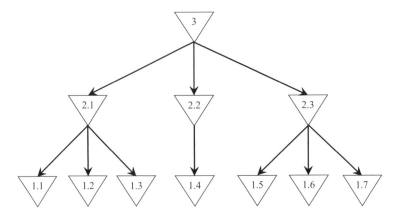

Fig. 6.3   A divergent or distribution supply chain.

Example 6.7 In the supply chain in figure 6.3, if the manager chooses a nested inventory policy, when warehouse 2.1 places an order, at least one of the warehouses 1.1, 1.2 and 1.3 orders as well. However, other warehouses (e.g., 3, 2.2, 1.4, 1.5 and so on) do not necessarily order at the same point in time, as they are not served by warehouse 2.1.   ▯

Under the Installation Stock logic in a linear supply chain, the inventory position at the warehouse n decreases (and thus can reach the reorder point R_n^i) if and only if warehouse $n-1$ places an order. Thus a necessary though not sufficient condition for warehouse n to cross the reorder point and place an order is that warehouse $n-1$ places an order. Obviously, for this to happen, warehouse $n-2$ must place an order, and so on, until we reach the warehouse in the first echelon.

On the contrary, the Echelon Stock logic is not bound to be nested. Indeed, when warehouse $n-1$ places an order to the warehouse n, the inventory position of the latter warehouse remains unchanged. The reduction in the inventory position at warehouse n is counterbalanced by the increase in the inventory position at warehouse $n-1$. So orders from direct customers drive replenishments under the Installation Stock logic, whereas they do not drive replenishments under the Echelon Stock logic (they are actually irrelevant fir the Inventory Position). So what drives orders in the Echelon Stock logic? The only event that reduces the Echelon Stock Inventory Position is final demand; that is, the exit of inventories from the supply chain. Orders and deliveries within the (downstream) supply chain do not matter from an Echelon Stock perspective. So in general, warehouse n can reach the Echelon Stock reorder point even when warehouse $n-1$ is not ordering. This, obviously, does not mean that the Echelon Stock policy cannot be nested, if parameters are selected properly. It simply means that it can also be nonnested, while the Installation stock policy is bound to be nested.

Concept 6.4 *While under the Installation Stock logic orders are driven by the internal demand, under the Echelon stock they are driven by final (external) demand.*

In our following analysis we assume that the reorder levels are such that $IP_{n,0}^i - R_n^i$ is an integer multiple of Q_{n-1}, so that we exactly reach the reorder level R_n^i (orders from the customer $n-1$ as well as deliveries from the supplier $n+1$ are multiple of this quantum of demand). Any reorder level $R_n^i + y$ with $0 \le y \le Q_{n-1}$ creates exactly the same pattern of orders. The one difference is that in this latter case the order is placed when the inventory position is y units below the reorder point rather than equal to the reorder point. So in any way the minimum inventory position is R_n^i, and we can make this assumption for the sake of clarity without any loss of generality.

Example 6.8 Let us consider a warehouse that receives weekly orders for 10 units at a time and places orders for 40 units at a time. Also, let us assume that the initial inventory position is 20 units. If we set the reorder point to 10 units, then we place an order in week 1. The inventory position increases by 40 units, reaching 50 units at time 1. Then it drops by 10 units each week, so that a second order is placed at time 5. Following the same logic, we can draw the conclusion that orders are placed at time 1, 5, 9, 13, 17, and so on. Now let us check what happens if we set the reorder point to 13 (or any level greater than 10 and lower than 20). The inventory position drops from 20 to 10 at time 1. So it crosses the reorder point and an order is placed exactly at time 1. So at time 1 the inventory position increases up to 50. Then again it drops by 10 units at time, 2, 3, 4, and finally at time 5 it drops from 20 to 10 units, crossing the reorder point. So even when we set the reorder point to 13 units, we place orders in periods 1, 5, 9, 13, 17, and so on. As we can see, setting the reorder point to 10 or 13 (as well as any other value between 10 and 20) is basically the same.   ⏹

Property 2: An Installation Stock policy can be replaced by an appropriate Echelon Stock policy that creates the same pattern of orders. The Installation logic is nested and thus we know that if warehouse n is placing an order at time t_0, then all warehouses 1 to $n-1$ are doing the same. So at time t_0^+, that is, just after all orders have been placed, the Installation Stock inventory position of the generic warehouse k with $1 \le k \le n$ is $R_k^i + Q_k$. With this information we can compute the Echelon inventory position at warehouse n at time t_0^+ that is the sum of the Installation inventory position of all warehouses from 1 to n:

$$IP_{n,t_0^+}^e = \sum_{k=1}^{n} (R_k^i + Q_k). \tag{6.1}$$

If the Echelon Stock logic wants to mimic the Installation Stock one, then it needs to order the same quantities at the same points in time. To make sure that the Echelon Stock logic orders the same quantities the Installation Stock logic orders, we just need to use the same lot size. To make sure that the Echelon Stock logic orders at the same points in time the Installation Stock logic orders, we shall make sure that it reaches the echelon reorder point R_n^e at time t_0 and thus that the inventory position at time t_0^+ is just equal to $IP_{n,t_0^+}^e = R_n^e + Q_n$.[7] Also, from equation (6.1), we know the Echelon Stock at this point in time. So we just need to make sure that

$$IP_{n,t_0^+}^e = R_n^e + Q_n = \sum_{k=1}^{n}(R_k^i + Q_k); \qquad (6.2)$$

hence

$$R_n^e = R_n^i + \sum_{k=1}^{n-1}(R_k^i + Q_k). \qquad (6.3)$$

So an appropriate selection of a parameters can make sure that the Echelon Stock logic leads to order the same quantities the Installation Stock logic orders at the same points in time the Installation Stock logic orders them. So the Echelon Stock can mimic the Installation Stock logic (in a linear system).

Property 3: A nested Echelon Stock policy can be replaced by an appropriate Installation Stock one. Not all Echelon Stock policies are nested, but those that are nested can always be replaced by an appropriate Installation Stock policy. Indeed, non-nested Echelon Stock policy cannot be possibly imitated by Installation Stock policies. In non-nested policies, warehouse n can place an order even when warehouse $n - 1$ does not place an order. This cannot possibly happen under Installation Stock policies simply because they are nested.

But now let us turn our attention to nested policies and let us try to select the parameters of an Installation Stock policy in such a way that it can mimic a nested Echelon Stock one. Let us start with the first warehouse. In the case of the first warehouse we just need to make sure that $R_1^i = R_1^e$. Indeed, in the case of the first warehouse there is no downstream inventories to add and thus the two logics are basically the same thing.

As to other warehouses, the generic warehouse n orders at time t_0 when the Echelon inventory position reaches the reorder point R_n^e and immediately bounces back to $IP_{n,t_0^+}^e = R_n^e + Q_n$. Since the Echelon Stock policy is nested, we know the same equation holds for all $1 \leq k < n$ ($IP_{k,t_0^+}^e = R_k^e + Q_k$) and in particular for $n - 1$.

This means that when we look at the distribution system from an Installation Stock perspective, the inventory position of warehouse n just after the

---

[7]Actually, to write this equation, we shall make an additional assumption. We shall assume that final demand is a continuous process or $IP_{n,0}^e - R_n^e$ is an integer multiple of Q_{n-1} to make sure that the echelon policy reaches exactly the reorder level at time t_0.

order is placed in t_0 is

$$IP^i_{n,t_0^+} = IP^e_{n,t_0^+} - IP^e_{n-1,t_0^+} = R^e_n + Q_n - \left(R^e_{n-1} + Q_{n-1}\right), \qquad (6.4)$$

that is, the inventory position of warehouse n minus the sum of all downstream inventory positions.

Now to make sure that the Installation Stock logic places orders exactly when the Echelon logic places them, we just need to select an Installation Stock reorder point such that the inventory position right after the order is placed in t_0 is equal to (6.4). Hence we just need to set

$$IP^i_{n,t_0^+} = IP^e_{n,t_0^+} - IP^e_{n-1,t_0^+} = R^e_n + Q_n - \left(R^e_{n-1} + Q_{n-1}\right) = R^i_n + Q_n,$$

thus the Installation reorder point

$$R^i_n = R^e_n - \left(R^e_{n-1} + Q_{n-1}\right). \qquad (6.5)$$

makes sure that the Installation Stock policy perfectly mimics the nested Echelon Stock one.

Properties 2 and 3 show that the Installation policy with a continuous final demand and a linear supply chain structure is basically a special case of the Echelon logic. However, when the two logics lead to the same decisions (how much and when to order) and thus to the same performance, we definitely prefer the Installation Stock logic since it only requires local information while the Echelon logic requires each warehouse to have prompt and precise information on the inventory position in all $n - 1$ downstream warehouses.

Example 6.9 Let us consider a linear distribution network with 3 echelons, where lot sizes are 2, 4, and 8 units, respectively, for warehouses 1, 2, and 3. Also, let us assume that the replenishment LT is 1 period for all three warehouses and the Installation reorder points are 2 units for all three warehouses ($R^i_k = 2$, $\forall k$). We assume the initial inventory levels are 4, 6, and 10 units, respectively, and final consumer demand is one unit per period.

- Warehouse 1 orders two units at time 2, 4, 6, 8, etc. (see figure 6.4).

- The Installation Stock Inventory Position in warehouse 2 drops to 4 units at time $t = 2$. Then it further drops to 2 units at time $t = 4$. So at time $t = 4$ we reach the reorder point and we immediately place an order for 4 units that take the inventory position back to 6. Given the demand pattern, the warehouse reorders in periods 4, 8, 12, 16, etc. (see figure 6.5).

- The inventory position in the third warehouse drops to 6 units at time 4, then to 2 units at time 8 when it reaches the reorder point, and we place an order for 8 units, and the inventory position is taken back to 10 units (see figure 6.6).

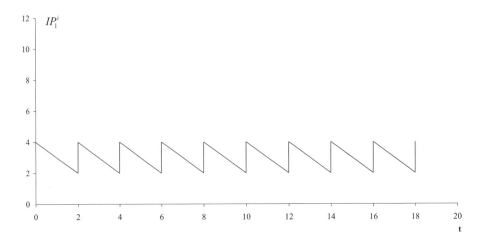

Fig. 6.4   Installation Inventory Position in the first warehouse.

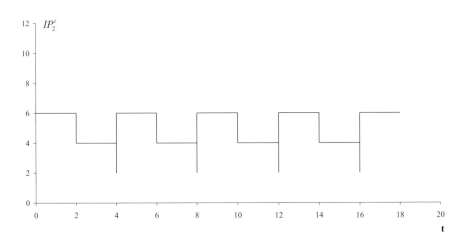

Fig. 6.5   Installation Inventory Position in the second warehouse.

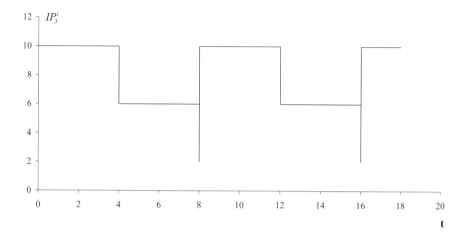

Fig. 6.6 Installation Inventory Position in the third warehouse.

Let us now check how the Echelons Stock logic can re-create the same ordering pattern. Lot sizes in this case too are 2, 4, and 8 units for warehouses 1, 2, and 3 respectively. The initial inventory position is 4, 10 $(= 6 + 4)$, and 20 $(= 10+6+4)$. The Echelon reorder point for the first warehouse equals the Installation reorder point, as there is no downstream inventory in the case of the first echelon so the Echelon and Installation Stock are basically the same: $R_1^e = R_1^i = 2$.

As to the second warehouse, we would like this warehouse to place an order in period 4. Equation (6.3) suggests that we set the Echelon reorder point for the second warehouse to $R_2^e = R_2^i + R_1^i + Q_1 = 2 + 2 + 2 = 6$. Actually, we know that only final demand decreases the Echelon inventory position. In this case demand is continuous, constant, and equal to one-unit-per-period (see figure 6.7). Thus, the initial Echelon inventory position drops at a one unit per period rate and reaches the reorder point $R_2^e = 6$ right at time $t = 4$, just like in the case of Installation Stock logic. Obviously, once the inventory position reaches the reorder point, we place an order for 4 units and the Echelon inventory position bounces back to 10 units. We can follow a similar logic to draw the conclusion that the Echelon Stock logic, too (with $R_2^e = 6$), leads us to place orders for 4 units at times 4, 8, 12, 16, etc. (see 6.8).

Finally, we can use equation (6.3) again to set the reorder point for warehouse 3: $R_3^e = R_3^i + R_2^i + R_1^i + Q_2 + Q_1 = 2 + 2 + 2 + 4 + 2 = 12$. The initial Echelon inventory position is 20 units. In this case, too, the final demand is the one variable that reduces the inventory position at a one-unit-per-period rate. So the inventory position reaches the reorder point only at time $t = 8$, when an order for 8 units is placed and the inventory position is taken back to 20 units. The Echelon Stock logic (with $R_3^e = 12$) leads us to order at time

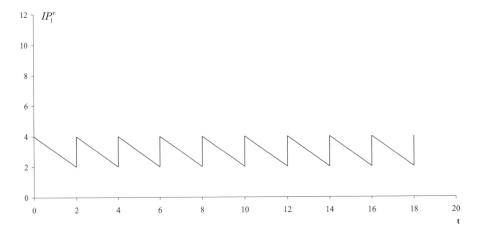

Fig. 6.7   Echelon inventory position in the first warehouse.

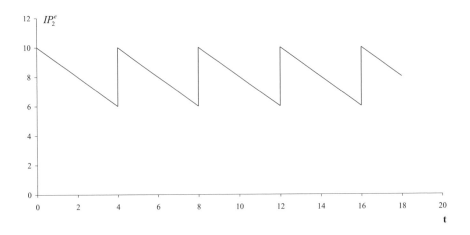

Fig. 6.8   Echelon inventory position in the second warehouse.

8, 16, 24, etc. (see 6.9). We were able to design an Echelon stock (Q, R) that mimics an Installation stock (Q, R) policy.   ⬜

So far we have underlined the similarities between the two logics. Let us now focus on the differences.

The Echelon policy can create non nested patterns of orders that in some instances might be superior to the nested ones, as example 6.10 shows.

Example 6.10 Let us consider a two-echelon distribution system with deterministic, continuous, and constant demand for one unit per period and a delivery LT of two periods and one period for the first and second ware-

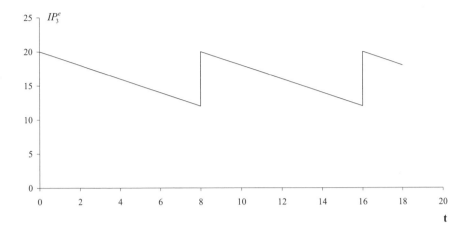

Fig. 6.9 Echelon inventory position in the third warehouse.

house, respectively. Also, let us assume that the lot size is 5 units for the first warehouse and 10 units for the second warehouse.

Given our assumptions, the first warehouse shall place an order when only two units are left as they are needed to cover demand over the replenishment LT (see chapter 4). The case of the second warehouse is slightly different, as we shall order one period in advance of the actual shipment to warehouse 1. For example, if initial Installation inventories are 7 and 5 units respectively, warehouse 1 orders 5 units at time $t = 5, 10, 15, 20$, etc. (see figure 6.10). So the 5 units initially in warehouse 2 are shipped to warehouse 1 at time $t = 5$. Warehouse 2 will get ready to ship 5 more units at time $t = 10$. So warehouse 2 should not order up to time $t = 9$. If we order 10 units at time $t = 9$ they are delivered at time $t = 10$ when 5 of them are also forwarded to warehouse 1. Thus the optimal policy we have just designed is actually non-nested since warehouse 2 orders at time $t = 9$ when warehouse 1 does not place any order (see figure 6.11). So no Installation policy (no matter which parameters we select) can generate an optimal pattern of orders, as all Installation Stock policies are nested and the optimal solution is not nested.

On the contrary, we can find the parameters of an Echelon Stock policy to place orders at time $t = 5, 10, 15$, etc. in the first warehouse, and $t = 9, 19, 29$, etc. in the second warehouse.

The initial Echelon inventory position is obviously 7 for the first warehouse and 12 ($IP_{2,0}^{e} = 7 + 5 = 12$) for the second one. The Echelon inventory position decreases by one unit per period in both warehouses (both inventory positions are driven by the final demand). The right reorder quantity for the first warehouse is obviously $R_1^i = 2$ (see comments above). In the case of the second warehouse, we know we want to place an order at time $t = 9$ and we know that at that time the Echelon inventory position has reached 3

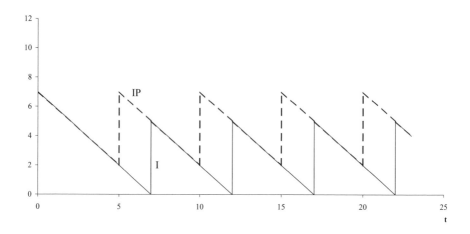

Fig. 6.10   On hand inventory and inventory position in the first warehouse.

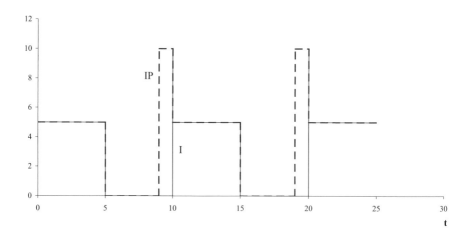

Fig. 6.11   On hand inventory and inventory position in the second warehouse.

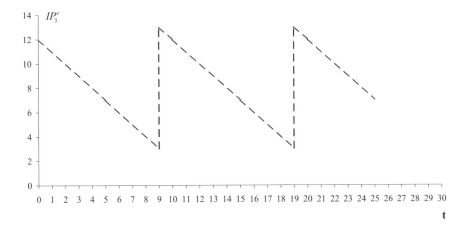

Fig. 6.12 Echelon Inventory Position for the second warehouse with $R_2^e = 3$.

$(IP_{2,9}^e = IP_{2,0}^e - 9 \cdot d = 12 - 9 \cdot 1 = 3)$. So we just need to set the Echelon reorder point of the second warehouse to 3 (R_2^e). Also, notice that this result is fairly logical: the system suggests that we reorder when in the distribution system consisting of 2 warehouses, only 3 units are left. These 3 units are actually needed to cover demand over the total LT of three periods; that is, the time it takes to move products from the external supplier to warehouse 2 (one period) and from warehouse 2 to warehouse 1 (two periods). ⬚

Finally, the Echelon Stock logic has another significant advantage: managers of the upstream warehouses can immediately observe any increase (or decrease) in demand. Indeed, an increase in demand immediately reduces the Echelon inventory position of all warehouses.

Concept 6.5 *The Echelon Stock logic is more flexible as it can generate both nested and non-nested policies. Also, the Echelon stock logic can immediately spot any change in final demand, since final demand has a direct impact on the Echelon stock inventory position.*

In an Installation Stock system, if inventories in the lower echelons of the supply chain can initially absorb the increase in demand, then managers of upstream echelons hardly notice the change. Managers of upstream echelons notice the change only when the orders coming from downstream warehouses increase. So the whole supply chain might figure out that demand has changed quite slowly. The tighter informative requirements of the Echelon Stock logic also mean that this logic tends to give decision makers in the upstream portion of the supply chain real-time information. Also, while the Echelon Stock logic provides decision makers with first-hand information on current demand, in the Installation Stock logic the information on final demand only comes through orders from downstream warehouses. This information can be both delayed and distorted by the ordering policies (see next section for a deeper analysis of this phenomenon) of downstream warehouses.

Concept 6.6 *In the Installation Stock policy the information on any change in final consumer demand is delayed and possibly made noisy by the stocking decisions of the downstream warehouses.*

In summary, in the case of linear (and convergent) distribution systems, the Echelon Stock logic is more flexible than the Installation Stock logic and thus can offer better performance. However, in the case of divergent distribution chains the benefits of (i) a broader perspective on the inventory level of the whole chain, (ii) better information on final demand, and (iii) a more flexible ordering system are counterbalanced by the inability to capture and manage unbalances in the distribution of the inventories among warehouses in the same echelon (see example 6.6).

6.3   COORDINATION IN THE SUPPLY CHAIN: THE BULLWHIP EFFECT

One of the major issues in multiechelon supply chains is the coordination of decisions among planners of the various warehouses (for a more general discussion of coordination among decision makers, see next chapter).

As we have just discussed, one approach to coordination is to simplify the planning problem and manage each single warehouse as if it were independent. This simplistic approach is often used when various warehouses in the supply chain belong to different companies or different organizations. However, even within a company or institution different parts of the organization might have partially contrasting objectives or lack of coordination.

A first apparent effect of such a simplistic solution is that each decision maker looks for locally optimal solutions and practices that, however, might turn into global inefficiencies.

A second, even more pervasive effect is the so-called *Bullwhip effect* also known as the *Forrester effect* (Forrester is an MIT professor that first investigated the effect through *Industrial Dynamics*). The bullwhip effect produces an increase in demand variability as we move upstream in the supply chain. This effect makes the demand for components more variable than the demand for the finished product at the retail stores or at the distributors.

Classic examples of this phenomenon are Pampers diapers and Barilla pasta. For both products, end-consumer consumption is quite flat (there is no sharp seasonality nor sharp trend, at least in most developed countries), as both products meet basic physiological needs. This relatively flat consumption drives purchases at the retail stores. However, these purchases show some fluctuations due to trade promotions, i.e., price promotions at retail stores.

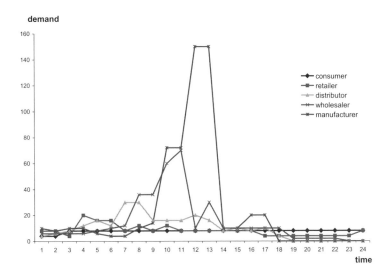

Fig. 6.13 Pattern of orders at various stages of the supply chain in the Beer Game.

Such promotions tend to shift demand among brands, as many consumers are willing to switch from one brand to the other, and over time since many consumers forward-buy their best-preferred brands during promotions (at least for durable goods). Such variability induced by promotions is even wider upstream. Orders from the retailer to the manufacturer and finally orders from the manufacturer's warehouse to the production plant are even more variable.

This phenomenon is described and studied through the *Beer Game*, which was designed in the 1960 at the MIT to re-create in a simple simulation the dynamics that create unpredictable demand variability, which creates uncertainty in supply chains (see [4]). Figure 6.13 shows the typical pattern of orders in a supply chain with four players: the retailer, the distributor, the wholesaler, and the manufacturer.

Demand variability in the upstream portion of the supply chain creates several inefficiencies. First, in any inventory policy an increase in unpredictable variability (i.e., uncertainty) requires an increase in safety stocks (or, more in general, in *slacks* such as spare capacity) required to gain a given service level.

In the remainder of this section we identify the root causes of this phenomenon and suggest some actions to remove them or mitigate their effects.

Lot sizing and planning. In lower echelons of the distribution chains, orders are small. For example, an average family buys a few kilograms of pasta per week (at least in Italy). Various families make purchase decisions rather independently, and the sum of all their orders tends to be a relatively stable random variable. However, in the upper echelons of the supply chain, single stores, grocery chains, distributors, and manufacturers have quite sizable incentives to buy in large quantities. For example, a (Q, R) policy suggests that we buy in large quantities to reduce the ordering costs.

Correlation among orders coming from different players in the same echelon of the supply chain (say, correlation among consumers or correlation in orders coming from various stores) can further increase variability. For example, consumers tend to spend more on grocery products right after salaries are paid. In Italy most companies pay the salary either at the end of the month or at the beginning of the following month. So we have the so-called "effect of the fourth week"; that is, many consumers run out of money toward the end of the month and postpone some purchases until the beginning of the following month.[8] Also, retailers, distributors or manufacturers might have some sort of incentive to reduce their inventories or increase their turnover toward the end of the quarter, thus creating an anomaly in the collection of orders.

Example 6.11 In the distribution of grocery products to small retail stores (so-called "mom and pop" stores), salespersons have some incentives to reach quarterly sales targets that often push them to collect as many orders as they can toward the end of the quarter, thus creating an anomalous peak in demand toward the end of the quarter. On the other hand, some salespersons that have already reached their target tend not to collect orders in the last few days of the quarter (actually some of them tend to collect the orders but do not key them into the system). The idea is that sales targets for future quarters are often based upon sales in past quarters; thus the higher the sales in the past, the higher the targets for the future. Also, if they keep some orders for the next quarter, they can have a jump start and thus are more likely to meet the sales target for the next quarter.   ☐

Example 6.12 A distributor of electronic components in Italy stops ordering and shipping products towards the end of the year (they basically do not place orders in November and December and do not ship goods in December) because of odd incentives. This company is the national distributor for a major Danish producer of components. The distributor has signed a contract with the manufacturer with a significant bonus based on sell-in targets. In other words, the distributor gets a bonus if it orders more than a minimal monetary amount per year. The sell-in target for 2007 is based upon sales for 2006. So this creates an incentive not to place orders toward the end of the year (in case, like in recent years, they have already ordered enough to meet the current year's sales target).

---

[8]Notice that this effect is particularly strong in time of recession or limited economic growth and for grocery products, whereas it is far less sensible for other product categories such as luxury products bought by customers with no major financial constraints.

Also, this privately held company postpones many deliveries to the beginning of the following year to postpone turnover and taxes.

Other companies in this business might have a similar behavior, thus generating odd end-of-year effects. Finally, we shall notice that other companies have similar behaviors for opposite reasons. Some public companies try to postpone purchases at the end of the year to reduce the inventories and thus the working capital so that they look leaner and thus can get more favorable evaluations from the market.   ⬜

We can address this issue in several ways.

1. *Reduce the fixed ordering cost.* As we have already discussed in chapter 4, fixed ordering costs give an incentive to buy in large quantities. Thus reducing fixed costs reduces the optimal lot size. So automation of all order-related activities such as planning, quality inspections, and administrative activities (e.g., one can substitute traditional tools such a fax machines and data entry with business to business online orders) can reduce the fixed ordering fees and reduce the lot sizes.

2. *Consolidate transportation.* Part of the fixed ordering costs comes from transportation. Some manufacturers give customers that place orders for full truckloads some additional discounts, since transportation of full trucks is more efficient. The flipside of this policy is that it tends to create large and infrequent orders that contribute to demand variability and uncertainty. A solution to this tradeoff between transportation efficiency and demand variability is to consolidate in a single truck various products from various suppliers (as we have discussed in chapter 2) or deliveries to several customers (see chapter 8). In the first case, the customer might consolidate the transportation and collect the goods from various suppliers. On the contrary, in the latter case the third party provider of transportation services can consolidate transports to fully load the truck.

3. *Reduce correlation among order patterns of various customers.* To reduce demand variability, we shall make sure that customers place orders at different points in time. So we shall reduce any incentive to place orders at specific points in time. Obviously, it is relatively easy to reduce the salespersons incentives to collect orders at the end of the year. On the contrary, it is relatively hard to work on financial problems of many families that postpone purchases at the end of the month. Service industries with high fixed costs and a relatively inflexible capacity such as telecom and electric energy have designed specific tariffs to make sure that demand fluctuations are smoothed out. Even in more traditional services, we have similar patterns with some furniture retailers reducing prices during low demand periods. Notice that furniture is not subject to sudden changes in fashion, so these initiatives do not aim at selling

goods left over; instead they aim at increasing demand in low demand periods (such as the end of December, oddly) to cover fixed costs such as personnel, energy, and real estate.

4. *Increase homogeneity in customer and order sizes.* Finally to reduce demand variability we shall have a large number of independent sources of demand; that is, a fairly large number of customers served from a single warehouse. There are several ways to reach this objective. Obviously, one can increase the number of customers, but this has little to do with logistics, though we have to acknowledge that this simplifies supply chain management. Moving to more supply-chain-related issues, one can reduce the number of warehouses in each echelon of the supply chain to make sure that the number of "customers" served by each warehouse increases and variations in the ordering patterns smooth out. One important prerequisite for this to happen is that all customers have a comparable size, otherwise any variability in the orders of one customer can hardly be counterbalanced by orders from any other customer (metrics for market concentration such Herfindal's or Gini's indexes in the economics literature can be very effective for this purpose; e.g., see [2]).

Forecasting   In a multiechelon supply chain the instability of forecast can contribute to increase variability in the upstream stages of the supply chain. Let us consider a situation where each decision maker only "sees" orders from direct customers with no information whatsoever on the final consumer demand (see the Installation Stock policy). In these circumstances, orders from direct customers is the one and only relevant piece of information to make a demand forecast. This can create significant distortions in the flow of information in the chain. Indeed, as we have learned in the previous chapters, planners place orders to optimize their inventories. So they are basically inventory decisions of the customer that, nevertheless they are interpreted as demand signals by the supplier, simply because he/she has no other information on demand. Let us try to understand what goes on in a supply chain and let us try to understand how information on final demand gets distorted as it is transmitted in the chain. Let us assume that final demand increases by Δd. Such an increase can lead the retailer to increase orders by far more than Δd. He/she increases orders for the following reasons:

1. He/she Replenishes inventories that went down more than he/she expected.

2. The recent increase in demand leads him/her to increase the demand forecast and target inventory levels.

3. The recent unpredicted increase in demand might lead him/her to update the estimate of demand uncertainty and thus the need for safety stocks (if, as it is often the case, the economics of the business suggest him/her to overstock).

Notice that in these contexts the retailer makes locally rational decisions (as we have seen, there are several good reasons to increase orders more than demand has increased) that the distributor can read as a signal on final consumer demand. Obviously, a similar process goes on between the distributors and manufacturers and between manufacturers and suppliers of raw materials.

Also, often forecasters (or the tools they use) tend to read a temporary increase in demand as a permanent trend. So, while the increase in demand Δd can be temporary or a one-time-only increase, forecasting tools might interpret that as a sign of a constant trend. So the forecasting method can (at one extreme) project a growth of Δd units per period. Such a reading of the demand signal would further increase the order size and would lead upstream decision makers, in their turn, to overstate demand.

Such an excess demand at the end of the day turns into inventory holding costs. Indeed, when each player overstates the orders he/she receives, we can easily get into a situation where the upstream portion of the supply chain (say manufacturers) produces far more than final consumers require. In these circumstances, inventories build up rather quickly.

Let us now check what we can do to reduce the impact of demand forecasting on the Bullwhip effect:

1. *Share information on final consumer demand.* First, all players in the supply chain can receive information on the final consumer demand. This way we decouple information flows and inventory decisions. Though new technologies are making such exchange of information cheaper and cheaper, very often signing contracts for the exchange of information might be hard. Indeed, downstream players (e.g., retailers) own the information and they shall release it to the upstream players. Unfortunately, such an information directly only solves the problems of the upstream players that are confronted with the high variability created by the Bullwhip effect. In the long run, a more stable demand can make production more effective and lead to a reduction in price, but the link is very weak and the retailer might wonder who is going to take advantage of the increase in efficiency. Is this going to increase the manufacturer's profits? Is it leading to a reduction in wholesale price that competing retailers are going to enjoy as well? Costs and risks of such exchanges of consumer data are very clear and often benefits for the retailers are too unclear. So, a key to the success of these initiatives is a clear plan to share the benefits they certainly create. For example, some retailers get paid for the information, others receive a better service and so on. Also, such an exchange of information raises some confidentiality issues. For example, a manufacturer might use the data he/she received from a retailer to prove to another retailer that a new product is in high demand in a given area. Clearly, this is a secret the retailer would like to preserve.

A further step beyond the simple exchange of information is to share forecasts in such a way that various players in the supply chain make their decisions independently, but at the very least they share the same vision of what the future will bring. These practices are often called *Collaborative Forecasting* (also known as CPFR, that is Collaborative Planning Forecasting and Replenishment). With these practices the retailer can leverage on some information the producer only has, such as advertising campaigns, while the producer can leverage on some information that the retailer only has, like promotions of competing products or planned retail price during the promotion (a 50% price reduction is much more effective than a 20% one). Indeed, a retailer that plans a promotion for fresh filled pasta of brand A might suggest the manufacturer of brand B that he/she expects a short-term reduction in sales volume.

2. *Vendor Managed Inventory* (VMI). An alternative solution to sharing consumer demand data or forecasts is to give a single decision maker the decision rights on all echelons in the supply chain. In particular, we call this practice Vendor Managed Inventory in case the supplier controls inventories at the distributor's or retailer's warehouses. These practices obviously make sure that all decisions in the supply chain are based on a single and consistent forecast. However, we shall also notice that these agreements might face a couple of problems.

First, they might be subject to incentive problems (see next chapter for a deeper discussion on incentives problems). In all inventory models for uncertain demand we have analyzed so far, the optimal quantity depends on the inventory (holding) cost and the cost of the stockout. However, the cost of a stockout might be substantial for the manufacturer while it might be negligible for the retailer since margins are often different and many consumers are very willing to substitute stocked out items in many categories of grocery product. So the manufacturer has a greater incentive to reduce stockouts than the retailer has. This partially explains why decision rights are allocated to the manufacturer rather than to the retailer. Also, this partially explains why the manufacturers increase the service level and make sure that retailers' warehouses have a very high service level (99%+ service level is not uncommon at the warehouses under VMI) but then the retail stores fail to turn this into a high service level on the shelves, simply because it is not as important for them. To make sure that these systems work we need to make sure that all inventory-related costs are on the shoulders of the decision maker. For example, when a manufacturer manages the distributor's inventories through VMI, he/she shall be held accountable for the inventory investment he/she generates; otherwise the manufacturer might have an insane incentive to increase inventories up to unreasonable levels simply because they are free for him/her. A second major concern actually regards information. When we move decision rights in the chain, we also want to make sure that the one decision maker has all relevant pieces of information (as well as skills and systems) to make the best possible decision. So, as we move decision rights in the chain we shall also make sure that relevant pieces of information are moved to the right point in a timely fashion. The next chapter discusses in some more detail economic reasons why it might make sense to adopt VMI.

3. *Reduction in LT.* Long LTs lengthen the forecasting horizon. Hence:

- Long LTs increase the uncertainty in a single time bucket, because a forecast for the near future tends to be more accurate than a forecast for the far future.

- Long LTs increase the out of control period that safety stocks shall cover (think of (Q, R) or S systems). So a reduction in LTs reduces the forecasting horizon and thus the need to invest in safety stocks.

Pricing policies.   Price promotions can create demand variability upstream. Trade promotions perturb the consumers' purchasing process. For nonperishable products such as canned food or dry pasta, consumers tend to forward buy (and to some extent increase the consumption of these products). Also, such price promotions generate further disturbances within the supply chain. Indeed, during trade promotions, producers reduce the wholesale price since an increase in demand benefits both the producer and the retailer and thus both contribute to the reduction of the final consumer price. However, such a temporary price reduction prompts the retailer to forward buy to stock inventories at a relatively low price and sell them at a full price once the promotion is over. By doing so, the distributor further increases the peak in demand caused by the trade promotion.

A rather radical solution to this cause of the Bullwhip effect is the *Every-Day-Low-Price* (EDLP) policy. Some retailers keep a constant and relatively low price for all of their assortment rather than periodically (usually promotions last a couple of weeks) reduce the selling price of some items. The price these retailers charge is lower than the standard price other retailers charge in off-promotion periods, though it is higher than the promotional price. While price promotions try to drive traffic into the store by advertising some items with a very low price for a limited period of time,[9] the EDLP policy tries to attract customers by promising a low ticket for the average shopping basket. The key idea is that a more stable demand is easier to manage, and such ease of management can increase efficiency and reduce costs. So basically EDLP rewards customers for the stability of their purchases with a lower price (of an

---

[9]Some discount retailers in Italy have started to offer deep discounts on a couple of items at a time (say an electric drill and a specific kind of sneakers) for one day only to drive traffic on specific low demand days.

average shopping basket, though other retailers might have a lower price of an item on promotion). The best-known case of EDLP is Wal-Mart, today the largest retailer in the world. These strategies do not always work. For example, in Italy, Barilla tried such strategies with the help of some US managers and reduced both price promotions and gadgets. However, they discovered that EDLP works only for products and purchases that are planned rationally like dry pasta. On the contrary, this strategy does not work for impulse purchases such as cakes and snacks for kids. For these products, motivations are hardly rational and have a lot to do with the fun content and emotions such as the feeling that we give our children the "best" food available.[10]

Allocation of capacity   Customers actually place orders to control the quantity the supplier delivers them. So in some instances they overstate their needs to get a higher priority and thus a better service. Indeed, in case the supplier does not have enough goods to meet the demand from all customers, the supplier very often allocates the limited amount of goods (or the limited capacity to manufacture/distribute them) proportionally to the orders that customers have placed.

Such a policy might look reasonable since it allocates more inventories or capacity to the customers that requested more goods and thus are more relevant and/or need more products. But this allocation policy creates an insane incentive to overstate demand in periods of high demand (and product scarcity), thus contributing to the Bullwhip effect. For this to happen, customers (say retailers) must realize that the supplier (say manufacturer) is running short of inventories. Once this vicious circle starts, it is really hard to stop it. It is basically a self-fulfilling prophecy. When customers expect a stockout, they overstate their orders to get a larger share of the limited inventories (capacity). But this actually creates a stockout.

This practice can distort final consumer demand since customers use orders to signal they want a large share of the limited quantity available rather than to signal the optimal quantity, let alone final consumer demand. Sooner or later, the supplier recovers from the stockout, often after investing in extra capacity for a demand that is actually not there. Once the crisis is over, the supplier can actually deliver the whole quantity customers have ordered. All of a sudden deliveries are excessive since the customer has overstated orders presuming that the supplier could not deliver the whole quantity. So either the customer cancels excessive orders and the supplier is left with excess inventories (or spare capacity), or the customer accepts the deliveries and his/her inventories go through the roof. So to reduce the inventory level the

---

[10]Notice that on page 327 we suggest the use of pricing strategies. However, the objective and the net effect in that case is just the opposite of standard price promotions. In that case we change price over time to flatten demand, whereas price promotions in the grocery business boost demand variability.

supplier stops ordering for a while and after a peak in demand we have zero orders for a while (see figure 6.13).[11]

We can manage allocation problems in several ways:

1. *Choose different allocation criteria such as past sales or sell-out.* A first option we might have is to destroy the incentive to overstate orders to get a larger share of the limited capacity. For example, one can use past sales or *sell-out* (that is, units or dollars sold to the final consumer or the customer's customer). These variables still give larger customers higher priority but do not distort their ordering policies.

2. *Cancelation policies and less flexibility in order changes.* Often customers can modify orders even shortly before they are delivered. From a marketing and sales perspective, such practice persuades customers to place orders with more confidence and disposes them to select suppliers that offer this sort of flexibility. However, this greater willingness to place orders shall be contrasted with their informative content, that is, with the information we can extract for planning purposes. If a customer knows that he/she can reserve inventories or capacity for free by placing an order, the customer has all kinds of incentives to place an order that exceeds his/her actual need by far. So the supplier can hardly understand what actual future demand is going to look like. So one shall always carefully consider the tradeoff between the flexibility we give to our clients and the ability to promptly collect reliable information on future demand.

Finally, we can dramatically reduce the Bullwhip effect by removing some of the echelons in the supply chain. Such an alternative shall be actually quite carefully considered as we shall ponder, What are the functions each warehouse is performing, and who else in the supply chain can perform them in case we remove the warehouse? One example of how effective such a decision can be is Dell computers, today one of the largest manufacturers of computers in the world. Dell sells directly through the Internet and catalogues to avoid the expensive inventory buildup in the distribution chain (in this business, goods lose value very quickly, so the holding cost is substantial). At each single store the demand for each single product variant is relatively low and thus demand at the item/store level tends to be fairly variable, unsold inventories tend to be high and thus the holding/obsolescence cost is substantial.

The success of Dell inspired many companies to follow suit in other industries such as furniture or grocery. But many of them simply failed. Indeed,

---

[11] A related problem regards the penalties for lack of service. Many suppliers agree to pay the customer for lack of service. So if orders are not fulfilled, the supplier must pay a given fee. This again gives the retailer an insane incentive to overstate the orders in peak times, as he/she can basically get a discount for large orders that can cause a stockout.

they tried to remove the so-called middle-man, that is, the intermediary between the consumer and the producer. But the success of these initiatives depends on several issues, among which:

- *The value of the product* that shall make the delivery at the customer's place economically sound (i.e., not too high as compared to the cost of the good) and quick enough;

- *Touch and feel content of the product* that shall be bought without trying, touching, and feeling it; certainly a PC can be easily described (to a person that knows the basic variables that describe a PC) through an Internet screen with some pictures and a list of technical features while a piece of furniture can hardly be fully described through a computer screen even to the most expert person (even companies in the business only buy goods once they see a sample);

- *Variety of the assortment at various levels of the supply chain.* When we remove one echelon of the supply chain, we save the inventories held in that echelon. Clearly, the greater the variety of items held in that echelon, the greater the savings. The case of Dell really tells a long story on this variable. Though Dell is considered to be a make-to-order manufacturer with no inventories, it actually carries (and/or lets suppliers carry) inventories. The one big difference is that other manufacturers carry (or used to carry) finished goods in the distribution chain while Dell carries components in the central warehouses that feed the production plant. So what is the advantage? The number of different components is relatively small, while the number of finished products one can generate by combining these components is basically infinite. Dell carries inventories only at the production plant while other manufacturers carry (or let the retailers carry) inventories in many locations. While Dell needs to plan actually a few item/location combinations with a relatively high and stable demand, most competitors need to plan thousands of item/location combinations with relatively low and variable demand.

Also, the case of online furniture retailing is telling. Often traditional furniture retailers do not carry large quantities of inventories. Products are often displayed in the store in a specific style/material/color combination, while the material/color variants (and in some cases slight variations in design such handles) are only presented through samples or catalogues. Products purchased by the consumer are then made to order. So taking the retailers out of the supply chain saves very little inventories and does not reduce the Bullwhip effect. Obviously, we can save on retailers' margins, but we shall check what their functions are. They guarantee a very efficient primary transportation to the retailer's warehouse. They manage the secondary transportation to the consumer's place. Retailers make this secondary transportation much more efficient, as they often transport goods coming from various suppliers. Retailers often repair small damages (such as scratches on leather sofas) that have occurred during the transportation and handling. This latter function is crucial in a business where products are rather frail, and it is hard to pack them effectively. They often make sure the product is repaired before the customers can see it and thus make sure that customers are 100% satisfied while they actually receive a product that has been repaired. Hence selling furniture directly through the Internet is just not as effective as selling PCs on the Internet.[12].

6.4 A LINEAR DISTRIBUTION CHAIN WITH TWO ECHELONS AND CERTAIN DEMAND: THE TWO-STAGE ECONOMIC ORDER QUANTITY

So far in our analysis of a multiechelon supply chain, we have considered lot sizes as a given. We now wonder how to set the lot sizes, and we consider a fairly simple supply chain consisting of two echelons and one warehouse per echelon. We assume that demand is deterministic, continuous, and constant, like in the case of the EOQ model. Also, we assume that the unit price is constant (no price discounts), and thus the only relevant costs for our purposes are ordering and holding costs for the two warehouses. To model our problem, we introduce the following notations:

- A_1 and A_2 are the fixed ordering costs in the downstream and upstream warehouse, respectively;

- h_1 and h_2 are the unit inventory holding cost in the downstream and upstream warehouse respectively; we assume $h_1 > h_2$, as, in general, downstream warehouses are smaller and closer to the final consumer (thus usually in areas with higher costs of real estate) and the downstream warehouse stocks inventories with more value added (e.g., transportation or bulk lots have been broken down in the upper echelons). Anyway, in case $h_2 \geq h_1$, clearly the optimal policy is to keep inventories only in the downstream warehouse where they are both cheaper and closer to the final consumer (and thus can offer a better service).

Moreover, we assume, without any loss of generality, that LTs are zero. As we have shown in the case of a single warehouse (see section 4.4), the only difference between zero LTs and nonzero LTs is that in the latter case we need to place an order before inventories drop to zero. But the actual holding cost

---

[12] [8] and [6] provide an extensive analysis of the root causes of the bullwhip effect and this section is partially based on these references. For more quantitative models on the bullwhip effect see, for example, [7] and [3]

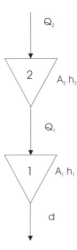

Fig. 6.14   Structure of the two-echelon linear supply chain.

and ordering costs (as well as the optimal lot sizes) do not change, and thus our results can be applied in the case of nonzero LTs too. Our assumption just simplifies the model, since the inventory position equals inventories physically on hand.

We first write the cost functions for the two warehouses separately, so that we can compare two scenarios:

- In the first scenario the cost functions of the two warehouses are optimized separately.

- In the second scenario the cost functions of the two warehouse are optimized jointly.

The cost function of the first (downstream) warehouse $C_{tot,1}$ is exactly the cost function we have developed for the EOQ model:

$$C_{tot,1} = A_1 \cdot \frac{d}{Q_1} + h_1 \cdot \frac{Q_1}{2}. \tag{6.6}$$

Clearly, in this case the optimal order quantity is

$$Q_1 = \sqrt{\frac{2A_1 d}{h_1}}. \tag{6.7}$$

For the second (upstream) warehouse, things are slightly more complex. We cannot use the EOQ model, since in this case demand is not continuous but rather consists of lots of size Q_1.

Just like in section 6.2, we assume that $Q_2 = jQ_2$, $j \in \mathbb{Z}^+$; that is, we assume that the lot size of the upstream warehouse is an integer multiple of the lot size for the downstream warehouse.

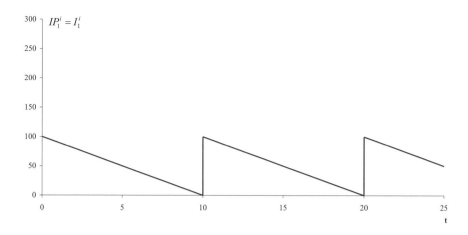

Fig. 6.15 Inventory position and inventories on hand in the first warehouse.

To build some intuition on the inventory dynamics in the second warehouse, we refer to example 6.13.

Example 6.13 Let us consider a product whose demand is 10 units per period, $Q_2 = 300$, and $Q_1 = 100$. If we assume that initial inventories are 100 and 200 units, respectively, the first warehouse places orders at time $t = 10, 20, 30$ and so on (see figure 6.15). Warehouse 2 runs out of inventories at time $t = 20$ but there is a need for additional inventories only at time 30, when warehouse 2 orders and receives 300 units, 100 of which immediately go to warehouse 1 (see figure 6.16). So inventories in the upstream warehouse do not fluctuate between 0 and 300 unit but vary between 0 and $Q_2 - Q_1 = 300 - 100 = 200$ with an average of 100 units (rather than 150). ⬜

More in general, the upstream warehouse places an order to receive the quantity Q_2 when it needs to ship a quantity Q_1 to the first warehouse. So inventories vary between 0 and $Q_2 - Q_1 = (j-1)Q_1$. The inventory level in the second warehouse remains constant at each level $0 \cdot Q_1, 1 \cdot Q_1, \ldots, (j-1) \cdot Q_1$ for Q_1/d periods. Thus the average inventory level is $\dfrac{(j-1)Q_1}{2}$.

As a consequence, the cost function for the second warehouse is

$$C_{tot,2} = A_2 \cdot \frac{d}{j \cdot Q_1} + h_2 \cdot \frac{(j-1)Q_1}{2}. \tag{6.8}$$

This cost function is convex in j. So the manager of the second warehouse can relax the problem and find a solution in $\mathbb{R}^+$ so that a (generally) non-integer solution is found as follows:

$$j^* = \frac{1}{Q_1}\sqrt{\frac{2A_2 d}{h_2}}. \tag{6.9}$$

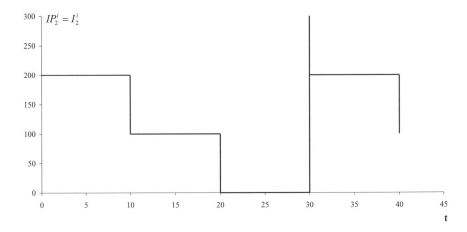

Fig. 6.16 Inventory position and inventories on hand in the second warehouse.

Once we have found this optimal solution (that in general is not integer and thus not feasible), we shall just consider the two closest integers and check which one minimizes the cost function. Indeed, we can leverage on the convexity of the cost function to rule out any other integer solution.

Example 6.14 Let us consider an European company that imports goods from the Far East and has a bonded warehouse where goods can be held before import duties are paid. The fixed cost of ordering to the Asian supplier (A_2) is 8.000€. Once duties are paid and goods clear customs, they are shipped to the warehouse nearby the city of Turin. A delivery from the bonded warehouse to the warehouse nearby Turin costs roughly 2.000€ (A_1). This cost covers both administrative costs and transportation. Monthly demand for the product is 8000 pcs. Holding one unit in the bonded warehouse for a month costs 4€, while holding it in the warehouse nearby Turin costs 5€ because of the greater capital investment due to import duties.

The manager of the first warehouse buys in lots of size

$$Q_1 = \sqrt{\frac{2A_1 d}{h_1}} = \sqrt{\frac{2 \cdot 2000 \cdot 800}{5}} = 800 \, \text{pcs},$$

and thus the total cost of the first warehouse is

$$
\begin{aligned}
C_{tot,1} &= \sqrt{2A_1 d h_1} = \sqrt{2 \cdot 2000€ \cdot 800 \, \text{pcs/month} \cdot 5€/(\text{pcs} \cdot \text{month})} \\
&= 4000€/\text{month}.
\end{aligned}
$$

Given this decision of the manager of the first warehouse, the manager of the second warehouse sets the parameter j through equation (6.9):

$$j^* = \frac{1}{Q_1}\sqrt{\frac{2A_2 d}{h_2}} = \frac{1}{800\,\text{pcs}}\sqrt{\frac{2 \cdot 8000€ \cdot 800\,\text{pcs/month}}{4€/(\text{pcs} \cdot \text{month})}} \simeq 2.24.$$

So the two candidate integer values are 2 and 3. We can identify the best solution through equation (6.8):

$$C_{tot,2}(2) = 8000\frac{800}{2 \cdot 800} + 4 \cdot \frac{1 \cdot 800}{2} = 5600€/\text{month},$$

$$C_{tot,2}(3) = 8000\frac{800}{3 \cdot 800} + 4 \cdot \frac{2 \cdot 800}{2} = 5866€/\text{month}.$$

Thus, the solution is $Q_1 = 800$ and $Q_2 = 2 \cdot 800 = 1600$, and total cost is 9600€/month. □

Now we can check that minimizing the two cost functions separately leads to suboptimal results. We write the cost function for the whole supply chain and find the optimal solution:

$$\begin{aligned}
C_{tot} &= A_1 \cdot \frac{d}{Q_1} + h_1 \cdot \frac{Q_1}{2} + A_2 \cdot \frac{d}{j \cdot Q_1} + h_2 \cdot \frac{(j-1)Q_1}{2} \\
&= \left(A_1 + \frac{A_2}{j}\right)\frac{d}{Q_1} + (h_1 + (j-1)h_2)\frac{Q_1}{2}.
\end{aligned} \qquad (6.10)$$

The cost function resembles the EOQ problem where ordering and holding costs are:

$$\begin{aligned}
A &= A_1 + \frac{A_2}{j} \\
h &= h_1 + (j-1)h_2.
\end{aligned}$$

Each j orders from warehouse 1, warehouse 2 places 1 order. Thus each ordering cycle in warehouse 1 costs A_1 but it also implies an indirect cost that is a fraction $1/j$ of the ordering cost A_2 of the second warehouse. As to holding costs, the average inventory level in the first warehouse is $Q_1/2$, while the average inventory level in the second warehouse is $(j-1)Q_1/2$, that is, $j-1$ times higher. This means that for each unit held in the first warehouse we hold $j-1$ units in the second warehouse, and thus increasing the inventory level of the first warehouse by one unit causes an increase of $j-1$ units in the second warehouse. So we do not need any further math and simply resort to the EOQ model to derive the optimal quantity Q_1^*:

$$Q_1^* = \sqrt{\frac{2\,(A_1 + A_2/j)\,d}{h_1 + (j-1)h_2}}. \qquad (6.11)$$

We can take the derivative of equation (6.10) with respect to j:

$$\frac{\partial C_{tot}}{\partial j} = -\frac{A_2 \cdot d}{Q_1 \cdot j^2} + \frac{Q_1 \cdot h_2}{2} = 0.$$

hence, rearranging terms and substituting the generic quantity Q_1 with the optimal quantity Q_1^*:

$$\frac{\partial C_{tot}}{\partial j} = \frac{1}{Q_1^*} \left(-\frac{A_2 \cdot d}{j^2} + \frac{h_2 \cdot Q_1^{*2}}{2} \right)$$

$$= \frac{1}{Q_1^*} \left\{ -\frac{A_2 \cdot d}{j^2} + \frac{2 h_2 \left(A_1 + A_2/j \right) d}{2 \left[h_1 + (j-1) h_2 \right]} \right\} = 0;$$

hence

$$-A_2 \cdot [h_1 + (j-1) h_2] + j^2 h_2 \left(A_1 + A_2/j \right)$$
$$= -A_2 \cdot h_1 + A_2 \cdot h_2 + j^2 h_2 A_1 = 0.$$

Hence,

$$j^* = \sqrt{\frac{A_2 (h_1 - h_2)}{A_1 \cdot h_2}}. \tag{6.12}$$

Before we proceed and try to implement the solution, we want to restate the problem with an Echelon stock logic. From an echelon stock standpoint we consider units in the first warehouse to be in the second warehouse as well. Thus the Echelon stock holding cost e_1 for the first warehouse is just the incremental holding cost; that is, the difference between the cost of holding one unit in the first warehouse minus the cost of holding one unit in the second warehouse $(h_1 - h_2)$. Indeed, units that are physically in the first warehouse already pay a holding cost h_2, since they are part of the echelon stock of the second warehouse (though they are physically in the first warehouse). This variable has a clear economic reading. It is the incremental cost of holding one unit in the downstream warehouse (2) rather than in the upstream one (1):

$$e_1 = h_1 - h_2,$$
$$e_2 = h_2.$$

We can restate the cost functions with echelon holding costs as follows:

$$C_{tot,1}^e = A_1 \cdot \frac{d}{Q_1} + e_1 \frac{Q_1}{2},$$
$$C_{tot,2}^e = A_2 \cdot \frac{d}{j \cdot Q_1} + e_2 \frac{j \cdot Q_1}{2},$$

where the average echelon inventory level for the second warehouse is $j \cdot Q_1/2$. Indeed, the Echelon Stock logic implies that the inventory level is not affected by the transfer of Q_1 units to the first warehouse. Also, inventories are only influenced by the final demand and vary linearly between 0 and $Q_2 = j \cdot Q_1$.

Hence the total cost is

$$C_{tot}^e = C_{tot,1}^e + C_{tot,2}^e = \left(A_1 + \frac{A_2}{j} \right) \cdot \frac{d}{Q_1} + (e_1 + j \cdot e_2) \frac{Q_1}{2}.$$

We can use the EOQ model to derive the optimal lot size for the first echelon:

$$Q_1^* = \sqrt{\frac{2d \left(A_1 + \dfrac{A_2}{j} \right)}{e_1 + j \cdot e_2}}, \tag{6.13}$$

and for the total cost when the optimal lot Q_1^* is selected:

$$C_{tot} = \sqrt{2d \left(A_1 + \frac{A_2}{j} \right) (e_1 + j \cdot e_2)}. \tag{6.14}$$

In this case too, the total cost function is convex in j and thus we can relax the problem, find the optimal j in $\mathbb{R}^+$, and finally check which of the two closest integers minimize the total cost. The cost function reaches its minimum where the function under the square root reaches its minimum. Thus the unconstrained optimal solution j is

$$j^* = \sqrt{\frac{A_2 \cdot e_1}{A_1 \cdot e_2}}. \tag{6.15}$$

Equation (6.15) is just a reformulated version of equation (6.12) (this must be the case given that it is still the same problem) where we use Echelon holding costs (e_1 and e_2) instead of Installation holding costs (h_1 and h_2). Moreover, here the result is derived more elegantly. The final step is to round j^* and check whether the floor of j^* rather than the ceiling of j^* are the best integer solution.

Example 6.15 Going back to example 6.14, we now check whether the minimization of total costs for the whole chain suggests a different and more efficient solution. If we minimize total costs, we can choose j through equation (6.15):

$$j^* = \sqrt{\frac{8000 \cdot (5 - 4)}{2000 \cdot 4}} = 1;$$

thus the lot size for the two echelons is basically the same and is equal to

$$Q_1 = Q_2 = \sqrt{\frac{2 \cdot 800 \left(2000 + \dfrac{8000}{1} \right)}{1 + 1 \cdot 4}} = 1788.9 \, \text{pcs}$$

for a total cost of 8944.3€/month, well below the cost of 9600€/month for the disjoint optimization.

We shall notice, though, that this solution involves higher costs for the first warehouse. Indeed, if we choose a lot size $Q_1 = 1788.9$ pcs, the cost for the first warehouse reaches $2000 \cdot 800/1788.9 + 5 \cdot 1788.9/2 = 5366.6$€/month, definitely more than the 4000€/month for the disjoint optimization. In this latter case the cost for the first warehouse is higher since when we look for a solution we account for the effects that the lot sizing decisions of the first warehouse generates on the second one. This increase in the costs of the first warehouse raises some issues on how the saving of a joint optimization should be split, in case the two warehouses belong to different organizations.   ⬜

Concept 6.7 *Integrated management of a supply chain increases the efficiency of the whole supply chain. However, this does not automatically turn into a benefit for each and every player in the chain. This is why we often shall carefully consider the incentives and potential gains for each player when we suggest a greater degree of integration* (see chapter 7).

6.5   ARBORESCENT CHAIN WITH TWO ECHELONS: TRANSIT POINT WITH UNCERTAIN DEMAND

As we have discussed in the second chapter, one of the key roles of distribution warehouses (also known as distribution centers) is to consolidate traffic and increase the efficiency (e.g., through a greater utilization of cargo space or the usage of larger trucks) of transportation from points of production to points of consumption. In addition, in distribution warehouses we can prepare assorted deliveries with tens of different products for a single store (or local warehouse) and increase the delivery frequency of each single product while still keeping the cost of transportation under control.

Distribution warehouses also play other functions. For example, they hold inventories that can be immediately delivered where demand is higher than we initially expected. Also they manage to deal with very large purchase quantities, since any inventories exceeding the immediate need of the lower echelons can be stored in the warehouse for a while.

However, some warehouses are designed not to perform these other functions and only perform the first one. They are designed to receive large quantities (say a full truck or a container) of a given product (or a limited assortment). We call the transportation of goods from the production site to the distribution center primary transportation. These warehouses immediately use these deliveries to feed smaller trucks (typically with a broader assortment since we load them with goods coming from various suppliers) that then deliver the products to stores and customers. We call this secondary transportation.

Example 6.16 For example, the Carrefour group, like most other grocery chains, uses transit points for fresh food, while it uses traditional warehouses for packaged goods. In the transit point for fresh food, they receive goods in the late afternoon full trucks from their suppliers. They might receive a full truck of vegetables, one full truck of fresh fish, one full truck of meat, and so on. In a few hours these products are mixed and ready for delivery to stores. Each store receives only one delivery of a single truck that carries vegetables, fresh fish, meat, and so on.   ▯

These nodes of the network are called *transit points, transshipment points* or *cross-docking centers*. In these centers, inventories are not stocked and thus usually there are no shelves. However, the large lots received are broken down into smaller quantities that are then used to prepare a mixed cargo for secondary transportation. The term transit point really means that goods only transit in the warehouse.

In this section we discuss how one can manage a transit point and show how a transit point can help us manage demand uncertainty. We investigate the case of a single product and thus we cannot fully capture the savings on transportation costs due to consolidation of a mix of products. As to savings on transportation, we refer to chapter 2.

For the sake of simplicity we investigate a two echelons supply chain where the transit point serves I stores (local warehouses). Each store has a stochastic demand with expected value m_i and standard deviation σ_i. We call LT_2 the delivery lead time to the transit point, and we call LT_1 the delivery lead time from the transit point to the stores. Finally, we assume a review period τ for the transit point and a S Echelon Stock inventory policy (notice that the transit point has basically no inventories and thus the Installation Stock policy would hardly make sense).

Under these assumptions we first describe how the system works and then try to build a model of it and prescribe how to run it. The first question we need to answer is, What is the out-of-control-period for the transit point? And what is the out-of-control-period for the local stores?

Let us consider an order placed by the transit point at time t_0. This order is delivered to the transit point at time $t_0 + LT_2$. At the transit point the goods are allocated to single stores where they are delivered at time $t_0 + LT_2 + LT_1$ (see figure 6.17).

In a system like this one, it does not make sense to set the inventory levels at the stores, as all goods received by the transit point must be immediately shipped to the stores. So the real question becomes how we allocate the goods we have just received to the stores.

To manage this supply chain we have to make two decisions.

- First, we have to choose the inventory policy for the transit point, in this specific case we have to select the appropriate level of the Echelon Stock order up to level S.

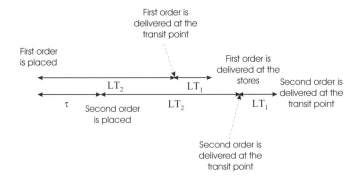

Fig. 6.17 Timeline of the inventory planning problem in case of a two-echelon supply chain with a transit point.

- Second, we have to define a policy to allocate to stores the inventories we receive at the transit point.

Selecting the Echelon Stock order up to level S.   As to the first problem, we shall first understand what demand inventories need to cover and thus what demand we shall use to set the order up to point S. The order we place in t_0 is delivered at the stores at time $t_0 + LT_2 + LT_1$. The following order placed at time $t_0 + \tau$ is then delivered at the stores at time $t_0 + \tau + LT_2 + LT_1$. So inventories ordered in time t_0 shall cover demand up to the next delivery in time $t_0 + \tau + LT_2 + LT_1$ and thus shall cover the whole out-of-control period $\tau + LT_2 + LT_1$.

The question then becomes, What is the relevant demand over the out of control period? The problem is all but trivial. While the expected value is obviously $(\tau + LT_2 + LT_1) \cdot \sum_{i=1}^{I} m_i$, understanding what is the relevant standard deviation is a more complex issue. We shall understand whether inventories shall (i) cover demand uncertainty for each single store or (ii) cover it at the chain level. In other words, we shall understand whether we want to (i) take the sum of all I standard deviations rather than (ii) take the standard deviation of the sum of all I demands (see chapter 2).

To answer this question, we shall carefully study the flow of goods in the chain and understand:

- When and where a unit of inventories can be used at all stores (is fungible for all stores) and thus can cover demand fluctuation wherever they happen;

- When and where, on the contrary, inventories can only be used to meet demand in a given store and thus can cover demand fluctuations at that store only.

Goods ordered at time t_0 are allocated to a specific store only when they are received in the transit point at time $t_0 + LT_2$. Up to this point in time, goods can be used to rebalance inventories among the stores as we can allocate more units to stores that have experienced a peak in demand. Up to $t_0 + LT_2$ we really do not care about the distribution of demand among the I stores but rather are interested in the overall demand for the network of I stores. Indeed, when one of the I stores faces an unpredicted peak in demand while another store experienced a period of low demand, the two variations compensate and cancel out. We simply allocate more inventories to the former store and less inventories to the latter. What might cause some concerns is a demand above the initial expectations for the whole chain of I stores. In this case we cannot allocate more inventories to one store and less to another. All stores (say) need more than we expected and the incoming order is just not enough.

At time $t_0 + LT_2$, inventories are allocated to each single store; and from this point in time, up to $t_0 + \tau + LT_2 + LT_1$ each single store has a limited and fixed quantity of goods to meet demand. During this time interval $(t_0 + LT_2, t_0 + \tau + LT_2 + LT_1)$ an increase in demand at one store cannot be possibly be counterbalanced by a decrease at another store since excess inventories in the latter cannot be used to meet demand in the former (if horizontal shipments are not allowed, like in all our models).

Hence, the order up to level S shall cover demand for the whole chain over a period LT_2 and shall cover demand from each single store over a period $LT_1 + \tau$.

Let us now introduce some notations. We call σ the standard deviation of demand for the whole chain. When demands are independent across stores, we have

$$\sigma = \sqrt{\sum_{i=1}^{I} \sigma_i^2}.$$

We call $\tilde{\sigma}$ the sum of standard deviations, that is, the sum of demand uncertainties for each single store:

$$\tilde{\sigma} = \sum_{i=1}^{I} \sigma_i.$$

So the relevant demand distribution to set the order-up-to-level S is a random variable with expected value

$$(\tau + LT_2 + LT_1) \cdot \sum_{i=1}^{I} m_i,$$

and, if we assume demand is not correlated over time, with a standard deviation

$$\sqrt{LT_2 \cdot \sigma^2 + (LT_1 + \tau)\tilde{\sigma}^2}.$$

Once we have identified the demand distribution, we can set the order-up-to-level S using the heuristics we have designed in 5.6. In this case too,[13]

$$1 - F(S) = \frac{h \cdot \tau}{p_u}, \qquad (6.16)$$

where p_u is the penalty cost for stockouts.[14]

Example 6.17 Let us consider a company with a review period τ of one week, a delivery lead time to the central transit point LT_2 of 2 weeks, and a lead time of 1 week to transfer goods from the transit point to the 9 stores LT_1 of 1 week. Each of the 9 stores has a normally distributed demand with expected demand 100 and standard deviation 25. We also assume that the margin of the product is 1€ and the holding cost for 1 week is 0.1€.

In this case, the demand we use to set the order-up-to-level has an expected value equal to the out-of-control period $(\tau + LT_2 + LT_1)$ times the sum of expected demands for the I stores $\sum_{i=1}^{I} E(d_i)$, i.e., $(1 + 2 + 1) \cdot 9 \cdot 100 = 3600$.

If we assume that stores are independent and there is no autocorrelation, the standard deviation of demand is

$$\sqrt{LT_2 \cdot \sigma^2 + (LT_1 + \tau) \tilde{\sigma}^2} = \sqrt{2 \cdot (9 \cdot 25^2) + (1+1)(9 \cdot 25)^2} = 335.41.$$

We look at the economics of the company to choose the optimal type I service level through formula (6.16):

$$1 - F(S) = \frac{0.1 \cdot 1}{1} = 0.1;$$

---

[13]Other authors suggest to set S in such a way that $F(S) = \frac{m}{m+h \cdot \tau}$. This solution traces back to the newsvendor problem. The above solution and the one we suggest make basically different assumptions on the holding cost. We assume that we incur the full (see note 20 on page 274) stock holding cost even if the marginal unit is sold before the end of the planning cycle. Other authors assume (implicitly or explicitly) that the company incurs the holding cost only if the product is left unsold at the end of the planning cycle. In this situation the inventory problem resembles the newsvendor one. More formally, under our simplifying assumptions the expected marginal return is $(1 - F(S)) \cdot m$ and the marginal cost of inventories is deterministic $(h \cdot \tau)$. In the alternative model the holding cost is stochastic and has an expected value $F(S) \cdot h \cdot \tau$. Obviously the truth lies in between these two extremes. We hold inventories for the whole cycle and thus pay the full holding cost $h \cdot \tau$ when we do not sell the marginal unit, while we carry the item for a portion of the planning cycle when we sell it by the end of the cycle (see note 20 on page 274). Clearly, one approximation might work better than the other according to the service level we provide. The higher the service level the more frequently the marginal unit is carried for the whole planning cycle; and even when it is sold, it gets sold towards the end of the planning cycle and thus our assumption works better.

[14]Notice that the transit point "per se" does not guarantee that we can postpone the allocation of inventories among stores up to time $t_0 + \tau$. Some US retail chains use transit points to consolidate transports of products imported from Asia but make firm decision on inventory allocation when they place the purchase order. This way they lose flexibility but gain efficiency. The expensive handling operations can be performed in Asia rather than in the USA with substantial savings. So the usage of the transit point to postpone decisions on allocation of inventories rather than the transit point per se guarantees a reduction in the inventories required to gain a given service level target.

thus $F(S) = 0.9$, and if the probability distribution is normal we can use tables for the standardized normal distribution and choose $z = 1.28$ and thus set S to $S = 3600 + 1.28 \cdot 335.41 = 4029.32$.

Notice that if we allocate the inventories to stores when we place the order in t_0 (say, we import in pre-set boxes of goods tagged to the final destination), rather than when we receive the goods (say in $t_0 + LT_2$), the inventories required to gain a 90% increase. Indeed, in this case each store is basically independent and we shall set the inventory level for each store and simply add up the requirements for each store. So, each store sets an order-up-to-level S_i to cover demand over the out-of-control period of 4 weeks. So each store i needs and order-up-to-level $S_i = (1 + 2 + 1) \cdot 100 + 1.28\sqrt{4 \cdot 25^2} = 464$, for a total of 4176 ($9 \cdot 464$) units for the network of 9 stores.

The transit point creates efficiency, since inventories shall only cover the uncertainty on the aggregate demand for the network of I sores and thus a peak of demand in a store can be counterbalanced by a low demand in another store. We simply allocate more inventories to the former store and less to the latter one. On the contrary, once products are allocated to single stores, a peak in demand at one store is not counter-balanced by a low demand at another store, as each single store can only rely on the inventories that have been allotted to it.   ⬜

Concept 6.8 *In a supply chain with a transit point, we can postpone decisions on the allocation of inventories to stores. Thus we can set inventory levels according to the aggregate demand and, thanks to* risk pooling, *reduce the uncertainty and thus reduce the need for safety stocks.*

Allocation of inventories to stores   We have set the Echelon Stock inventory target for the transit point, but we still have one open question: How do we allocate goods among the stores?

Before we get into the details of how to allocate inventories, we shall understand what is the relevant demand distribution.

Figure 6.17 shows that the order placed in time t_0 is allocated to stores at time $t_0 + LT_2$, it reaches the store at time $t_0 + LT_2 + LT_1$, and the successive order is delivered to stores at time $t_0 + \tau + LT_2 + LT_1$. Inventories we allocate to stores at time $t_0 + LT_2$ shall cover demand up to $t_0 + \tau + LT_2 + LT_1$ and thus they shall cover the out of control period $LT_1 + \tau$.

To solve the allocation problem, we refer to section 5.4.1 and assume that the size of the stockout, rather than its occurrence, matters. The same logic we have developed in that case applies to this case, as we face an allocation of a limited amount of common resources (in this case, inventories delivered to the transit point).

If we assume that all stores share similar costs of stockout (say margins) and holding costs (this assumption is very reasonable in many retail chains), we shall allocate the incoming order in such a way that all stores share the same type I service level. Unit $Q + 1$ is sold with a probability $1 - F(Q)$. If store 1 has a higher service level than store 2, then we have $1 - F_1(Q_1) < 1 - F_2(Q_2)$ and thus we would rather allocate a marginal unit of inventories to store 2 (with a lower service level) than to store 1 (with a high service level).

So, we just want to use the allocation process suggested in section 5.2.1 for the multi-item newsvendor problem. In this case we allocate inventories of one item to stores, rather than a common resource to various products, but still the logic is the same.

The budget, i.e., the total amount of common resources, we must allocate is the overall inventories level k in the downstream supply chain (on hand plus any goods in transit, i.e., the sum of inventory positions for all stores) plus inventories entering the transit point, while $r_i = 1; \forall i$. Shipments to store i are equal to the optimal inventory level for store i minus initial inventory position for the store.

If expected demand in a store drops sharply while others experience a significant increase in demand and horizontal shipments are not allowed, the optimal allocation might turn out to be unfeasible. Some stores might indeed have already more inventories than they should have under the optimal allocation and if horizontal shipments are not allowed, we cannot reduce the inventory position immediately but simply have to wait for demand to progressively consume inventories. This means that some of the inventories that other stores would need are stuck in the store with excess inventories. This really means that the whole allocation plan is unfeasible and shall be reviewed, as example 6.18 shows.

Example 6.18 Let us assume store 1 has 1000 units, while the other 2 stores in the chain only have 100 units on hand. Let us assume that the 3 stores face a similar demand distribution and the same economics. Also let us assume there is no in transit inventory between the transit point and the stores. Finally, the transit point is receiving 600 units and shall allocate them to the stores.

Obviously, the optimal solution is to evenly distribute inventories among the stores: $(1000 + 100 + 100 + 600)/3 = 600$ units. However, in the first store we already have more than 600 units (1000 units) and horizontal shipments are not allowed, so this solution in unfeasible. Still, it provides us with a precious piece of information: We shall not ship any goods to store 1. We just allocate the 600 units the transit point is receiving to stores 2 and 3, so we take their inventory position to 400 units. The service level for stores 2 and 3 is still significantly below the service level of store 1 (whose inventory position is 1000 units), but we would need horizontal shipments to improve our solution. ▯

If the current inventory position of any store exceeds the optimal inventory position, we would need to ship a negative quantity (i.e., ship some units from that store to other stores). If that is not possible, we exclude stores that should receive a negative shipment from our allocation and reallocate inventories only among stores that in the previous run had a positive shipment.

In other words, we rule out stores that quite obviously should not receive any goods and we allocate incoming goods only among the stores that might need some inventories.

To illustrate an allocation algorithm, we introduce the following notations:

- Sh_i is the shipment to store i.

- O_i is a dummy variable $(0/1)$ that captures whether we want to ship any good to store i.

- V is the total amount of inventories that is entering the transit point and shall be allocated to stores.

A possible allocation policy is based on Lagrangian multipliers (i.e., opportunity cost) λ of the budget constraint. Notice that in this case we have to allocate all goods we receive in the warehouse, so the multiplier λ can be both positive and negative. Indeed, in some instances we deliver to the stores more goods that would be required simply because we need to allocate all goods we have received.

1. $O_i = 1; \forall i$, so initially we consider all stores as potential candidates for the shipment of a portion of the goods we are receiving at the transit point.

2. $\lambda = 0$, so initially we assume that inventories have a zero opportunity cost.

3. The total amount of inventories to allocate is $k = \sum_{i=1}^{I} IP_i \cdot O_i + V$. Notice that we only consider the inventory position of the stores that we consider to be part of the allocation. Once a store i is removed from the allocation process $(O_i = 0)$ its inventories are removed as well. The store basically exits the allocation game.

4. Find $Q_i = F_{d_i, LT_1+\tau}^{-1} \left(\dfrac{E(d_i) - \lambda \cdot r_i}{E(d_i) + c_i} \right)$.

5. If current solution is such that $\sum_{i=1}^{I} Q_i \leq k$, increase the opportunity cost of inventories (capacity) λ; otherwise reduce λ.

6. If the solution meets (with a given tolerance) the constraint $\sum_{i=1}^{I} Q_i = k$ proceed; otherwise goto 4; this step guarantees that we can find an optimal allocation among the active stores $(O_i = 1)$, but still there is no guarantee that the solution is feasible.

7. We calculate the shipment for each store, $\forall i$, $Sh_i = Q_i - IP_i$.

8. IF the shipment is negative, then we draw the conclusion that we shall not ship any goods to the store that is excluded from the allocation: $\forall i$, if $Sh_i < 0$, then $O_i = 0$.

9. If in the previous step any store was excluded from the allocation process taking O_i from 1 to 0, goto step 2, otherwise stop.

In the specific case of normal demand distribution and cost of inventories h_i and cost of stockout $E(d_i)$ equal for all stores, we want all stores to reach the same type I service level; thus, all stores shall reach the same quantile in their respective demand distribution and thus use the same standardized value z. Hence

$$\sum_{i=1}^{I} \left((LT_1 + \tau) E(d_i) + z \cdot \sqrt{LT_1 + \tau}\sigma_i \right) O_i = k$$

and thus

$$z = \frac{k - \sum_{i=1}^{I} O_i (LT_1 + \tau) E(d_i)}{\sum_{i=1}^{I} O_i \sqrt{LT_1 + \tau}\sigma_i}, \qquad (6.17)$$

where $O_i = 1$ for stores that actually receive some goods from the central transit point.

Example 6.19 Let us consider a network of 11 stores with a normally distributed monthly demand. The expected monthly demand is 100 units, while the standard deviation is 20 units. The inventory position for stores 1–5 is 50 units, the inventory position for stores 6–10 is 100 units, and the inventory position for the 11th store is 200 units. Stores share the same prices, purchasing costs, and holding costs. The LT to replenish the transit point is 2 months, while the LT to deliver stores from the transit point is 1 month. Goods are ordered to the suppliers once a month. We are receiving 350 units at the transit point and we shall allocate them to the 11 stores. The 11 stores face the same demand distribution, and thus we know that the optimal solution is to provide them with the same inventory position. So we simply divide the sum of all inventories in the system or entering the system by 11. Overall inventories currently in the downstream portion of the supply chain are $50 \cdot 5 + 100 \cdot 5 + 200 + 350 = 1300$, that is 118.18 units per store. Unfortunately the 11th store already has 200 units and, when horizontal shipments are not allowed, the solution is unfeasible. This really means that we do not want to ship any goods to store 11 and thus we allocate the incoming goods to the first 10 stores only. In this case the overall inventories in the set of stores we currently consider for allocation is $50 \cdot 5 + 100 \cdot 5 + 350 = 1100$. Thus we want to have 110 units in each of the first 10 stores and we ship 60 $(110 - 50)$ units in each of the first 5 stores, and ship 10 $(110 - 100)$ for stores 6 to 10. Notice that the total shipment is 350 units, which is exactly the quantity we are receiving at the transit point.   ⬜

Example 6.20 Let us consider a network of 3 stores. Stores have a normally distributed demand with an expected value of 100, 200, and 100, respectively, and standard deviations 20, 20, and 30, respectively. The delivery lead time to the stores is 2 weeks. The company places orders (and thus receives goods) every two weeks. Stores share prices, purchase, and holding costs. The inventory position of the 3 stores is currently 60, 100, and 40 units, respectively. The transit point is receiving 270 units.

When goods are received, the total amount of inventories in the (downstream[15]) distribution chain is $60 + 100 + 40 + 270 = 470$. This amount of inventories is used to reach a common service level in the 3 stores. We use equation (6.17) to find the optimal solution:

$$z = \frac{470 - (100 + 200 + 100)}{20 + 20 + 30} = 1.$$

The optimal inventory positions for the 3 stores are $100 + 1 \cdot 20 = 120$, $200 + 1 \cdot 20 = 220$, and $100 + 1 \cdot 30 = 130$ respectively. So we shall ship $120 - 60 = 60$, $220 - 100 = 120$, and $130 - 40 = 90$ units to the three stores respectively. Once again notice that the total shipment to the three stores is $70 + 120 + 80$, that is exactly the 270 units that are entering the transit point. □

6.6 A TWO-ECHELON SUPPLY CHAIN IN CASE OF STOCHASTIC DEMAND

Let us consider a stochastic extension of the problems addressed in the previous section. [16]We now assume that the warehouse in the second echelon actually carries inventories and is not just a transit point. Though this problem resembles the previous one, inventories in the upstream warehouse increase the complexity of the system and, as a consequence, the complexity of our model.

Under uncertain conditions, there is no guarantee that when the downstream warehouse (echelon 1) places an order, it can be actually fulfilled by the upstream warehouse (echelon 2). Hence, even if we assume deterministic lead times for materials handling and transportation, we might still face a delay in the delivery to the downstream warehouse simply because the central warehouse might be stocked out. So demand uncertainty creates an uncertainty in delivery lead times.

To show this concept, we analyse a simplistic case where the downstream warehouses use a *one-for-one* logic; that is, when they sell one unit, they order one unit. In other words, they use an S policy with continuous review. Under this assumption, the demand at the central warehouse is simply the sum of

---

[15]Notice that we do not account for any outstanding order the transit point has already sent to the supplier. It is part of the inventory position of the transit point but it is irrelevant for the allocation problem.
[16]This section was inspired by [5].

demands from all stores. Indeed, orders from store i are just equal to demand at store i. This simplifies our model substantially, as we do not need to model the ordering policy of the stores.

Finally, we assume that the central warehouse adopts a (Q, R) policy and an Installation Stock logic.

We adopt the following notations:

- l_i is the (deterministic) transportation and handling time to deliver to store i; that is, the Lead Time in case the central warehouse has enough inventories to immediately fulfill the order coming from store i; to build our intuition we can assume that this is the time required to handle the goods, load the truck, and deliver. We basically start our clock when goods are available in the central warehouse for delivery.

- L_i is the (stochastic) delivery lead time of an order placed by store i; it is the time elapsed from the time the order is placed up to the delivery of goods at the warehouse; so it includes any waiting time at the central warehouse in case the product is currently stocked out (this is actually the random portion of this delivery lead time).

- $B(Q, R)$ is the average backorder in the central warehouse and thus is the queue (waiting line) an order finds (on the average) at the central warehouse.[17]

- $W(Q, R)$ is the average time the order waits at the central warehouse because the product is stocked out.

In order to proceed, we need to find a relationship between the average backorder $B(Q, R)$ and the waiting time $W(Q, R)$. We shall use Little's law, a fundamental equation in queuing theory. It links the average waiting line Λ, the average waiting time W, and the throughput θ (average number of customers served per unit of time, e.g., hour):

$$\Lambda = \theta W.$$

In our case, backorders are basically a waiting line and the throughput is demand at the central warehouse, that is, the sum of demands from all stores.

Hence, applying Little's law to our problem, we find that the average waiting time is the ratio between the average Backorder and demand, as it is the

---

[17]Notice that the definition of average backorder depends on both R and Q, while the backorder at the end of the cycle $n(R)$ only depends on the reorder point R. In this case we try to capture the average number of customers (units ordered) that are waiting for their order to be fulfilled on the average rather than at the end of the planning cycle (that is, the point in time where we expect the longest queue). $B(Q, R)$ looks at the average queue during the whole planning cycle and thus depends on the duration of the cycle as well. Clearly, Q determines the average duration of the planning cycle.

average rate of arrival of inventories and, thus, the rate at which customers are served:

$$W(Q, R) = \frac{B(Q, R)}{E(d)}.$$

On the average , the delivery lead time for orders placed by store i is equal to the deterministic component l_i plus the stochastic time we wait at the central warehouse. So the expected delivery lead time is equal to l_i plus the expected waiting time $W(Q, R)$:

$$E(L_i) = l_i + W(Q, R). \tag{6.18}$$

We can use this delivery lead time to store i to set the order up to point S_i. One could be tempted to treat this random delivery lead time as if it was deterministic and equal to $E(L_i) = l_i + W(Q, R)$. This implicitly means that we assume that all orders wait $W(Q, R)$ units of time at the central warehouse, while some orders do not wait at all since the product is available immediately, and others wait much more than $W(Q, R)$. We know that a fraction of orders equal to the type II service level, β, is immediately fulfilled. For these orders, the delivery lead time equals l_i. On the contrary, orders that are not fulfilled immediately might wait much more than $W(Q, R)$. So the distribution of delivery lead times is definitely complex and can hardly be modeled properly. In other words, the waiting time is zero with a probability β and follows an unknown distribution with a probability $1 - \beta$. However, we can build a simplified model of this distribution of waiting times. We can assume that the waiting times at the central warehouse are zero when the product is in stock, but are equal to a constant a when the product is stocked-out. We basically assume that the waiting time follows a binomial distribution (with one single draw, which is also called Bernoulli distribution see A.3.1 on page 446). The waiting time is zero with a probability β and equals a constant a with a probability $1 - \beta$.

Now we need to set the parameter a of this simplified distribution. We can select a in such a way that the expected value of the binomial distribution (see figure 6.19) equals the expected value of the actual distribution of waiting times (see figure 6.18). This process is called *moment matching*. We basically replace the actual and complex distribution with a simplified one, but we make sure that the moments (in this case the first moment, that is, the expected value) of the simplified distribution match those of the more complex and realistic one.

So we set a in such a way that

$$E(L_i) = l_i + W(Q, R) = \beta \cdot l_i + (1 - \beta) \cdot (l_i + a);$$

hence

$$a = \frac{W(Q, R)}{1 - \beta}.$$

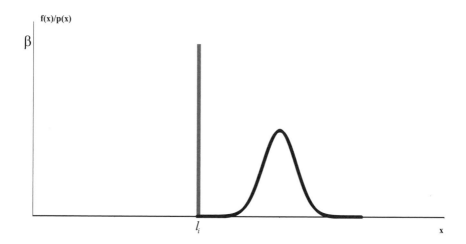

Fig. 6.18   Actual distribution of the LT_1.

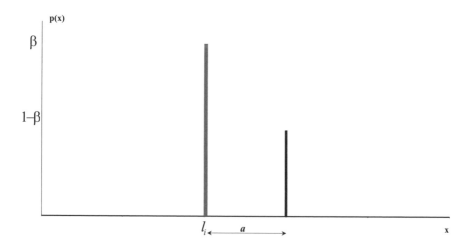

Fig. 6.19   Simplified distribution of LT_1.

Once we have selected a, we can use it to estimate the variability of the time stores have to wait for the delivery. Basic statistics tells us that the binomial distribution with one draw has a standard deviation

$$\text{Var}(L_i) = \beta\,(1 - \beta)\,a^2 = \frac{\beta}{1 - \beta}W^2(Q, R); \qquad (6.19)$$

hence

$$\sigma(L_i) = \sqrt{\frac{\beta}{1 - \beta}} \cdot W(Q, R)$$

Concept 6.9 *The simplified model shows that stockouts at the central warehouse increase both the expectation (6.19) and the standard deviation (6.20) of delivery lead time to the stores. Indeed, some orders are fulfilled immediately, while others shall wait at the central warehouse for the product to be replenished.*

This situation poses a new challenge: We have to model uncertain LTs. Indeed, when we plan inventories at the stores, we do not exactly know what period of time the inventory should cover and thus we do not know the relevant demand distribution. For example, inventories might cover demand for two weeks or maybe three weeks. Statistically, this means that the relevant demand distribution is the sum of an uncertain number of random variables (demands for each period of time). The following example illustrates this rather complex concept.

Example 6.21 Let us consider a store that receives goods from a central warehouse. The central warehouse delivers in one week (with a 70% probability) or two weeks (with a 30% probability) according to the availability of the product. Also we assume that the retailer follows a continuous review policy so that inventories shall only cover the LT (e.g., he uses a (Q, R) policy or the S policy with continuous review; that is, he orders as soon as a customer buys one unit). We assume that weekly demand is a normal distribution with expected value 100 and standard deviation 20. Demand is not correlated over time.

Under these assumptions the probability distribution of demand is bimodal (see figure 6.20); that is, demand is drawn from a normal distribution with expected value 100 and standard deviation 20, with a 70% probability, and it is drawn from a normal distribution with expected value 200 and standard deviation $\sqrt{2} \cdot 20$, with a 30% probability.

This means that the expected value of demand during the LT really depends on the duration of the LT. Nevertheless, we can compute the overall expected value across the two scenarios of $70\% \cdot 100 + 30\% \cdot 200 = 130$. Now, to calculate the standard deviation, we shall resort to the law of transport of moments. Here we derive the final result in full detail for the sake of completeness:

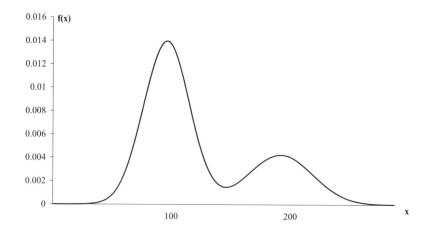

Fig. 6.20   Bimodal demand distribution of over the LT.

$$\mathrm{Var}(d_{LT}) = \int_{-\infty}^{+\infty} f_{d_{LT}}(x)\,(x-130)^2\,dx$$

$$= \int_{-\infty}^{+\infty} \left(0.7 \cdot f_{N(100,20)}(x) + 0.3 \cdot f_{N(200,\sqrt{2}\cdot20)}(x)\right)(x-130)^2\,dx$$

$$= 0.7 \int_{-\infty}^{+\infty} f_{N(100,20)}(x)\,(x-130)^2\,dx$$

$$+\,0.3 \int_{-\infty}^{+\infty} f_{N(200,\sqrt{2}\cdot20)}(x)\,(x-130)^2\,dx,$$

where

$$0.7 \int_{-\infty}^{+\infty} f_{N(100,20)}(x)\,(x-130)^2\,dx$$

$$= 0.7 \int_{-\infty}^{+\infty} f_{N(100,20)}(x)\left[(x-100)-(100-130)\right]^2\,dx$$

$$= 0.7 \int_{-\infty}^{+\infty} f_{N(100,20)}(x)\left[(x-100)^2 + 2\cdot30(x-100)+(100-130)^2\right]\,dx$$

$$= 0.7\left(20^2 + 0 + 30^2\right) = 910,$$

since:

- $\int_{-\infty}^{+\infty} f_{N(100,20)}(x)(x-100)^2\,dx$ is by definition the standard deviation of the demand that follows the distribution $N(100,20)$;

- $\int_{-\infty}^{+\infty} f_{N(100,20)}(x)\cdot30(x-100)\,dx$ is zero because the expected value of the distribution is 100;

- $\int_{-\infty}^{+\infty} f_{N(100,20)}(x) \cdot (100 - 130)^2 \, dx = (100 - 130)^2 = 30^2$ since it is a constant times the integral of a probability density over its support.

Similarly,

$$0.3 \int_{-\infty}^{+\infty} f_{N(200,\sqrt{2}20)}(x) \left[(x - 200)^2 - 2 \cdot 70 \cdot (x - 200) + 70^2 \right]^2 \, dx$$

$$= 0.3 \cdot \left(2 \cdot 20^2 + 0 + 70^2 \right) = 1710,$$

and thus the variance of demand over the LT is $910 + 1710 = 2620$ and standard deviation is $\sqrt{2620} = 51.2$, which is actually far greater than the variance of demand in a week, as we might face two very different demand scenarios. ◻

We can reach the same result following the same process described in example A.14, on page 474. If the duration of the LT and demand are independent random variables and demand is not autocorrelated, the variance of demand over the LT is

$$\text{Var}(d_{LT}) = \text{E}(LT) \cdot \text{Var}(d) + \text{Var}(LT) \cdot \text{E}(d)^2, \tag{6.20}$$

where $\text{E}(LT)$ and $\text{E}(d)$ are, respectively, the expected value of the LT and of demand in the unit of time. The first term models demand uncertainty, while the second term models LT uncertainty. Equation (6.20) can be applied whenever LT is uncertain. So equation (6.20) can be applied both (i) when the uncertainty is an endogenous variable and the manager can control through the service level β of the central warehouse and (ii) when it is an exogenous variable due to unreliable suppliers or uncertainty in transportation lead times (say weather or strikes).

Example 6.22 Going back to example 6.21 we can check that equation (6.20) correctly estimates the variability of demand over an uncertain LT. Indeed, LT has a binomial distribution and thus its variance is $0.7 \cdot (1 - 0.7) \cdot 1 = 0.21$ and its expected value is $1 \cdot 0.7 + 2 \cdot 0.3 = 1.3$ (see appendix A.3.1 on page sec:exdiscretedistrib. Thus the variability of demand over the LT is

$$\text{Var}(d_{LT}) = 1.3 \cdot 20^2 + 0.21 \cdot 100^2 = 2620,$$

which confirms the above calculations. ◻

Once we have obtained the relevant demand distribution we can set the order up to level for the stores through the heuristics we have designed in section 5.6.

REFERENCES

1. S. Axsater. *Inventory Control*. Kluwer Academic Publishers, Dordrecht, 2000.

2. E. Bartezzaghi, R. Verganti, and G. Zotteri. A Simulation Framework for Forecasting Uncertain Lumpy Demand. *International Journal of Production Economics*, 59:499–510, 1999.

3. F. Chen, Z. Drezner, J.K. Ryan, and D. Simchi-Levi. Quantifying the Bullwhip Effect in a Simple Supply Chain: The Impact of Forecasting, Lead Times, and Information. *Management Science*, 46:436–443, 2000.

4. R. Croson and K. Donohue. Impact of POS Data Sharing on Supply Chain Management: An Experimental Study. *Production and Operations Management*, 12:1–11, 2003.

5. W. Hopp and M. Spearman. *Factory Physics (2nd Ed.)*. McGraw-Hill, New York, 2000.

6. H.L. Lee, V. Padmanabhan, and S. Whang. The Bullwhip Effect in Supply Chains. *Sloan Management Review*, pages 93–102, Spring 1997.

7. H.L. Lee, V. Padmanabhan, and S. Whang. Information Distortion in a Supply Chain: The Bullwhip Effect. *Management Science*, 43:546–558, 1997.

8. D. Simchi-Levi, P. Kaminsky, and E. Simchi-Levi. *Designing and Managing the Suppy Chain (2nd Ed.)*. McGraw-Hill/Irwin, New York, 2002.

7

$$\text{\Large\itshape 7}$$

Incentives in the Supply Chain

7.1 INTRODUCTION

In the previous chapter we introduced the issue of multiechelon supply chains. As we have said, there are two major topics in multiechelon supply chains:

- On the one hand, in multiechelon supply chains, planning problems are quite complex, so we need rather complex planning processes and tools: this is what we called the engineers' perspective of the multiechelon supply chain problem. It is a hard problem and so we need "engineers" to find a solution that then all the players in the chain will be willing to implement, simply because it is a "good solution."

- On the other hand, decision makers in the supply chain might have at least partially different objectives, so they might not care about a good solution that minimizes the total cost for the whole chain if it contrasts with their own objectives. This is a very common case in a supply chain where different players belong to different companies. In this case, understandably, the various decision makers consider the profitability (or value creation) of their company as the primary objective. What is even more interesting is that these problems arise even among organizational units of the same given company. Indeed, decision makers might control a fraction of the overall company. Large companies have rather complex organizational structures with various organizational units. Each of them is usually controlled on a given set of performance metrics; the bonus, tenure in the current position, future career, and income of managers usually depend on such metrics. So, the managers of the various

organizational units tend to have different objectives rather than a con-
current interest in the profit of their company (e.g., see footnote 6.1 on
page 305). This is what we called in chapter 6 the economists' perspective: people are bright and can find solutions even to rather complex
problems. However, they are selfish and tend to do what is in their best
interest. Economists call these incentive problems. This is basically the
perspective we adopt in this chapter. For the sake of simplicity, in our
models we assume that the objective the various decision makers try to
maximize is the profit of their organization. In other words, we assume
that either organizations are independent and profit maximizing companies or they are different organizational units within the same company
but each of them is judged on profit.

In the models we present in this chapter we basically ignore inventory planning or transportation issues. In particular, in all our models the time variable
does not play a role, we basically assume the problem to be static. This really
means that the models presented here are simplistic and are designed to be
thought-provoking and informative rather than to provide tools or solutions.

In the remainder of this chapter we present models that show a variety of
incentives problems; that is, we describe the conflicts among decision makers
on several decisions. These conflicts lead decision makers to make locally
optimal decisions that turn into a suboptimal performance. In section 7.2 we
show the contrast of incentives between a producer and a retailer when setting
the final price of a product. In section 7.3, we investigate how things change
when a single producer provides the product to various retailers. In section
7.4 we discuss the contrast of interest on stocking decisions. In section 7.5 we
finally discuss the incentive to deploy an effort to increase demand through
better product design, additional product features, or better service at the
retail outlet.

In each of these sections we basically compare the performance (profit) of a
fully integrated chain where a single decision maker interested in the profit of
the supply chain makes all decisions with those of a nonintegrated chain where
each decision maker makes his/her own decisions to maximize the profit of
his/her own organization. The one difference between the situations we contrast is that in the latter case we only reach local optima. So the vertically
integrated case is bound to lead to superior (or at least equal) performance.
We show that under standard conditions (e.g., standard contracts among the
organizations in the supply chain) the vertically integrated supply chain out-
performs the disintegrated one. This actually does not mean that we suggest
that the vertically integrated solution is always the first best solution. There
are several reasons why a company might want to focus on core activities and
thus resort to partners to distribute or manufacture products. Actually, over
the past decades we have seen a general trend towards outsourcing, and thus
the issue of suppliers' and distributors' management has become more and
more relevant. We acknowledge that our models do not fully capture vari-

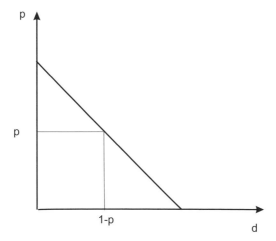

Fig. 7.1 Demand function.

ables that lead companies to outsource some activities and thus reduce the degree of vertical integration. In practice, the disintegrated chain (and decentralized decision making) can lead to substantial technical and/or economic advantages (such as lower cost of labor, specialized know how etc.). In our models we deliberately ignore these issues to highlight the decision making issues that arise in a vertically disintegrated chain. So we do not suggest using the models to make a decision on vertical integration. In our perspective the degree of vertical integration (integrated vs. disintegrated chain) is a given. Our models are simply meant to show that in the disintegrated chain some incentive issues arise and shall be accounted for. Also, in each section we show what are the basic remedies to the problems we highlight. We basically show that there are counterintuitive solutions that give all players (manufacturers and retailers in our models) a common interest in the global performance of the chain. This leads to a global optimum even in a disintegrated chain.

7.2 DECISIONS ON PRICE: DOUBLE MARGINALIZATION

Let us consider a one-product supply chain that consists of two stages. In the first stage the product is manufactured, while in the second stage it is sold to consumers at a retail store. We assume that the marginal (variable) production cost is c, while selling the product at the retail store has a zero marginal unit cost. Also, let us assume that the demand for the product is deterministic and linear. In particular, we assume that (see figure 7.1)

$$d = 1 - p \tag{7.1}$$

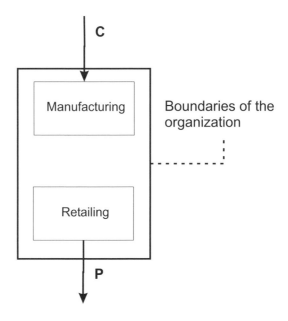

Fig. 7.2 Structure of the supply chain in the case of vertically integrated organization.

where d is the demand per unit of time and p is the price the final consumer pays. We also assume that the decision makers perfectly know this demand curve. Finally, this supply chain is a monopolist. The manufacturing stage is the only one production plant for the product, and the retailing stage is the only store selling the product. In other words, in this simple initial model we do not have any strategic interaction among competing manufacturers or competing retailers (see next section for the case of competing retailers).

7.2.1 The first best solution: the vertically integrated firm

In the first scenario the two operations (manufacturing and retailing) are performed by a single company that is vertically integrated (see figure 7.2). Also, we assume that within this company no incentive problem arises, so that either there is one decision maker or all decision makers share the same objective: maximization of the company profits.

In our scenario the company needs to set the price to maximize profit. The profit function is

$$\pi = d \cdot (p - c) - FC = (1 - p) \cdot (p - c) - FC. \tag{7.2}$$

We look for optimal conditions and thus fixed costs FC are irrelevant, if they do not make the optimal profit negative. For the sake of simplicity we assume $FC = 0$. To find the optimal price, we simply take the derivative of the profit function with respect to p:

$$\frac{\partial \pi}{\partial p} = c - p + 1 - p = 0;$$ (7.3)

hence

$$p^* = \frac{1 + c}{2}$$ (7.4)

so by substituting p with $p^* = \frac{1+c}{2}$ in (7.1) we can derive the optimal quantity and by substituting in 7.2 we can derive the optimal profit:[1]

$$d^* = 1 - p^* \quad = \quad \frac{1 - c}{2}$$

$$\pi^* \quad = \quad (1 - p^*)(p^* - c) = \frac{1 - c}{2} \cdot \frac{1 - c}{2}$$

$$= \quad \frac{(1 - c)^2}{4}.$$

In our simplistic model this is the *first best solution* where the single relevant decision (final price) is taken with a global perspective on the whole supply chain. In the next section we discuss how the situation changes as the supply chain in disintegrated.[2]

7.2.2 The vertically disintegrated case: independent manufacturer and retailer

In this section we investigate a slightly different scenario. We assume that manufacturing and retail distribution are performed by two separate organizations (e.g., two companies) with separate objectives. Each of the two players wants to optimize the profit of his/her organization (e.g., his/her company), and the product is sold to the retailer at a wholesale price p_w, so the cost function of the retailer is linear.

In this case, we shall describe the decisions the two decision makers face. Both the retailer and the wholesaler need to set a price. The retailer sets the

---

[1]In the more general case where manufacturing faces a marginal cost c and the retailer faces a marginal cost r, the marginal cost for the company becomes $c + r$ and thus the optimal price is $p^* = \frac{1+c+r}{2}$, optimal demand is $d^* = \frac{1 - (c+r)}{2}$ and the optimal profit is $\pi^* = \frac{[1 - (c+r)]^2}{4}$.

[2]Notice that to make the model work properly we need to assume that the marginal cost c is lower than one; otherwise this market simply does not exist as the production cost of one unit is larger than the maximum value of the product for the single consumer that values the product the most. This entails that $c < p < 1$. Notice that, as one would expect, the price is an increasing function of c, while the demand as well as the profit is a decreasing function of c.

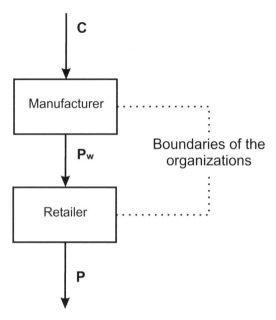

Fig. 7.3 Structure of the supply chain in the case of vertically disintegrated organizations: one manufacturer and one retailer.

final price p, while the manufacturer sets the wholesale price p_w. When setting the wholesale price the manufacturer shall understand what the reaction of the retailer will be: The manufacturer shall understand how many units of the product the retailer is going to purchase at various levels of the wholesale price p_w. To do so, the manufacturer shall anticipate the behavior of the retailer and understand how he/she reacts to any change in the wholesale price p_w (economists call this process *backward induction*, meaning that we shall start our analysis from the player that makes the decisions at a later stage, since the player that makes the first decision shall understand the reactions to his/her moves before he/she can make the optimal decision). So we start our analysis by investigating the decisions of the retailer to understand how he/she reacts to any change in the wholesale price p_w.

The decisions of the retailer   Given our assumptions, the profit function of the retailer π_r is

$$\pi_r = d \cdot (p - p_w) = (1 - p)(p - p_w); \qquad (7.5)$$

so the structure of the profit function resembles (7.1), where c is replaced by p_w. Indeed, the only one difference between the retailer in this case and the vertically integrated firm in the previous case is that the retailer faces a marginal cost p_w rather than a marginal cost c. This really means that we can derive optimal price, quantity, and profit by substituting c with p_w:

$$p^* = \frac{1 + p_w}{2},$$

$$d^* = 1 - p^* = \frac{1 - p_w}{2},$$

$$\pi_r^* = \frac{(1 - p_w)^2}{4};$$

the decision (p^*) and the profit (π_r) of the retailer depend on the wholesale price (p_w) the manufacturer charges. Oddly, the manufacturer can "control" the behavior of the manufacturer through p_w.

The decisions of the manufacturer   The manufacturer should set the wholesale price to maximize his/her profit while keeping in mind that the decisions of the retailer change according to the price he/she decides to charge. In particular, the manufacturer should estimate the demand curve. Actually, given that the final consumers' demand is deterministic, the retailer buys exactly the quantity that he/she can sell. So the quantity $d^* = \frac{1 - p_w}{2}$ is the demand curve for the manufacturer. Hence, the profit function of the manufacturer π_m is

$$\pi_m = \left(\frac{1 - p_w}{2}\right)(p_w - c). \tag{7.6}$$

Once again this function resembles (7.2) where p is replaced by p_w (the factor $1/2$ is irrelevant when we look for the optimal price p_w) and thus the optimal price for the manufacturer is

$$p_w^* = \frac{1 + c}{2} \tag{7.7}$$

and the optimal profit for the manufacturer is:

$$\pi_m^* = \frac{1 - c}{4} \cdot \frac{1 - c}{2} = \frac{(1 - c)^2}{8}. \tag{7.8}$$

Understandably, the profit of the manufacturer is lower than the profit of the vertically integrated firm. In the next subsection we investigate whether the profits the manufacturer fails to make are made by the retailer or, vice versa, are simply lost.

Performance of the vertically disintegrated chain.   At this stage, we can put together the information about the manufacturer's pricing policy and the retailer's reaction to draw conclusions on the performance of this supply chain (structure and contracts).

   We know that the manufacturer charges $p_w^* = \frac{1 + c}{2}$ and the retailer charges the final consumer a price $p^* = \frac{1 + p_w}{2}$, so the final consumers' price is

$$p^* = \frac{1 + p_w^*}{2} = \frac{3 + c}{4}. \tag{7.9}$$

.

Notice that this final consumer price is higher than in the vertically integrated case $((1+c)/2)$ since $c < 1$. Also, a higher price means that consumers buy a smaller quantity: They buy $(1 - c)/4$ in this case, while they buy $(1 - c)/2$ in the vertically integrated one.

This is a first relevant finding. The *consumers are definitely better off with a vertically integrated firm*, as they buy twice the quantity at a lower price. This result might look somehow odd. Indeed, in basic courses in economics we learn that monopolists (like the vertically integrated firm) take advantage of consumers, while this finding suggests that one monopolist is better than two firms. Indeed, in the model both companies are monopolist in their respective position: The model simply suggests that two monopolists are even worse than a single one. When we see this result we are tempted to draw the conclusion that consumers are worse off since the industry (in our case the supply chain) is making more profit (basically, the consumers have to feed two companies rather than just one). This is actually a zero sum game perspective. As we shall see, the industry is making less profits as well, so this is a negative sum game! We have already seen that the manufacturer makes a profit

$$\pi_m^* = \frac{(1 - c)^2}{8}. \tag{7.10}$$

Given the price the manufacturer charges, the retailer makes a profit

$$\pi_r^* = \frac{(1 - p_w)^2}{4} = \frac{(1 - c)^2}{16}. \tag{7.11}$$

Hence, the total profit π_{tot} for the supply chain (manufacturer and retailer) is

$$\pi_{tot}^* = \frac{(1 - p_w)^2}{4} = \frac{3(1 - c)^2}{16}, \tag{7.12}$$

which is lower than the profit for the vertically integrated firm $\left(\frac{(1 - c)^2}{4} \right)$.

So the disintegration reduces the profits for the industry, raises final price, and reduces the quantity bought, reducing the welfare of the consumers.[3] So

---

[3] Economists use the expression *surplus of the consumer*. The meaning of this economic concept is basically the following. The aggregate demand curve for a given product shows that different consumers value the product differently. In our example, some consumers value it 1 and others value it 0. Consumers that buy the product at a given price p are those that value the product at the very least p. Others simply prefer to keep their money. Among those that buy the product, some value it more than they paid for it and thus enjoy a surplus. When a company cannot price discriminate like in our example, the sum of all

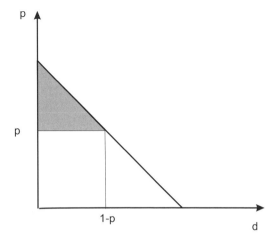

Fig. 7.4   Consumer surplus (see footnote 3 on page 366).

who is better off? Basically, no one is better off; this is a second best solution for all players. Then the question becomes, How is this happening? What is driving this?

This is the so-called *double-marginalization problem*. In other words, two players do a marginal analysis and find optimal prices. The retailer perceives a variable cost p_w that is higher than the actual variable cost of the chain c and then tends to charge a final price that is higher than the optimal one. On the other hand, by doing so the retailer reduces the demand for the whole chain and thus the manufacturer sees a demand that is smaller than it could potentially be. The two decision makers make locally optimal decisions that, however, turn into a global disaster for all parties.

Concept 7.1 *The presence of intermediaries in a supply chain can lead to distortion of incentives: In a disintegrated supply chain each player tries to maximize its own profit and disregards the effects of his/her decisions on other players (as well as on the consumer). Such misalignment of incentives can lead to a relatively high price that reduces the surplus of the consumer. Also, the misalignment of incentives reduces the profits of the firms in the supply chain. So these incentive issues tend to have negative effects both for companies and consumers.*

Also, it is rather interesting to focus on the tradeoff the manufacturer faces. On the one hand, the manufacturer can easily solve the double marginalization

these surpluses is simply the integral over all consumers that actually bought the item of the demand curve (actual value of the product for a given consumer) minus the price actually paid for the item. In the case of a linear demand like in our example, it is the area of the triangle highlighted in figure 7.4.

problem by reducing the wholesale price p_w down to c. In this case actually the double marginalization basically disappears. The retailer sees a marginal cost that is just the marginal cost for the whole supply chain, and as a result he/she charges an optimal price:

$$p^* = \frac{1 + p_w}{2} = \frac{1 + c}{2}, \tag{7.13}$$

which is actually the optimal price that leads to the maximum profit for the whole supply chain. So the manufacturer can lead the retailer to choose the optimal price and maximize the profits for the whole supply chain. However, there is a major issue: In this case the unit margin for the manufacturer is zero and thus the profit is just zero. So the manufacturer can make the supply chain behave optimally at the expenses of his/her own profit, which is actually a very unlikely scenario.

On the other hand, the manufacturer can maximize the unit margin $p_w - c$ and, at one extreme, set $p_w = 1$. In this case, however, the demand simply fades away and drops to zero.

The manufacturer would like to (i) take a large share of the total profit of the chain and (ii) make sure that profits are maximized. Unfortunately, he/she tries to achieve these contrasting objectives with a single lever – that is, the wholesale price. While to maximize total profit for the chain the manufacturer shall keep the wholesale price low to avoid the double marginalization problem, to take a large share of the total profit the manufacturer shall increase p_w up to a point where $p = p_w$, that is, $p = p_w = 1$.

So the manufacturer basically cannot have the cake and eat it too. With a single lever p_w, he/she can either have the cake or eat it. The best solution is actually a compromise where the price generates a good profit for the chain but at the same time guarantees that a significant portion of it is gained by the manufacturer rather than by the retailer.[4]

---

[4]Notice that in the more general case of a production cost c and a distribution cost r, the retailer sees a marginal cost $p_w + r$ and thus charges a price $p^* = \frac{(1+p_w+r)}{2}$, so that the demand from the consumer is $d^* = \frac{1-p_w-r}{2}$. Then the profit for the manufacturer is

$$\pi_m = (p_w - c)(\frac{1 - p_w - r}{2}),$$

and thus the optimal wholesale price is

$$p_m^* = \frac{1 + c - r}{2},$$

and the optimal final consumer price is

$$p^* = \frac{1 + p_w^* + r}{2} = \frac{3 + c + r}{2}.$$

Notice that we can rewrite the above final price as

$$p^* = \frac{3 + c}{2} + \frac{r}{2},$$

7.2.3 A way out: designing incentive schemes

When one encounters the double-marginalization problem, intuition suggests that there must be a way out of it, since none is better off in the second scenario we have investigated.

Actually, there are several ways out. A common feature of all solutions we are going to discuss is that they re-design incentive schemes. In other words, we want to redesign the rules of the game so that all players (decision makers) change their behavior and we can reach an optimal (or at the least better) solution. The key idea is that the manufacturer and the retailer choose high prices and end up with relatively low profits simply because they have an incentive for this suboptimal behavior and if incentives are changed their objectives and behavior are going to change as well.

Let us now discuss several ways out.

Change the structure of the supply chain   The first and obvious solution is to *change the structure of the supply chain* and go back to the vertically integrated case. Clearly this solution solves the problem structurally. In many industries, many companies are going toward this solution, as they are opening more and more single-brand stores such as (a) Bulgari, Cartier, etc., in luxury and (b) Adidas, Nike, etc., in the sportswear business. However, as we have discussed in the introduction to this chapter, this is often a hardly viable solution. Empirical evidence tells us that retailers exist and often they are not part of a manufacturing company. This is because often retailers perform functions that our model fails to capture such as:

- *product selection*; that is they act as agents for the customers and select products that customers can hardly judge on their own. This is a typical function performed in some multibrand specialty stores such as wine bars, where the owner of the wine bar has the technical ability to judge the product and repetitively purchases from a series of wine producers and sells to a series of loyal customers. This way he acts as an intermediary that builds reputation both upstream and downstream and reduces the incentive of the producers to free ride.

- *creation of a wide assortment*; in many instances the value of a single product is limited as compared to the search cost – that is, the cost to look for, locate, purchase, and transport the product. A classic example of this phenomenon is grocery. Customers rather pay the retailer to create a wide assortment and transport products from various suppliers

and compare it with the price of the vertically integrated solution $p^* = \frac{1+c+r}{2} = \frac{1+c}{2} + \frac{r}{2}$. As we can see, the double marginalization has an effect only on the marginal costs of the manufacturer, whereas it has no effect whatsoever on the marginal costs at the retailer. Indeed, for these latter costs there is no double marginalization, since the retailer "sees" the actual marginal cost r rather than an overstated wholesale price $p_w > c$.

than buy single products from single suppliers at a lower cost. Simply because shopping at many single-brand stores – one selling Palmolive products, the second selling only P&G products and another one selling only Barilla products – might be very expensive (in terms of cost of transport and time consumed). Think of visiting 15 stores during your weekly shopping trip rather than a single supermarket!

Our model fails to capture such problems, since we have a single product model. So, in the model, we ignore issues that might make the vertical integration a hardly viable solution. This really means that there is a need for a retailer. In principle the retailer could be just a consortium of manufacturers. However, one can easily understand how hard it would be for all consorted manufacturers to control the retailing consortium and make sure the consortium acts in the best interest of each single producer. Writing a contract between the consortium and each single manufacturer is basically impossible. Also, even if we assume a contract was written, controlling it and enforcing it in a court is actually quite hard and definitely very expensive. Finally, we should notice that even in those cases where other variables not included in our simplistic model lead us to choose a vertically disintegrated supply chain, still the problem of double marginalization stays and we need to find a solution to this issue. When the solution cannot possibly be structural, it must be contractual, that is, we shall rewrite the contract between the manufacturer and the retailer in such a way that they are led to set the optimal price and can gain optimal results. In other words, when we still have two decision makers, we would better make sure their incentives lead them in the right direction, otherwise performance drops.

A second solution is to adopt more complex and subtle contracts between the supplier and the manufacturer to make sure they act in the best interest of the supply chain and make sure that the "cake is as big as it can possibly be." As we have already discussed, a single lever p_w is just not enough to achieve two contrasting objectives (unit margins and large quantities sold). So we need a more complex pricing strategy. This is what economists call the *two-part tariff*; that is, a pricing strategy with two parts. Basically, these contracts use two parameters to achieve the two contrasting objectives.

Franchising contracts   A first example of these two-part tariffs is the so-called *franchising contract*. Under this contract the manufacturer, called *franchisor*, sells the retailer, called *franchisee*, both the goods and the right to sell them.

The franchisor can sell products at marginal cost $p_w = c$ to make sure that the retailer has the right incentives to set the retail price appropriately at $p = (1 + c)/2$ and maximizes the profit for the chain $((1 - c)^2/4)$. Once such a profit is generated, the franchisor can use a second parameter in the contract that is a fixed fee F; that is, the right to sell the products of the franchisor. We just need to set $F = (1 - c)^2/4 - \epsilon$ to move a large portion of the optimal profit upstream. In real-life applications the fixed fee is actually just a mean to split the supply chain profit between the two parties. The actual split depends obviously on the bargaining power of each party. But in general we can keep the marginal cost at a minimum to reduce the double marginalization problem and then use a second lever to move the maximized profit within the supply chain appropriately.

Quantity discounts Another example of two-part tariffs are *quantity discounts.*[5] Under such a contract the cost of marginal quantities is very low (at one extreme equal to the marginal production cost) to provide the retailer with an incentive to increase the quantity and reduce the selling price.[6] On the other hand, the first units sold can have a fairly high price to move the profits upstream.

Let us consider a case where the producer asks for a price p_w for the first Q_1 units (with $Q_1 < (1-c)/2$, that is lower than the optimal quantity for the vertically integrated supply chain) and then charges only the marginal cost c for any unit on top of this. In this case the profit function of the retailer is[7]

$$
\begin{aligned}
\pi_r &= p \cdot d - p_w \cdot Q_1 - c(d - Q_1) \\
&= (p - c)d - Q_1 \cdot (p_w - c) \\
&= (p - c)(1 - p) - Q_1 \cdot (p_w - c).
\end{aligned}
$$

The above equation basically resembles equation (7.2) other than for a basically fixed cost $Q_1 \cdot (p_w - c)$. This really means that when fixed cost are not excessive,[8] the retailer simply selects the optimal price $p^* = (1+c)/2$ and the optimal quantity $d^* = (1-c)^2/2$. This means that the retailer makes a profit

$$
\pi_r = \frac{(1+c)^2}{4} - Q_1 \cdot (p_w - c), \tag{7.14}
$$

and the manufacturer can reduce the profit of the retailer and increase its profit by appropriately selecting Q_1 and p_w in such a way that the retailer's profit is reduced to a small quantity ϵ:

$$
\pi_r = \frac{(1+c)^2}{4} - Q_1 \cdot (p_w - c) = \epsilon. \tag{7.15}
$$

---

[5] In this context we consider the marginal units price discount while the all units is ineffective in this scenario.
[6] Notice that from the incentive standpoint franchising contracts and marginal unit quantity discounts are basically the same. $F + c$ can be interpreted as the very high cost of the first unit.
[7] Notice that this is the profit function for $d > Q_1$, to be precise we shall also consider the option to purchase a quantity below Q_1.
[8] Indeed, the retailer might have the option to exit the market in case of negative profits by buying zero units.

Resale price maintenance   Finally, another solution is to contractually set the final price p; that is, sell the product to the retailer if and only if he/she charges the optimal price. This is the so-called *resale price maintenance*. Under these contracts the manufacturer sells to the retailer at a price $p_w = \frac{1+c}{2}$ and the contract also forces the retailer to sell at a price $p = p_w$ or a price $p \leq p_w$. Finally, an equivalent solution is to contractually set a minimum quantity $q \geq \frac{1-c}{2}$. This solution is basically equivalent to a large all unit discount. Notice that in many countries fixing the final price contractually might be considered illegal (it is a restraint to price competition in the retailing stage of the supply chain). Manufacturers can only "suggest" a price the retailer should charge the final consumer. Though the final price cannot be included in the contract, many manufacturers check the final price of the product and at times take the decision not to sell the product to the retailers that set a price that significantly differs from the one they suggest.

So as intuition suggests there are several fixes to the double marginalization problem.

Concept 7.2 *Once we acknowledge that the presence of intermediaries can raise incentive problems, we can design contractual solutions, such as franchising contracts, quantity discounts, or fixed consumer prices that can re-build an incentive for all parties to set an optimal price and make sure that the supply chain gains optimal profits (and increases the consumer surplus). In a sentence a supply chain can be profitable only if we make sure that every company on it has reasons to pull in the same direction.*[9]

7.3   DECISION ON PRICE IN A COMPETITIVE ENVIRONMENT

In this section we still discuss the pricing decisions in a supply chain. However, we make a small change to the model we have investigated. In this section we assume that in the supply chain there is just one manufacturer of the homogeneous product and a large number n of retail outlets (see figure 7.5). We assume zero search cost and homogeneous products. Thus we assume that consumers select retailers solely on price. In this case too, we compare the vertically integrated case with the vertically disintegrated one. Obviously, we do not need to recalculate the optimal policy in the case of the vertically integrated firm. The company still chooses to charge the same final price $p^* = \frac{1+c}{2}$ in all n stores.

---

[9]The last sentence is adapted from [5] that also provides some interesting real life examples on incentives issues in supply chain management.

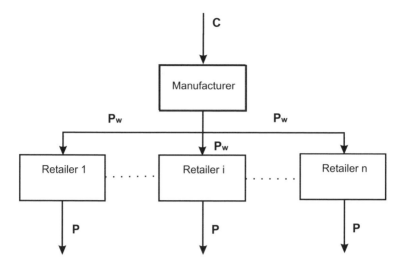

Fig. 7.5   A supply chain with a competitive retailing sector

7.3.1   The vertically disintegrated supply chain: independent manufacturer and retailer.

On the contrary, we shall investigate the vertically disintegrated chain. In this case, indeed, the situation changes, since the retail market is competitive. To understand this supply chain we shall resort to basic economic theory.

Basic economic theory is outside the scope of this book. For example, we refer to [3] for a more precise analysis of perfect markets. Here we simply recall the basic intuition behind the perfect market model. The model assumes that a large number n of competing firms sells an homogeneous product and use a homogeneous technology and thus share a common cost structure. Consumers have a perfect information and easy access to all suppliers. So they select the supplier on price, since it is the one differentiating variable. Under these assumptions any competing firm that reduces the price below competitors takes all the demand, and any company that increases the price above competition loses all the demand. Thus all companies charge the same price. This really means that all companies cut the price to the very minimum to increase demand, immediately followed by competition.[10] At steady state, in such markets the final price is equal to the marginal cost; that is, companies

---

[10]Notice that this is just to build intuition, as economists might challenge this statement. They might argue that in this model, companies are pure price takers and are not aware of strategic interaction with their competitors. So they are not aware that competitors will react to their actions.

make no profit[11]. They cannot reduce the price below the marginal cost, as they would have a negative profit. But they also cannot increase it, because they would lose all the demand and would be basically out of business. Economists say that in such a business the companies are just *price takers*, as they have no latitude on price, they simply charge the standard price to the consumer.

This downstream competition has several consequences for our supply chain. Like in the previous case, we shall start the analysis from the final stage of the supply chain (backward induction). In this case the analysis is really simple. All retailers charge a price p equal to their marginal cost p_w. So each of the n retailers gets a portion $1/n$ of the overall aggregate demand $d = 1 - p = 1 - p_w$ and makes no profit whatsoever. Thus the profit for the manufacturer is

$$\pi_m = (p_w - c)(1 - p_w). \tag{7.16}$$

We know that this equation resembles equation (7.2) and thus the optimal wholesale price is

$$p_w = \frac{1+c}{2}. \tag{7.17}$$

Hence also the final price p is optimally set to $p^* = p_w^* = \frac{1+c}{2}$, and so the aggregate demand for the industry is $d^* = (1 - p^*) = \frac{1-c}{2}$ and the profit for the manufacturer (and the whole supply chain as the retailers make no profit) is $\pi_m^* = (1 - p^*)(p^* - c) = \frac{(1-c)^2}{4}$.

The results deserve a couple of comments. As the reader has probably already noticed, in this case even a very simple and linear pricing policy leads to an optimal price. So when there is a large number of retailers, the vertically disintegrated supply chain performs just as well as the vertically integrated firm. What are the economic interpretations of this apparently odd result?

A first, technical reading is that in a perfect market the retailers do not make any margin and thus though there are two echelons in the supply chain, one actually does not increase the price. A second, more managerial reading is that there are two players but actually one is simply a *price taker* that makes no actual decision. So in this case, coordination of decisions is rather trivial, since there is only one actual decision maker.

Concept 7.3 *The double marginalization problem depends on the market structure. If the retailing stage is a competitive market (with no search costs) then the double marginalization problem simply fades away. More in general, incentive issues might depend on the structure of the supply chain and of the markets.*

---

[11] Again an economist might argue that companies make no extra-profit, that is the make no profit on top of those minimal profits that reward the capital investment and the efforts of the entrepreneur

7.4   DECISION ON INVENTORIES: THE NEWSVENDOR PROBLEM

In this section we investigate the impact of the supply chain structure on inventory decisions.[12] To isolate this issue from other decisions, we assume that the final price is fixed (say by law or there's a standard price in the market, like for newspapers that in Italy basically share the same common price). Also, to make inventory decisions relevant we assume that demand is uncertain.

In this case we shall use three variables. First we have the exogenous demand y, then we have the production quantity Q, and finally we have sales V. In the previous sections, we had a single variable since we had assumed a deterministic demand and thus purchase quantity was exactly equal to demand and sales.

In our model we assume that the expected demand is still $1 - p$, but in this case the demand is variable and is uniformly distributed between 0 and $2 - 2p$ ($f(x) = 1/(2 - 2p)$). To keep the problem simple, we assume the product expires like newspapers, so that we can consider our inventory problem to be static; that is, the one purchase decision on a single period (day) has no impact whatsoever the next period (day). In other words we face a multi-echelon newsvendor problem (see 5.2 for the classic newsvendor problem). Actually, we have already discussed the issues that arise while setting prices in a supply chain and thus ignore them in this section by fixing the price p.

7.4.1   The first best solution: the vertically integrated firm

The vertically integrated firm has basically one decision to make: how many units Q to purchase.

The newsvendor problem suggests balancing the cost of a stockout and the cost of inventories in such a way that the type I service level is $SL = \frac{m}{m+c}$, where m is the cost of a stockout and c is the cost of excess inventories.

In our simple case the cost of a stockout for the vertically integrated firm is $p - c$ and the cost of inventories is c, as we assume that inventories have no salvage value. This means that the vertically integrated firm shall provide a service level $SL = \frac{p-c}{p}$. Under our simplistic assumptions the company shall stock a percentage $\frac{p-c}{p}$ of the maximum potential demand $2 - 2p$. So the optimal stocking quantity is

$$Q^* \frac{p - c}{p} \cdot (2 - 2p).$$

---

[12]Notice that even issues discussed in section 6.4 can fall under this umbrella. While in this section we present inventory planning issues in a static and uncertain environment, in section 6.4 we present the same issue in a dynamic and certain environment.

7.4.2 The vertically disintegrated case: independent manufacturer and retailer

In this case the retailer purchases copies of the newspaper in the early morning from the manufacturer and tries to sell them during the day. So, the retailer makes the stocking decision and runs the inventory risks. This is actually a very important issue. In our model we assume that decision makers are risk neutral, so this allocation of risks in the supply chain has no effect, but when players show different attitudes towards risks, we shall carefully consider the allocation of risks in the supply chain.

In the vertically disintegrated case, the retailer faces a decision problem that resembles the one of the vertically integrated company. The one difference is that the marginal cost of the retailer is p_w. So the quantity the retailer decides to stock is $Q_r^* = \frac{p - p_w}{p} \cdot (2 - 2p)$. This quantity is also the demand for the manufacturer.

Given the stocking decision of the retailer, the profit function for the manufacturer is

$$\pi_m = \frac{p - p_w}{p} \cdot (2 - 2p) \cdot (p_w - c). \tag{7.18}$$

So the manufacturer charges a wholesale price p_w that maximizes his/her profit:

$$\frac{\partial \pi_m}{\partial p_w} = \frac{2 - 2p}{p} \cdot (p - p_w - p_w + c) = 0, \tag{7.19}$$

and the optimal wholesale price is

$$p_w^* = \frac{p + c}{2}. \tag{7.20}$$

The manufacturer charges a wholesale price that is halfway between the final consumer price p and the marginal cost. This finding deserves some comments. Equation (7.18) suggests that once again the wholesale price performs two functions. On the one hand, the manufacturer would like to increase the wholesale price to increase the margin $p_w - c$ he/she makes. On the other hand, the wholesale price performs a second function: It sets the retailer's incentives to stock large quantities. So the manufacturer would be tempted to reduce the wholesale price p_w to increase the stocking quantity $Q_r^* = \frac{p - p_w}{p} \cdot (2 - 2p)$.

Like in the case of decisions on price, the manufacturer tries to achieve contrasting objectives with a single lever p_w. He/she can set $p_w = c$ and lead the retailer to stock the optimal quantity but the manufacturer has no margin and thus makes no profit. On the other hand, the manufacturer can set $p_w = p$ and maximize margins, but in this case the retailer has no incentive to stock and this reduces demand to zero. The optimal solution mediates between these two contrasting objectives and actually is halfway between the two extreme (and ineffective) solutions we just discussed ($p_w = p$ and $p_w = c$).[13]

Given this wholesale price p_w^*, the retailer stocks a quantity

$$Q_r^* = \frac{p - p_w^*}{p} \cdot (2 - 2p)$$

$$= \frac{p - \dfrac{p+c}{2}}{p} \cdot (2 - 2p)$$

$$= \frac{p-c}{p}(1-p).$$

In other words, in this situation the stocking quantity is 50% lower than the optimal one, and the service level is reduced by 50% as well. Clearly, this is not good news for the consumers that are very likely not to find the item they want (at the least at the end of the day). Is this more profitable for the industry? Actually it is not. Indeed, we can show that for the whole supply chain it would be appropriate to increase the stocking quantity above the current level $Q_r = \frac{p-c}{p}(1-p)$.

We can go back to the intuition behind the basic newsvendor problem and show that the chain profit increases if we increase the stocking quantity marginally. The probability of not selling a marginal unit is $(p - c)/2p$ while the probability selling it is $(p+c)/2p$. The cost of lost sales is $(p-c)$ while the cost of excess inventories is c. So what we gain by increasing the inventory level marginally is $p - c$ times the probability that we actually sell the additional marginal quantity. What we lose is the cost of inventories c times the probability $(p - c)/2p$ that we are not going to sell the additional quantity. Hence, the marginal profit of an increase in inventories is

$$\frac{p+c}{2p} \cdot (p - c) - \frac{p - c}{2p} \cdot c = \frac{p - c}{2} > 0. \tag{7.21}$$

Increasing the stocking quantity increases the profit for the chain and the solution Q^* is suboptimal. Unfortunately, what is good for the supply chain is not good for the one decision maker, the retailer.

Like in section 7.2 the vertically disintegrated supply chain leads to suboptimal performance, since it provides bad incentives to the players. The manufacturer charges a wholesale price p_w above the marginal cost to gain a profit. However, such a wholesale price reduces the retailer's margin and thus reduces the incentive for the retailer to stock. This leads to a reduction in the retailer's stocking quantity, in service level for the consumers, and in profits for the chain. Like in previous cases, this is a bad situation for all players, and intuition suggests there must be a way out.

---

[13] Actually, the optimal wholesale price depends on the demand and cost functions. So the fact that the optimal solution is halfway between marginal cost and consumer price is just a pure chance. In general though, it is within this range.

7.4.3   A way out: designing incentives and reallocating decision rights

As we have seen in section 7.2, a first solution can be vertical integration of companies. As we already discussed, this solution has several consequences our simplistic models fail to capture. So we consider the structure of the supply chain as a given and we try to find other, contractual rather than structural, solutions.

Like in the case of decisions on prices, we discovered that the manufacturer is trying to achieve two contrasting objectives (make margins and give the retailer incentives to stock) with a single lever (the wholesale price p_w).

So a first reasonable solution is to adopt two-part tariffs so that both objectives can be sought.

Franchising contracts.   Franchising contracts again can be a very good option. The key idea is again to sell the product at marginal cost c to give the retailer the incentive to stock the optimal quantity for the whole chain $Q^* = \frac{p-c}{p}(2 - 2p)$. This solution leads to optimal profits for the whole chain. Then a fixed fee F that the retailer pays to the manufacturer can be used to move profits upstream. So in this case one lever (the wholesale price $p_w = c$) is used to give the retailer the incentive to stock the right quantity and a second lever (the fixed fee F) is used to distribute the profits in the chain.

Buy-back contracts.   A second option is to use the wholesale price p_w to make a profit and use a second lever to give the retailer an incentive to stock the optimal quantity. The intuition behind these contracts is that we increase the willingness of the retailer to invest in inventories by decreasing the cost of excess inventories. So, while in the franchising contract we increase the willingness to invest by increasing the cost of a stockout (i.e., increasing the margins) in this case we act on the other economic parameter of the newsvendor model: the cost of excess inventories. Under these contracts, basically the manufacturer offers to buy back the inventories left over at the end of the day (or selling period more in general). Basically in this case too, the pricing policy consists of two parameters. The manufacturer sells the newspaper to the retailer at a wholesale price p_w early in the morning and buys back the units unsold at the end of the day at a buy-back price p_b. Basically, we reduce the cost of a stockout to $p_w - p_b$. So any pricing policy (p_w, p_b) such that the optimal type I service level for the retailer equals the optimal type I service level for the vertically integrated firm leads to an optimal stocking quantity Q^* and thus to optimal profits for the chain:

$$(p_w, p_b)| \frac{p - c}{p} = \frac{p - p_w}{(p - p_w) + (p_w - p_b)} = \frac{p - p_w}{(p - p_b)}.$$

Thus, the buy-back price is a function of the wholesale price p_w:

$$p_b = p - p \cdot \frac{p - p_w}{p - c}$$

$$= p\left(1 - \frac{p - p_w}{p - c}\right)$$
$$= p \cdot \frac{p_w - c}{p - c}.$$

While an infinite number of pricing strategies (p_w, p_b) is just as good to fix the retailer's incentives to stock, they are very different from the standpoint of the manufacturer's profit. Clearly among these couples the manufacturer wants to choose a couple with a very high p_w to increase unit margins and thus profits.

So the optimal policy for the manufacturer is to charge a price $p_w = p - \epsilon$ and a buy back price $p_b = p \cdot \frac{p - \epsilon - c}{p - c}$. Such a pricing policy builds the right incentive to stock and at the same time creates a profit for the manufacturer. So, quite interestingly, the manufacturer has an incentive to pay for the unsold goods. Intuition would suggest that the manufacturer would rather not pay for the unsold inventories at the retailer. Why should one want to pay for goods that at the end of the day lost most of their value? A deeper analysis shows that buying the unsold goods is actually in the best interest of the manufacturer.

Finally, we shall conclude this section with a note on contracts that do not solve this issue. In section 7.2 we suggested solving the double marginalization issue by fixing the final price p contractually. We shall acknowledge that this solution is actually completely ineffective in the case of uncertain demand. Indeed, in our simple model we assume the price to be fixed, but still the stocking problem has arisen.

Example 7.1 These contracts are often used in several industries where the cost of inventories is substantial to give the retailer an incentive to stock appropriately. A classic example is the electronics business, where products lose value month by month. In this business, the OEM share the cost of excess inventories with the retailers and refund a portion of the products' loss of value. A similar practice is fairly common in the fresh food business. Large companies such as Nestlé partially refund the cost of expired food to the retailers. These practices aim at providing the retailers with an appropriate incentive to stock inventories. Manufacturers could give up some margin to reach the same objective but they rather buy products back.   □

Revenue sharing contracts. Another partial solution to the double marginalization problem is to adopt profit sharing contracts. In these two-part tariffs the manufacturer charges a low wholesale price $p_w \approx c$ and get paid a percentage of the final revenues. In other words the manufacturer is paid partially on the sell-out and partially on the sell-in.

These contracts can generate a better incentive for the retailer to stock. However, they never re-create the whole incentive to stock the optimal quantity, since a portion of the final consumer price (that exceeds the marginal cost of the manufacturer) is given to the manufacturer, thus reducing the retailer's incentive to stock. However, these contracts have two other beneficial effects

First these contracts distribute the risk among the players in the supply chain. In these contracts uncertainty on the final demand has an effect on both parties rather than on just the retailer. So, these contracts can adapt to the inclination toward risk of the two players. So these contracts can be an effective mean to re-allocate risks in the supply chain according to the ability/willingness to accept risk.

Also, demand fluctuations can be partially random, but they might partially be due to actions of both the retailer and the manufacturer. Such efforts might be hardly contractable, meaning that it might be hard to write a contract that (i) fully describes the efforts of the two parties and (ii) can be enforced before a court. In these fairly common circumstances, these contracts give both the manufacturer and the retailer a partial incentive to increase demand the way they can.[14] For example, the retailer might provide more selling effort, allocate more space in the store, or put the item in the front window. On the other hand, the manufacturer can spend more on advertising, add extra features, or add extra contents.

Section 7.5 discusses the issue of efforts in supply chains.

Example 7.2 These contracts are used in the media business. Blockbuster noticed very frequent stockouts of top movies (premieres) during the weekends. Nevertheless, purchasing more copies of the movies was not profitable for Blockbuster. DVDs have a very low marginal production cost, since the vast majority of costs are fixed (think of the production cost of the first unit). Thus Blockbuster has signed a revenue sharing agreement with the movies' majors to share the revenues of DVD rents. This new contract gives the majors an incentive to reduce stockouts in order to increase the revenues shared. So they reduce the wholesale price to increase Blockbuster's incentive to stock. As a consequence, this new contract changes the incentives of Blockbuster to stock DVDs, and it increases the availability of copies for the consumers (even during times of peak demand such as the first few weekends after the launch of the DVD); also, the turnover increases in a business where the marginal cost is basically negligible, thus increasing the profits for the chain. It is interesting to notice that only technology made this deal a viable option. Indeed, the two parties need to make sure that total revenues is a certified number so that revenues can be shared fairly. A third party provider of technology certifies the revenues and thus makes the total revenues observable to the manufacturer (they are obviously observable to the retailer) and thus the revenue sharing contract a viable option. In economics we would say that technology is making revenues a contractable variable, that is a variable that can be included in a contract (see also [5]).

---

[14] In case the wholesale price exceeds the marginal cost of the manufacturer, as it is very often the case.

Re-allocation of decision rights.   Another, more radical solution is to reallocate decision rights in the supply chain. In other words, we keep the supply chain structure and thus we keep the number of players constant. However, the rights to make decisions are changed. One option in this context is to move the right to make inventory planning decisions upstream. This is the so-called Vendor Managed Inventories. Under such contracts the manufacturer holds the right to make decisions on inventories. A simple reallocation of decision rights would lead to a scary incentive scheme without a reallocation of the cost of inventories. Think of what would happen if the manufacturer makes decisions on retailers' inventories. The manufacturer would be in a position to freely increase deliveries and thus turnover at the expenses of the retailer that might carry unnecessary inventories. Usually these agreements require a change in the allocation of cost of inventories among players. Often retailers give the manufacturers a maximum budget for the inventory investment to make sure they do not over-invest in inventories as they do not pay for the holding cost. Another solution is to agree that the retailer pays for the sell-out (units sold to the consumer or delivered from the central distribution center to the stores) rather than for the sell-in (units delivered to the retailer's DC) so that the manufacturer pays for excess inventories and thus has no insane incentive to overstock.

These contracts are very popular nowadays, and some companies such as Procter and Gamble have proven that they can be very effective, if properly deployed. However, there are several *caveats*.

First, these contracts work only if the retailer provides the manufacturer with the information required to make inventory decisions effectively. Some of these pieces of information are easy to transfer such as sell-out data. Others are somehow more problematic. For example, information about the promotion of a competing substitute product can be crucial to plan inventories, but the retailer might not want/should not give the manufacturer this relevant piece of information. Also, a retailer might have some qualitative information that the producer lacks, such as roadworks around the store, new openings of competing stores, etc. On the other hand, the manufacturer might have some other pieces of private information such as the launch of new products in the same category of new advertising campaigns.

More in general, when we allocate decision rights in the supply chain, we shall always wonder who has the right information (and skills) to make better decisions. Second, we shall wonder who has the right incentives to make appropriate decisions. One example can show some problems these contracts can have. Retailers often sell products from competing manufacturers. Often consumers are quite willing to switch among brands and substitute stocked-out products (see [1]). However, such substitution reduces the cost of a stockout for the retailer, but it does not change (or might even increase) the cost of a stockout for the manufacturers (see [4]). This really means that the manufacturers tend to overinvest in inventories, as they target a service level that is greater than the optimal one for the whole supply chain. More in general, when allocating decision rights, we should carefully study incentives to over-invest/underinvest in inventories as well as the availability of relevant pieces of information and skills.

Concept 7.4 *In a disintegrated chain the inventory and service levels drop below the optimal point since the retailer only sees a fraction of the total margin and thus is less willing to invest in inventories. Franchising, buy-back and VMI contracts rebuild an incentive to invest in inventories and gain a better service.*

Example 7.3 [15] To make the problem simple, let us think about sales of a newspaper in a small town where this is the only newspaper sold. The final consumer price for the newspaper is 1€. For newspapers, usually the marginal production cost is rather low, and in our example we assume that the marginal production cost c is 0.2€.

Let us assume that demand is uniformly distributed between 0 and 200, with an expected level of 100 units ($f(x) = 1/200$).

The first best solution: the vertically integrated firm.

As we have learned in section 5.2 the newsvendor has an 80-cent cost of a stockout and a 20-cent cost of inventories (we assume that the salvage value is zero), and thus the optimal type I service level is 80%. Given the uniformly distributed demand, the optimal stocking quantity Q^* is 160 units, as in 80% of the cases the demand is lower than or equal to 160 units.

Given the demand distribution and the stocking quantity, we can compute the expected sales E(V):

$$\begin{aligned}
\text{E}(V) &= \int_0^{160} y \frac{1}{200} \, dy + \int_{160}^{200} 160 \frac{1}{200} \, dy \\
&= \frac{1}{200} \left[\left(\frac{y^2}{2} \right)_0^{160} + 160 \, (200 - 160) \right] \\
&= 96.
\end{aligned}$$

Expected sales are 96 units, and given the stocking quantity of 160, the expected inventories left over is 64 (160−96). This means that the total expected profit for the vertically integrated chain is 96units · 0.80€/unit − 64units · 0.20€/unit = 64€.

The vertically disintegrated chain: independent manufacturer and retailer

Now let us assume that the manufacturer (editor) figures out that retailing is just a different business and that he/she is not interested in it or is not good at it. So the manufacturer (editor) starts selling through a retailer. Let

---

[15]This example is adapted from the Hamptonshire express case (see [4]).

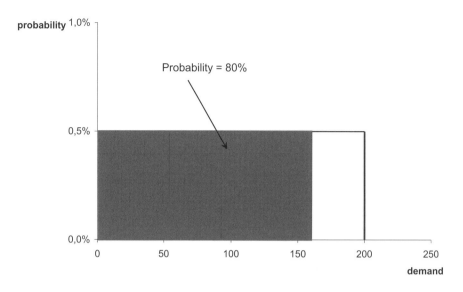

Fig. 7.6   Demand distribution and optimal stocking quantity.

us assume that the manufacturer charges the retailer a price $p_w = 0.8$ and the retailer makes the inventory planning decision before the newspaper starts selling. Under this contract the retailer has a low cost of stockout (he/she loses 0.2€ for each unit of demand lost) and a high cost of inventories (he/she loses 0.8€ for each unit of excess inventories). This really means that the retailer has all incentives to understock. The optimal target service level for the retailer is 20% and thus the optimal stocking quantity is just 40 units, way below the optimal quantity for the chain and far less than the expected demand (100 units).

With such a low stocking quantity the profit for the manufacturer is very low:

$$\pi_m = 40 \, \text{units} \cdot (0.8€/\text{unit} - 0.2€/\text{unit}) = 24€. \qquad (7.22)$$

Let us check the profit for the retailer. Expected sales are

$$
\begin{aligned}
\mathrm{E}(V) &= \int_0^{40} y \, dy + \int_{40}^{200} 40 \frac{1}{200} \, dy \\
&= \frac{1}{200} \left[\left(\frac{y^2}{2} \right)_0^{40} + 40 \, (200 - 40) \right] \\
&= 36
\end{aligned}
$$

Given the stocking quantity of 40 units, 4 are left unsold, on the average, and the expected profit for the retailer is[16]

$$\mathrm{E}(\pi_r) = 36\,\mathrm{units} \cdot (1€/\mathrm{unit} - 0.8€/\mathrm{unit}) - 4\,\mathrm{units} \cdot 0.8€/\mathrm{unit} = 4€.$$

The total profit for the chain drops to 28€ (compare it to the 64 units for the integrated channel). Also consumers have a lower service level (20% instead of 80%) and experience very frequent stockouts.

Let us check the efficacy of the buy back contracts. Let us assume that the manufacturer charges 0.992€ and buys the units left unsold at 0.99€. In this case the retailer's margin is 0.008€ and the cost of inventories is 0.002€. Hence the optimal service level for the retailer is 80%(0.008/(0.008+0.002)). Thus the stocking quantity is 160 units and the whole supply chain makes optimal profits. Also, the manufacturer is the one that makes the bulk of the profits. The retailer's expected profit is

$$\mathrm{E}(\pi_r) = 96\,\mathrm{units} \cdot (1 - 0.992)€/\mathrm{unit} + 64\,\mathrm{units} \cdot (0.992 - 0.99)€/\mathrm{unit} = 0.64€,$$

while the manufacturer sells 160 units at 0.9992 €/unit, 96 are actually sold and thus entail only the manufacturing cost of 0.2 €/unit, and the remaining 64 entail a buy back cost of 0.99 €/unit on top of the manufacturing cost:

$$\pi_m = 160\,\mathrm{units} \cdot (0.992 - 0.2)€/\mathrm{unit} - 64\,\mathrm{units} \cdot 0.99€/\mathrm{unit} = 63.36€.$$

Thus the total profit for the supply chain is 64, and it is actually equal to the profit of the vertically integrated firm. ▯

7.5 DECISION ON EFFORT TO PRODUCE AND SELL THE PRODUCT

As we briefly discussed in the previous section, demand depends not only on price but also on other decisions both at the retailing stage and at the manufacturing stage. As we mentioned on the one hand, one can improve the store look and size, hire more salespersons, train the salespersons, add services such as home delivery, and include customization of the products (e.g., assembling the add-on peripherals of a PC or sizing the sleeves of shirts to customers). On the other hand, one can make more/better advertising, add new features to the product, update the design of the product more frequently, and so on.[17]

---

[16]Notice that the retailer runs the inventory risk so the retailer's profit is uncertain, while the manufacturer's profit is certain. So in this case we take the expectation, while we had no expectation in the case of the manufacturer.

[17]Notice that even the inventory investment can be interpreted as a sort of effort to increase sales. However, given the focus of this book, we think it is worth investigating the issue of the inventory investment separately.

We can call these actions efforts to improve sales at the retailing and manufacturing stage of the supply chain. In our discussion we first focus on the efforts at the retailing stage and later discuss a situation where we have both efforts at the manufacturing and retailing stage.

Also, the cost of effort can be both a variable cost or a fixed cost.[18] For example the assistance at the retail store is an effort from the retailer (s_r) and it generates a variable cost, as it is proportional to the number of customers served. On the contrary the look and size of the store tends to be a fixed cost, as one does not need to increase the size of the store or the number of lights as the number of visitors and customers increases.[19] Similarly, we can have both fixed and variable costs of effort at the manufacturing stage. Examples of fixed costs are advertising campaigns, sponsorships, efforts in the design stage. Examples of variable costs are new product features and optionals (say air conditioning in a car or a camera in a cellular phone).

In our initial model we only have promotional effort from the retailing stage. We can formally build a model on promotional effort at the downstream stage of the supply chain through the variable s_r. Also let us assume that the cost $e_r(s_r)$ is variable rather than fixed. For example, let us assume that this is the cost of the time the salespersons spend with each customer. The effort increases demand $d(p, s_r)$, which is a function of both final consumer price p and promotional effort s_r. Finally, let us assume that $d(p, s_r)$ is concave in s_r, as we have decreasing returns of the sales effort. Also, let us assume that the cost function is linear or convex. So there are diminishing marginal returns of the effort s_r and constant or increasing marginal cost of effort e_r.

7.5.1 The first best solution: the vertically integrated firm

Given our assumptions, the profit for the vertically integrated chain is

$$\pi_{tot} = (p - c - e_r(s_r)) \cdot d(p, s_r). \tag{7.23}$$

So we can find the optimal effort for the vertically integrated firm by taking the derivative of the above equation with respect to s_r:

$$\frac{\partial \pi_{tot}}{\partial s_r} = (p - c - e_r(s_r))\frac{\partial d(p, s_r)}{\partial s_r} - e'(s_r)d(p, s_r) = 0;$$

therefore

$$(p - c - e_r(s_r))\frac{\partial d(p, s_r)}{\partial s_r} = e'(s_r)d(p, s_r).$$

---

[18] Here by variable we mean that the unit cost depends on the effort, but such cost is incurred for each unit sold. So the cost varies proportionally to demand.
[19] This holds unless the space in the store is the binding constraint that censors sales, which is a fairly rare situation.

The above equation suggests that the vertically integrated firm would increase the promotional effort up to a point where the marginal increase in demand $[\partial d(p, s_r)/\partial s_r]$ is counterbalanced by the marginal cost of the effort $[e'(s_r)d(p, s_r)]$.

7.5.2   The vertically disintegrated case: independent retailer and manufacturer

Now let us check what happens when the retailer and the manufacturer are independent organizations that maximize their own profit (and use simple linear contracts).

The manufacturer charges a wholesale price $p_w > c$ to make some margins and thus some profits. The drawback of this policy is that the retailer sees only a portion $(p - p_w)$ of the total chain margin $(p - c)$ and thus has a lower incentive to promote the item.[20] Indeed, the profit function for the retailer is

$$\pi_r = (p - p_w - e_r(s_r)) \cdot d(p, s_r),$$

and thus the optimal solution for the retailer is

$$(p - p_w - e_r(s_r)) \frac{\partial d(p, s_r)}{\partial s_r} = e'(s_r)d(p, s_r).$$

The above equation suggests that the marginal return for the effort is smaller for the retailer $[(p - p_w - e_r(s_r))]$ than for the vertically integral firm $[(p - c - e_r(s_r))]$, and thus, under the very reasonable assumption of diminishing returns of the promotional effort, the retailer chooses a level of effort that is lower than in the vertically integrated case. The side effects of this decision are lower demand, lower service for the consumers (and thus a reduction of their welfare), and lower profits for the chain. Why is this all happening? Once again the decision maker fails to fully capture the benefits of his decisions for the chain. Basically, the retailer overlooks the beneficial effects of the increase in demand he/she can cause on the manufacturer profit. He/she only sees a benefit $p - p_w - e_r(s_r)$ rather than the actual benefit $p - c - e_r(s_r)$.

A very easy solution to the above problem would be to contractually set this level of effort deployed by the retailer. Unfortunately, these variables can hardly be set in a contract. How do you measure the time a salesperson spends with a consumer? How do you measure the quality of a salesperson? And even if you write these variables in a contract, it can hardly be enforced in a court and thus it tends to be ineffective. Incentives are much more effective means to control efforts.

---

[20] Notice that this resembles the issues we have discussed in the previous section. Indeed, inventory investment can be interpreted as one of the means to increase sales. We have discussed it as it is a very relevant issue for the specific focus of this book.

Example 7.4 One way out of this is to make sure that the manufacturer pays a portion of the promotional effort in the stores. Large grocery stores (supermarkets and hypermarkets) tend to be self-service environments with limited sales service. However, when a new product is launched, some manufacturers pay for salespersons that invites the potential consumers that are visiting the store to try the product (usually at a launch price lower than the full price). So, quite interestingly, manufacturers in this business recognize that retailers make a suboptimal selling effort and thus they supplement the retailer's effort with their own personnel.   ☐

So far we have investigated a case where the retailer's effort influences demand and the cost of effort is variable. Let us now investigate the case where of the retailer's effort generates a fixed rather than a variable cost through an example.

Example 7.5 Let us consider a supply chain where the final consumer demand depends on both the final price p and the promotional effort s_r at the retail store. We assume the demand function to be

$$d = 1 - p + \sqrt{s}. \tag{7.24}$$

We assume that the marginal cost of production is c. Also, we assume that the promotional effort entails a cost $0.5 s_r$ (imagine that the cost is related to the size and look of the store rather than the time spent assisting each single customer).

The first best solution: the vertically integrated chain

Given the above assumptions, the profit function for the vertically integrated firm is

$$\pi = (1 - p + \sqrt{s_r}) \cdot (p - c) - 0.5 \cdot s_r. \tag{7.25}$$

Hence we can find the optimal price and promotional effort by taking the derivative with respect to p and s_r.

$$\frac{\partial \pi}{\partial p} = c - p + 1 - p + \sqrt{s_r} = 0, \tag{7.26}$$

$$\frac{\partial \pi}{\partial s_r} = \frac{(p - c)}{2 \cdot \sqrt{s_r}} - 0.5 = 0;$$

thus

$$\sqrt{s_r} = p - c$$

and by substituting in (7.26), $p^* = 1$ and thus $\sqrt{s_r^*} = 1 - c$.

We can now find the optimal profit by substituting p^* and s_r^* in equation (7.25):

$$
\begin{aligned}
\pi^* &= (1 - p^* + \sqrt{s_r^*}) \cdot (p^* - c) - 0.5 \cdot s_r^* \\
&= (1 - c)^2 - 0.5(1 - c)^2 \\
&= 0.5(1 - c)^2.
\end{aligned}
$$

The vertically disintegrated case: independent manufacturer and retailer.

Now let us investigate the case of the vertically disintegrated supply chain. As we have learned in previous sections we start by analyzing the decisions of the retailer.

The profit function for the retailer resembles the profit function for the whole supply chain with one minor difference: the marginal cost is p_w rather than c.

$$\pi_r = (1 - p + \sqrt{s_r}) \cdot (p - p_w) - 0.5 \cdot s_r. \tag{7.27}$$

Thus the optimal final price and promotional effort for the retailer are $p^* = 1$ and $\sqrt{s_r^*} = 1 - p_w$ and demand is $d^* = 1 - p_w$. This is then the demand curve for the manufacturer that tries to set the wholesale price to generate demand but at the same time he/she tries to generate margins $(p_w - c)$. The profit function of the manufacturer (which in this example has no promotional effort and thus no effort-related cost) is

$$\pi_m = (1 - p_w)(p_w - c), \tag{7.28}$$

and the optimal wholesale price for the manufacturer is $p_w^* = \frac{1+c}{2}$.

Now we can go back to the retailer and check how he reacts to this level of the wholesale price. The retailer charges an optimal price $p^* = 1$, but he/she makes less effort to sell the product ($\sqrt{s^*} = 1 - p_w^* = \frac{1-c}{2}$) than the vertically integrated company does ($\sqrt{s^*} = 1 - c$).

Let us now check the profits of the retailer and the manufacturer by substituting p^*, p_w^*, and s_r^* in equations (7.27) and (7.28).

$$
\begin{aligned}
\pi_r &= (1 - p + \sqrt{s_r}) \cdot (p - p_w) - 0.5 \cdot s_r \\
&= \frac{1-c}{2} \cdot \frac{1-c}{2} - 0.5 \left(\frac{1-c}{2}\right)^2 \\
&= \frac{(1-c)^2}{8}, \\
\pi_m &= (1 - p_w)(p_w - c) \\
&= \frac{1-c}{2} \cdot \frac{1-c}{2} \\
&= \frac{(1-c)^2}{4},
\end{aligned}
$$

so the total profit for the whole chain is $\frac{3(1-c)^2}{8}$.

Once again we shall now sit back and read the results carefully. As usual, things get worse with the vertically disintegrated chain. Consumers get a lower service (in our example they shop in smaller and less fancy stores), the profit for the chain is reduced, and the final price remains unchanged. It is rather interesting to compare these findings with those of section 7.2. In both cases the profit for the chain drops by 25%. However, in this latter case this is not due to an inappropriate increase in final consumer price (both in the vertically integrated and in the vertically disintegrated case the price is 1),[21] but rather to a lower than optimal promotional effort. So quite interestingly the suboptimal decision variable is different in the two cases, but the basic dynamics are the same. The party that makes the decision sees only a portion $(p - p_w)$ of the overall margin $(p - c)$ and thus makes decisions that are in his/her best interest but reduce the size of the overall cake.   ▯

7.5.3   A way out: designing incentive schemes.

As usual we have a situation where nobody wins, the consumer gets a worse service, and the manufacturer and the retailer make relatively less profits. There must be a way out.

Franchising contracts   A solution is to re-create an incentive for the retailer to deploy an optimal service by giving the retailer all the margins. Again like in the previous cases we can design a franchising contract where the retailer pays a variable price $p_w = c$ and a fixed fee F. The fixed fee distributes the profit in the chain while the low variable wholesale price re-creates an appropriate incentive to provide the optimal sales effort. Again, like in previous cases, these contracts leave all uncertainties on the shoulders of the retailer. So, these contracts raise major concerns when the retailer is risk adverse (or at the least more risk adverse than the manufacturer). [22]

Example 7.6 Let us now check whether the franchising contracts can solve the problem we have discussed in example 7.5. Let us assume that the manufacturer charges a wholesale price $p_w = c$. This re-creates the retailer's incentive to make an optimal effort, as the retailer's profit function resembles equation (7.25). So both the final price and the retailer's effort are optimal: $p = 1$ and $s_r = \sqrt{1 - c}$. Clearly such a policy makes sure that the retailer's profit is just equal to the optimal profit for the whole chain (as we said, retailer's profit function is just the whole chain's profit function in case $p_w = c$). However, the manufacturer's profit is zero, since the unit margin is zero $(p_w - c = 0)$. So we use the fixed fee F to move (some) profits upstream by setting $F = 0.5(1 - c)^2 - \epsilon$.

---

[21] Notice that this is a rather obvious result; indeed, all fixed costs have no impact on the price that is set through a marginal analysis. On the contrary, variable costs of service tend to have an impact on price.

[22] Notice that when the manufacturer as well has an ability to influence the demand this solution completely destroys his/her incentives to work hard to increase demand. So actually we might want to balance the incentives to increase demand according to the parties' ability to influence demand (see next subsection).

Quantity discounts Another option is to give quantity discounts, that is, charge a low wholesale price $p_w = c$ only for marginal quantities. The first few units bought make sure that the manufacturer makes a profit, while the last few units sold create a marginal incentive for the retailer to deploy an optimal sales effort (see section 7.2). Notice that the *minimum order quantity contract* where the manufacturer refuses to sell less than a given minimum quantity is basically a variant of quantity discounts, where first few units have a very high price that make the purchase of small quantities economically not sound.[23] Finally, we shall notice that other solutions do not work. Again fixing the final consumer price contractually does not solve this problem. As example 7.5 shows, setting the final price in the vertically disintegrated case is no guarantee of optimal solution. Actually, the key issue is to make sure that the retailer (more in general, the party that can influence the demand through its efforts) has the right incentive to work hard to increase demand. Setting the final price contractually gives no guarantee that this is going to happen.

Concept 7.5 *In a disintegrated chain the promotional effort might be suboptimal. The retailer might hire less salespersons and less experienced ones than in a vertically integrated chain. Stores might be less appealing and the manufacturer might spend relatively little on, say, advertising. This reduces both the profits of the industry and the welfare of the consumer. Well designed contracts can, at the least partially, solve this problem and give all parties an incentive to deploy a greater sales effort.*

7.5.4   The case of efforts both at the upstream and downstream stage

So far we have discussed the case where the retailing stage can increase demand through a promotional (sales) effort. More in general, we can improve demand both at the retail outlet and at the manufacturing stage (e.g., we can add new features, improve conformance quality etc.). So we slightly change our model and introduce a second variable s_m that captures the effort the manufacturer makes to increase demand by adding extra features to the product, by improving the product quality or through an advertising campaign. In our example, we assume that the effort at the retail store creates a variable cost (e.g., think of the sales assistance in a store) $e_r(s_r)$ while the effort from the manufacturing organization is a fixed cost (e.g., think of an advertising campaign or an effort to improve the look of the product) $e_m(s_m)$.[24] Again we

---

[23]Notice that even the franchising contracts can be interpreted as a specific kind of quantity discounts, where the first unit costs $F + p_w$, while successive units cost p_w to the retailer.
[24]Notice that the basic issue we raise in this section does not change when we change the assumption on whether the cost of the effort is variable or fixed. We suggest that the reader checks how the model changes when the cost of effort at the retail outlet is fixed

assume that demand function is concave in s_r and s_m, while the cost functions are linear (or increasing)

The first best solution: the vertically integrated chain   In this case the profit for the whole chain is

$$\pi_{tot} = [p - c - e_r(s_r)] \cdot d(p, s_r, s_m) - e_m(s_m). \tag{7.29}$$

We can find the optimal efforts by taking the derivative with respect to s_r and s_m:

$$(p - c - e_r(s_r^*))\frac{\partial d(p, s_r, s_m^*)}{\partial s_r} = e_r'(s_r^*)d(p, s_r^*, s_m^*); \tag{7.30}$$

$$(p - c - e_r(s_r^*))\frac{\partial d(p, s_r^*, s_m)}{\partial s_m} = e_m'(s_m^*). \tag{7.31}$$

The vertically disintegrated chain: independent manufacturer and retailer   Let us now investigate the case of independent manufacturer and retailer. In this scenario, the manufacturer makes two decisions: (i) the wholesales price p_w and (ii) the manufacturer's effort e_m. The retailer makes two decisions: (i) the final consumer price p and (ii) the retailer's effort e_r.

In this case, the two parties have the following profit functions:

$$\pi_m = (p_w - c) \cdot d(p, s_r^*, s_m) - e_m(s_m);$$
$$\pi_r = (p - p_w - e_r(s_r)) \cdot d(p, s_r, s_m^*).$$

Therefore they find the optimal efforts s_r^* and s_m^* by taking the derivative with respect to e_r and e_m:

$$\frac{\partial \pi_r}{\partial s_r} = (p - p_w - e_r(s_r))\frac{\partial d(p, s_r, s_m^*)}{\partial s_r} - e_r'(s_r)d(p, s_r, s_m^*) = 0 \tag{7.32}$$

$$\frac{\partial \pi_m}{\partial s_m} = (p_w - c) \cdot \frac{\partial d(p, s_r^*, s_m)}{\partial s_m} - e_m'(s_m) = 0 \tag{7.33}$$

Basically, the profit function of retailer is equal to the profit function of the previous case. The retailer does not pay for the effort the manufacturer makes. So like in the previous case the retailer makes a suboptimal effort since the reward he/she gets is only a portion $[p - p_w - e_r(s_r)]$ of the chain's benefit $[p - c - e_r(s_r)]$.[25] As we have learned, franchising contracts with $p_w = c$

(e.g., store size and look) and the cost of effort at the manufacturing stage is variable (e.g., more product features or optionals and thus more variable costs).

[25] Notice that while the retailer makes a suboptimal effort the manufacturer might make a less-than-optimal but also a more-than-optimal effort. On the one hand, he ignores a portion of the marginal revenues since his/her profit depends on the wholesale price rather than on the final price p. On the other hand, he also ignores part of the marginal cost $e_r(s_r)$.

are very effective means to re-create the retailer's incentive to make an optimal effort. Unfortunately, while this policy maximizes the retailer's incentive to promote the item, it completely destroys the manufacturer's incentive to make an effort to improve the product or invest in an expensive advertising campaign [see equation (7.33)]. Actually, in this case there seems to be no way out. We need a very low wholesale price to give the retailer incentives to make an effort in the store, and we need a high wholesale price to give the manufacturer an incentive to invest in advertising and product design. Clearly, if the demand function is much more sensitive to one of the two efforts (say the manufacturer's effort) rather than to the other, there is a fairly easy way out: We can set the wholesale price to give the right incentive to the decision maker (in our example the manufacturer) that matters the most. In this case we deliberately ignore one of the two efforts (in our example the retailer's effort). So this solution works only to the extent that one of the two efforts is actually negligible. In the more general case, we need a wholesale price that is at the same time very high $[p_w = p - e_r(s_r^*)]$ to give the manufacturer the right incentive [i.e., make sure that equation (7.33) is equal to equation (7.31)] and very low $(p_w = c)$ to give the retailer the right incentive [i.e., make sure that equation (7.32) is equal to equation (7.30)]. In this case clearly two-part tariffs do not work. Indeed, in this case the wholesale price shall perform three functions: (i) it moves profits in the chain, (ii) it sets the retailer's incentives to deploy an effort, and (iii) it sets the manufacturer's incentives to deploy an effort. And two parameters cannot possibly enable us to achieve three objectives at once. This apparently unsolvable problem actually has a solution. We just need to be creative. We just need a third party that buys from the manufacturer at a marginal price $p_{wm} = p - e_r(s_r^*)$ and sells to the retailer at a marginal price $p_{wr} = c$ (see figure 7.7). To make ends meet, the third party shall also charge a fixed fee F to the retailer. So the third party pays $p_{wm} \cdot d = [p - e_r(s_r^*)] \cdot d$ to the manufacturer and is paid $p_{wr} \cdot d = F + c \cdot d$ by the retailer. Finally, the fixed fee F moves the profits upstream and makes sure all parties have a nonzero profit (in particular the retailer's profit sets an upperbound to F and the third party's profit sets an lower bound to F).[26]

---

[26][2] shows that this solution might still face some problems as the manufacturer and the retailer might take advantage of the intermediary. Indeed, the intermediary's marginal profit is negative as his/her margins are negative and this means he/she is basically giving the other two companies an incentive to increase the quantity above the optimal quantity, simply because the retailer and the manufacturer get some margin from the consumer and from the intermediary. Indeed, if the intermediary has a negative margin, the retailer and the manufacturer are gaining a positive margin as demand increase. This means that at the optimum quantity, the retailer and the manufacturer are not gaining any margin from the consumer, but are still gaining some margin from the intermediary. Thus they have an insane incentive to further increase the quantity.

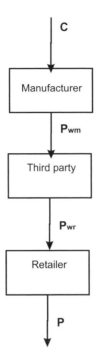

Fig. 7.7 Supply chain with a third party providing right incentives for all players.

7.6 CONCLUDING REMARKS

Though this is not a book on economics and incentives[27], in this chapter we provide some basic information on the issue of incentives in supply chains. The issue of incentives is very important, as it drives behavior both within companies and among companies. So, the issues presented in this chapter are very relevant in all multiechelon supply chains no matter whether they cut the borders of companies or lie within the boundaries of a company.

This chapter does not advocate vertical integration of supply chains. We acknowledge that there are very good reasons why companies decide to outsource and/or to split the overall company into independent organizational units. This chapter simply suggests that this wise decision raises some issues that deserve some attention. The disintegration of the supply chain implies that several independent players make their decisions (pricing, inventory, or effort) independently, to optimize their own profit (or, more in general, performance metrics).

Under these circumstances, using standard linear prices can be fairly ineffective, as single players only see a fraction of the benefit of their actions (e.g.,

[27]For a comprehensive book on these topics we refer to [6]

the retailer sees a unit margin $p - p_w$ rather than an unit margin $p - c$, with $p > p_w > c$). Economists call these externalities, meaning that the decisions of one player have side effects on other players and he/she fails to account for that. For example, in our model with promotional effort, the retailer makes a suboptimal effort because he does not account for the increase in the manufacturer's profit that the increase in demand creates. In other words, each player tries to maximize his/her slice of the cake, but by doing so, he/she is shrinking the cake.

On the contrary, we shall think about the incentives that contracts give to all players (managers of organizational units of a company rather than owners and managers of independent companies), since these incentives tend to drive people's behavior. As we have seen in all our sections on pricing, inventory planning, and promotional effort, the behavior of decision makers depends on the rules of the game. Appropriately designed contracts align incentives and make sure that all players contribute to making the cake as large as possible.

Also, we shall acknowledge that monetary/formal incentives schemes are not the only way to drive behavior. In recent weeks in an Italian newspaper, a manager was suggesting that the "clan" attitude can help the effectiveness of a company. The idea is that a feeling of belonging to a group of people (in our case a supply chain) can drive behaviors and make sure that all decisions are made to maximize the wealth of the group (in our case the profit of the supply chain). While we do acknowledge the role of these "soft" issues, in this chapter we only discuss how economic incentives drive behavior.

While in other chapters we provide tools and methods, in this chapter we present only stylized models that are designed to provide insights, intuitions, and guidelines rather than tools. This is at least partially due to the fact that this body of knowledge is relatively recent and thus robust models are still far from real life applications. Nevertheless, the basic ideas and concepts that these models provide are proven to be effective, as some of the examples discussed in the chapter show. A manager needs to adjust and fine-tune these concepts to his/her specific environment where efforts on both parties, uncertainties, and pricing might all play a role. Finally, the design of incentives through contracts actually depends on laws that might be country-specific, though the basic economic rules we have discussed in this chapter are general.

REFERENCES

1. D. Corsten and T. Gruen. Desperately Seeking Shelf Availability: An Examination of the Extent, the Causes, and the Efforts to Address Retail Out-of-Stocks. *International Journal of Retail & Distribution Management*, 31:605–617, 2003.

2. B. Holmstrom. Moral Hazard in teams. *Bell Journal of Economics*, 10:324–340, 1982.

3. A. Koutsoyiannis. *Modern Microeconomics*. The Macmillan Press Ltd., Hong Kong, 1980.

4. V.G Narayanan and A. Raman. *Hamptonshire Express, case 9-698-053*. Harvard Business School Publishing, Boston, MA, 2002.

5. V.G Narayanan and A. Raman. Aligning Incentives in Supply Chains. *Harvard Business Review*, November 2004.

6. J. Tirole. *The Theory of Industrial Organization*. MIT Press, Boston, MA, 1988.

8

Vehicle Routing

In this chapter we consider a typically operational problem, i.e., the optimal planning of routes for a set of vehicles; each vehicle is used for multiple deliveries within its route. Such a problem has a lot of variations and is known as VRP (*Vehicle Routing Problem*). In the simplest version of the problem, we have a set of customers located over some geographic region; each customer should be delivered a given amount of goods. Each customer is associated with a point in the region of interest; we know the distances between any pair of customer locations. Another point of interest is the deposit from which goods must be transported by a fleet of vehicles with limited capacity; the departure point of these vehicles is the deposit, and we also know the distance between the deposit and any customer location. We would like to deliver the required amount to all of the customers at minimum cost; the total cost function can depend, e.g., on the total miles traveled by the vehicles, on the total travel time, or on a combination of both. For the sake of simplicity, in most of the chapter we assume that only mileage is relevant. We are facing a twofold problem: On the one hand, we must assign a subset of customers to each vehicle, subject to capacity constraints; on the other one, we should plan a route for each vehicle, i.e., a sequence of customers, in order to minimize the traveled distance. Typically, such a problem makes sense over relatively short distances and time spans. The amount demanded by each customer is small enough, with respect to vehicles' capacity, to accommodate multiple deliveries; otherwise we would resort to a point-to-point transportation mode.

What we have outlined is just the basic VRP, as there are many complications in practice, in terms of both costs and constraints. Costs can be linked to both space and time; there can be a fixed cost for using a vehicle; as to constraints, delivery might be subject to time windows; the vehicle fleet may be heterogeneous, and capacity can be multidimensional (volume and weight). Still, even the basic VRP is hard to solve to optimality, unless very sophisticated approaches are used. Hence, we will just describe basic principles that can be used for the development of heuristics. These principles should be regarded as building blocks for heuristics aimed at more realistic versions of VRP. Optimization modeling can also be used, but naive mixed-integer models have weak continuous relaxations; hence, use of commercial branch and bound packages is ineffective and ad hoc strategies must be employed, which are definitely outside of the scope of an introductory book. Still, optimization models can be used to address *parts* of a VRP within clever decomposition strategies (see section 8.3.2).

Since we are interested in distribution, we just deal with deliveries, but the VRP is formally equivalent to a problem in which we want to *collect* goods; a more complicated task pops up when we have a mixed delivery/collection problem, as is the case with some postal services offering package collection to subscribers. Yet another related problem deals with *fixed routing*, in which we have to determine a set of routes which will then be followed regularly. This is more of a tactical than an operational problem. As an example of a more strategic issue, we may consider fleet sizing problems.

VRP is a classic among network routing problems. In section 8.1 we give an introduction to routing problems. If we have one vehicle with infinite capacity, VRP boils down to the classical Traveling Salesperson Problem (TSP). Solution methods for TSP can be somehow adapted to deal with VRP; indeed, TSP is a component of VRP. This is why we devote section 8.2 to illustrate some basic heuristics for solving TSP. Then we use these heuristics as building blocks to cope with basic VRP in section 8.3. Finally, in section 8.4 we illustrate a few complications arising in more realistic versions of VRP.

As a general remark, for the sake of simplicity, in this chapter we assume *deterministic* problems; we do not associate any uncertainty with demand, as we consider short-term operational problems, whereby customers have placed orders and we must just deliver the required goods. However, demand uncertainty can play a role in more tactical problems such as fixed routing. Demand uncertainty may play a role even in the short-term; in fact, there are goods which are not ordered from the warehouse, but it is the driver himself which receives orders on the spot, when visiting retailers (as a practical example, consider how fresh milk and butter are delivered to small retail stores). By the same token, we do not consider uncertainty in the traveling time; in urban transportation, delivery may be heavily affected by traffic jams or accidents.

8.1 NETWORK ROUTING PROBLEMS: THE TSP

Network routing is a general header for a very wide class of problems. Within distribution logistics, we typically adopt network routing models to tackle service scheduling problem, aimed at finding the optimal use of transportation resources (e.g., trucks) to deliver some goods to a set of customers located on a region, which is modeled as a network.[1]

From section 2.2, we recall that a network is a graph with additional information. A graph consists of a set of nodes and arcs. In our case, nodes correspond to locations (retail stores or vehicle deposits). An information which may be associated with each node corresponding to a customer is the amount of demand. Formally, arcs are ordered pairs of nodes; they can be used to represent the possibility of traveling from one node to another one, and the information associated with the arc can be distance, traveling time, or cost. We also recall that a graph can be directed or undirected. In a directed graph, we have oriented arcs, i.e., node pairs are ordered. An oriented arc is typically represented as an arrow, whereas a line is used when the orientation is irrelevant. We should also mention that for undirected graphs we should use the term *vertex*, rather than node, and *edge*, rather than arc, since latter terms are reserved to directed graphs. However, we will use just one pair of terms to keep it simple.

In vehicle routing problems, arcs are oriented if the distance (or traveling time, or cost) from node i to node j does not equal the distance from j to i. This may sound odd, but in a urban transportation problem one-way street may have that effect. On a geographical scale, if the nodes represent Los Angeles and Boston, we may argue that the distance is symmetric.[2] In this chapter we only deal with symmetric problems for the sake of simplicity.

To make things concrete, let us consider the five points depicted in the left part of figure 8.1. Think of those points as cities, or points within a city, that must be visited in order to deliver goods to customers. In the right part of the figure, we give the coordinates of each point, with respect to an arbitrary point of reference. The essential information is the distance between nodes. The real-life distance between two points may be hard to compute, because of roads, natural obstacles, etc.; if we assume that the plain Euclidean distance is a good proxy for distance, we get the distance matrix illustrated in the right part of figure 8.2 (distances have been rounded to the nearest integer). A distance matrix is a handy way to collect distance information. In our case, the distance matrix is symmetric by construction; hence, we may just show the upper triangle of the matrix, as we did in figure 8.2. The left part of the figure illustrates the corresponding network, with undirected arcs depicted as lines joining nodes; in a sense, this representation is abstract, in that node placement in the figure has no physical interpretation.

---

[1] Actually, network routing can refer to quite different problems, such as optimizing the layout in VLSI (Very Large Scale Integrated) circuits.

[2] It is worth noting that costs in transportation problems might not be symmetric even if distances are. If there is more goods flow in one of the two directions, e.g., from Detroit to New York, demand/offer mechanisms may induce asymmetric transportation fares.

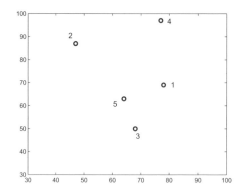

Point	X	Y
1	78	69
2	47	87
3	68	50
4	77	97
5	64	63

Fig. 8.1   Map and coordinates of five points on a region.

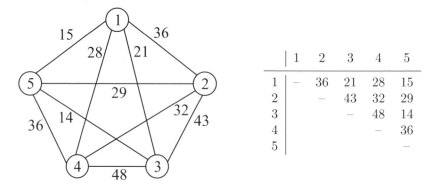

	1	2	3	4	5
1	–	36	21	28	15
2		–	43	32	29
3			–	48	14
4				–	36
5					–

Fig. 8.2   Network and distance matrix for a symmetric network with five nodes.

Given a network, a **node routing** problem calls for finding the optimal way of visiting nodes, with respect to a given criterion, subject to certain restrictions. A prototypical node routing problem is the *Traveling Salesperson Problem*, or TSP for short. In this problem, the salesperson lives in a city and must visit all of the other cities in the network before coming back home; each city (or customer) must be visited once. The traveling route is closed, and it is typically referred to as a tour. Among the many possible tours, she would like to find one with a minimal total traveled distance. The network in figure 8.2 may represent represents a simple TSP, whereby the salesperson lives in city 1 and must visit the other four cities in a clever sequence. The distance matrix can be interpreted literally, but it could also represent travel times; whatever the case, we interpret the labels associated to arcs as *costs*; the cost may also depend on both time and space. The cost for going from city i to city j is c_{ij}, and we have already noted that the matrix in the figure

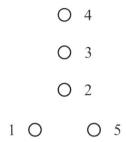

Fig. 8.3 Simple network to illustrate issues in representing network distances.

is symmetric. When $c_{ij} = c_{ji}$, for all i and j, we have a symmetric TSP; otherwise we have an asymmetric TSP, also denoted by ATSP.

As we pointed out, one of the cities should be regarded as the city where the salesperson lives and must get back to; however, due to the cyclical nature of the problem, which city is the starting point is irrelevant. For instance, the tours $(1, 2, 3, 4, 5, 1)$ and $(2, 3, 4, 5, 1, 2)$ have the same total length. As a consequence, if there are n cities, there are $n!$ permutations of them, but only $(n - 1)!$ possible solutions for the TSP; actually, in the symmetric case, there are only half of that, because we may travel any tour in two ways, obtaining the same mileage. Formally, a cycle visiting all of the nodes of a graph exactly once is called a *Hamiltonian cycle*. So, TSP calls for finding the shortest Hamiltonian cycle.

The restriction that a city must be visited exactly once may sound illogical. After all, if three cities (say i, k, and j) are geographically arranged on a line, it may be advantageous to travel from the first one, to the second one, then to the third one, and finally travel back to the first city passing through the second one. In more concrete terms, if there is a convenient freeway joining three cities, it might well be the case that we travel twice through a city. Consider for instance the network of figure 8.3, and assume for simplicity that all of the distances between neighboring cities are 1. The optimal solution of the TSP is obviously to start from city 1; go through cities 2, 3, and 4; then go to city 5, going through 3 and 2 again; finally, get back home. This may not look like a Hamiltonian cycle, but it is if we build the network in a more abstract way. From the point of view of an abstract network, like the one in figure 8.2, we travel from city 4 to city 5 "directly," along an arc of length 3, which is the sum of the distances between cities 4 and 3, 3 and 2, and finally 2 and 5. This may be the only way of reaching city 5 from city 4, or maybe just the optimal way. The bottom line is that the abstract network representation includes a "full" distance matrix, with no empty entries even though some cities are not directly linked. The distance matrix consists of *optimal* distances between pairs of nodes.

The distance matrix, since it is an "optimal" distance matrix rather than the direct translation of a map, must satisfy a rather obvious requirement, which is called **triangularity** property:

$$c_{ij} \leq c_{ik} + c_{kj}. \tag{8.1}$$

The distance from i to j cannot be larger than the distance from i to k plus the distance from k to j. We may have an equality in the aforementioned case of three cities arranged in sequence on a line. If we are dealing with a full matrix of triangular distances, we may look for a Hamiltonian cycle in the associated network; a city will be visited twice if it is optimal in the real world, but we do not see this on the abstract network model, which should not be taken too literally.

We consider solution methods for the symmetric TSP in section 8.2. Then, in section 8.3, we generalize the TSP by associating a demand information with each node. If customers must be served by finite-capacity vehicles, it is unlikely that all of the customers may be served by just one tour. If the overall demand cannot fit one vehicle, we must use multiple tours or multiple vehicles. This generalization leads to the Vehicle Routing Problem, which is the core of this chapter. However, we should at least mention the existence of other network routing problems, while referring to [3] for a full account of network routing.

8.1.1   Other network routing problems

In this chapter we only consider very basic symmetric node routing problems, but it is worth noting that node routing problems have lots of applications outside the logistics field. A symmetric TSP can be used to find the optimal path planning for a robot which has to visit a set of points in space to take measurements or to carry out spot welding operations. An asymmetric TSP can be used to model sequence-dependent setup times in a machine scheduling problem; if you produce black paint after a batch of white paint, maybe you do not need to wash the machine too accurately; going the other way around is not that easy, as producing white paint after a batch of black one requires a thorough setup. Similar considerations apply when producing vermouth or, in the textile industry, when we deal with both cheap wool and cashmere. A few concepts we use in solving symmetric problems may also be used to cope with asymmetric problems, but the latter typically require more care, depending on the solution algorithm we use.

It should also be mentioned that sometimes we have to cope with *arc* (or edge) routing problems. Consider a postman in charge of visiting all houses within a portion of a city. Since houses are arranged linearly along streets, it may be much better to represent his problem as the one of visiting all of the arcs at least once, rather than the nodes (which are used in this setting to represent crossroads). Ideally, the postman should visit all of the arcs once, along what we call a *Eulerian* cycle. Actually, a strictly Eulerian cycle may not exist.[3] The prototypical arc routing problem is the Chinese Postman Problem, i.e., the problem asking for the shortest tour of a graph which visits each arc (actually, edge) at least once.

Finally, we associate arcs with either time, space, or cost information. When dealing with very complex transportation scheduling problems, one may develop a space–time network. On such a network, some arcs represent movement in space and other arcs represent movement in time. This modeling framework is important if we want to manage, e.g., the flow of freight wagons on a railroad network (see [16]). Further complications arise when you also consider the many constraints you may have on the crews to be scheduled on trains or aircrafts.

8.2 SOLUTION METHODS FOR SYMMETRIC TSP

In this section we describe basic heuristic principles for the solution of symmetric TSP. The principles we illustrate are not the most advanced ones, but they are useful to build intuition and pave the way for the development of heuristics aimed at VRP. Conceptually, TSP is a trivial problem: Find the best sequence of stops in a set of cities. Mathematically, we have to find the best solution within a finite set of permutations of "cities." We could simply enumerate all of them and spot the best one. Unfortunately, such a simple-minded approach is not practically feasible but for very small problem instances. If we have 25 cities, there are $24!/2 \approx 3.1 \cdot 10^{23}$ alternative solutions. Assuming that we are able to generate and evaluate one billion solutions per second, it would take something like 9.84 million years to get the optimal tour. If you have to dispatch a fleet of vehicles each and every morning, you need a seriously faster decision approach.

In section B.6.1 we illustrate the branch and bound method as a way to solve optimization problems with a combinatorial component, without resorting to complete enumeration. In principle, we could build a mixed-integer linear programming model with binary variables modeling the sequencing decisions and use a good commercial solver implementing LP-based branch and bound. However, we have also pointed out that the efficiency of these methods relies on the quality of lower bounds; simple TSP model formulations have very weak relaxations, and unless very sophisticated and ad hoc modeling frameworks and solution methods are used, finding the optimal solution is very hard. We will not pursue such approaches, which are hardly available in commercial software, as we prefer to illustrate some *principles* which lend themselves to generalizations when coping with additional constraints that are important for a real-life VRP. Anyway, we should keep in mind that we

---

[3]Many of us have checked this as children, trying to draw certain geometric figures always keeping the pencil in contact with the paper, without passing twice on the same segment.

could ask someone to come up with an algorithmic black box able to solve a TSP to optimality for not-too-large problems; this can be handy in devising decomposition-based methods.

There is a huge literature on solution methods for TSP, but the methods we consider here can be broadly classified into two categories:

1. **Constructive** methods aim at building a tour by expanding a partial route according to some reasonable criterion; such methods build *one* solution directly. We illustrate two basic constructive approaches in sections 8.2.1 and 8.2.2.

2. **Iterative** methods start from a given solution and try to improve the initial tour by generating a *sequence* of alternative solutions; clearly, iterative methods are more time-consuming and require a constructive method to get a starting point. Nevertheless, the resulting gain in solution quality may be remarkable. We outline iterative methods based on local search in section 8.2.3.

8.2.1   Nearest-neighbor heuristic

The *nearest-neighbor* heuristic is arguably the simplest heuristic that may come to mind to solve TSP. We select a city acting as a starting point, and we grow a partial sequence by appending cities at the end of it. To select the next city to visit, we always choose the closest one to the last city we visited (ruling out those we have already visited). Then, after visiting all of the cities, we close the route by going back to the starting point.

The procedure can be formally stated as follows:

Step 0: initialization. Let $\mathcal{N} = \{1, 2, 3, \ldots, n\}$ be the set of cities we want to visit. Choose a starting point $i^\circ \in \mathcal{N}$; let $\mathcal{V} = \mathcal{N}\backslash i^\circ$ be the set of cities we still have to visit and let $\mathcal{S} = (i^\circ)$ the current partial sequence.[4]

Step 1: choose the next city. Let i^l be the last city in the partial sequence $\mathcal{S}$. Find the closest city j^* in $\mathcal{V}$, i.e., solve $\arg\min_{j \in \mathcal{V}} c_{i^l, j}$. If there are alternative optima, break ties arbitrarily.

Step 2: expand partial sequence. Append city j^* at the end of the partial sequence ($\mathcal{S} \leftarrow (\mathcal{S}, j^*)$) and cancel it from the set of cities yet to visit ($\mathcal{V} \leftarrow \mathcal{V}\backslash j^*$).

Step 3. If $\mathcal{V} = \emptyset$, i.e., there is no city left to visit, close the route by appending the initial city at the end of the sequence ($\mathcal{S} \leftarrow (\mathcal{S}, i^\circ)$); otherwise, go to step 1.

---

[4]We recall that the \ operator denotes set difference.

Example 8.1 Let us apply the nearest-neighbor heuristic to the problem of figures 8.1 and 8.2, starting from city 1. The closest city to 1 is city 5, and the partial sequence so far is $(1, 5)$. Among the remaining cities, the closest one to city 5 is 3; the partial sequence is expanded to $(1, 5, 3)$. From city 3 we should go to city 2, and finally we have to terminate the sequence with city 4. The complete tour is $(1, 5, 3, 2, 4, 1)$, with total length 132.

Actually, it is easy to see from the figures that this is not the optimal solution. From the map, the tour $(1, 3, 5, 2, 4, 1)$ looks more sensible; indeed, its length is 124, and it turns out that this is really the optimal tour. In this trivial case, we see quite clearly what is wrong with the nearest-neighbor: We should have gone from node 1 to node 3, but we were too greedy. We may also see that the method might yield different solutions, depending on the starting point. If we start from city 5, we get the tour $(5, 3, 1, 4, 2, 5)$, which is equivalent to $(1, 3, 5, 2, 4, 1)$. We could try all of the possible starting points and keep the best result. However, even this cannot guarantee the optimality of the solution we get.   ▯

The nearest-neighbor heuristic is conceptually simple, easy to implement, and quite fast. The bad news is that it is a *greedy* heuristic, and there is no guarantee on the optimality of the solution we get. Choosing what looks best for the current decision we have to make (select the next city) does not ensure the optimality of the whole tour. A clear danger is disregarding some inconvenient city, leaving it to the last steps of the procedure. The quality of the overall solution can thus deteriorate significantly, as the inconvenient city (which may demand a substantial cost to visit) is going to be inserted at the last step of the procedure; this means that the most critical city is practically inserted in a random position in the tour.

8.2.2   Insertion-based heuristics

The nearest-neighbor approach has many obvious limitations, which we have already mentioned. An additional one is the fact that it allows us to append a city only at the *end* of the current sequence. We could allow insertions in any point in the sequence. Since we must get back to the starting point, it would be even better to expand a closed route, rather than an open sequence that we close at the last step of the procedure. This idea leads to insertion heuristics, which are still very simple. At each step of the algorithm, we have a set $\mathcal{V}$ of residual cities to visit and a partial tour $\mathcal{T}$; what we need is to select an arc (i, j) in $\mathcal{T}$, which should be "opened" to allow insertion of a new city between i and j, leading to a subsequence (i, k, j). Actually, given a partial route, we have to make two decisions:

1. which city $k \in \mathcal{V}$ to insert in $\mathcal{T}$;

2. the insertion point, i.e., between which cities i and j already in $\mathcal{T}$ we should insert k.

Since we assume that the triangularity property holds, inserting a new city can only increase the total length of the current partial route. Hence, a reasonable criterion is to make decisions in such a way as to minimize the incremental cost of the insertion. The incremental cost of inserting city k between i and j is

$$c_{ik} + c_{kj} - c_{ij}. \tag{8.2}$$

This additional length is typically called **extra mileage**.

The first point we must take care of is how to find the initial partial route. One possibility is selecting the shortest arc (i, j) and let $\mathcal{T} = (i, j, i)$ be the initial partial route. To find the next city to insert in the partial route, we may search $\mathcal{V}$ for the closest city to $\mathcal{T}$, i.e., we may solve

$$\min_{i \in \mathcal{T}, k \in \mathcal{V}} c_{ik}.$$

Then, given the new city (breaking ties arbitrarily), we may look for the best insertion point by minimizing extra mileage. The procedure is repeated until we have the complete tour.

Example 8.2 Let us consider the TSP of figure 8.2 again. The are two cities in the initial route. Choosing the shortest arc in the network, we set the initial route as $\mathcal{T} = (3, 5, 3)$. The closest city to those included in $\mathcal{T}$ is city 1. For now, there is no substantial degree of freedom in choosing the insertion point, and we update the partial route $\mathcal{T} = (3, 5, 1, 3)$. This route is equivalent to $\mathcal{T} = (3, 1, 5, 3)$, since the problem is symmetric and the way we travel the tour is irrelevant.

Now the closest city to those in $\mathcal{T}$ is city 4, since its distance from city 1 is 28, whereas the distance between cities 2 and 5 is 29. Now we must find the optimal insertion point among the three following possibilities:

$$c_{34} + c_{45} - c_{35} = 48 + 36 - 14 = 70,$$
$$c_{54} + c_{41} - c_{51} = 36 + 28 - 15 = 49,$$
$$c_{14} + c_{43} - c_{13} = 28 + 48 - 21 = 55.$$

Hence, we set $\mathcal{T} = (3, 5, 4, 1, 3)$. Note that there is no need to reevaluate the whole tour after insertion, as only the incremental cost of the insertion is needed to make the decision. Finally, we have to accommodate city 2:

$$c_{32} + c_{25} - c_{35} = 43 + 29 - 14 = 58,$$
$$c_{52} + c_{24} - c_{54} = 29 + 32 - 36 = 25,$$
$$c_{42} + c_{21} - c_{41} = 32 + 36 - 28 = 40,$$
$$c_{12} + c_{23} - c_{13} = 36 + 43 - 21 = 58.$$

The final route we get is $(3, 5, 2, 4, 1, 3)$, with total length 124.   □

In this case, we get the optimal solution, but this is not guaranteed in general, as the insertion-based heuristic is still a greedy heuristic. We could represent our basic constructive procedures as a greedy way to explore a search tree, a concept that we introduce in section B.6.1 on branch and bound methods. In a branch and bound method, we prune a branch of the search tree only if we are sure that it cannot lead to an optimal solution. In greedy heuristics, we basically select the most promising branch, forgetting about the others. However, we could reduce the myopic behavior of greedy heuristics by adopting a look-ahead strategy, whereby we explore the consequence of a choice by examining its consequences a few steps further. A further issue concerns breaking ties when we have to make a decision. In insertion-based procedures, we might have two insertion points with the same extra mileage; in the nearest-neighbor heuristic, we may have two or more cities with the same distance from the last one in the partial sequence. In such a case, we could explore the consequences of each alternative a bit deeper in the search tree, rather than breaking ties arbitrarily and take a basically random branch.

In the specific case of the insertion-based approach above, we may also try to improve results, at some additional computing cost, by considering all possible pairs consisting of a new city to insert and its insertion point. In fact, in the procedure above we select a city, and then we explore possible insertion points; we could find the optimal insertion point for each city, and only after evaluation of the result we make a decision. Another variation on the theme is the choice of the initial two-city tour; we could start from the two farthest cities, rather than from the closest pair.

8.2.3   Local search methods

The two approaches we have just considered are constructive, in that they directly build one solution, with a possibly greedy logic. An alternative consists of examining a *sequence* of solutions. The basic idea is trying to improve a given solution using some simple recipe. We can perturb the solution according to a predefined set of rules, which define a *neighborhood* of the current solution; the name stems from the fact that we just apply small changes to the current solution. For instance, since a TSP solution is basically a permutation of cities, we could consider swapping pairs of cities in the tour. Having defined the neighborhood structure, we may look for the best solution within the neighborhood of the current tour. This new candidate solution may be an improvement or not. In the first case, we set the candidate as the new current solution and we repeat the procedure; otherwise we stop.

This very simple approach is called **iterative improvement** and is the simplest example of a large family of methods collectively called **local search** methods. Since we only search locally in the neighborhood of the current solution, we might well get stuck in a locally optimal solution that is far less performing than the globally optimal one. We should note that "locally" means "with respect to the neighborhood structure." In figure 8.4 we illustrate the issue conceptually. If we are minimizing a nonconvex cost function $f(x)$, and we are at point x_L, there is no way to escape from this local optimum and

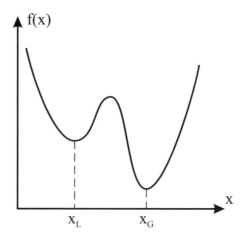

Fig. 8.4 Getting stuck in a local minimum.

get to the global optimum x_G, if we just look to the left and to the right and accept only improving steps. In local search, we cannot really draw a picture like that, because we are moving within a space whose *discrete* points are, e.g., tours on a network; nevertheless, with respect to some "weird" topology, we may have lack of convexity in the cost function, possibly leading a local improvement procedure into bad local optima. Clearly, there is a tradeoff between computational requirements and the richness of the neighborhood structure (in the limit, a somewhat expensive neighborhood could require the complete enumeration of the feasible solutions). On the one hand, defining a small neighborhood is very efficient computationally, but it can leave us in a very bad local optimum. On the other hand, a very rich neighborhood structure opens many more search paths, but it can be too demanding from a computational point of view.

Indeed, the art of local search consists of devising a parsimonious, yet effective neighborhood structure. For instance, in the TSP case we could swap pairs of consecutive cities in the sequence, which is a rather limited neighborhood structure. A richer, and quite effective, neighborhood structure is known as 2−opt. Given a complete tour, we consider all pairs of nonconsecutive arcs. They are canceled and substituted by two alternative arcs in such a way that we obtain another tour. The idea is illustrated in figure 8.5. We see that the two canceled arcs are substituted by arcs "crossing" each other (remember that the network we draw need not be taken as a pictorial representation of the underlying geography). The idea can be generalized by canceling k arcs and replacing them in all possible ways. The k−opt approach, for $k > 2$, tends to get more complex and time-consuming, and significant advantages in terms of quality are not guaranteed.

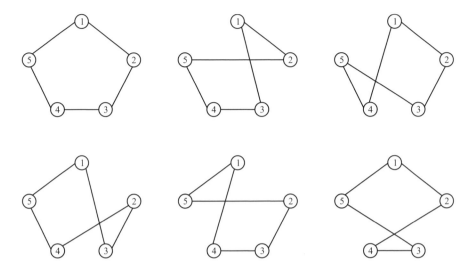

Fig. 8.5 An example of neighborhood generated by the 2−opt rule; the current solution is depicted in the upper-left corner.

Example 8.3 We consider once more the TSP of figure 8.2, and we tackle it by a 2−opt approach starting from the tour $(1, 2, 3, 4, 5, 1)$, whose total length is 178. We must compute the total length of each neighboring tour, as depicted in figure 8.5:[5]

$$(1, 2, 5, 4, 3, 1) \rightarrow 170,$$
$$(1, 2, 3, 5, 4, 1) \rightarrow 157,$$
$$(1, 3, 2, 4, 5, 1) \rightarrow 147,$$
$$(1, 4, 3, 2, 5, 1) \rightarrow 163,$$
$$(1, 2, 4, 3, 5, 1) \rightarrow 145.$$

The best tour in this set is $(1, 2, 4, 3, 5, 1)$, which gets to be the new current tour. Then we evaluate the new neighborhood:

$$(1, 2, 5, 3, 4, 1) \rightarrow 155,$$
$$(1, 2, 4, 5, 3, 1) \rightarrow 139,$$
$$(1, 4, 2, 3, 5, 1) \rightarrow 132,$$
$$(1, 3, 4, 2, 5, 1) \rightarrow 145,$$
$$(1, 2, 3, 4, 5, 1) \rightarrow 178.$$

[5]From an implementation point of view, this task can be made extremely efficient by proper use of data structures, also avoiding the recomputation of total length from scratch and just evaluating an incremental cost; see [18].

Note that the last solution in this neighborhood is just the initial tour. The new current solution is $(1, 4, 2, 3, 5, 1)$, with total length 132. Repeating the procedure one more time, we get

$$(1, 4, 5, 3, 2, 1) \rightarrow 157,$$
$$(1, 4, 2, 5, 3, 1) \rightarrow 124,$$
$$(1, 2, 4, 3, 5, 1) \rightarrow 145,$$
$$(1, 3, 2, 4, 5, 1) \rightarrow 147,$$
$$(1, 4, 3, 2, 5, 1) \rightarrow 163.$$

We leave to the reader the task to verify that no further improvements can be obtained. Since we cannot find any improving tour, the algorithm stops.   ⬜

In this lucky example, we actually end up with the optimal solution, but we do not know that (in this small case, we may prove that 124 is the optimal length by complete enumeration). In general, this does not happen, and the solution we stop at may depend on the initial tour. The difficulty is that the search process may get stuck into a local optimum, and there is no way out because we only accept improving perturbations (see figure 8.4). There are a couple of ideas that may come to our mind to overcome this difficulty:

- We could start the search from different initial tours, possibly generated by alternative constructive heuristics or by random generation. The idea of generating multiple starting points randomly leads to GRASP (**Greedy Randomized Adaptive Search Procedure**) methods.

- We may try to overcome the tendency to get stuck in local optima by allowing nonimproving perturbations according to a sensible strategy. In fact, looking back at figure 8.4, we see that in order to travel from x_L to x_G, we must accept a temporary increase in cost.

The last idea has lead to a fairly wide family of local search approaches, which we just outline below, referring the interested reader to references at the end of the chapter.

- In **simulated annealing**, optimization is interpreted as an energy minimization process. In classical mechanics, a physical system evolves in such a way as to minimize its energy: A ball subject to gravity force will roll into a hole, minimizing its potential energy, and will stay there. There is no way a ball can pop up from the hole all by itself. In optimization terms, this means that if the ball rolls into a local minimum, it gets stuck there. In Statistical Mechanics, under the effect of thermal noise, there is some probability that a system will find itself in a higher energy state without external intervention. The probability of this upward jump increases with temperature and decreases with the size of the jump, i.e., the energy difference between the two states. Annealing is a technological process whereby a material is slowly cooled, allowing it to escape from local minima and to reach a lower energy level. If we cool the material too fast, we get a glass; if the cooling process is slow, we get a good crystal structure when the final temperature is so low that the system cannot change configuration anymore. Simulated annealing exploits this idea for optimization, allowing nonimproving perturbations according to a stochastic mechanism. Given a current solution with cost C_{old}, we randomly sample an alternative solution in its neighborhood, with cost C_{new}. The alternative solution is accepted with probability given by

$$\min\left\{1, \exp\left[\frac{-(C_{\mathrm{new}} - C_{\mathrm{old}})}{T}\right]\right\},$$

where T is a control parameter acting as a temperature, which is decreased according to a cooling schedule. We see that at high temperatures, the search process is free to wander and explore the solution space, whereas at low temperatures it works just like local improvement. When the algorithm freezes, the best solution visited will be reported.

- Another idea for a stochastic search mechanism is mimicking biological evolution, rather than statistical mechanics. In **genetic algorithms**, unlike other local search mechanisms, we work on a *population* of solutions. Only the best members within the current population have a high chance of surviving: The current population evolves by crossover (offspring are created from two parents) and mutation (a random perturbation is applied) mechanisms, whereby probability of selection and survival depends on the quality of each solution. In this case, we need a way to map a solution to a data structure, which works like a chromosome, whose genes are the features of a solution (or the parameters of an algorithm to build a solution). The mechanisms for crossover and mutation define the neighborhood structure for this stochastic search algorithm.

- Maybe the most widely applied local search mechanism, as far as TSP and VRP are concerned, is **tabu search**. This approach, unlike the previous two, need not be stochastic. The rationale is that the best solution in the neighborhood of the current one should be accepted, in order to escape from local minima, while biasing the search process towards good solutions. The trouble with this simple idea is that cycling is most likely to occur: When escaping from a local minimum, we accept a nonimproving alternative, but the best solution in the neighborhood of this new solution may well be the previous local optimum. To avoid cycling, we use a data structure to store some attributes of each solution we visit, or some feature of the perturbations we apply to get them. This data structure works as a tabu list, which forbids revisiting solutions or applying perturbations undoing what we have just accomplished. The tabu list is a sort of short-term memory, as only the most recent tabu attributes are kept there, in order to avoid restricting the search process too much. Long-term memory mechanisms have been proposed to improve the ability of diversifying search by exploring new regions of the solution space.

Local search algorithms look conceptually simple, but in fact getting them to work properly requires a fair amount of skill and ingenuity, not to mention experience. Defining a good neighborhood structure, as well as setting the parameters governing the algorithm, is not trivial. To get a feeling for the subtle issues we may have to face, consider the application of the 2−opt neighborhood structure to an asymmetric TSP. If we cross arcs, like we did in figure 8.5, the consequence is that we actually invert part of the sequence; in other words, part of the tour is traveled clockwise rather than counterclockwise (and vice versa). This is not relevant in the symmetric case, but when the distance matrix is not symmetric, the new solution may be radically different from the previous one. We face a similar issue when dealing with time-windows in a VRP; even if distances do not change, changing the time instants at which we visit customers may have adverse effects. In practice, some knowledge of network and graph optimization may be needed in order to find a good heuristic for a complex case; common sense is not always enough.

8.3   SOLUTION METHODS FOR BASIC VRP

VRP is a generalization of TSP, accounting for multiple vehicles whose routes are subject to additional constraints. There is a set of n customers; each customer is located on a node in a network. To serve customers, we have a fleet of vehicles located in node 0. We consider a fleet of homogeneous vehicles, each featuring the same capacity, and one deposit; real-life problems may require relaxing such assumptions. A known demand d_i, $i = 1, \ldots, n$, is associated with customer i. Demand need not be necessarily associated to one item type; what is really important is that demand is measured in the same units as vehicle capacity. Just like in TSP, we would like to minimize distance traveled (or time, or cost); but unlike TSP, each vehicle has a finite capacity, in terms of volume and/or weight. This is what creates the need for multiple vehicles and/or multiple routes, because we cannot serve all of the customers with one route. We have to develop a set of routes, starting and terminating at the deposit, which can be carried out sequentially by one vehicle, or in parallel by a set of vehicles.

Given such assumptions, our input data are a symmetric distance (or travel cost) matrix, the demand per customer, and the vehicles' capacity; the number of vehicles may be given or not, depending on the specific assumptions about the way routes are carried out. We want to find a set of routes minimizing total distance traveled, subject to vehicle capacity constraints. In the

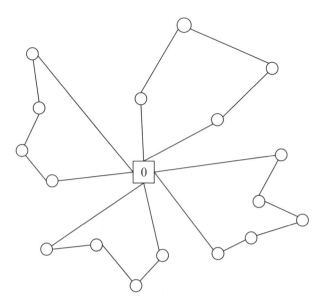

Fig. 8.6   Example of solution of a VRP.

basic VRP, we do not consider additional constraints such as the maximum route duration, which is just another capacity constraint, or time windows for serving customers. Another important simplification is that we sequence customers within each route, but we do not really schedule routes. For instance, suppose that early in the morning we devise five routes. Any route can be executed within the current working day, but we have just four vehicles. Hence, we should decide which routes should be carried out today, and which one will be carried out tomorrow. Clearly, this may depend on priorities associated with customers; alternatively, we could try to devise two routes that can be carried out by the same vehicle within one working day, by returning to the deposit between the two routes. We see that such timing issues might be rather complicated. In the basic VRP, we either assume that the number of vehicles is unlimited, or we try to find a solution serving all of the customers with a given number of vehicles, reporting infeasibility otherwise.

Despite all of these severe limitations, the basic VRP is a tough problem, and tackling it paves the way for solution of more realistic versions. Figure 8.6 illustrates one solution of a VRP. The figure points out the twofold nature of VRP. The solution consists of two elements, since each route consists of a subset of customers and the sequence according to which they are visited by the vehicle. Given the first element, we have one TSP per vehicle. This can be exploited in decomposition strategies; it also suggests that TSP heuristics can provide some basic principles to tackle VRP as well. VRP heuristics, too, can be constructive or iterative. We do not consider local search methods for VRP, because devising neighborhood structures coping with both dimensions of the problem (i.e., allocation of customers to routes and sequencing within each route) is not trivial, even though the effort in doing so can be quite rewarding.

8.3.1   Constructive methods for VRP

Constructive methods for VRP are based on the idea of growing routes according to various patterns and based on various criteria.[6] To classify constructive methods, we should begin by drawing the line between

- sequential algorithms, in which one route is grown at a time, until all customers have been routed, and

- parallel algorithms, in which several routes are grown together.

Parallel algorithms, in turn, can be classified into two subcategories:

1. We may start from a set of small routes, one per customer, and we proceed by merging routes. The procedure stops when vehicles' capacities prevent us from coalescing routes. One clear disadvantage of this approach is that we have no control over the number of routes we end up with, which may be larger than the number of available vehicles.

2. In order to overcome the aforementioned disadvantage, we may fix the number of routes a priori, say m. The number of routes can be the number of vehicles we plan to use. Typically, we use m well-selected customers to devise an initial set of "seed" routes, each one consisting of one customer. Then we proceed by selecting one customer at a time, which is inserted in one of the m growing routes.

Finally, we have to specify the criteria we use in growing routes. There are many of them, but we illustrate the two fundamental ones by referring to figure 8.7.

- The **savings** criterion. The rationale behind the savings criterion is that if two customers, say i and j, are served by two vehicles along separate routes, the two vehicles have to drive from the deposit to the customer and back. Hence, the total traveled distance amounts to $c_{0i} + c_{i0} + c_{0j} + c_{j0}$. If the two routes are merged and the two customers are served by the same vehicle, the new total length will be $c_{0i} + c_{ij} + c_{j0}$, with a saving $s_{ij} = c_{i0} + c_{0j} - c_{ij}$. Referring to figure 8.7, we cancel the two dashed arcs, replacing them with arc (i, j). Actually, the argument, as it is stated, applies only to routes consisting of one customer visit. In fact, it can be applied more generally, provided that customers i and j are the first or the last on their respective routes (see figure 8.8; remember that

---

[6]This section relies heavily on material from [7].

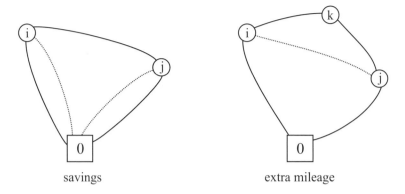

Fig. 8.7 Illustrating savings and extra-mileage criteria.

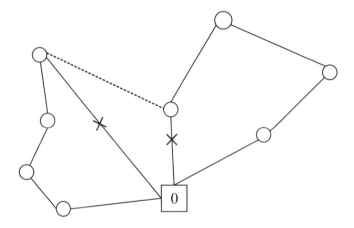

Fig. 8.8 Merging two partial routes by the end points.

we are just considering symmetric problems; hence, being last or first in the route is actually the same thing). This idea can be used to merge partial routes together, provided that capacity constraints are satisfied; according to this metric, we should give priority to the merger with the largest saving. Another relevant point is that by merging routes, we decrease the number of required vehicles.

• The **extra-mileage** criterion. We have already met the extra-mileage criterion when discussing insertion-based heuristics for TSP (see section 8.2.2). Here the idea is inserting customer k on the path from customer i to j, incurring an increase of the route length given by $e_{ikj} = c_{ik} + c_{kj} - c_{ij}$. Referring again to figure 8.7 (right side), we get rid of the dashed arc and insert two arcs, in such a way that extra mileage e_{ikj} is minimal.

Table 8.1 Distance matrix and customer demand for example 8.4

c_{ij}	0	1	2	3	4	5	6	7
0	–	4	2	4	3	3	5	6
1		–	2	7	4	6	5	3
2			–	5	4	4	3	4
3				–	3	1	6	9
4					–	3	7	7
5						–	5	8
6							–	5
7								–

i	1	2	3	4	5	6	7
d_i	9	6	14	8	9	6	5

Table 8.2 Savings matrix for example 8.4

r_{ij}	1	2	3	4	5	6	7
1	–	4	1	3	1	4	7
2		–	1	1	1	4	4
3			–	4	6	2	1
4				–	3	1	2
5					–	3	1
6						–	6
7							–

To illustrate the concepts above in a concrete setting, we may describe an early algorithm for VRP, known under the names of Clarke and Wright.[7] The method is based on the savings criterion, and it is a parallel algorithm of the first type, i.e., it is based on the coalescence of smaller routes.

Example 8.4 Clarke–Wright's algorithm is best illustrated by a small example, whose input data are displayed in table 8.1. The distance matrix is symmetric and we assume that vehicles' capacity is 20. To begin with, we may compute a savings matrix, with an entry for each pair of customers; the result is reported in table 8.2. Since we always join customers when they are placed at an endpoint of a route, this savings matrix can be computed once for all. Actually, not all of its entries are relevant: For instance, customers 1 and 3 cannot be served by the same vehicle, because their total demand is $9 + 14 = 23$, which exceeds vehicle capacity. The starting set of routes is

$$(0, 1, 0); \quad (0, 2, 0); \quad (0, 3, 0); \quad (0, 4, 0); \quad (0, 5, 0); \quad (0, 6, 0); \quad (0, 7, 0).$$

---

[7]See the original reference [8].

We see from table 8.2 that 7 is the largest saving, and it is obtained by joining customers 1 and 7, leading to the new set of routes:

$$(0, 1, 7, 0); \quad (0, 2, 0); \quad (0, 3, 0); \quad (0, 4, 0); \quad (0, 5, 0); \quad (0, 6, 0).$$

Then the table shows that two savings amount to 6, but the one associated with customers 3 and 5 is not compatible with vehicle capacity. We should check if joining customers 6 and 7 is feasible, since the latter customer is already on the same route as customer 1; the total demand for these three customers equals the vehicle capacity; hence, we may get rid of customers 1, 6, and 7. Now, current routes are

$$(0, 1, 7, 6, 0); \quad (0, 2, 0); \quad (0, 3, 0); \quad (0, 4, 0); \quad (0, 5, 0),$$

The best feasible option is merging customers 4 and 5, with a saving of 3. Joining them, we get

$$(0, 1, 7, 6, 0); \quad (0, 2, 0); \quad (0, 3, 0); \quad (0, 4, 5, 0).$$

Now, neither customer 2 nor customer 3 fits the route $(0,4,5,0)$; all we can do is merging customers 2 and 3, which yields the final set of routes:

$$(0, 1, 7, 6, 0); \quad (0, 2, 3, 0); \quad (0, 4, 5, 0).$$
                                                              ▯

Clarke and Wright's algorithm is conceptually quite easy, and it played a prominent historical role, but it suffers from a few limitations. To begin with, when we merge routes, we do so only by joining a pairs of customers at the endpoints of their respective route (see figure 8.8). Maybe, inserting new customers in arbitrary points of a route could be advantageous. Furthermore, there is no control over the number of routes we end up with; in the example above, we could not use less than three vehicles anyway, but in general, if we have a given number of vehicles and we have to serve all of the customers in parallel, we would like to make sure that the number of routes is kept under control.

We can also exploit the ideas behind the insertion-based heuristic for TSP (see section 8.2.2) to come up with a sequential algorithm based on extra mileage. The idea is growing one route by inserting one customer at a time; the customer and its insertion point are determined by minimizing extra mileage, provided that the vehicle capacity constraint is satisfied. The current route may be closed when there is no way to insert any other customer; then we start again with a new route. A potential weakness of such an idea is that, in order to saturate the current route, we could be forced to add a very distant customer. This may happen if there is a small residual capacity on the truck and the only customer with a small demand, fitting the residual capacity, is really far from the cluster of customers in the current route. This may be a good reason to prefer a parallel approach, in which we select which customer to insert, on which route, and at which point. If we wish to use m vehicles, it is natural to start from m seed routes, each consisting of one customer. We can grow the routes using the extra-mileage criterion; this way, we can control the number of vehicles we use (assuming we can serve all of the customers with that number of vehicles). If there is a fixed cost associated with each vehicle, we may change the number of seed routes, trading off the number of vehicles against total distance traveled.

A common issue with parallel approaches of the second kind is the selection of seed customers. A sensible rule is that they should be distant from the deposit and distant from each other. The rationale behind the first requirement is that distant customers are an inconveniency, but they must be served anyway; it is better to include them in a route immediately, in order to avoid late insertions that may generate a large increase of the route length. Furthermore, it is natural to think that if customers are far away from each other, they are best served by separate routes. Hence, let us denote by σ_j, $j = 1, \ldots, m$, the seeds to initialize the desired m routes. The first seed is selected by maximizing its distance from the deposit. Then, after having selected the first k seeds, the next seed σ_{k+1} is found by solving

$$\max_i \min \{c_{i0},\ c_{i,\sigma_1},\ c_{i,\sigma_2}, \ldots,\ c_{i,\sigma_k}\}. \tag{8.3}$$

In plain terms, the new seed maximizes the minimum distance between itself and the deposit and the other seeds. The idea is illustrated in figure 8.9, under the assumption of Euclidean distances. Customer 1 is the first seed we would select, since it is the farthest one from the deposit. Customers 3, 4, and 5 do not make good seeds because they are close to the deposit. The next farthest node in the network is associated with customer 2. However, this node is close to customer 1; it is reasonable to assume that they will be served by the same route. The second seed we should select is customer 6, which is far from both node 0 and node 1. Of course, this is just a sensible heuristic, which only considers distance. One could also consider demand size with respect to vehicle capacity: It may be not advisable to leave customers with large demand to late insertions, as they may be hard to fit to residual capacity.

We see that there is room for a large variety of combinations of heuristics principles. Since we may grow routes using extra-mileage or savings criteria, a natural question is whether one of them performs best. As expected, there is no easy answer, and the result may depend on the problem instance. To get an intuitive feeling for the underlying issues, we may have a look at picture 8.10, which illustrates a rather artificial but instructive example (see [7]). We have four customers, located on an equilateral triangle; here we consider Euclidean distance, i.e., we assume that the distances we see in the drawing correspond to the real ones. Each customer demand is 1, and vehicle capacity is 2; so, each route should serve two customers. If we start with four separate one-customer routes, and we merge them in parallel using a savings criterion, we end up with

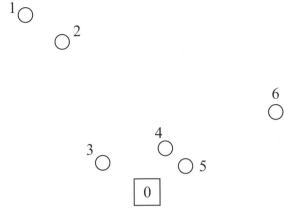

Fig. 8.9   Selecting seeds for parallel constructive heuristics.

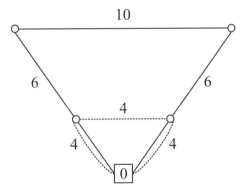

Saving criterion: total length 42

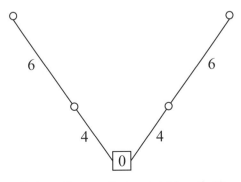

Extra-mileage criterion: total length 40

Fig. 8.10   Alternative criteria to merge routes (case 1).

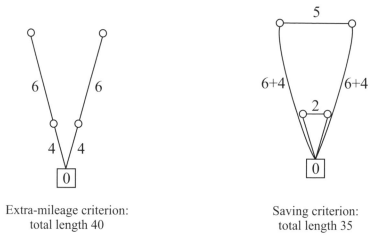

Extra-mileage criterion: Saving criterion:
total length 40   total length 35

Fig. 8.11   Alternative criteria to merge routes (case 2).

the solution illustrated in the upper part of figure 8.10, with total length 42. Note that the largest saving $(10 + 10 - 10 = 10)$ is obtained by joining the two farthest customers; this leads to a long an "circumferential" route. If we use extra mileage, e.g., starting from two seed routes associated to the farthest customers, we get the solution in the lower part of the figure, with total length 40; this happens because we can serve the two closer customers with no extra mileage. In this problem instance, the extra-mileage criterion performs better, but if we shrink the bottom angle and reduce horizontal distances in the triangle, as illustrated in figure 8.11, we get a different conclusion. In this second case, we still get total length 40 when using extra mileage, whereas the saving criterion yields a solution with total length 35. By the way, this second solution has lower total length, but it could be unsatisfactory in terms of workload balance: One driver gets a much easier task than the other one, an issue that we do not consider here, but may play a very important role. This example is clearly artificial, but it helps in building intuition about the qualitative properties of routes developed using the two criteria. We see from figures 8.10 and 8.11 that saving yields "circumferential" routes. This happens because the savings criterion may consider joining far customers attractive. In fact, Clarke and Wright's algorithm was included in an early software tool for VRP, developed by IBM in the 1970s. This package, called VPSX, was sometimes criticized by practitioners just because of the circumferential nature of proposed routes. On the contrary, when looking for small extra-mileage, it is natural to get more "radial" patterns.

The bottom line of the discussion so far is that it may be difficult to devise a *robust* method based on a single heuristic principle. Occasionally, any heuristic may yield a very poor solution. One way out of this difficulty is combining heuristic principles, possibly introducing one or more parameters which may be adjusted as needed. One such idea is introducing a modified saving

$$\tilde{s}_{ij} = s_{ij} - \theta c_{ij},$$

where the parameter θ tends to penalize the inclusion of long arcs in the route, even if they yield large savings. This may prevent the creation of too circumferential routes. This is a simple example of parameterized criterion, and quite complex criteria have been proposed in the literature. Choosing the right value of one or more parameters is a tough task, but since constructive heuristics are quite fast, probably the best idea is simply using brute force and running the heuristic for several values of the parameter, keeping the best solution. Then, the solution can be further refined by local search.

8.3.2   Decomposition methods for VRP: cluster first, route second

VRP is a twofold problem with a *clustering* component (i.e., assigning a group of customers to each vehicle) and a routing component (i.e., finding the best route for each vehicle). Since the second problem dimension boils down to a set of TSP problems, which we may deal with rather effectively, the idea of decomposing the overall problem into two subproblems is quite natural. Various decomposition methods have been proposed and can be classified into two broad categories:

1. In **route-first, cluster-second** methods we first find one tour covering all of the customers, e.g., using some TSP solution method; then, we partition the resulting tour into routes compatible with vehicles' capacity.

2. In **cluster-first, route-second** methods we first assign customers to vehicles, subject to capacity constraints, and then we solve one TSP per vehicle.

Here we outline a couple of possible implementations of the second principle.

An early and intuitive sequential decomposition method, due to Gillett and Miller,[8] is called the **sweep** method. The approach has a strong geometric motivation, which is illustrated in figure 8.12. We draw a ray from the deposit, and we rotate the ray clockwise or counterclockwise; in doing so, we "sweep" customers in an order depending on their location. Whenever the ray passes over a customer, this is included in the current cluster and the process continues until we find a customer which cannot be fitted to the residual vehicle capacity. Then we form a cluster by grouping the "swept" customers; we route this subset of customers by solving the corresponding TSP, and we proceed by forming and routing the next cluster. We see that this approach is sequential in nature, which means that we have no control over the number

---

[8]See the original reference [11].

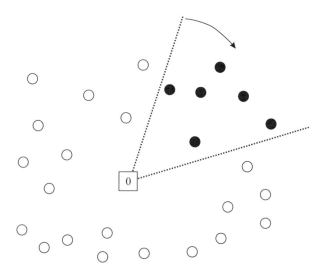

Fig. 8.12   Applying the *sweep* method to a Euclidean VRP.

of routes we will end up with; maybe by swapping and reassigning customers between clusters, we could reduce the number of routes. However, even if there exists a set of feasible routes involving a given number m vehicles, there is no guarantee that we will be able to find it. A further limitation of the approach is that it relies on a geometric argument; figure 8.12 should be really interpreted as a map of customer locations. If distances between nodes are strongly related to the Euclidean distances between points on the map, the result may be satisfactory. However, if the nature of terrain and roads is such that Euclidean distances are not closely related to actual distances, the quality of results could be low.

More recent and sophisticated decomposition-based methods have been proposed to overcome the limitations above, by exploiting partial mathematical modeling of VRP. Modeling the VRP by integer programming is possible, but not effective, unless nontrivial modeling and solution approaches are adopted. However, we may build a partial model, e.g., in order to assign customers to vehicles, leaving the routing task to a TSP solver. We illustrate here an idea due to Fisher and Jaikumar [10], which exploits a prototypical combinatorial optimization model known as *generalized assignment*. In this problem we are given a set of n jobs which must be carried out on a set of m machines (typically, $m < n$). Machines need not be identical: For each pair consisting of job $i = 1, \ldots, n$ and machine $k = 1, \ldots, m$, we have a processing time p_{ik} and a cost c_{ik}. We would like to carry out the whole set of jobs at minimal cost, but we might not be able to assign each job to the cheapest machine, because of capacity constraints: Each machine k is available for R_k time units. We can build an integer programming model by introducing a set of binary decision variables y_{ik}, set to 1 if job i is assigned to machine k, 0 otherwise:

$$\min \quad \sum_{i=1}^{n}\sum_{k=1}^{m} c_{ik}y_{ik} \tag{8.4}$$

$$\text{s.t.} \quad \sum_{k=1}^{m} y_{ik} = 1, \qquad i = 1,\ldots,n, \tag{8.5}$$

$$\sum_{i=1}^{n} p_{ik}y_{ik} \le R_k, \qquad k = 1,\ldots,m, \tag{8.6}$$

$$y_{ik} \in \{0,1\}.$$

The objective function (8.4) is total cost; constraint (8.5) makes sure that each job is assigned to exactly one machine, and (8.6) is the capacity constraint for each machine. This literal description of generalized assignment leaves room to many interpretations. In the VRP case, we may interpret jobs as customers to be served and machines as vehicles. So far, we have assumed that vehicles are identical, and we stick to this case denoting the vehicle capacity by R; however, we see an immediate advantage of this approach, which helps in getting rid of many limitations. We may also include restrictions on the type of vehicle that can be used to serve a customer (e.g., because large trucks cannot be used in old town centers). Denoting the demand from customer i by d_i, *in principle* we can write the following model:

$$\min \quad \sum_{k=1}^{m} f(\mathbf{y}_k)$$

$$\text{s.t.} \quad \sum_{k=1}^{m} y_{ik} = 1, \qquad i = 1,\ldots,n,$$

$$\sum_{i=1}^{n} d_i y_{ik} \le R, \qquad k = 1,\ldots,m,$$

$$y_{ik} \in \{0,1\}.$$

This is just a generalized assignment problem with a weird objective function. Here vector $\mathbf{y}_k$ consists of all of the decision variables y_{ik} associated with vehicle k. From a conceptual point of view, we may imagine a function $f(\mathbf{y}_k)$ which yields the optimal tour length obtained by solving to optimality a TSP restricted to the customers assigned to vehicle k. If we were really able to write such a function analytically, the model above would be a working model for VRP. Of course, we are not that lucky, but we can try to approximate function $f(\cdot)$ in a way which is suitable to solution by linear integer programming:

$$f(\mathbf{y}_k) \approx \sum_{i=1}^{n} g_{ik}y_{ik}.$$

The parameter g_{ik} should be an approximation of the cost of assigning customer i to vehicle k. Clearly, such a linear function cannot really capture the interactions among various assignments, which influence each other when solving the TSP. However, we can try to find a suitable value if we assume that a set of m customer seeds is given. Such seeds play the same role as in constructive parallel heuristics of type 2 (i.e., those based on growing a given number of routes), and they could be selected by the logic behind expression (8.3) on page 418; we select m seeds by finding a subset of m customers which are far from the deposit and far from each other. Given the seeds, which are associated with the m routes, we may estimate the cost of inserting customer i in any route by computing the extra mileage with respect to the deposit 0 and the seed σ_k ($k = 1, \ldots, m$) of that route:

$$g_{ik} = c_{0i} + c_{i,\sigma_k} - c_{0,\sigma_k}.$$

Now that we have a linear approximation of the TSP cost, we may solve the generalized assignment problem by branch-and-bound, or by ad hoc methods if problem size precludes using a commercial integer programming package; then we solve one TSP for each vehicle. A noteworthy feature of the approach is that if there exists a feasible solution using m vehicles, we will find one; constructive heuristics do not offer such a guarantee. Another important remark is that the generalized assignment formulation, as we have already pointed out, can be extended to cope with heterogeneous vehicles and to model some additional constraints on the vehicles that can be used to serve a customer (some goods need freezer trucks, or separate sections because of mutual incompatibility; for instance, think of food and chemicals).

A later extension of the generalized assignment approach was proposed by Bramel and Simchi-Levi [6], in order to avoid the *a priori* selection of seeds. In order to integrate seed selection with customer clustering, they proposed a *concentrator location* formulation. Let us introduce the following decision variables:

$$z_j = \begin{cases} 1 & \text{if customer } j \text{ is selected as a seed,} \\ 0 & \text{otherwise;} \end{cases}$$

$$y_{ij} = \begin{cases} 1 & \text{if customer } i \text{ is assinged to a route, whose seed is customer } j, \\ 0 & \text{otherwise.} \end{cases}$$

We also need the cost coefficients:

$$g_{ij} = c_{0i} + c_{ij} - c_{0j}, \qquad v_j = 2c_{0j}.$$

The first cost is a familiar extra mileage, whereas the second one is associated to the selection of customer j as a seed; the cost is approximated by the distance of a round trip from deposit to customer j and back. The resulting optimization model is

$$\min \quad \sum_{i=1}^{n} \sum_{j=1}^{n} g_{ij} y_{ij} + \sum_{j=1}^{n} v_j z_j \qquad (8.7)$$

$$\text{s.t.} \quad \sum_{j=1}^{n} z_j = m, \tag{8.8}$$

$$\sum_{i=1}^{n} d_i y_{ij} \leq R, \qquad j = 1, \ldots, n, \tag{8.9}$$

$$\sum_{j=1}^{n} y_{ij} = 1, \qquad i = 1, \ldots, n, \tag{8.10}$$

$$y_{ij} \leq z_j, \qquad i, j = 1, \ldots, n, \tag{8.11}$$

$$y_{ij}, \; z_j \in \{0, 1\}.$$

The objective function (8.7) is total cost; constraint (8.8) enforces the selection of a given number m of seeds; constraints (8.9) and (8.10) are essentially the same as the generalized assignment formulation; finally, constraint (8.11) states that if customer j is not selected as a seed, we cannot assign any customer i to it (i.e., to the route associated with seed j).

8.4 ADDITIONAL FEATURES OF REAL-LIFE VRP

In the previous section we have considered solution approaches for the basic VRP. Typical routing problems have several additional features that make their solution a bit tougher, even though the heuristic principles we have just outlined can be adapted. In the following list, we illustrate a few of these complications.

- In the basic VRP, given a set of customers along with their demand, we have to build a brand new set of routes; if the demand pattern changes, the set of routes may change as well. From an organizational point of view, this may be an inconveniency. Hence, at a more tactical level, one may try to come up with a set of fixed routes which are traveled several times. Such **fixed routing** problems may also be formulated and solved in the case of uncertain demand.

- We have taken the number of vehicles as given (or irrelevant). In **fleet planning** problems, the aim is sizing a fleet of vehicles. This type of problem is also relevant in point-to-point transportation.

- We have considered a static, deterministic, and single-period problem. Depending on the practical context, uncertainty may affect travel times and/or demand. One simple approach to tackle the first source of uncertainty would be introducing slack time by judiciously overestimating travel times. Demand uncertainty can be tackled by similar means. However, the real issue is arguably the real-time management for such a problem; this calls for efficient real-time data collection and an effective organization in order to adapt routes and delivery on-the-fly. Dealing with uncertainty may result in a tough, dynamic and stochastic problem. Even if we rule out uncertainty, we may have to cope with a multiperiod problem, whereby we have to develop routes for a few consecutive periods; as we have already noted in the book, multiperiod problems need not be dynamic in the sense of adapting to uncertain events.

- We have considered a symmetric distance matrix, whereas sometimes the underlying TSP structure is asymmetric. Apart from adjustments in solution algorithms, a difficult issue is filling the matrix with reliable data. In the past, one possibility was computing plain Euclidean distances and then inflating them by coefficients modeling the difficulty of the terrain. Given technological advancement, geographic information systems are now typically exploited to this aim. This can also be done at the single customer address level, e.g., analyzing the ZIP code, by a process called *geocoding*.

- The maximum tour length in terms of time and space may be a constraining factor, and not only capacity. Capacity is actually multidimensional, involving both weight and volume, potentially for separate parts of the truck, such as refrigerated and nonrefrigerated. Exploiting the available volume is an optimization problem in itself, for which software packages have been developed. We should emphasize however, that such an optimization is desirable for point-to-point service, but it may get into the way of unloading stuff in multiple delivery problems such as VRP: Having fully loaded the truck is of little use, if the parcel of the first customer in the route lies at the unreachable bottom of the truckload.

- The objective function of a real VRP may involve multiple costs, and not only distance. Some desirable features of a route may hardly be expressed in monetary terms. If not all of the customers can be accommodated for delivery today, we must decide which ones will be served tomorrow. Moreover, overtime driver cost may by quantified, but workload balance issues in human resource management may be hardly turned into a cost. The flexibility of local search in dealing with complex objectives may be used to advantage.

- We may have multiple deposits and heterogeneous vehicles (possibly with separate sections). Cluster-first, route-second methods may be adapted in some cases.

- In **inventory routing** problems, vehicle routing is coupled with inventory management. In basic VRP, we consider customer demand as given; but if inventories are taken into account, a brand new dimension is open, offering both degrees of freedom and additional complexity. If inventory is available at customer nodes, we may better manage vehicle capacity, by delivering flexible quantities depending on current inventory state. We may also better cope with demand uncertainty. The price we pay for such opportunities is the difficulty of the integrated problem.

- Last but not least, delivery may be subject to **time windows**, linked to traffic conditions, customer's requirement, or to the availability of an unloading bay. This places additional constraints on solution methods.

Given all of these complexity factors, the staggering amount of scientific literature on VRP is no surprise. One way to cope with real-life VRPs is to extend and adapt the heuristic principles we have briefly illustrated. Local search approaches are certainly an interesting way to tackle generalized versions of the basic VRP. However, sometimes constraints are so tight that even finding a *feasible* solution is difficult. In this case, sophisticated mathematical modeling and solution approaches can offer some advantage. Since this level of sophistication requires advanced optimization concepts, in the next section we just offer some clues on how constructive heuristics can be extended to cope with time windows. Whatever solution approach we take, we should bear in mind that real-life VRPs may be subject to significant uncertainty and ill-defined objectives linked to human factors; hence, solution approaches must be cast within a well-designed decision support system.

8.4.1 Constructive methods for the VRP with time windows

In the VRP with time windows,[9] each customer $i = 1, \ldots, n$ is associated with an interval $[e_i, l_i]$, whose endpoints are the *earliest time* and the *latest time* for the start of service (in our case, unloading the vehicle). In practice, multiple time windows may be associated with a single customer. Let s_i be the duration of service and t_{ij} be the time to travel from customer i to customer j. If the vehicle arrives early with respect to the time window, then it must wait. Hence, if b_i denotes the start of service for customer i, and the vehicle visits customer j after customer i a j, we have

$$b_j = \max\{e_j,\ b_i + s_i + t_{ij}\}.$$

In the basic VRP we were deliberately ambiguous in using a "cost" c_{ij} which could be related to space, or time, or a mixture of both. For the VRP with time windows, we must take both time and space into account; we will denote the distance between customers i and j by q_{ij}.

Based on our knowledge of constructive heuristics, one of the first ideas that may come to our mind is to build routes sequentially, using an extension of the nearest-neighbor TSP approach. In this case, "nearest" mixes both

---

[9]This section is based on [17], to which we refer for a full treatment.

space and time considerations. We recall that in this algorithm customers are always appended at the end of the growing route, which may be a limitation. Let i be the last visited customer; we must define a hybrid measure c_{ij} of "closeness" between i and j. One possibility is the following:

$$c_{ij} = \lambda_1 q_{ij} + \lambda_2 T_{ij} + \lambda_3 v_{ij},$$

where:

- the weights λ_i are non-negative and sum up to one (actually, we must just give two weights);

- q_{ij} is the distance between customers i and j;

- the quantity
$$T_{ij} = b_j - (b_i + s_i)$$
takes into account the time difference between the end of service at i and the beginning of service at j (if j follows i on the route);

- the quantity
$$v_{ij} = l_j - (b_i + s_i + t_{ij})$$
measures the time slack we still have for service at customer j, i.e., how much time remains to the end of its time window (the smaller v_{ij}, the more urgent it is serving j after i; hence, this factor works in the same way as the previous two in making service of j after i desirable).

This generalized metric is used as a simple priority rule to append customers to the current route. When no more customers can be appended, because the vehicle is full or no time window is compatible, we close the current route and start a new one. Clearly, we have no direct control on the number of routes we build, and the algorithm looks quite greedy. In any metric depending on weights, parameter fine-tuning is an issue. However, in this case we just have to select a combination of two parameters ranging between 0 and 1; since a greedy procedure is very fast, we may simply carry out a grid search, trying several weight combinations and plucking the best solution found.

Given the already familiar limitations of nearest-neighbor, we may consider adapting insertion-based heuristics. Consider a partial route $0, i_1, i_2, \ldots, i_m, 0$, starting and terminating at deposit 0. We may work at two levels:

1. For each unrouted customer u, we compute the best insertion point in the partial route (provided vehicle capacity is not exceeded), according to some metric.

2. We select the best customer to insert, applying some metric, which need not be the same as in the previous point.

We should note that inserting a customer my imply a time shift for all of the following customers along the route.

To evaluate the opportunity of inserting customer u between i and j, we can adapt the following metric (which should be minimized to find the best insertion point):

$$c_1(i, u, j) = \alpha(q_{iu} + q_{uj} - \mu q_{ij}) + (1 - \alpha)(b_{j_u} - b_j),$$

where α and μ are parameters to be chosen. The parameter α must be selected in the range $[0, 1]$ and controls the relative weight we assign to space vs. time considerations. In fact, the first term in the sum is linked to extra mileage; indeed, it is be the familiar extra mileage if $\mu = 1$. The additional parameter μ allows for extra fine-tuning of the heuristics; actually, such coefficients are common in variations of insertion heuristics for TSP. The second term includes the difference between the new start time of service at customer j, if we insert u, denoted by b_{j_u}, and the current start time b_j. This term tries to capture the time shift effect due to insertion, provided that the insertion is feasible with respect to time windows.

To select the customer u to insert, given the best insertion point above, we may consider the following metric (to be maximized):

$$c_2(i, u, j) = \lambda q_{0u} - c_1(i, u, j),$$

where λ is a parameter to be chosen. If we select parameters $\mu = \alpha = 1$ and $\lambda = 2$, this metric measures the saving in terms of traveled distance if we serve customer u between i and j rather than serving u directly from the deposit. An alternative choice is

$$c_2(i, u, j) = \beta R_d(u) + (1 - \beta)R_t(u),$$

where $R_d(u)$ and $R_t(u)$ are total distance and total time of the current route if we insert u, respectively; the parameter β must be selected in the range $[0, 1]$ and has essential the same meaning as the parameter α above. This metric, which should be minimized, tries to capture more fully the effect of the insertion.

These very simple rules, and related variants, have a definite advantage in terms of CPU time and conceptual simplicity. They might not be very effective in tightly constrained problems. If we do not want to resort to complex mathematics, we could also consider local search algorithms, but there are additional complications when we try to apply something like a $2-$opt neighborhood structure to a VRP with time windows. When we delete and cross a pair of arcs, we invert the direction of some part of the route; this is irrelevant in terms of distance, provided that the problem is symmetric, but it is definitely relevant in terms of time windows. Nevertheless, many clever approaches have been proposed over the years, which are described in the references listed at the end of the chapter.

8.5   FINAL REMARKS

In this chapter we have considered basic VRP as a straightforward extension of classical TSP. The interest of these network routing problems has spurred a significant amount of work, which is documented by a vast literature where a wide array of methods is presented. It is hard to tell if there is one best approach. On the one hand, very sophisticated mathematical approaches have been developed, and despite technical intricacy, their potential for economic impact must be carefully considered. On the other hand, the variety of constraints and complicating features has led software developers to privilege simpler and possibly more adaptable approaches. What we tried to accomplish in this chapter is just to get the reader acquainted with the conceptual foundations of these approaches.

We should also raise a couple of general points, whose practical importance cannot be overemphasized. The first one is that we have considered VRP as an off-line scheduling problem. In practice, disruptions and uncertainty are a way of life; hence, it is essential to develop suitable user interfaces to manage such situations. Such decision support systems must also rely on proper data collection from the field; new satellite-based technologies are being exploited for this task. Last, but not least, we have just considered cost minimization. Environmental issues should remind us that proper transportation management and organization has a significant impact, which goes beyond the bottom line of a single firm.

8.6   FOR FURTHER READING

- We did not consider mathematically sophisticated approaches to solve the TSP, which include branch and bound methods based on Lagrangian relaxation (i.e., the relaxation of complicating constraints by Lagrangian multipliers; see section B.4) and branch and cut methods (i.e., branch and bound methods in which constraints, i.e., cuts, are added to strengthen the lower bound we get from the continuous relaxation; see section B.6.1). A not-so-recent, but still relevant reference book is [14].

- An overview of local search methods for combinatorial optimization can be found in [1]; a recent survey is [5]. A specific reference on tabu search is [12]; see [15] for an application of GRASP to the TSP.

- An excellent survey on early approaches to VRP can be found in [7], from which we have taken part of section 8.3 (in particular, the examples and the discussion associated with figures 8.10 and 8.11).

- A more recent survey, which also includes approaches based on mathematical programming, can be found in [9]. See also [19].

- For VRP under uncertain demand see, e.g., [4].

- For a complete description of simple heuristics for VRP with time windows, see [17]; we have just hinted at a few basic concepts in section 8.4.1.

- Some commercially available software packages for VRP are just based on principles we have outlined, but the complexity of real-life VRP can only be appreciated by having a look at the data requirements to define a problem. Many complicating constraints must be addressed, and the data we have taken for granted, such as the distance matrix, may require a link to a geographic information system. For instance, you can have a look at the web site http://www.bestroutes.com/

- Many commercial tools of VRP are "closed" products. There is also the possibility of using software components to tailor a specific application. This is the approach taken by the ILOG Dispatcher library (see http://www.ilog.com/)

- For routing applications in transportation by aircraft or railway, see [2] and [13], respectively.

REFERENCES

1. E. Aarts and J.K. Lenstra. *Local Search in Combinatorial Optimization.* Wiley, NewYork, NY, 1997.

2. A.P. Armacost, C. Barnhart, K.A. Ware, and A.M. Wilson. UPS Optimizes Its Air Network. *Interfaces*, 34:15–25, 2004.

3. M.O. Ball, T.L. Magnanti, C.L. Monma, and G.L. Nemhauser, editors. *Network Routing (Handbooks in Operations Research and Management Science, Vol. 8).* Elsevier Science, Amsterdam, 1995.

4. D. Bertsimas and D. Simchi-Levi. The New Generation of Vehicle Routing Research: Robust Algorithms Addressing Uncertainty. *Operations Research*, 44:286–304, 1996.

5. C. Blum and A. Roli. Metaheuristics in Combinatorial Optimization: Overview and Conceptual Comparison. *ACM Computing Surveys*, 35:268–308, 2003.

6. J. Bramel and D. Simchi-Levi. A Location-Based Heuristic for General Routing Problems. Working Paper, Department of Industrial Engineering and Operations Research, Columbia University, 1992.

7. N. Christofides, A. Mingozzi, and P. Toth. The Vehicle Routing Problem. In N. Christofides, A. Mingozzi, P. Toth, and C. Sandi, editors, *Combinatorial Optimization.* Wiley, Chichester, 1979.

8. G. Clarke and J.W. Wright. Scheduling of Vehicles from a Central Depot to a Number of Delivery Points. *Operations Research*, 12:568–581, 1964.

9. M.L. Fisher. Vehicle Routing Problem. In M.O. Ball, T.L. Magnanti, C.L. Monma, and G.L. Nemhauser, editors, *Network Routing (Handbooks in Operations Research and Management Science, Vol. 8)*. Elsevier Science, Amsterdam, 1995.

10. M.L. Fisher and R. Jaikumar. A Generalized Assignment Heuristic for Vehicle Routing. *Networks*, 11:109–124, 1981.

11. B.E. Gillett and L.R. Miller. A Heuristic Algorithm for the Vehicle-Dispatch Problem. *Operations Research*, 22:340–349, 1974.

12. F.W. Glover and M. Laguna. *Tabu Search*. Kluwer Academic Publishers, Dordrecht, 1997.

13. P. Ireland, R. Case, J. Fallis, Carl Van Dyke, J. Kuehn, and M. Meketon. The Canadian Pacific Railway Transforms Operations by Using Models to Develop Its Operating Plans. *Interfaces*, 34:5–14, 2004.

14. E.L. Lawler, J.K. Lenstra, A.H.G. Rinnooy Kan, and D.B. Shmoys. *The Traveling Salesman Problem: A Guided Tour of Combinatorial Optimization*. Wiley, NewYork, NY, 1985.

15. Y. Marinakis, A. Migdalas, and P.M. Pardalos. Expanding Neighborhood GRASP for the Traveling Salesman Problem. *Computational Optimization and Applications*, 32:231–257, 2005.

16. W.B. Powell. Dynamic Models of Transportation Operations. In A.G. de Kok and S.C. Graves, editors, *Supply Chain Management: Design, Coordination, and Operation*. Elsevier, Amsterdam, 2003.

17. M.M. Solomon. Algorithms for the Vehicle Routing and Scheduling Problems with Time Window Constraints. *Operations Research*, 35:254–265, 1987.

18. R.E. Tarjan. *Data Structures and Network Algorithms*. SIAM, Philadelphia, 1983.

19. P. Toth and D. Vigo. *The Vehicle Routing Problem*. SIAM, Philadelphia, 2001.

Appendix A
A Quick Tour of
Probability and Statistics

The tools of Probability Theory and Statistics are essential in formulating and solving several problems in Distribution Logistics. The main reason for this need is uncertainty in demand, even though other uncertain factors that may affect a supply chain are lead time, price, and exchange rates in an international context. This appendix aims at recalling, in a reasonably concise manner, the fundamental concepts that we use in the main body of the book. It goes without saying that we do not intend the following treatment as a substitute for a serious study of the matter, which is often subtle and requires nontrivial concepts for a deep and thorough exposition. Hence, we illustrate the main ideas by examples, including some counterexamples whose purpose is to point out some potential traps of intuitive thinking, to show some common misunderstandings, and to underline some pitfalls of the most used tools. We refer the interested reader to the references for a deeper and more rigorous treatment.

The probability of an event is a fairly intuitive concept; it may refer to the frequency at which a random event occurs, or it may stem from a subjective assessment. For instance, characterizing demand uncertainty for a brand new product may require a different approach than for a well-established one, with a long history of sales. In the axiomatic approach to probability theory, in order to provide a sound foundation, a sample space is defined and probabilities are associated with subsets corresponding to events. These concepts are introduced in section A.1. Given this conceptual framework, section A.2 introduces the fundamental ideas of conditional probability and independent events. Then, we proceed to treat random variables, both discrete (section A.3) and continuous (section A.4), and to describe some common probability distributions such as geometric, binomial, Poisson, exponential, and normal.

The generalization to the multivariate case of jointly distributed random variables is the subject of section A.5, which is then expanded in section A.6, where we deal with fundamental issues such as independence between random variables, conditional expectation, covariance, and correlation; we also introduce useful distributions which are built on the basis of the normal, such as chi-square and Student's t, as well as the central limit theorem.

When we consider jointly distributed random variables, we may think of random realizations of different phenomena at the same time (e.g., sales of different items in the same week) or successive realizations of the same phenomenon over time (e.g., sales of the same item over a range of time periods). The last idea leads us to the concept of a stochastic process, which is briefly described in section A.7.

In probability theory, we assume that a large body of knowledge is available, and we ask possibly complex questions about expected values, probabilities, etc. In practice, such knowledge is a scarce commodity, and we must extract it from empirical data. Then, on this basis, we may also try to come up with inferences about unknown parameters or forecasts. This leads us into the realm of Statistics, which rests on the theory of probability, but it is the empirical side of the coin. We will deal with the most relevant topics for our applications, such as parameter estimation and confidence intervals (section A.8), hypothesis testing (A.9), and simple linear regression (A.10). These concepts play a key role in demand forecasting. The aim of simple linear regression is to use one variable to explain the behavior of another variable of interest. Of course, one can use multiple explanatory variables; this leads to multiple linear regression, which is the topic of web section W.A.11.

In this appendix we illustrate more ideas than we actually use in the main body of the text. One good reason for doing so is to provide the reader with a stronger background. Another reason is to pave the way for more web supplements, that we will include over time in the book web page.

A.1 SAMPLE SPACE, EVENTS, AND PROBABILITY

The starting point in defining probability according to the axiomatic approach is a **sample space**, which we denote by Ω, representing the set of possible outcomes $\omega \in \Omega$ of a random experiment or of a sequence of random experiments. Intuitively, an **event** is something that, depending on the outcome of the random experiment, may happen or not. Formally, an event E is a subset of the relevant sample space; note that an event need not be a single element of the sample space, which is just a particular case.

Example A.1 The best-known example is dice throwing. When throwing a dice, the outcome will be an element of the following sample space:

$$\Omega = \{1, 2, 3, 4, 5, 6\}.$$

The definition of events depends on our purpose and on what we may know about the outcome of the random experiment, as our ability to observe results may be partial. If what we are interested in is just whether the outcome is an even or an odd number, we define the following events:

$$\mathsf{E_{ven}} = \{2, 4, 6\}, \qquad \mathsf{O_{dd}} = \{1, 3, 5\}.$$

We see that the two events are subsets of the sample space Ω. If we combine several experiments, by throwing the same dice repeatedly or a pair of dice, we may define more complex sample spaces and events. ▯

The next logical step is assigning a probability to events. In order to do so in a consistent way, we would like to be able to work with the following concepts:

- The probability that an event does *not* occur. To this aim, given an event $E \subseteq \Omega$, it is natural to consider its complement $E^C \equiv \Omega \backslash E$.[1] With reference to example A.1, $\mathsf{E_{ven}^C} = \mathsf{O_{dd}}$.

- The probability that *at least* one of two (or more) events occurs. In this case it is natural to exploit the concept of set *union*.

- The probability that two (or more) events occur jointly. To this aim we exploit the concept of set *intersection*.

This allows us to work on events by using elementary set operations, but in order to do so in a consistent way, we must require some additional conditions. In fact, we should require that by working with complements, differences, unions, and intersections of events we get other events. To formalize this requirement, given a sample space Ω, we define a family of events $\mathcal{F}$, i.e., a set of subsets of Ω. Such a family of events must satisfy the following properties.[2]

- If an event is an element of the family $\mathcal{F}$, then its complement is too:

$$E \in \mathcal{F} \rightarrow E^C \in \mathcal{F}.$$

---

[1] The $\backslash$ operator denotes the difference between two sets: The set $B \backslash A$ consists of the elements of B which do *not* belong to A.

[2] Technically, these requirements lead us to consider a *field* of sets.

Note that $\Omega^C = \emptyset$, and vice versa, which justify the inclusion of the empty set $\emptyset$ in $\mathcal{F}$.

- If two events belong to $\mathcal{F}$, then their union does too:

$$E_1, E_2 \in \mathcal{F} \rightarrow E_1 \cup E_2 \in \mathcal{F}.$$

- If two events belong to $\mathcal{F}$, the same holds for their intersection:

$$E_1, E_2 \in \mathcal{F} \rightarrow E_1 \cap E_2 \in \mathcal{F}$$

If the intersection of two events is empty, i.e., $E_1 \cap E_2 = \emptyset$, the two events are said **mutually exclusive** or **disjoint**.

We may extend the above requirements by considering the union and intersections of an arbitrary number of events.[3]

Finally, armed with a sample space Ω and a suitable family of events $\mathcal{F}$, we may associate a **probability measure** $P(E)$ to each event $E \in \mathcal{F}$. The probability measure is a function mapping an event to a real number in the interval from 0 to 1. The probability measure, together with the other ingredients, defines a probability space $(\Omega, \mathcal{F}, P)$. We should note that, given a sample space, we may define different families of events. The choice depends on our purpose and on the available information, i.e., what we may observe; hence, several probability spaces may be defined on the basis of the same sample space.

The probability measure must satisfy the following conditions:

1. $0 \le P(E) \le 1$, for any event $E \in \mathcal{F}$;

2. $P(\Omega) = 1$;

3. for each sequence of mutually exclusive events $E_1, E_2, E_3, \ldots$, i.e., such that $E_i \cap E_j = \emptyset$, for $i \ne j$, we have

$$P\left(\bigcup_{i=1}^{\infty} E_i\right) = \sum_{i=1}^{\infty} P(E_i).$$

The first and second conditions are fairly self-explanatory: The probability of an event can be neither negative nor larger than 1, or 100%; the probability of the sample space is 1, because "something must occur anyway." To get a feeling for the third one, let us refer to a pair of *disjoint* sets E_1 and E_2. In

---

[3]From a mathematical point of view, passing from an arbitrary but finite number of events to an infinite (but countable) family of sets requires some care. We go from a field to a σ-algebra, which is beyond our scope.

such a case, it sounds reasonable to say that the probability of the union of the two sets is the sum of the two probabilities:

$$P(E_1 \cup E_2) = P(E_1) + P(E_2).$$

The third property generalizes the idea and allows us to express the probability of any event by decomposing it into a set of disjoint events.

Given these basic properties of a probability measure, we can prove all of the properties, which we intuitively associate to the concept of probability, as well as some less intuitive ones. We now give a few illustrative examples.

- Given the probability of an event E, what is the probability of its complement E^C? Since $E \cup E^C = \Omega$ and the two sets are disjoint, applying property 3 yields

$$P(E) + P(E^C) = P(\Omega) = 1,$$

which implies $P(E^C) = 1 - P(E)$. This is a rather intuitive property, but it may be very useful. If computing the probability of an event is difficult or time-consuming, it may be convenient to compute the probability of its complement, and then to use the property to get what we are interested in.

- From elementary set theory, we know that intersection can be expressed on the basis of union and complement:

$$(E_1 \cap E_2) = (E_1^C \cup E_2^C)^C.$$

Hence, the three properties enable us to work with intersection too, i.e., the probability that pairs of events occur together. Sometimes, instead of notation $P(E_1 \cap E_2)$, the joint probability of two events is denoted by $P(E_1 \cdot E_2)$ or $P(E_1 E_2)$.

- For two overlapping events, we cannot say that the probability of the union event is the sum of the two probabilities. To see this, consider a deck of poker cards, and imagine drawing a card at random. What is the probability of getting a king or a spade? Out of the 52 cards, we have four kings and thirteen spades, but the answer is not $(4 + 13)/52$; we should not count the king of spades twice. In other words, we should not count the intersection of the two sets twice. Indeed, it is easy to show that

$$P(E_1 \cup E_2) = P(E_1) + P(E_2) - P(E_1 \cap E_2).$$

We leave this as an exercise. (Hint: Note that $E_1 \cup E_2 = E_1 \cup (E_2 \backslash E_1)$.)

- Finally, if $E_1 \subseteq E_2$, then $P(E_1) \leq P(E_2)$ (if E_1 occurs, then E_2 occurs too for sure, but the converse does not hold). Indeed, in such a case we may write $E_2 = E_1 \cup (E_2 \backslash E_1)$. Since the two sets are disjoint, $P(E_2) = P(E_1) + P(E_2 \backslash E_1)$, which implies $P(E_2) - P(E_1) = P(E_2 \backslash E_1) \geq 0$, proving the claim.

A.2   CONDITIONAL PROBABILITY AND INDEPENDENCE

When one considers a pair of events, a natural question is whether information about one of them tells us something about the probability that the other one occurs. For instance, in the dice throwing case, we know that $P(1) = P(2) = 1/6$ if the dice is fair. But if we have partial knowledge, e.g., we know that event $\mathsf{E_{ven}}$ occurred, we also know for sure that the outcome cannot be 1, whereas there are increased chances that the result has been 2. As a practical example, knowing that a customer purchases a certain product may tell us something about the probability that she will buy another one, particularly if they are complements or substitutes. A striking example of how past purchase information can be exploited is familiar to customers of online sellers such as Amazon.com: After you buy a few books, they are able to send rather accurate recommendations for related titles, based on observed patterns of other customers. The formalization of such an idea leads us to the concept of conditional probability of an event E given the occurrence of another event G.

DEFINITION A.1 (Conditional probability) *The probability of event E, conditional on the occurrence of event G, is denoted as $P(E \mid G)$ and is given by*

$$P(E \mid G) = \frac{P(E \cap G)}{P(G)}.$$

As an intuitive justification, if we know that event G has occurred, then we also know that E occurs if and only if the joint event $E \cap G$ occurs. Moreover, G becomes the new sample space, and we have to renormalize the probability dividing it by $P(G)$ (since this probability is generally less than 1, this amounts to increasing the ratio). We should note that the definition above makes sense only if $P(G) > 0$.

As a simple example, we may compute the conditional probability

$$P(2 \mid \mathsf{E_{ven}}) = \frac{P(2 \cap \mathsf{E_{ven}})}{P(\mathsf{E_{ven}})} = \frac{1/6}{1/2} = \frac{1}{3}.$$

In this case, the conditional probability is larger than the unconditional probability $P(2)$. In other cases, the information on G does not tell us anything about the probability of E. If throw a pair of dice, the number shown by one of them does not tell us anything about the other one. In this case $P(E \mid G) = P(E)$, i.e., conditional and unconditional probabilities are the same. This leads us to the property characterizing independent events.

DEFINITION A.2 (Independent events) *Two events E and G are said independent if*

$$P(E \cap G) = P(E) \cdot P(G).$$

We see that the joint probability of two independent events is simply the product of the individual probabilities. This idea generalizes fairly easily to an arbitrary number of events, but we believe it is also useful to see the real meaning of independence in terms of *information*. If E and G are independent events, we obtain

$$P(E \mid G) = \frac{P(E \cap G)}{P(G)} = \frac{P(E) \cdot P(G)}{P(G)} = P(E),$$

$$P(G \mid E) = \frac{P(G \cap E)}{P(E)} = \frac{P(G) \cdot P(E)}{P(E)} = P(G).$$

The following examples aim at reinforcing the understanding of this concept.

Example A.2 Two disjoint events *cannot* be independent. In fact, if we know that one of them has occurred, we have quite a good amount of information about the other one, because its occurrence can be ruled out. Formally, if $E \cap G = \emptyset$, we know $P(E \cap G) = 0 \neq P(E) \cdot P(G)$; moreover, $P(E \mid G) = 0$.

By the same token, if $G \subset F$, the two events cannot be independent, since if G occurs, then F occurs as well. Formally,

$$P(E \mid G) = \frac{P(E \cap G)}{P(G)} = \frac{P(G)}{P(G)} = 1.$$

⬜

Independence is a concept that can be extended to several events.

DEFINITION A.3 *Events* $E_1, E_2, \ldots, E_N$ *are said independent if, given any arbitrary subset* $E_{j_1}, E_{j_2}, \ldots, E_{j_m}$ *of this family of events, with* $m \leq N$, *we have*

$$P(E_{j_1} \cap E_{j_2} \cap \ldots \cap E_{j_m}) = P(E_{j_1}) \cdot P(E_{j_2}) \cdots P(E_{j_m}).$$

We should stress that the meaning of the above definition is that knowledge about *any* subset of events does not tell us anything about the remaining ones. We should also stress that the definition requires that joint probabilities may be factored into the product of individual probabilities for *any* subset of the given family. Intuition may be misleading. For instance, it is tempting to think that if all events are pairwise independent, then they are independent, but this is false in general. The example below illustrates this point.

Example A.3 Let us consider three pairwise independent events. A simple example is provided by the draw of an integer number between 1 and 4, assuming that the four outcomes are equally likely. We see that the events $A \equiv \{1, 2\}$, $B \equiv \{1, 3\}$, $C \equiv \{1, 4\}$ have the same probability, $1/2$. It is also easy to see that these events are pairwise independent:

$$P(A \cap B) = 1/4 = P(A) \cdot P(B),$$
$$P(A \cap C) = 1/4 = P(A) \cdot P(C),$$
$$P(B \cap C) = 1/4 = P(B) \cdot P(C).$$

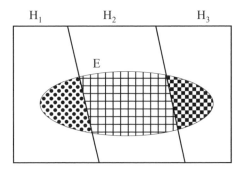

Fig. A.1 Partitioning an event into disjoint subsets.

However,

$$P(A \cap B \cap C) = P(\{1\}) = 1/4 \neq P(A) \cdot P(B) \cdot P(C).$$

To really get the point, it is useful to reason in terms of information and conditional probabilities. For instance, $P(A \mid B) = P(A) = 1/2$, because knowing that B occurred does not provide us with any additional information about occurrence of event A. However, $P(A \mid (B \cap C)) = 1 \neq P(A) \cdot P(B \cap C)$, because if we know that the event $(B \cap C)$ occurred, then necessarily the number 1 has been drawn, so A occurred for sure.   ⬚

Consider a finite partition of a sample space Ω, i.e., a set of events

$$\{H_1, H_2, H_3, \ldots, H_n\}$$

mutually exclusive (empty intersections) and collectively exhaustive (their union yields Ω); formally,

$$H_i \cap H_j = \emptyset, \quad \text{for } i \neq j, \quad \text{and} \quad \bigcup_{i=1}^{n} H_i = \Omega.$$

Any event E can be partitioned too, on the basis of the partition of Ω, by putting together intersections of the form $E \cap H_i$. Clearly,

$$E = \bigcup_{i=1}^{n} (E \cap H_i).$$

This idea is illustrated in figure A.1. Since the elements of the partition are disjoint, given the third property of a probability measure, we have

$$P(E) = \sum_{i=1}^{n} P(E \cap H_i).$$

If we rewrite this sum in terms of conditional probabilities, extending the idea to an infinite countable partitioning, we get the following useful theorem.

THEOREM A.4 (Theorem of total probabilities) *Consider a partition of a sample space Ω, i.e., a family of mutually disjoint and collectively exhaustive subsets $H_1, H_2, H_3, \ldots\ldots$ Then, for any event $E \subseteq \Omega$, we have*

$$P(E) = \sum_{i=1}^{\infty} P(E \mid H_i)P(H_i).$$

The following example shows how the theorem can be used to compute probabilities by conditioning.

Example A.4 A lazy professor, rather than administering serious exams, adopts a multiple choice quiz. Actually, even if a student gives the correct answer, there is still no guarantee that he really knows the subject, because he may try a random answer and succeed by sheer luck. Let m be the number of multiple answers and let p be the probability that the student knows the exact answer (hence, $1 - p$ is the probability that he will take his chances with a random answer). Let K be the event "the student knows," and let OK be the event "correct answer." Assuming that when the student tries at random, any choice is equally likely, we may write

$$P(K) = p,$$
$$P(OK \mid K) = 1,$$
$$P(OK \mid K^C) = 1/m.$$

However, in order to understand if the test is reliable enough, what we would like to assess is the conditional probability $P(K \mid OK)$. To this aim, we may use the theorem of total probabilities as follows:

$$
\begin{aligned}
P(K \mid OK) &= \frac{P(K \cap OK)}{P(OK)} \\
&= \frac{P(OK \mid K)P(K)}{P(OK)} \\
&= \frac{P(OK \mid K)P(K)}{P(OK \mid K)P(K) + P(OK \mid K^C)P(K^C)} \\
&= \frac{p}{1 \cdot p + (1/m)(1 - p)} = \frac{mp}{1 + (m-1)p},
\end{aligned}
$$

where the first two lines follow by definition, and the third line results from applying the theorem of total probabilities. Note that we assume $P(OK \mid K) = 1$, i.e., we disregard the impact of emotional factors. For instance, if $p = 1/2$ and $m = 3$, then $P(K \mid OK) = 3/4$. ⬚

In tackling the previous example, we have actually proved a particular case of the following Bayes' theorem.

THEOREM A.5 (Bayes' theorem) *If $H_1, H_2, H_3, \ldots, H_n$ is a partition of the sample space Ω and E is an event, then the following holds:*

$$P(H_i \mid E) = \frac{P(E \mid H_i)P(H_i)}{\sum_{j=1}^{n} P(E \mid H_j)P(H_j)}.$$

A.3 DISCRETE RANDOM VARIABLES

When working with events, we basically ask questions whose answer may be "yes" or "no" (e.g., did we meet all customers' demand this week?) and we reason about the probability of each answer (what is the probability that we will not lose customer orders during the next couple of months?). Sometimes, we may also need more quantitative answers; this is the case, e.g., when we want to forecast future demand in order to properly plan inventories and we also need some measure of confidence in such a forecast.

Formally, we should associate numerical values to events. The case of dice throwing is very simple, since the outcome of the random experiment is naturally linked to the number shown by the dice. More generally, a **random variable** is a function mapping the events within a probability space to numbers. If the possible numerical values are a discrete finite set or the set of integer numbers, we speak of **discrete** random variables. In logistics, discrete random variables yield suitable models to capture the variability of demands for items which are naturally sold in discrete units and relatively low volumes (e.g., expensive spare parts for large equipments or slow-moving items). When the demand volume is large, even if the goods are sold in discrete units, a continuous model may be a reasonable approximation; we deal with continuous random variables, which are functions mapping events to real numbers, in the next section.

In mathematically inclined books, the notation $X(\omega)$ is often used in order to point out that a random variable is a function mapping events to numbers. We will not be that rigorous, but the least we can do is to distinguish very clearly a random variable from its realization, i.e., the numerical value taken by the random variable. To this aim, we mostly follow the typical notation whereby capital letters such as X are used to denote a random variable, whereas the corresponding lowercase x denotes a numerical value assumed by the variable. To get the point, assume that we are interested in the probability that a random variable takes a value less than 10; the notation $P\{X < 10\}$ should be used, whereas $P\{x < 10\}$ makes no sense. When using Greek letters, it may be convenient to adopt the notation $\tilde{\epsilon}$ (random variable) and ϵ (realization); this notation is quite common in Economics.

Now we may start wondering how we may reason on random variables in probabilistic terms. From a theoretical point of view, a probability measure is not really associated with a random variable, but rather with the events in the underlying probability space. However, in relatively simple applications, this point of view is left implicit. An event that is naturally linked to the realization of a random variable is $\{X \leq x\}$. The probability of this event defines the **cumulative distribution function** (or CDF for short) of the random variable X:

$$F_X(x) \equiv P\{X \leq x\}.$$

We have seen before that if event F is included in event G, then $P(F) \leq P(G)$. An immediate consequence is that the cumulative distribution function is a nondecreasing function with respect to its argument. Furthermore, if we denote by x_i, $i = 1, 2, 3, \ldots$, the possible values of a discrete random variable, assuming that they are labeled in an increasing order, so that $x_i < x_{i+1}$, we also see

$$F_X(x) = 0 \qquad \text{if } x < x_1$$

and

$$\lim_{i \to +\infty} F_X(x_i) = 1. \tag{A.1}$$

The cumulative distribution function for a discrete random variable is a non-decreasing piecewise constant function, with discontinuities corresponding to possible values of the realization (see example A.5 below).

In the discrete case, we may assign a probability to the event $\{X = x_i\}$, which results in a **probability mass function**, or PMF for short, associating a probability with each possible outcome:

$$p_X(x_i) \equiv P\{X = x_i\}.$$

We should note again that a probability measure is actually associated with the events $\{X = x_i\}$, and only through the function mapping events to numbers we may speak of the "probability of a value."[4]

The link between PMF and CDF is two-way. Given the latter, we may recover the former:

$$p_X(x_i) = P\{X \leq x_i\} - P\{X < x_i\} = F_X(x_i) - F_X(x_{i-1}).$$

In other words, the jumps in the distribution function are exactly the probabilities of the corresponding values. On the other hand, given the PMF, we may build the CDF:

$$F_X(a) = P\{X \leq a\} = \sum_{x_i \leq a} p_X(x_i).$$

Example A.5 Consider the cumulative distribution and the probability mass functions for a random variable directly linked to dice throwing. They are shown in figure A.2. The mass function assigns the same probability $(1/6)$

---

[4]This point may get a bit thorny when dealing with continuous random variables or with sequences of random variables over time. From a rigorous point of view, additional technical conditions are required.

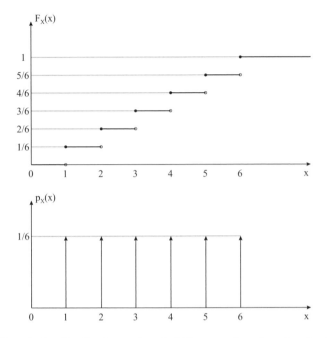

Fig. A.2   Cumulative distribution and probability mass functions for dice throwing.

to each value, and by calculating the cumulative sum we get the distribution function, which jumps on the six possible values. The jump height is exactly given by the probabilities. We may also note that the distribution function is continuous from the right, whereas it has discontinuities from the left.   ⬚

An immediate consequence of the conditions we require for a probability measure is

$$\sum_{i=1}^{\infty} p_X(x_i) = 1. \tag{A.2}$$

Here and in the following, if there is a finite set of N possible values, the sum should be rewritten as $\sum_{i=1}^{N}$. Of course, this property is just a rewriting of (A.1).

Both the CDF and the PMF contain the whole knowledge about a discrete random variable. In applications, it may be more natural to work on the mass function, as it is intuitively related to the phenomenon we wish to model. We would also like to come up with a few numbers summarizing some basic features of the random variable, such as which value we could use as a prediction of the outcome and how values are dispersed. The concept of **moment** of a random variable is used to this aim.

DEFINITION A.6 (Moment of a discrete random variable) *Given a discrete random variable X, with probability mass function $p_X(\cdot)$, its moment* *of order k is defined by*

$$\mu_X^{(k)} \equiv \sum_{i=1}^{\infty} x_i^k p_X(x_i).$$

Often, to ease notation, we use p_i instead of $p_X(x_i)$. The first few moments of a random variable allow us to capture some (not all) of its essential features. The **expected value** $E[X]$ is the first-order moment:

$$E[X] \equiv \sum_{i=1}^{\infty} p_i x_i.$$

Intuitively, the expected value is a location measure of a probability distribution and it is what we should expect "on the average." An important property of expected value is linearity:

$$E[\alpha X + \beta] = \alpha\,E[X] + \beta,$$

where α and β are arbitrary numbers.

Another very important quantity is **variance**:

$$\text{Var}(X) \equiv E[(X - E[X])^2].$$

Variance is often denoted by σ^2; its square root σ is called **standard deviation**. It is important to note that variance is non-negative by definition; it can be zero in the "degenerate" case of a constant random variable, which is quite predictable. In fact, variance and standard deviation are dispersion measures, since they are related to deviations with respect to the expected value. Deviations are squared to avoid cancelation between deviations with different signs. Standard deviation has the advantage of being expressed in the same unit of measurement as expected value. In practice, it may be difficult to get a true feeling for a deviation in absolute terms: Can we say that a standard deviation of 10 is large or small? Not really, since it makes a big difference whether the expected value is 5 or 1000 . To overcome this difficulty, the **coefficient of variation** may be used, which is defined as

$$c_X \equiv \frac{\sqrt{\text{Var}(X)}}{|E[X]|},$$

and is often given squared, c_X^2. A value close to zero suggests a low variability. Unlike variance, the coefficient of variation does not depend on the chosen unit of measurement. A pair of important properties of variance are shown in the following example.

Example A.6 Quite often, to calculate variance, we use the following property:

$$\text{Var}(X) = E[X^2] - E^2[X].$$

To show this, we just need to rewrite the definition of variance:

$$
\begin{aligned}
\mathrm{Var}(X) &\equiv \mathrm{E}[(X - \mathrm{E}[X])^2] \\
&= \mathrm{E}[X^2 - 2X \cdot \mathrm{E}[X] + \mathrm{E}^2[X]] \\
&= \mathrm{E}[X^2] - \mathrm{E}[2X \cdot \mathrm{E}[X]] + \mathrm{E}[\mathrm{E}^2[X]] \\
&= \mathrm{E}[X^2] - 2 \cdot \mathrm{E}[X] \cdot \mathrm{E}[X] + \mathrm{E}^2[X] \\
&= \mathrm{E}[X^2] - \mathrm{E}^2[X].
\end{aligned}
$$

Here it is important to realize that the expected value $\mathrm{E}[X]$ is a *number* and it can be taken outside the expectation operator.

By the same token we can see that

$$
\begin{aligned}
\mathrm{Var}(\alpha X + \beta) &= \mathrm{E}\left[(\alpha X + \beta - \mathrm{E}[\alpha X + \beta])^2\right] \\
&= \mathrm{E}\left[(\alpha X + \beta - \alpha\mathrm{E}[X] - \beta])^2\right] \\
&= \mathrm{E}\left[\alpha^2 (X - \mathrm{E}[X])^2\right] \\
&= \alpha^2 \, \mathrm{Var}(X).
\end{aligned}
$$

This second property shows that shifting a random variable by a given amount has no influence on its dispersion (it has on its location, of course) and that variance, unlike expected value, is a nonlinear operator.   ⬚

In a similar way, we may consider a function of a random variable. The expected value of the function $g(X)$ of the random variable X is

$$
\mathrm{E}[g(X)] = \sum_{i=1}^{\infty} p_i g(x_i).
$$

For instance, the moment of order k is actually the expected value of X raised to the corresponding power: $\mu_X^{(k)} \equiv \mathrm{E}[X^k]$.

A.3.1 A few examples of discrete distributions

In this section we recall some common distributions of discrete random variables, limiting ourselves to those which are more commonly applied in distribution logistics. It is very important to realize that a practically relevant issue is to analyze available data to figure out *which* distribution can reasonably model the phenomenon of interest. In this textbook we will not dwell deeply in such a fundamental issue.

Uniform distribution   The uniform distribution assigns the same probability mass to all possible values, $p(x_i) = p$, like in dice throwing. Clearly, this is possible only if a finite number N of values are considered. From the condition

$$
\sum_{i=1}^{N} p_i = 1,
$$

we immediately see $p = 1/N$.

Bernoulli distribution   A Bernoulli random variable stems from the idea of a random experiment, which can result in a success with probability p or in a failure with probability $1 - p$. If we assign to the random variable X the value 1 in case of a success and 0 in case of a failure, we get the PMF:

$$p_X(0) \equiv P\{X = 0\} = 1 - p,$$
$$p_X(1) \equiv P\{X = 1\} = p.$$

We can calculate expected value

$$E[X] = 1 \cdot p + 0 \cdot (1 - p) = p$$

and variance

$$\text{Var}(X) = E[X^2] - E^2[X] = [1^2 \cdot p + 0^2 \cdot (1 - p)] - p^2 = p(1 - p).$$

This makes sense: The variance is zero for $p = 1$ and $p = 0$ (there is no uncertainty on the outcome of each experiment), and it is maximized for $p = 1/2$.

Geometric distribution   The geometric distribution is a straightforward extension of the Bernoulli distribution; it is obtained by thinking of repeating a sequence of identical and independent experiments until we get the first success. The number of experiments we carry out is a random variable, whose PMF is

$$p_i = P\{X = i\} = (1 - p)^{i-1}p, \tag{A.3}$$

where p is the probability of success. To understand (A.3), we note that $X = i$ if we have $i - 1$ failures before getting the first success, which stops the experiment. We should also note that this distribution has an infinite support, i.e., there are infinite values that the variable can take with strictly positive probability. Figure A.3 shows the PMF of a geometric variable with parameter $p = 0.2$.

To compute the expected value of a geometric variable, it is useful to recall a couple of properties of the geometric series, which hold for $\alpha \in (0, 1)$.[5] The first property,

$$\sum_{i=0}^{\infty} \alpha^i = \frac{1}{1 - \alpha}, \tag{A.4}$$

can be justified by writing

$$S = \sum_{i=0}^{\infty} \alpha^i = 1 + \sum_{i=1}^{\infty} \alpha^i = 1 + \alpha \sum_{i=0}^{\infty} \alpha^i = 1 + \alpha S,$$

[5]This condition is needed to ensure convergence of the series to a finite value.

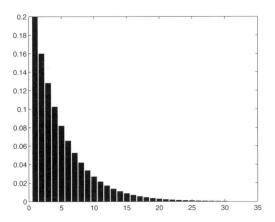

Fig. A.3 Probability mass function of a geometric random variable with $p = 0.2$.

and rearranging to obtain S. The second property is obtained by taking the derivative of the series with respect to α, term-by-term:

$$\sum_{i=1}^{\infty} i\alpha^i = \alpha \sum_{i=1}^{\infty} i\alpha^{i-1} = \alpha \frac{d}{d\alpha}\left(\sum_{i=0}^{\infty} \alpha^i\right) = \alpha \frac{d}{d\alpha}\left(\frac{1}{1-\alpha}\right) = \frac{\alpha}{(1-\alpha)^2}. \quad (A.5)$$

The first property of the geometric series allows us to prove that the (A.3) makes sense, i.e., the sum of the probabilities is 1 (this is left as an exercise to the reader). The second one is useful to compute the expected value:

$$E[X] = \sum_{i=1}^{\infty} i(1-p)^{i-1}p = \frac{p}{1-p} \sum_{i=1}^{\infty} i(1-p)^i = \frac{p}{1-p} \cdot \frac{1-p}{[1-(1-p)]^2} = \frac{1}{p}.$$

It is worth noting that the expected number of trials grows when the success probability is decreased. We will consider again the variance of a geometric random variable in example A.13 on page 473, as an example of computing moments by conditioning.

The binomial distribution   The idea, as in the case of the geometric variable, is to repeat independent experiments, but here we carry out a given number n of experiments and we count the number of successes. The PMF is

$$P\{X = r\} = \binom{n}{r} p^r (1-p)^{n-r},$$

where we use the binomial coefficient

$$\binom{n}{r} \equiv \frac{n!}{(n-r)!r!}.$$

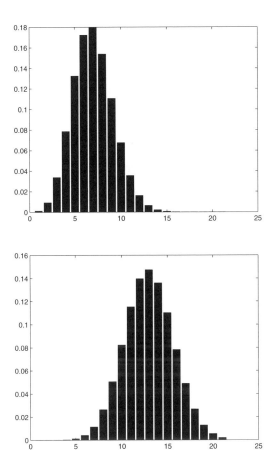

Fig. A.4   Probability mass function of two binomial random variables.

To see why this coefficient should be used, we should consider that there are many ways in which we may have r successes and $n - r$ failures; which experiment succeeds or fails is irrelevant. We have $n!$ permutations of the experiment,[6] $(n - r)!$ permutations of failures, and $r!$ permutations of successes; but the specific order of each permutation is irrelevant. This distribution has finite support and depends on two parameters. Figure A.4 shows the PMF for $n = 30$, and $p = 0.2$, $p = 0.4$. Using properties of sums of independent random variables (see section A.5), it is easy to show that

$$\mathrm{E}[X] = np, \qquad \mathrm{Var}(X) = np(1 - p).$$

---

[6]We recall the definition of the factorial of an integer number, $n! \equiv n \cdot (n-1) \cdot (n-2) \cdots 2 \cdot 1$; by convention, $0! = 1$.

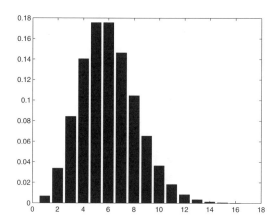

Fig. A.5 Probability mass function of a Poisson random variable.

The binomial variable, just like the geometric variable, has a "physical" motivation; we may disregard this motivation and, given empirical data on the demand for an item sold in small volumes, we may find the parameters n and p yielding the best fit of the theoretical model against the real data.

Poisson distribution   The Poisson random variable is characterized by a parameter λ, and it may take values on the set $\{0, 1, 2, 3, \ldots\}$. The PMF is

$$p_i = e^{-\lambda}\frac{\lambda^i}{i!}, \qquad i = 0, 1, 2, \ldots.$$

Despite the apparent complexity of this function, it is straightforward to check that it meets the fundamental requirement of a PMF:

$$\sum_{i=0}^{\infty} p_i = e^{-\lambda}\sum_{i=0}^{\infty}\frac{\lambda^i}{i!} = e^{-\lambda}e^{\lambda} = 1.$$

Now we may also compute the expected value:

$$
\begin{aligned}
\mathrm{E}[X] &= \sum_{i=0}^{\infty} i e^{-\lambda}\frac{\lambda^i}{i!} = \lambda e^{-\lambda}\sum_{i=1}^{\infty}\frac{\lambda^{i-1}}{(i-1)!} \\
&= \lambda e^{-\lambda}\sum_{k=0}^{\infty}\frac{\lambda^k}{k!} = \lambda.
\end{aligned}
$$

It is a bit more tedious to prove that variance has the same value, $\mathrm{Var}(X) = \lambda$. Figure A.5 shows the PMF of a Poisson random variable for $\lambda = 5$.

  In order to get a grasp of why this distribution is relevant, it is essential to understand the physical meaning of the parameter λ, which is also the expected value. Suppose that we want to model the number of customers or orders arriving during a certain time interval. Any customer arrival is an event, and it is natural to measure the flow intensity of requests by an arrival rate, which is the average number of customers arriving per unit time. The Poisson distribution is a possible model for such a phenomenon, which is suitable when arrivals are independent among them and uniform over time with average rate λ. In practice, we often consider a parameter λt, where λ is the event rate in the time unit and t is a time span of interest. We will see that the Poisson distribution is strictly linked to the exponential distribution and to the Poisson stochastic process, which are described in the following. The Poisson distribution can also be thought as the limit of a binomial variable for $p \to 0$ and $n \to +\infty$. In other words, if we have a large number of customers and each one of them orders an item, over a time interval, with a very small probability, the aggregate demand will be Poisson distributed.

Empirical distributions So far, we have considered "theoretical" distributions, which have their roots in a simple random process and are characterized by one or two numerical parameters; more complicated distributions, depending on more parameters, may be devised. Sometimes, however, none among the known theoretical distributions seems to fit the available data. In such a case, a possible alternative is to settle for an empirical distribution, whereby the mass function is given by a vector of probabilities p_i, $i = 1, \ldots, N$, which are obtained by analyzing the empirical frequencies of the observed outcomes. This is fairly straightforward to do, but it is essential to keep in mind a couple of limitations and pitfalls.

- Often, theoretical distributions do not fit empirical data, because these are affected by several underlying phenomena. The possibility of using an empirical distribution should not prevent us from understanding the root causes of the lack of fit, and sometimes the random phenomena should not be confused. A typical example is demand affected by promotional sales. If time period with full and discounted prices are alternated, this is likely to have a significant impact on the demand distribution.

- Another issue is that a simple-minded approach to devising an empirical distribution does not consider at all the possibility of having realizations of the random variable *outside the range* observed so far. In some cases, to account for extreme realizations, a mixture of empirical and theoretical distributions may be used to add some "tail" to the distribution. A similar problem occurs if we build a distribution on a subjective basis, e.g., considering demand forecasts by several experts for a brand new product, lacking a demand history suitable for statistical analysis. If we denote by D_k the forecast of expert k, $k = 1, \ldots, M$, we could consider a crude distribution where each discrete value has probability $1/M$. This makes sense if we believe that the experts are equally reliable and each guess is as good as the other ones. However, this automatically rules out any value outside the minimum and maximum of D_k.

A.4 CONTINUOUS RANDOM VARIABLES

If a random variable takes values on a continuous set, such as the real line $\mathbb{R} = (-\infty, +\infty)$ or a bounded interval $[a, b]$, we speak of a continuous random variable. From a logical point of view, introducing continuous random variables follows the same steps as in the discrete case, but there are some technical complications due to the fact that the range now is noncountable. For instance, we cannot say that there is a finite probability that the random variable takes a specific value, as this probability is zero. The key issue is that the probability mass is not concentrated on discrete points, but it is distributed on a continuous range, and this requires using slightly different concepts. As usual, we will rely on intuition without bothering too much about rigorous arguments.

A good starting point is the cumulative distribution function, $F_X(x) \equiv \mathrm{P}\{X \leq x\}$, which can be defined just as in the discrete case. Actually here, unlike the discrete case, there is no practical difference between a strict inequality or not, since $\mathrm{P}\{X \leq x\} = \mathrm{P}\{X < x\}$; however, it is preferable to maintain conceptual uniformity. As we said, we cannot speak of a probability mass, but we may define a related concept. Consider a bounded interval $[a, b]$; then, given the CDF, we may write

$$\mathrm{P}\{a \leq X \leq b\} = \mathrm{P}\{X \leq b\} - \mathrm{P}\{X \leq a\} = F_X(b) - F_X(a).$$

Under some conditions, we may come up with a **probability density function**, or PDF for short, such that we may also write

$$\mathrm{P}\{a \leq X \leq b\} = \int_a^b f_X(z)\, dz. \tag{A.6}$$

Now we may see why, in the continuous case, we have

$$\mathrm{P}\{a \leq X \leq b\} = \mathrm{P}\{a < X \leq b\} = \mathrm{P}\{a \leq X < b\} = \mathrm{P}\{a < X < b\}.$$

The reason is that including an extreme point of the integration interval, i.e., a set of zero measure, does not change the integral. More generally, the density function $f_X(\cdot)$ allows us to associate a subset B of the real line with a probability

$$\mathrm{P}\{X \in B\} = \int_B f_X(x)\, dx.$$

As usual, note that X is a random variable and that x is an irrelevant integration variable.

The PDF is non-negative and, in strict analogy with condition (A.2), the following condition must hold:

$$\int_{-\infty}^{+\infty} f_X(x)\, dx = 1.$$

The **support** of the distribution is the subset of $\mathbb{R}$ where the density is strictly positive.

To get an intuitive feeling for the meaning of the PDF, let us consider a small interval $(a, a + \delta x)$. Then we have

$$P\{X \in (a, a + \delta x)\} = \int_{a}^{a+\delta x} f_X(y)\, dy \approx f_X(a)\, \delta x.$$

This probability goes to zero when $\delta x \to 0$. Hence, we see that the PDF does tell us where the realization of a continuous random variable is more or less likely to happen, but the probability of getting a single real number is always zero.

To see the link between PDF and CDF, we may consider

$$P\{a \le X \le b\} = \int_{a}^{b} f_X(z)\, dz = F_X(b) - F_X(a)$$

and take the limit $a \to -\infty$:

$$F(b) = P\{X \le b\} = \int_{-\infty}^{b} f(x)\, dx.$$

Going the other way around,

$$\frac{dF(x)}{dx} = f(x),$$

which is a consequence of the fundamental theorem of Calculus.

Intuitively, the main difference between discrete and continuous variables is that in the former case we have discrete sums involving a mass function, whereas in the latter we have integrals (i.e., the limit of a sum) involving a density. Given a CDF, in the discrete case we get a PMF by taking differences, i.e., by considering increments. By the same token, in the continuous case we get the PDF by taking the derivative of the CDF, which is the limit of an incremental ratio.[7]

Keeping the intuition above in mind, we may follow the same path we have seen in the discrete case and define the **moment** of order k for a continuous random variable:

$$\mu_X^{(k)} \equiv \mathrm{E}[X^k] = \int_{-\infty}^{\infty} x^k f_X(x)\, dx.$$

---

[7] A bridge between these two worlds can be built by resorting to "impulse functions," which concentrate the probability mass on a discrete set of points. This is necessary to deal with mixed distributions, partly discrete and partly continuous, which are not used in this book.

By the same token, we may define the following concepts:

- The **expected value** of a continuous random variable X,

$$\mathrm{E}[X] = \int_{-\infty}^{+\infty} x f_X(x)\, dx.$$

- The expected value of a function of a continuous random variable,

$$\mathrm{E}[g(X)] = \int_{-\infty}^{+\infty} g(x) f_X(x)\, dx.$$

With reference to the second concept, it is important to stress that, in general, the expected value of the function is *not* the same as the function of the expected value:

$$\mathrm{E}[g(X)] \neq g(\mathrm{E}[X]). \tag{A.7}$$

The two values are the same for an affine function, given the linearity of the expected value operator, but nothing can be said in general for nonlinear functions. We will stress this concept again in counterexample A.8 on page 457.

Variance is defined as a (central) moment of second order, just like in the discrete case:

$$\mathrm{Var}(X) \equiv \int_{-\infty}^{+\infty} (x - \mathrm{E}[X])^2 f_X(x)\, dx = \mathrm{E}[X^2] - \mathrm{E}^2[X].$$

We should not take for granted that variance always exists, as this integral may diverge for certain heavy-tailed distributions. It is worth recalling that there are other quantities summarizing some features of a distribution, such as mode and median.

DEFINITION A.7 *The **mode** of a distribution is defined as the point where the density $f_X(x)$ is maximized (if the maximum exists).*

Given this definition, it is tempting to say that the mode is the most likely value among those the random variable may assume. To begin with, this idea makes no sense for a continuous variable, as all values have the same probability (zero), whereas in the discrete case it could sound more convincing. However, the idea of "most likely" value should be made more precise, as shown in the following example.

Example A.7 Given a density function $f_X(x)$, say that we must provide a good forecast α for the value that the random variable X will take. To rationalize the problem, we must first clarify what a "good" forecast is. One possible criterion to measure the quality of a forecast is the **mean square error**, or MSE for short:

$$\mathrm{E}[(X - \alpha)^2],$$

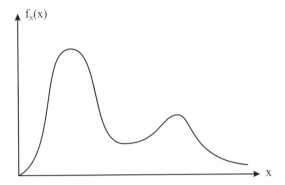

Fig. A.6 An example of multimodal density.

i.e., the expected value of the error in the forecast, which we square to avoid compensation between errors by excess or by defect. Hence, we should solve the following optimization problem:

$$\min_{\alpha} \mathrm{E}[(X - \alpha)^2].$$

Rewriting MSE as

$$\mathrm{E}[X^2 - 2\alpha X + \alpha^2] = \mathrm{E}[X^2] - 2\alpha \mathrm{E}[X] + \alpha^2$$

and setting the derivative with respect to α to zero, we get[8]

$$\alpha^* = \mathrm{E}[X].$$

We see that the best forecast, according to the MSE criterion, is not the mode but the expected value. □

As we will see in the following, all of the common theoretical distributions have a single mode, in the sense that the density has a single global maximum, rather than multiple local maxima. Sometimes, when analyzing real data, we may find multimodal distributions such as the one depicted in figure A.6, which features a local maximum as well. A typical task, given a set of empirical data, is finding a theoretical distribution fitting the data in some "best" way. Of course, we could do without this effort by building an empirical density, but this could result in some undesirable overfitting. Hence, quite often spurious modes are "leveled off" by using a theoretical single-mode distribution. However, sometimes a more careful analysis is needed, as different modes can be linked to different elements of a phenomenon, which deserve

---

[8]We are taking for granted that this first-order condition is sufficient. For a proper treatment of optimization methods, see appendix B.

careful modeling. One significant case in distribution logistics is the distribution of lead time. When ordering items from a supplier, we may observe a typical value of lead time, with some fluctuations due, e.g., to transportation delays. Occasionally, a much larger lead time can be observed, leading to a secondary mode. This may be the result of occasional stockouts suffered by the supplier, which result in unusually long lead times and need proper modeling.

DEFINITION A.8 *The* **median**, *for a continuous distribution, is a value* m_X *such that*

$$F_X(m_X) = \frac{1}{2}.$$

Mode, median, and expected value *may* be the same for a distribution characterized by a symmetric density, but they are different in general.[9] The median is a specific case of quantile.

DEFINITION A.9 *Given a continuous random variable* X, *with cumulative distribution function* $F_X(x)$, *the* α-**quantile** *is the smallest number such that*

$$F_X(x_\alpha) = \alpha,$$

with $\alpha \in [0, 1]$.

In other words, to the left of x_α we have an area α below the graph of the density. In the definition we account for pathological cases, but for typical distributions the CDF is invertible and there is a unique quantile satisfying the equation in the definition: $x_\alpha = F_X^{-1}(\alpha)$.

Some more care is needed for discrete distributions, since we might not find a possible realization x_i such that

$$\sum_{k=1}^{i} p_X(x_k) = \alpha.$$

To see this, consider a distribution over the set $\{0, 1, 2, 3, 4\}$, with probability mass function $p_0 = 0.1$, $p_1 = 0.4$, $p_2 = 0.3$, $p_3 = 0.1$, and $p_4 = 0.1$, and try to find the quantile with probability level 0.85. Hence, we should modify the definition, so that the quantile x_α is the smallest number such that $F_X(x_\alpha) \geq \alpha$. In the example, we have $x_{0.85} = x_{0.90} = 3$. To get a feeling for the rationale behind this definition, think of this distribution as a demand distribution, and ask yourself which inventory level would guarantee a 85% probability of meeting demand from stock.

---

[9]It is interesting to note, with reference to example A.7, that the median is the best forecast for X if we take the mean absolute deviation $E[|X - \alpha|]$, rather than MSE, as a metric.

By the same token, if we want to define the median in the discrete case, we should find a number such that the following inequalities are satisfied:

$$P\{X \leq m_X\} \geq \frac{1}{2} \quad \text{and} \quad P\{X \geq m_X\} \leq \frac{1}{2}.$$

A.4.1 Some continuous distributions

Uniform distribution A uniform variable on the interval $[a, b]$ features a constant density function on this support:

$$f_X(x) = \begin{cases} 1/(b-a) & \text{if } x \in [a, b], \\ 0 & \text{otherwise.} \end{cases}$$

A commonly used notation to say that a random variable X has this distribution is $X \sim U[a, b]$, or $X \sim U(a, b)$. It is easy to see that

$$E[X] = \int_a^b \frac{x}{b-a} \, dx = \frac{b^2 - a^2}{2(b-a)} = \frac{b+a}{2}$$

and

$$\begin{aligned} \text{Var}(X) &= E[X^2] - E^2[X] = \int_a^b \frac{x^2}{b-a} \, dx - \left(\frac{a+b}{2}\right)^2 \\ &= \frac{b^3 - a^3}{3(b-a)} - \frac{(b+a)^2}{4} = \frac{(b-a)^2}{12}. \end{aligned}$$

Example A.8 We show here a case to illustrate what we pointed out concerning relationship (A.7), i.e., that the expected value of a function is not, in general, the function of the expected value. Consider the function $g(x) = x^2$ and a random variable X uniformly distributed over the interval $[-1, 1]$. Its PDF is $f_X(x) = 1/2$, on the support, and $E[X] = 0$. Hence, $g(E[X]) = 0$. However,

$$E[g(X)] = \int_{-1}^1 x^2 \cdot \frac{1}{2} \, dx = \int_0^1 x^2 \, dx = \frac{1}{3}.$$

□

The uniform distribution is commonly used in computer simulation, since pseudorandom numbers $U \sim U(0, 1)$ are the basis to sample from an arbitrary distribution. In practice, its use is often justified by lack of knowledge. When we just know a range for an uncertain quantity, the uniform distribution could be our only choice. Triangular distributions are often used for the same reason; in this case, we basically give the support (a, b) and the mode c, i.e., a lower and an upper bound on possible values, as well as the "most likely" value (in a very loose sense, as we have already clarified).

Exponential distribution   An exponential random variable can only take non-negative values, i.e., its support is $[0, +\infty)$. The name stems from the functional form of its density,

$$f_X(x) = \begin{cases} \lambda e^{-\lambda x} & \text{if } x \geq 0, \\ 0 & \text{if } x < 0, \end{cases}$$

where $\lambda > 0$ is a given parameter, and the notation $X \sim \exp(1/\lambda)$ is often used.[10] The CDF is

$$F_X(x) = \int_0^x \lambda e^{-\lambda t}\, dt = 1 - e^{-\lambda x}, \tag{A.8}$$

and the expected value is

$$\mathrm{E}[X] = \int_0^\infty x \lambda e^{-\lambda x}\, dx = \frac{1}{\lambda}.$$

It is worth noting that the expected value is quite different from the mode, which is zero. It can be shown that variance for the exponential distribution is $1/\lambda^2$, implying that the coefficient of variation is $c_X = 1$.

Unlike the uniform distribution, there are typically good physical reasons to adopt this distribution to model a random quantity. A common use is to model interarrival times, e.g., the time elapsing between two consecutive arrivals of customer orders. Note that λ is, within this interpretation, the order arrival rate and that the mean interarrival time is $1/\lambda$. There is a strong link with the Poisson distribution: When the interarrival times are exponentially distributed with rate λ, the number of orders received in a time interval of length t is a discrete random variable following a Poisson distribution with parameter λt. Furthermore, we will see later that this phenomenon corresponds to a common stochastic process, which is unsurprisingly known as the Poisson process.

In example A.15 on page 475 we will see that the exponential distribution enjoys a very peculiar property known as "lack of memory." Roughly speaking, this says that whatever time interval we observe without any arrival occurring, the distribution of the time we must wait until the next arrival is always the same. To get the point, imagine that we use the exponential distribution to model time between failures of an equipment. Lack of memory implies that even if the machine has been in use for a long time, this does not mean that it is more likely to have a failure in the near future. We should note the big difference with a uniform distribution. If we know that time between failures is uniformly distributed between, say, 50 and 70 hours, and we also know that 69 hours have elapsed since the last failure, we must expect the next failure within one hour. If the time between failures is exponentially

---

[10]We are assuming that the parameter used in the notation is the expected value.

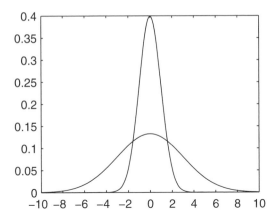

Fig. A.7 PDF of normal random variables with $\mu = 0$ and $\sigma = 1$, $\sigma = 3$.

distributed and 69 hours have elapsed, we cannot conclude anything, since from a probabilistic point of view the machine is brand new. If we think of purely random failures, due to bad luck, the exponential distribution may be a plausible model, but definitely not if wear is a factor.

Normal (Gaussian) distribution   The normal distribution characterizes what is arguably the best-known type of random variable. Its support is the whole real line and its PDF is the bell-shaped function

$$f_X(x) = \frac{1}{\sqrt{2\pi}\,\sigma} e^{-(x-\mu)^2/2\sigma^2}, \qquad -\infty < x < +\infty,$$

with parameters μ and σ; the density is symmetrical with respect to the point with abscissa μ. A few calculations show the meaning of the two parameters:

$$\mathrm{E}[X] = \mu, \qquad \mathrm{Var}[X] = \sigma^2.$$

Often, the notation $X \sim N(\mu, \sigma^2)$ is used.[11]

As we said, the parameter μ is a location measure of the distribution, whereas σ is a dispersion measure and tells how much the density is concentrated around the expected value. In figure A.7 we show the PDF for two normal distributions with $\mu = 0$ and $\sigma = 1, 3$, respectively. In this case mode, median, and expected value are the same.

In applications, a very special role is played by the **standard** normal distribution, characterized by parameters $\mu = 0$ and $\sigma = 1$. The reason of its

---

[11] We should note a potential ambiguity here. Sometimes, the notation $X \sim N(\mu, \sigma)$ is used. Indeed, the second parameter characterizing a normal distribution can equivalently be σ or σ^2.

importance is that if we are able to work with a standard normal in terms of quantiles and distribution function, then we are able to work with a generic normal variable. This is important, as the CDF is the integral of the bell-shaped function above, for which no analytical formula is known. The key point is that, given the properties of expected value and variance, if X is normal with parameters μ and σ, then $\alpha X + \beta$ has normal distribution with parameters $\alpha\mu + \beta$ and $\alpha\sigma$. In particular, $Z = (X - \mu)/\sigma$ is a standard normal. We may also go the other way around, starting from $Z \sim N(0,1)$, and obtain a generic normal variable $X \sim N(\mu, \sigma^2)$ by considering $X = \mu + \sigma Z$.

Although the general CDF for a normal variable is not known analytically, efficient numerical approximations (and fairly accurate tables) are widely available for the standard normal case:

$$\Phi(x) = \mathrm{P}\{Z \le x\} = \frac{1}{\sqrt{2\pi}} \int_{-\infty}^{x} e^{-z^2/2} \, dz.$$

These tables and numerical procedures also yield the quantiles z_q defined by

$$\mathrm{P}\{Z \le z_q\} = q,$$

for a probability level $q \in (0, 1)$. The tables are often given with different conventions, which may be confusing at first sight. However, given the symmetry of the standard normal distribution, any ambiguity is readily resolved. A common notation, which is ubiquitous in Statistics, is $z_{1-\alpha}$, where α is a relatively small number (say, 0.1 or 0.05) and it is intended that the quantile $z_{1-\alpha}$ leaves to its left an area $1 - \alpha$ under the graph of the density, whereas α is the area under the right tail.

Sometimes we are interested in a symmetric interval around the origin, such that α is the probability that a realization of the random variable will fall outside the interval. Then, we should cut two symmetric tails, each one with an area $\alpha/2$ under the CDF, to the left and to the right. Given the symmetry of the density, we have

$$\mathrm{P}\{-z_{1-\alpha/2} \le Z \le z_{1-\alpha/2}\} = 1 - \alpha.$$

The idea is illustrated in figure A.8. The statistical tables also give quantiles for other relevant distributions in Statistics, which are obtained from the standard normal, as we shall see in section A.6.3.

Example A.9 The knowledge of the CDF and the quantiles for a standard normal yields all the required values for a generic normal. In fact, if X is normal with parameters μ and σ, we know that $Z = (X - \mu)/\sigma$ is a standard normal. Hence, for instance,

$$\mathrm{P}\{X \le b\} = \mathrm{P}\left\{\frac{X - \mu}{\sigma} \le \frac{b - \mu}{\sigma}\right\} = \Phi\left(\frac{b - \mu}{\sigma}\right).$$

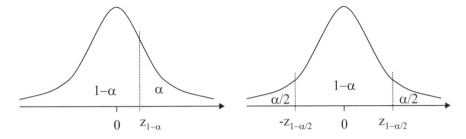

Fig. A.8 Quantiles of the normal distribution.

Suppose that $X \sim N(3, 16)$ and that we are interested in the probability $P\{2 < X < 7\}$. Looking at the statistical tables, we obtain

$$
\begin{aligned}
P\{2 < X < 7\} &= P\left\{ \frac{2-3}{4} \leq \frac{X-3}{4} \leq \frac{7-3}{4} \right\} \\
&= \Phi(1) - \Phi(-1/4) \\
&= \Phi(1) - [1 - \Phi(1/4)] \\
&= 0.8413 + 0.5987 - 1 \\
&= 0.4400.
\end{aligned}
$$

When we are interested in the quantiles, the relationship above implies that if we want the quantile x_q for $X \sim N(\mu, \sigma^2)$, all we have to do is find the corresponding quantile z_q for the standard normal and compute

$$
x_q = \mu + \sigma z_q.
$$ □

A.5 JOINTLY DISTRIBUTED RANDOM VARIABLES

So far, we have considered a single random variable, but in distribution logistics we typically work with several variables at a time. For instance, we may be interested in the sales of a given item over several days or weeks, or in the demand for several, possibly related items within the same period. It is very important to figure out which relationship, if any, may link these variables. For the sake of simplicity, we will only deal with the case of two random variables with a joint distribution, leaving the general case as a relatively straightforward extension.

The pathway to define all the relevant concepts for two jointly distributed random variables X and Y is similar to the case of a single variable. The starting point consists of joint events $\{X \leq x\} \cap \{Y \leq y\}$, to which a probability measure is associated. Given random variables X and Y, we must specify the joint CDF

$$
F_{X,Y}(x, y) = P\{X \leq x, Y \leq y\},
$$

where we use the simpler notation $\{X \leq x, Y \leq y\}$ for the joint event, instead of intersection. Just as in the case of a single random variable, this CDF collects all the relevant information; however, we may also find mass and density function very useful.

In the discrete case, we may define the PMF:

$$p_{X,Y}(x_i, y_j) = P\{X = x_i, Y = y_j\}.$$

Under some technical conditions, we may come up with a PDF $f_{X,Y}(x, y)$ in the continuous case, such that, given a region D in the two-dimensional plane, we have

$$P\{(X, Y) \in D\} = \iint_D f_{X,Y}(x, y)\, dy\, dx.$$

With respect to the single variable case, a new concept is the **marginal** distribution for the two variables, which is obtained as follows for the continuous case:

$$
\begin{aligned}
P\{X \in A\} &= P\{X \in A, Y \in (-\infty, +\infty)\} = \int_A \int_{-\infty}^{+\infty} f_{X,Y}(x, y)\, dy\, dx \\
&= \int_A f_X(x)\, dx,
\end{aligned}
$$

where

$$f_X(x) = \int_{-\infty}^{+\infty} f_{X,Y}(x, y)\, dy$$

is the marginal density for X. The marginal density $f_Y(y)$ is obtained by the same token, and the discrete case is similar as well. It is very important to realize that, given the joint density, we may find the two marginals, but we cannot really go the other way around. Quite different joint distributions may have the same pair of marginal distributions, and this depends on the relationship between the two variables, which we will investigate later.

The definition of expected value, variance, and moments is similar to the scalar case. Given a function $g(X, Y)$ of the two random variables, its expected value is

$$
E[g(X, Y)] = \begin{cases} \displaystyle\sum_i \sum_j g(x_i, y_j) p_{X,Y}(x_i, y_j) & \text{in the discrete case,} \\[2em] \displaystyle\int_{-\infty}^{+\infty} \int_{-\infty}^{+\infty} g(x, y) f_{X,Y}(x, y)\, dy\, dx & \text{in the continuous case.} \end{cases}
$$

Given the linearity of the sum and integral operators, we may see that the expected value of a linear combination of random variables is the same linear combination of the expected values. Formally, if we define a random variable

$$Z = \sum_{i=1}^{n} \lambda_i X_i,$$

where the numbers λ_i are the weights in the linear combination, its expected value is

$$E[Z] = \sum_{i=1}^{n} \lambda_i E[X_i].$$

This result does **not** apply, in general, to variance:

$$\text{Var}(X + Y) \neq \text{Var}(X) + \text{Var}(Y).$$

By the same token, in general,

$$E[g(X) \cdot h(Y)] \neq E[g(X)] \cdot E[h(Y)].$$

We may have an equality in some cases, which require the introduction of the concept of independence.

A.6   INDEPENDENCE, COVARIANCE, AND CONDITIONAL EXPECTATION

A.6.1   Independent random variables

Independence among random variables is directly related to the familiar concept of independence between events. Two random variables X and Y are independent if the two events $\{X \leq x\}$ and $\{Y \leq y\}$ are independent, i.e., if for any x and y we have

$$F_{X,Y}(x, y) = P\{X \leq x, Y \leq y\} = P\{X \leq x\} \cdot P\{Y \leq y\} = F_X(x) \cdot F_Y(y).$$

We see that independence allows us to factorize the joint CDF into the product of the individual CDFs. A consequence of this condition is that, in the discrete and continuous case, we may also factorize the PMF and the PDF into the product of marginals:

$$p_{X,Y}(x, y) = p_X(x)p_Y(y), \qquad f_{X,Y}(x, y) = f_X(x)f_Y(y).$$

This allows us to decompose double sums and double integrals, so that

$$E[g(X)h(Y)] = E[g(X)] \cdot E[h(Y)].$$

A further consequence is that, for a set of mutually independent random variables, we have

$$\text{Var}\left(\sum_{i=1}^{n} \lambda_i X_i\right) = \sum_{i=1}^{n} \lambda_i^2 \text{Var}(X_i).$$

We should note that a similar expression applies to the expected values of a linear combination of random variables; however, the formula for the expected value does not rely on any assumption about independence and holds in general. In particular, for independent variables X and Y we have

$$\text{Var}\,(X + Y) = \text{Var}\,(X) + \text{Var}\,(Y).$$

Note that we may sum *variances*, but *not* standard deviations:

$$\sigma_{X+Y} = \sqrt{\sigma_X^2 + \sigma_Y^2}.$$

The next example illustrates an interesting consequence of these properties.

Example A.10 Consider a set of i.i.d. random variables X_i, $i = 1, \ldots, n$, with expected value μ and variance σ^2. By "i.i.d." we mean independent and identically distributed. In Statistics (see section A.8), we are commonly interested in their average

$$Z = \frac{1}{n} \sum_{i=1}^{n} X_i,$$

which is a random variable as well. Let us compute the expected value and variance of Z. From the linearity of expectation, we immediately see

$$\text{E}[Z] = \text{E}\left[\frac{1}{n} \sum_{i=1}^{n} X_i\right] = \frac{1}{n} \sum_{i=1}^{n} \text{E}\,[X_i] = \frac{1}{n} \sum_{i=1}^{n} \mu = \mu.$$

As far as variance is concerned, given our assumption of independence, we have

$$\text{Var}[Z] = \text{Var}\left(\frac{1}{n} \sum_{i=1}^{n} X_i\right) = \frac{1}{n^2} \sum_{i=1}^{n} \text{Var}\,(X_i) = \frac{1}{n^2} \sum_{i=1}^{n} \sigma^2 = \frac{\sigma^2}{n}.$$

If we evaluate the squared coefficient of variation, we see

$$c_Z^2 = \frac{\text{Var}(Z)}{\text{E}^2[Z]} = \frac{\sigma^2}{n^2 \mu^2} = \frac{c_X^2}{n},$$

which shows an intuitive property of the average of independent variables: It is often "less uncertain" than the individual realizations.

It is very important to point out the difference between

$$\text{Var}(nX_i) = n^2 \sigma^2$$

and

$$\text{Var}\left(\sum_{i=1}^{n} X_i\right) = n\sigma^2.$$

If we were to simulate the first case, we would take one realization (or sample) of a random variable, and we would just multiply it by n. In the second case, we take n independent realizations and we sum them. Intuitively, the second case is less subject to random variability. As a practical illustration, consider the random daily demand for an item, assuming that demands in different days are independent. Weekly demand is obtained by summing daily demands and *not* by taking daily demand and multiplying it by the number of days in a week. ▯

A.6.2   Covariance and correlation

If two random variables are not independent, it is natural to investigate their degree of dependence, which means how can we measure it and how can we take advantage of it. The second task leads to statistical modeling, which we will investigate later in the simplest case of linear regression. The first task is not as easy as it may seem, as capturing dependence is a tricky issue. Nevertheless, we may settle for a less ambitious task and try to figure out a way to characterize the "concordance" between random variables. For instance, suppose that the random variable X tends to assume "large" values whenever Y does the same. More precisely, say that in most joint realizations (X, Y), both values tend to be either larger or smaller than the respective expected values. We could try to come up with a measure of this association.[12] One intuitive measure of this link is **covariance**:

$$\text{Cov}(X, Y) \equiv \text{E}[(X - \text{E}[X])(Y - \text{E}[Y])].$$

We have positive covariance when the events $\{X > \text{E}[X]\}$ and $\{Y > \text{E}[Y]\}$ tend to occur together, as well as the events $\{X < \text{E}[X]\}$ and $\{Y < \text{E}[Y]\}$, because the signs of the two factors in the product tend to be the same. If the signs tend to be different, we have a negative covariance.

For instance, if two products are complements, it is natural to expect positive covariance between their demands; negative covariance can be expected if they are substitutes. Similarly, if we observe over time the demand for an item whose long- or mid-term consumption is steady, a day of high demand should be typically followed by a day with low demand (as an example, consider the weekly demand of diapers if there is a promotional sale during one week).

From a computational point of view, it is very handy to express covariance as follows:

$$\text{Cov}(X, Y) \quad \equiv \quad \text{E}\Big[(X - \text{E}[X]) \cdot (Y - \text{E}[Y])\Big]$$

---

[12]Strictly speaking, we obtain a measure of concordance, rather than a measure of dependence; the latter should be something in the range $[0, 1]$, whereas as will shall see, correlation is in the range $[-1, 1]$; furthermore, a measure of dependence should meet some reasonable requirements which are beyond the scope of the book.

$$
\begin{aligned}
&= \quad \mathrm{E}\Big[XY - \mathrm{E}[X] \cdot Y - X \cdot \mathrm{E}[Y] + \mathrm{E}[X] \cdot \mathrm{E}[Y]\Big] \\
&= \quad \mathrm{E}[XY] - \mathrm{E}[X] \cdot \mathrm{E}[Y].
\end{aligned}
$$

We easily see that if two variables are independent, then their covariance is zero, since independence implies $\mathrm{E}[XY] = \mathrm{E}[X] \cdot \mathrm{E}[Y]$. However, the converse is *not* true in general, as we may see from the following example.

Example A.11 Let us consider a uniform random variable on the interval $[-1, 1]$; its expected value is zero and the density function is, on its support, constant and given by $f_X(x) = 1/2$. Now, let a random variable Y be given by

$$
Y = \sqrt{1 - X^2}.
$$

Clearly, there is a very strong dependence between X and Y, because, given the realization of X, the other one is perfectly predictable. However, their covariance is zero. We have seen that

$$
\mathrm{Cov}(X, Y) = \mathrm{E}[XY] - \mathrm{E}[X]\mathrm{E}[Y],
$$

but $\mathrm{E}[X] = 0$ and

$$
\mathrm{E}[XY] = \int_{-1}^{1} x\sqrt{1 - x^2} \cdot \frac{1}{2}\, dx = 0,
$$

because of the symmetry of the integrand function, which is an odd function, in the sense that $f(-x) = -f(x)$.

The key issue is that covariance is not really a good measure of dependence. It is only able to get a *linear* association between random variables, whereas in this case there is a very nonlinear link, since points with coordinates (X, Y) lie on the upper half of the unit circumference $X^2 + Y^2 = 1$. A more intuitive explanation is that if $Y > \mathrm{E}[Y]$, then we may have either $X > \mathrm{E}[X]$ or $X < \mathrm{E}[X]$. ▯

The following properties, whose proof is recommended as an exercise, are very useful:

- $\mathrm{Cov}(X, X) = \mathrm{Var}(X)$,

- $\mathrm{Cov}(X, Y) = \mathrm{Cov}(Y, X)$,

- $\mathrm{Cov}(aX, Y) = a\,\mathrm{Cov}(Y, X)$,

- $\mathrm{Cov}(X, Y + Z) = \mathrm{Cov}(X, Y) + \mathrm{Cov}(X, Z)$.

Using such properties, or the definitions, we may show

$$
\begin{aligned}
\mathrm{Var}(X + Y) &= \quad \mathrm{Var}(X) + \mathrm{Var}(Y) + 2\,\mathrm{Cov}(X, Y), \\
\mathrm{Var}(X - Y) &= \quad \mathrm{Var}(X) + \mathrm{Var}(Y) - 2\,\mathrm{Cov}(X, Y).
\end{aligned}
$$

By the way, we point out that variances are always summed, even if we take differences of random variables; otherwise, we could end up with a negative variance, which is impossible. More generally, for an arbitrary sum of random variables:

$$\text{Var}\left(\sum_{i=1}^{n} X_i\right) = \sum_{i=1}^{n} \text{Var}(X_i) + 2\sum_{i=1}^{n}\sum_{j<i} \text{Cov}(X_i, X_j).$$

We see that, in the case of mutually independent variables, the variance above boils down to the sum of variances.

A further issue with variance is that its value depends on how we measure the underlying quantities. We cannot say that a covariance of 100 is large or small. To define a measure which is independent on the unit of measure, we may introduce the **correlation** coefficient $\rho_{X,Y}$:

$$\rho_{XY} \equiv \frac{\text{Cov}(X,Y)}{\sqrt{\text{Var}(X)}\sqrt{\text{Var}(Y)}}.$$

It can be shown that the correlation coefficient takes values in the interval $[-1, 1]$. A value close to 1 shows a strong degree of positive correlation; a value close to -1 shows a strong degree of negative correlation. If correlation is zero, we speak of uncorrelated variables. We stress again that uncorrelated variables need not be independent. A notable case in which lack of correlation implies independence is the multivariable normal distribution: If two jointly normal variables are uncorrelated, they are independent too.

A.6.3   Distributions obtained from the normal and the central limit theorem

In general, if we sum identically distributed random variables, we do not get a random variable with the same distribution. For instance, if we sum two i.i.d. uniform random variables, we do not get a uniform random variable. By the same token, summing independent exponential variables, we do not get an exponential variable. If we allow the possibility of dependencies among the variables, the issue can get really complicated.

A most notable exception is the normal distribution. If we sum independent normal random variables, we still get a normal variable.[13] If we take a linear combination, with weights λ_i, of a set independent normal variables X_i, $i = 1, \ldots, n$, with parameters μ_i and σ_i, we get a normal random variable

$$X = \sum_{i=1}^{n} \lambda_i X_i,$$

---

[13]Proving this requires the introduction of moment generating function, which is beyond our scope.

with

$$E[X] = \sum_{i=1}^{n} \lambda_i \mu_i, \qquad \text{Var}(X) = \sum_{i=1}^{n} \lambda_i^2 \sigma_i^2.$$

It is important to note that this result does *not* require independence. If we have jointly normal random variables, any linear combination of them is still normal; the covariances affect the variance of the linear combination. If we group variances and covariances in a matrix $\boldsymbol{\Sigma}$, with elements $\sigma_{ij} = \text{Cov}(X_i, X_j)$, $\sigma_{ii} = \text{Cov}(X_i, X_i) = \text{Var}(X_i) = \sigma_i^2$, we may see that

$$\text{Var}(X) = \boldsymbol{\lambda}' \boldsymbol{\Sigma} \boldsymbol{\lambda},$$

where $\boldsymbol{\lambda}$ is a column vector grouping coefficients λ_i.

If we combine independent standard normal variables according to more complex patterns, we find some useful distributions.

Lognormal distribution   A random variable X is said to have a lognormal distribution if $\ln(X)$ is normal (note the use of natural logarithm). In other words, if Y is normal with parameters μ and σ^2, then $X = e^Y$ is lognormal. There is a link between the parameters of the lognormal variable and those of the underlying normal variable. For instance, it can be shown that

$$E[X] = e^{\mu + \sigma^2/2}.$$

Hence, if $Y \sim \mathcal{N}(-\sigma^2/2, \sigma^2)$, then $E[X] = 1$. This suggests using a lognormal variable with unit expected value to model random errors in a multiplicative model, whereas a normal variable with zero expected value would model random errors in an additive model (see section A.10.6). More generally, the normal distribution enjoys the nice property that by summing normal variables, we still get a normal variable; a similar property holds when we multiply lognormal random variables.

Chi-square distribution   Let Z_i, $i = 1, \ldots, n$, be standard and independent normal variables. The random variable X defined as

$$X = Z_1^2 + Z_2^2 + \cdots + Z_n^2$$

is certainly not normal, as it cannot take negative values. This variable is called chi-square with n degrees of freedom and is often denoted by χ_n^2.

We note that the expected value of a squared standard normal is

$$E[Z^2] = \text{Var}(Z) + E^2[Z] = 1,$$

from which we immediately see

$$E[X] = n.$$

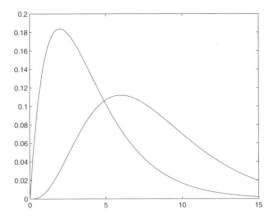

Fig. A.9 PDF of two chi-square random variables with 4 and 8 degrees of freedom, respectively; the variable with 4 degrees of freedom is less uncertain and has a higher mode.

It can also be shown that

$$\text{Var}(X) = 2n.$$

Quantiles of chi-square variables are tabulated for several degrees of freedom. Figure A.9 shows the PDF for a chi-square variable with 4 and 8 degrees of freedom.

Student's t distribution   If Z and χ_n^2 are a standard normal and a chi-square with n degrees of freedom, respectively, and they are also mutually independent, the random variable

$$T_n = \frac{Z}{\sqrt{\chi_n^2/n}}$$

has a Student's t distribution with n degrees of freedom. One could wonder why this weird ratio plays a practical role; we will see why later, when we consider parameter estimation and confidence intervals.

The density of the t distribution is bell-shaped and it looks much like a standard normal; the main differences lie in its heavier tail and in a lower mode. In figure A.10 we show the densities of T_1 and T_5 random variables, along a standard normal Z. We see that for increasing n, the tails of T_n get thinner and the density tends to a standard normal. In fact, for large n the two distributions are virtually identical.

It can be shown that

$$\text{E}[T_n] = 0\,, \qquad \text{Var}[T_n] = \frac{n}{n-2}.$$

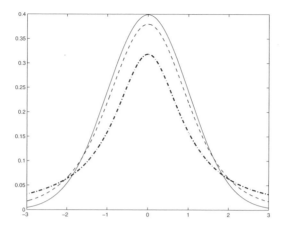

Fig. A.10 PDF of Student's t distribution, with $n = 1$ (dash-dotted line) and $n = 5$ (dashed line), compared with a standard normal (continuous line).

Statistical tables are available for quantiles $t_{1-\alpha,n}$, i.e., numbers satisfying the condition

$$P\{T_n \leq t_{1-\alpha,n}\} = 1 - \alpha.$$

F distribution   Finally, if we combine two independent chi-square variables with n and m degrees of freedom, respectively, to get a variable defined as

$$F_{n,m} = \frac{\chi_n^2/n}{\chi_m^2/m},$$

we obtain the F distribution with n and m degrees of freedom. This distribution too has applications in Statistics, which has motivated the compilation of tables yielding quantiles.

Central limit theorem   We have stressed that, by summing identical random variables, we do not get a similar distribution in general. However, if we sum a large number of i.i.d. random variables (please note *independence*), we obtain a distribution which gets closer and closer to a normal. This observation, which can be formalized as follows, contributes to explain the role of the normal distribution: When a phenomenon results from the sum of a large number of independent components, a normal distribution can make a good model.

This result is known as **central limit theorem**. A rigorous statement of the theorem requires some concepts of stochastic convergence, but we may try to clarify the sense of the theorem. Consider a sequence X_1, X_2, ..., X_n of i.i.d. random variables, with expected value μ and standard deviation σ. For n going to infinity, it can be shown that the sum

$$X_1 + X_2 + \cdots + X_n$$

has a distribution which is approximately normal with expected value $n\mu$ and standard deviation $\sqrt{n}\sigma$. In other words, the variable

$$\frac{X_1 + X_2 + \cdots + X_n - n\mu}{\sqrt{n}\sigma}$$

tends to a standard normal distribution. A little more precisely, we have

$$\mathrm{P}\left\{\frac{X_1 + X_2 + \cdots + X_n - n\mu}{\sqrt{n}\sigma} < x\right\} \approx \Phi(x),$$

where $\Phi(x)$ is the CDF of a standard normal. To illustrate the practical consequence of the theorem, if we have a large number of independent customers who order a certain item, we may assume that the overall demand can be modeled by a normal random variable. Such an assumption must be validated by proper statistical procedures, or at least checked in terms of basic tests. For instance, if the standard deviation is too large with respect to the expected value, there is a non-negligible chance of a negative demand, which makes no sense; in such a case, an asymmetric demand distribution would probably be a better model.

A.6.4   Conditional expectation

In section A.2 we have introduced conditional probabilities. When we deal with random variables, we may introduce the concept of conditional expectation, which is essentially the expected value of a variable X, given knowledge of the realization of another variable Y. As a practical example, we might wonder what is the expected demand for ice cream (random variable X) as a function of temperature (random variable Y). What we really get is a random variable denoted by $\mathrm{E}[X \mid Y]$. Conditional expectation is actually a very subtle concept in probability theory, which requires some nontrivial technical machinery when continuous random variables or stochastic processes (see next section) are involved. However, we may start from the discrete case to build intuition heuristically.

We would like to know how the event $\{Y = y\}$ influences the distribution of X. Working along the lines of conditional probability, we may introduce a conditional probability mass function:

$$p_{X|Y}(x \mid y) = \mathrm{P}\{X = x \mid Y = y\} = \frac{\mathrm{P}\{X = x, Y = y\}}{\mathrm{P}\{Y = y\}} = \frac{p_{X,Y}(x, y)}{p_Y(y)}. \quad \text{(A.9)}$$

Then, we may define conditional expectation:

$$\mathrm{E}[X \mid Y = y_j] = \sum_i x_i \mathrm{P}\{X = x_i \mid Y = y_j\} = \sum_i x_i p_{X|Y}(x_i|y_j). \quad \text{(A.10)}$$

It is a useful exercise to check that, if X and Y are independent, then $\mathrm{E}(X \mid Y = y) = \mathrm{E}(X)$. We cannot really extend this concept directly to continuous random variables, as the event $\{Y = y\}$ has zero probability. What we can do is to define a conditional density

$$f_{X|Y}(x \mid y) = \frac{f_{X,Y}(x, y)}{f_Y(y)},$$

which yields the conditional expectation

$$E[X \mid Y = y] = \int_{-\infty}^{+\infty} x f_{X|Y}(x \mid y) \, dx. \tag{A.11}$$

Example A.12 Suppose that the joint density of two continuous random variables is

$$f_{X,Y}(x, y) = K(x + y),$$

where $0 \leq x, y \leq 1$ and K is a normalization constant such that integrating the density over its support we get 1. To compute $E(X \mid Y)$, the first step is finding the marginal density of Y:

$$f_Y(y) = \int_0^1 K(x + y) \, dx = K \left(y + 1/2 \right).$$

Hence, the conditional density is

$$f_{X|Y}(x \mid y) = \frac{f_{X,Y}(x, y)}{f_Y(y)} = \frac{x + y}{\frac{1}{2} + y},$$

and we actually see that knowledge of the constant K is not needed. Applying the definition[14] (A.11), we get

$$E(X \mid Y = y) = \int_0^1 x f_{X|Y}(x \mid y) \, dx = \int_0^1 x \frac{x + y}{y + 1/2} \, dx = \frac{2 + 3y}{3 + 3y}. \qquad \square$$

From this example, we see that we get a function of y. The notation $E(X \mid Y = y)$ points out that if the realization of the random variable Y is y, then we have a number which is the conditional expected value of X. The notation $E(X \mid Y)$ actually shows that this is a random variable, which is actually the best forecast we can come up with, as a function of Y. Indeed, conditional expectation is all about the proper use of (partial) information.

If we interpret $E(X \mid Y)$ as a random variable, then it is natural to consider its expected value (expectation, in this case, is with respect to Y). A fundamental property of conditional expectation is

$$E[X] = E[E[X \mid Y]]. \tag{A.12}$$

---

[14]Readers with a background in measure-theoretic probability would object that this is not really the definition, which requires a way to model information by some σ-algebra, but please bear with us.

We may clarify the meaning of (A.12) by rewriting it explicitly:

$$
E[X] = \begin{cases} \displaystyle\sum_j E[X \mid Y = y_j] P\{Y = y_j\} & \text{in the discrete case,} \\[2mm] \displaystyle\int_{-\infty}^{+\infty} E[X \mid Y = y] f_Y(y)\, dy & \text{in the continuous case.} \end{cases}
$$

We will not prove this property, but at least in the discrete case we may see a link with the theorem of total probabilities theorem.

Equation (A.12) is very relevant from a practical point of view. Among other things, it may be exploited as a trick of the trade to compute expected values when a direct approach looks too difficult. In the following we will show a few applications of this technique.

Example A.13 In the previous sections we have considered the geometric distribution with parameter p and we have proved that its expected value is

$$
E[X] = \frac{1}{p}.
$$

To this aim, we have used some familiar properties of the geometric series, but there is a much more straightforward way to obtain the same result by conditioning on the outcome of the first trial (the reader should recall the physical motivation of this distribution). If the first trial is a success, and this occurs with probability p, we have $X = 1$ because we have just attained our success and we stop the sequence of trials immediately. Otherwise, we have already failed once, and we must try again. However, since experiments are independent, we are just back to square one, and the expected number of trials to go is the same as before. Formally:

$$
E[X] = E[X \mid OK] \cdot P\{OK\} + E[X \mid NOK] \cdot P\{NOK\} = 1 \cdot p + (1 + E[X])(1 - p),
$$

from which we immediately get $E[X] = 1/p$, which confirms our previous result. The real bonus, though, comes when computing variance. As a preliminary step, we have

$$
\begin{aligned}
E[X^2] &= E[X^2 \mid OK] \cdot P\{OK\} + E[X^2 \mid NOK] \cdot P\{NOK\} \\
&= 1^2 \cdot p + E[(1 + X)^2](1 - p) \\
&= p + (1 + 2E[X] + E[X^2])(1 - p) \\
&= p + (1 + 2/p)(1 - p) + E[X^2](1 - p),
\end{aligned}
$$

which yields

$$
E[X^2] = \frac{2 - p}{p^2}.
$$

Then we immediately obtain

$$
\operatorname{Var}(X) = E[X^2] - E^2[X] = \frac{2 - p}{p^2} - \frac{1}{p^2} = \frac{1 - p}{p^2}.
$$

☐

Example A.14 Here we generalize the result obtained in example A.10 on page 464, by considering a random variable defined as

$$Z = \sum_{i=1}^{N} X_i,$$

where now N is a discrete random variable, with known expected value and variance, rather than a number; we also assume that N is independent from variables X_i, which are mutually independent as well. In other words, we consider a sum of a random number of random variables, and we want to come up with its expected value and variance. This is practically relevant, for instance, when we want to sum the demand over a random number of time periods corresponding to a random procurement lead time.

By conditioning with respect to N, we have

$$\mathrm{E}\left[\sum_{i=1}^{N} X_i\right] = \mathrm{E}\left[\mathrm{E}\left[\sum_{i=1}^{N} X_i \,\middle|\, N\right]\right].$$

We start from the inner conditional expectation:

$$\mathrm{E}\left[\sum_{i=1}^{N} X_i \,\middle|\, N = n\right] = \mathrm{E}\left[\sum_{i=1}^{n} X_i \,\middle|\, N = n\right] = \mathrm{E}\left[\sum_{i=1}^{n} X_i\right] = n\mathrm{E}[X_i],$$

where in the second-to-last step we have used the independence between N and X_i. Hence, we have

$$\mathrm{E}\left[\sum_{i=1}^{N} X_i \,\middle|\, N\right] = N\mathrm{E}[X_i],$$

and, by computing the overall expectation,

$$\mathrm{E}\left[\sum_{i=1}^{N} X_i\right] = \mathrm{E}\left[N\mathrm{E}[X_i]\right] = \mathrm{E}[N] \cdot \mathrm{E}[X_i].$$

This result is fairly intuitive, actually, and one could wonder if we took unnecessary pains in using conditioning arguments. However, when we consider variance, things are a bit more difficult. Using the same technique, and some patience, we can arrive at the result:

$$\mathrm{Var}\left(\sum_{i=1}^{N} X_i\right) = \mathrm{E}[N] \cdot \mathrm{Var}(X_i) + \mathrm{E}^2[X_i] \cdot \mathrm{Var}(N).$$

We should note the *imperfect* symmetry of this formula, where expectations and variances of the involved variables are combined in a way that boils down to (A.10) when N is a number, in which case its variance is zero.   ☐

When we have introduced the exponential distribution, we have hinted at its "lack of memory" property. Now we may get a clear picture of what we meant.

Example A.15 Consider an exponential random variable X with parameter λ, and say that X models the random life of some equipment (or a light bulb) whose average life is $1/\lambda$. From the cumulative distribution function (A.8) we see that

$$P\{X > t\} = e^{-\lambda t}.$$

This makes sense, as this probability goes to zero when t increases, with a speed which is large when expected life is short. Now suppose that, after lighting the bulb, we notice that it is alive and kicking at time t; we could wonder what its expected residual life is, given this information. In general, after a long time of work, the death of a piece of equipment gets closer and closer.[15] To formalize the problem, we should consider the conditional probability that the overall life of the light bulb is larger than $t + s$:

$$
\begin{aligned}
P\{X > t + s \mid X > t\} &= \frac{P\{(X > t + s) \cap (X > t)\}}{P\{X > t\}} \\
&= \frac{P\{X > t + s\}}{P\{X > t\}} \\
&= \frac{e^{-\lambda(t+s)}}{e^{-\lambda t}} \\
&= e^{-\lambda s} \\
&= P\{X > s\}.
\end{aligned}
$$

We see a rather surprising result: The elapsed time t does not influence the residual life s and, after a time span of length t, the light bulb is statistically identical to a brand new one. This is why we speak of lack of memory in the exponential distribution, which makes it suitable to model certain "purely random" phenomena, but not situations such as failures due to wear.   ▯

A.7   STOCHASTIC PROCESSES

When we think of the joint distribution of random variables, we may naturally think of the realization of several phenomena at the same time. However, we may also be interested in the successive realizations of a single phenomenon over time, i.e., a collection of random variables indexed by time. The time index can take integer values, for instance, when we are interested in observing daily or weekly demand for an item, so that time is discretized in time buckets.

---

[15]Even more so in case of early burnout, which is typical of former Engineering students like the authors.

In such a case we will use notation like

$$X_t, \qquad t = 0, 1, 2, 3, 4, \ldots. \tag{A.13}$$

The sequence of random variables (A.13) is a **discrete-time stochastic process**. When the integer parameter does not represent time, we may speak of a discrete-parameter process. In some loose sense, the stochastic process is a generalization of deterministic functions of time, in that for any value of t it yields a random variable (which is a function itself) rather than a number. If we observe a sequential realization of the random variables over time, we get a **sample path** of the process.

Naive thinking would draw us to the conclusion that, in order to characterize a stochastic process, we should give the marginal distribution of X_T for all the relevant time instants t. Actually, this is not enough, as we should consider the *joint* distribution of the random variables. This may be very hard in general, and it is customary to look for relatively simple cases. The easiest one is arguably the case in which all the random variables are mutually independent.

Example A.16 (Gaussian processes) A common class of stochastic processes consists of sequences of random variables whose marginal distribution is normal, which is why they deserve the name of Gaussian processes. To be precise, we should say that a Gaussian process requires that the random variables $X_{t_1}, X_{t_2}, \ldots, X_{t_m}$ have a *joint* normal distribution for any possible choice of time instants $t_1, t_2, \ldots, t_m$, but for the sake of simplicity we will put in the same bag any process for which the marginal distribution of X_t is normal. However, it is important to realize that in doing so we are considering processes which may be very different in nature. Consider the stochastic process

$$X_t = t \cdot \tilde{\epsilon}, \qquad t = 0, 1, 2, 3, \ldots,$$

where $\tilde{\epsilon}$ is standard normal variable. In our loose sense, we may say that this is a Gaussian process, since X_t is normal with expected value 0 and variance t^2. However, it is a somewhat degenerate process, since uncertainty is linked to the realization of a *single* random variable. If we know the value of X_t for a single time instant, then we can figure out the whole sample path. Figure A.11 illustrates this point by showing a few sample paths of this process. A quite different process is obtained if all variables X_t are normal with parameters μ and σ^2 and mutually *independent*. Figure A.12 shows a sample path of the process $X_t = t \cdot \tilde{\epsilon}_t$, where $\tilde{\epsilon}_t \sim N(0, 1)$. However, the marginal distributions of the individual random variables X_t are exactly the same for both processes.

□

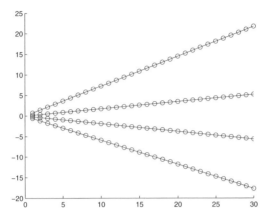

Fig. A.11   Sample paths of the stochastic process $X_t = t \cdot \tilde{\epsilon}$, $\tilde{\epsilon} \sim N(0,1)$.

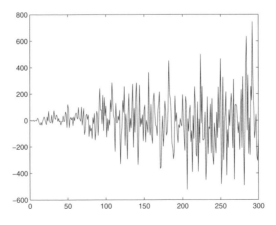

Fig. A.12   Sample path of process $X_t = t \cdot \tilde{\epsilon}_t$, where $\tilde{\epsilon}_t \sim N(0,1)$.

The example above deals with processes such that the expected value of X_t is constant.[16] In practice, we should expect trends (e.g., for recently developed products) and/or seasonality (e.g., for ice cream). A process with constant statistical properties is probably easier to deal with, but we should clarify what we mean exactly.

DEFINITION A.10 *A stochastic process* X_t *is said to be* **weakly stationary***, or second-order stationary, if the expected value is constant, i.e.,* $E[X_t] = \mu$, *and the covariance* $\text{Cov}(X_t, X_{t+s})$, *for all* $s = 0, 1, 2, 3, \ldots$, *does not depend on* t.

This definition deserves a few comments.

- To begin with, we speak of *weak* stationarity because we are only considering the first two moments of the relevant random variables, whereas stationarity in general has to do with joint distribution of every possible subset of random variables.

- The second condition has two implications.

  1. By setting $s = 0$, we see that variance is constant too, as $\text{Var}(X_t) = \text{Cov}(X_t, X_t) = \sigma^2$. Hence, the two processes in example A.16 are not really stationary, since variance changes with time.

  2. The second point is that the covariance between X_{t_1} and X_{t_2} depends only on the time distance $|t_1 - t_2|$, i.e., on the width of the time interval we consider, but not on *where* this interval is placed.

The second implication of definition A.10 suggests that we should reflect a bit on the link between two random variables X_t and X_{t+s}. In example A.16 we just considered two extreme cases: In the first one, knowledge of X_t implies perfect knowledge of X_{t+s} for any s; in the second one, because of independence, such a knowledge does not tell us anything. It is reasonable to guess that there are many interesting intermediate cases.

Example A.17 Consider a small shop with one clerk. Customers arrive according to some probabilistic law, and if the clerk is busy with another customer, they wait in a queue. To keep it simple, assume that the shop is open 24/7, so that there is no issue with closing periods. The service time is also a random variable, characterized by some suitable distribution. Let W_k be the waiting time of the kth customer; if we consider the sequence of waiting times for $k = 1, 2, 3, \ldots$, we obtain a discrete-parameter stochastic process. Can we say that the variables W_k are independent? Ruling out pathological

---

[16]We should stress that, for the "degenerate" process, it is the *unconditional* expectation which is constant and equal to zero; conditional expectation is quite different, since a very little knowledge results in a deterministic function of time.

cases, the general answer is no. If customer k undergoes a long waiting time, then we may conclude that this unlucky customer probably arrived at a time in which the system is congested and there is a long queue. Hence, we might expect that the waiting time for customer $k + 1$ will be large too. However, if the clerk, on the long term, is able to serve all of the customer, sooner or later this congestion will be resolved. Hence, we should expect that the waiting times of two faraway customers, say W_k and W_{k+1000}, are practically independent. In other words, intuition suggests that the random variables W_k and W_{k+s} should have some positive correlation and that this tends to fade out for increasing values of s. □

The last example motivates the following definition.

DEFINITION A.11 (Autocovariance and autocorrelation)
Given a weakly stationary stochastic process X_t, the function

$$C(s) = \text{Cov}(X_t, X_{t+s})$$

is called autocovariance of the process with delay s. The function

$$R(s) = \frac{C(s)}{\sigma^2}$$

is called autocorrelation function.

We should note that because of how it is defined, autocovariance depends only on s, which is justified for a stationary process. The definition of autocorrelation relies on the fact that variance is constant, which implies

$$R(s) = \rho(X_t, X_{t+s}) = \frac{\text{Cov}(X_t, X_{t+s})}{\sqrt{\text{Var}(X_t)}\sqrt{\text{Var}(X_{t+s})}} = \frac{C(s)}{\sigma^2}.$$

Even from this cursory and crude treatment, we may see that stochastic processes are a thorny object to deal with, since in general we should describe the joint distribution of all the involved random variables. For this reason, whenever it is practically acceptable, we should work with processes in which mutual dependence among random variables is at least limited to a simple structure, if not absent at all. We have a relatively easy case when the process "memory" is limited to its last value. Formally,

$$E[X_{t+1} \mid X_t, \ X_{t-1}, \ X_{t-2}, \ X_{t-3}, \ldots] = E[X_{t+1} \mid X_t].$$

A process meeting this condition is said a **Markov process**. A typical example of Markov process is

$$X_t = X_{t-1} + \tilde{\epsilon}_t,$$

where $\tilde{\epsilon}_t$ is a normal random variable and all variables $\tilde{\epsilon}_t$ are mutually independent. Figure A.13 shows two sample paths in the case $\tilde{\epsilon}_t \sim N(0, 1)$; a process like this is also called **random walk**.

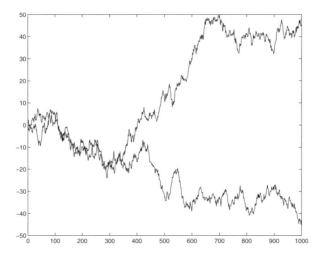

Fig. A.13 Two sample paths of the process $X_t = X_{t-1} + \tilde{\epsilon}_t$, with initial condition $X_0 = 0$ and $\tilde{\epsilon}_t \sim N(0,1)$.

So far, we have dealt with discrete-time processes, but the time parameter may also be represented by a real number. In such a case we have a **continuous-time stochastic process**

$$X(t), \qquad t \geq 0.$$

In principle, the definitions we have given above for stationarity, autocovariance, and autocorrelation have a straightforward extension to the continuous-time case. Actually, continuous-time processes require a much more complex machinery for a deep understanding. Here we limit ourselves to considering a very common and useful process, which is known as **Poisson process**.

Example A.18 The Poisson process is an example of counting process, i.e., a stochastic process $N(t)$ counting the number of events occurred in the time interval $[0, t]$. Such a process starts from zero and has unit increments over time. We may use such a process to model order or customer arrivals. The Poisson process is obtained when we make specific assumptions about the interarrival times of customers. Let X_k, $k = 1, 2, 3, 4, \ldots$, be the interarrival time between customer $k - 1$ and customer k; by convention, X_1 is the arrival time of the first customer after the start time $t = 0$. We obtain a Poisson process if we assume that variables X_k are mutually independent and all exponentially distributed with parameter λ, which is in this case the arrival rate, i.e., the average number of customers arriving per unit time. A sample path is illustrated in figure A.14; we see that the process "jumps" whenever a customer arrives, so that sample paths are piecewise constant.

We have already mentioned the link between Poisson and exponential distributions and the Poisson process. If we consider a time interval $[t_1, t_2]$,

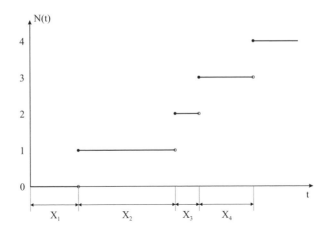

Fig. A.14 Sample path of the Poisson process.

with $t_1 < t_2$, then the number of customers arrived in this interval, i.e., $N(t_2) - N(t_1)$, has Poisson distribution with parameter $\lambda(t_2 - t_1)$. Furthermore, if we consider another time interval $[t_3, t_4]$, where $t_3 < t_4$, which is disjoint from the previous one, i.e., $(t_2 < t_3)$, then the random variables $N(t_2) - N(t_1)$ and $N(t_4) - N(t_3)$ are independent. We say that the Poisson process has *stationary* and *independent* increments.

The Poisson process is a useful model to represent the random arrival of customers who have no mutual relationships at all. This is a consequence of the lack of memory of the exponential distribution, which we have illustrated in example A.15 on page 475.

The model can be generalized to better fit reality. For instance, if we observe the arrival process of customers at a big retail store, we easily observe variations in the arrival rate. If we introduce a time-varying rate $\lambda(t)$, we get the so-called **inhomogeneous** Poisson process. Furthermore, if we consider not only customer (or order) arrivals, but the demanded quantities as well, we see the opportunity of associating another random variable, the quantity per order, with each customer. The cumulative quantity demanded $D(t)$ in the time interval $[0, t]$ is another stochastic process, which is known as **compound** Poisson process. The sample paths of this process would be qualitatively similar to those in figure A.14, but the jumps would be random variables. This is a possible model for demand, when sale volumes are not large enough to warrant use of a normal distribution. ☐

A.8 PARAMETER ESTIMATION

In this section we enter the realm of Statistics, which, in a sense, goes the other way around with respect to probability theory. In the latter, we assume perfect knowledge of essential properties of a phenomenon, e.g., encoded in the density function of a random variable, and we ask possibly complicated questions about the probability of occurrence of an event of interest. Whatever the questions, in probability theory we take for granted that we have knowledge of parameters such as expected value and variance. In Statistics, it is raw data we start from, and we would like to build a probability distribution; we *do not* really know expected values or variances, and we would like to come up with some sound procedure to estimate them, to qualify the reliability of such estimates, and to test hypotheses about them. This does not mean that we discard probability theory; on the contrary, this is the conceptual foundation for such "sound" procedures, but the mindset is completely different. In Statistics, we aim at squeezing information out of available data, and this may require some finesse in understanding which data are relevant, which ones should be discarded, and how they can be related.

The basic problem we consider in this section is parameter estimation, with specific reference to the estimation of an expected value. The starting point is a set of data; these can be obtained, e.g., by historical demand data, by a survey where customers have been interviewed, or by computer-based experiments with Monte Carlo simulation. A formalization of the vague idea of "a set of data" is needed to rely on a sound probabilistic foundation, and this yields the concept of a random sample.

DEFINITION A.12 (Random sample) *If $X_1, X_2, \ldots, X_n$ are independent random variables characterized by the same CDF F_X, then they are a random sample.*

In other words, the elements of the random sample are a sequence of i.i.d. random variables. It is very important to stress the role of *independence* in the definition above. All of the following concepts depend critically on this assumption. It may well be the case that there is correlation in a practical sample, but then a blindfolded application of naive statistical procedures may lead to erroneous conclusions and a possible business disaster. Furthermore, we also assume that the data are somewhat homogeneous, since they are identically distributed. Clearly, if the data have been observed under completely different settings, the conclusions we draw from their analysis may be severely flawed.

Given a random sample, we typically summarize the data by using some recipe. Formally, we compute a statistic.

DEFINITION A.13 (Statistic) *A **statistic** is a random variable whose value is determined by a random sample.*

In other words, a statistic is basically a function of a random sample. As a concrete example, the most common statistic is the **sample mean**:

$$\overline{X} = \frac{1}{n} \sum_{i=1}^{n} X_i. \tag{A.14}$$

The sample mean should not be regarded as a number, but as a random variable. Assume that the expected value of the random variables in the sample is μ. Then the expected value of the sample mean, as we know from example A.10 on page 464, is

$$\mathrm{E}\left[\overline{X}\right] = \mu.$$

This property justifies using the sample mean $\overline{X}$ as *one* possible **estimator** of the expected value μ. Hence, we are using realizations of a random variable to estimate the unknown value of a number, and a natural question is, Which properties make a good estimator? The first one is that an estimator should be *unbiased*, in the sense that its expected value is the parameter we wish to estimate. As we have seen, the sample mean is an unbiased estimator of the expected value. However, there are other issues in using an estimator, since this is a random variable with some variance; of course, we would like to have estimators with small variance. From example A.10, we know the variance of sample mean,

$$\mathrm{Var}\left[\overline{X}\right] = \frac{\sigma^2}{n}, \tag{A.15}$$

where σ^2 is the variance of each single element X_i in the random sample. As it is reasonable to expect, this variance decreases for an increasing size of the sample. However, we insist again that the last property relies on independence in the sample, whereas unbiasedness of the sample mean does not.

Equation (A.15) is useful in drawing some conclusions on how reliable an estimate is, but it relies on another parameter, σ, which is typically unknown as well. The typical estimator for variance is **sample variance**:

$$S^2 = \frac{1}{n-1} \sum_{i=1}^{n} \left(X_i - \overline{X}\right)^2. \tag{A.16}$$

This formula can be understood as a sample counterpart of the definition of variance: It is basically an average squared deviation with respect to sample mean. From a computational point of view, the following rearrangement can be useful:

$$
\begin{aligned}
S^2 &= \frac{1}{n-1} \sum_{i=1}^{n} \left(X_i - \overline{X}\right)^2 \\
&= \frac{1}{n-1} \left(\sum_{i=1}^{n} X_i^2 - 2\overline{X} \sum_{i=1}^{n} X_i + n\overline{X}^2 \right) \\
&= \frac{1}{n-1} \left(\sum_{i=1}^{n} X_i^2 - n\overline{X}^2 \right). \tag{A.17}
\end{aligned}
$$

The **sample standard deviation** is just S, the square root of sample variance. We note again that these estimators are random variables, and we should make sure that they are unbiased, i.e., $E[S^2] = \sigma^2$ This is the reason for the apparently odd denominator in (A.16), which is $n-1$ rather than n. From the rearrangement above we see

$$
\begin{aligned}
E\left[S^2\right] &= \frac{1}{n-1}\left(\sum_{i=1}^{n} E\left[X_i^2\right] - nE\left[\overline{X}^2\right]\right) \\
&= \frac{n}{n-1}\left\{\sigma^2 + \mu^2 - \left(\frac{\sigma^2}{n} + \mu^2\right)\right\} \\
&= \sigma^2.
\end{aligned}
$$

Actually, the need for the denominator $n-1$ stems from the fact that we are measuring deviations against $\overline{X}$ rather than μ. From an intuitive point of view, we could say that the need to estimate the unknown expected value implies that we "lose one degree of freedom" in the n available data in the sample. This point of view can be made rigorous, but we will settle for the intuitive sense.

Now that we know something about expected value and variance of the sample mean, we can dig deeper and ask questions about its distribution. Finding the distribution of an estimator can be a tricky issue since, as we have seen, summing random variables with a given distribution need not result in a random variable with the same distribution. However, we may at least look for some partial results. To begin with, the central limit theorem says that for a large sample, the distribution of the sample mean tends to a normal. We also know that if we are actually sampling from a normal distribution, then the sample mean will be normal too for any size of the sample.

Sample variance is a bit trickier, even if we assume normal samples. From an intuitive point of view, we see from equation (A.17) that it involves squares of normal variables. Given what we know about the chi-square and Student's t distribution, the following theorem, which summarizes basic results on the distribution of the estimators we have considered, should not come as a surprise.[17]

THEOREM A.14 Let $X_1, \ldots, X_n$ be a random sample from a normal distribution with expected value μ and variance σ^2. Then:

1. The sample mean $\overline{X}$ has normal distribution with expected value μ and variance σ^2/n.

2. The random variable $(n-1)S^2/\sigma^2$ has chi-square distribution with $n-1$ degrees of freedom.

3. Sample mean and sample variance are independent random variables.

---

[17]See, e.g., [4] for a proof.

4. The random variable

$$\frac{\overline{X} - \mu}{S/\sqrt{n}}$$

has t distribution with $n - 1$ degrees of freedom.

The third statement above is somewhat surprising, since sample mean and sample variance are statistics depending on the same random variables, but it is essential in establishing the last distributional result, which will play a fundamental role in the following. We should note that if the true variance were known, we could work with the statistic

$$\frac{\overline{X} - \mu}{\sigma/\sqrt{n}},$$

which is a standard normal. If the random sample is not normal, the results above do not hold for a small sample. We will rely on central limit theorem in justifying the application of statistical procedures, which are valid for a normal sample, to a large non-normal sample. This is enough for what we do in the main body of the book, but this heuristic approach should not be applied to other problems without due care.

A.8.1   Sample covariance and correlation

In applications, we are often interested in modeling the relationships among different variables: For instance, we would like to estimate the impact of advertising on sales, or of temperature on ice cream demand. The amount of ads and the temperature are treated as *explanatory variables*, in that they contribute to explain demand. In section A.10 we will discuss regression models in some detail, but it is useful to start discussing here the sample counterparts of covariance and coefficient of correlation, which we have introduced in section A.6.2 to investigate the link among random variables. Indeed, one test that we can carry out to check the impact of a variable on another one is estimating their coefficient of correlation. As with expected value and variance, we assume that we lack knowledge of the probability distributions involved, and we must analyze empirical data to come up with estimates. It is important to realize that our sample must consist of *joint* realizations of variables X and Y. If we want to investigate the impact of temperature on ice cream demand, we must have pairs of observation taken in the same place at the same time; clearly, mixing observations is no use.

Just as we have defined sample variance in equation (A.16), we may define **sample covariance** S_{XY} between random variables X and Y:

$$S_{XY} = \frac{1}{n-1} \sum_{i=1}^{n} (X_i - \overline{X})(Y_i - \overline{Y}), \qquad (A.18)$$

where n is the size of the sample, i.e., the number of *pairs*. This definition is also consistent with sample variance, since $S_X^2 = S_{XX}$.

To estimate the (coefficient of) correlation ρ_{XY} between X and Y, we may use **sample coefficient of correlation**, or sample correlation for short:

$$r_{XY} = \frac{S_{XY}}{S_X S_Y} = \frac{\sum_{i=1}^{n}(X_i - \overline{X})(Y_i - \overline{Y})}{\sqrt{\sum_{i=1}^{n}(X_i - \overline{X})^2} \cdot \sqrt{\sum_{i=1}^{n}(Y_i - \overline{Y})^2}}, \quad -1 \le r_{XY} \le +1,$$

(A.19)

where factors $n-1$ cancel each other. Once again, we stress that the estimators we have just defined are *random variables* depending on the random sample we take.

As in the case of sample variance, we have to include a term $n - 1$ to make the estimators unbiased. Checking unbiasedness is left as an exercise, but one could wonder why we should divide by $n - 1$ and not $n - 2$, given that we rely here on the estimates of *two* parameters, the expected values of both X and Y. Apart from a formal proof, which we omit, one intuitive check is that using $n - 1$ is required for consistency with sample variance. As another intuitive argument, we could note that the minimal size of the sample to get an estimate of variance, covariance, and correlation is two; otherwise, we cannot compute any deviation from the mean. We will see that in linear regression the minimal sample size is actually three, and there we will see a $n - 2$ factor come into play.

The sample coefficient of correlation, just as its probabilistic counterpart, is adimensional and it lies in the range $[-1, 1]$. We recall that this is not really a measure of dependence, but a measure of concordance in the deviations with respect to the means. The sample correlation tells us if a positive (negative) deviation of Y with respect to $\overline{Y}$ is associated with a positive (negative) deviation of X from $\overline{X}$. A positive coefficient suggests that, on average, when Y is larger than $\overline{Y}$, then also X is larger than $\overline{X}$; similarly, when Y is smaller than $\overline{Y}$, also X is smaller than $\overline{X}$ on average (see figure A.15). On the contrary, a negative coefficient of correlation suggests that when Y is larger than $\overline{Y}$, X is smaller than $\overline{X}$, and vice versa. A case of *negative correlation* is illustrated in figure A.16. When none of these patterns occurs, we have no correlation, like in figure A.17. Clearly, if we want to use these tools for forecasting, a very small correlation tells that, probably, the variable X is not very useful in predicting Y, since apparently there is no relationship. On the contrary, a strong (in absolute sign) correlation suggests that there is a strong link and that *maybe* X can be very useful in predicting Y.

Correlation analysis is very useful, but like any other tool we must be well aware of its pitfalls and limitations to use it properly.

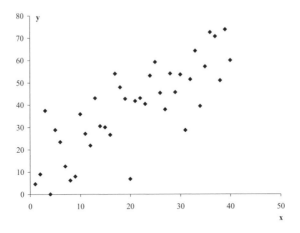

Fig. A.15   A case of correlation 0.8.

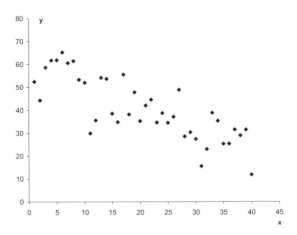

Fig. A.16   A case of correlation −0.8.

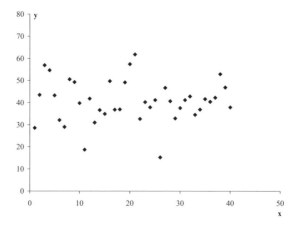

Fig. A.17   A case of correlation 0.

- In the first place, the concept of correlation is often confused with *causation*. When X and Y are correlated, it is tempting to conclude that X "causes" Y. This may be true, but it is knowledge of the phenomenon that allows to draw such a conclusion. Correlation, *per se*, does not measure anything but a *symmetric* association. In fact, the definition of covariance and correlation is symmetric: $S_{XY} = S_{XY}$. It may even be the case that there is a third variable, say Z, which is actually causing Y and is correlated with X; this *lurking variable* effect can lead us to an erroneous conclusion. As a well-known example, assume that X is the amount of spending in advertisements, Z is the amount of discount in promotional sales, and Y is demand. We might observe an increase in demand due to ads, leading us to conclude that advertisements are very useful. However, it *might* be the case that the real cause of the increase in demand is the reduction in price which is often associated to ads in order to boost sales.

- We have already pointed out that, in general, lack of correlation does not imply independence. When the relationship between X and Y is nonlinear, the coefficient of correlation could not reflect this link at all. An example is shown in figure A.18, where we may see that there is indeed a link between the two variables, but sample correlation is practically zero. This happens because when Y is larger than its mean, X can be larger or smaller than its mean (see also example A.11 on page 466). In the last part of this appendix we will outline some nonlinear transformations of data which *could* be useful to overcome, at least partially, this limitation.

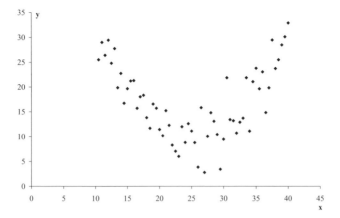

Fig. A.18   A nonlinear relationship between X and Y.

Table A.1   An example of negative correlation

i	1	2	3	4	5	6	7	8	9	10
x	9	6	10	2	7	3	2	5	6	2
y	7.9	7.5	13.6	16.8	8.1	15	9.8	12.8	15.8	17.7

- Finally, peculiar data can have a very large impact on the analysis. In fact, when we have a single observation (X_i, Y_i) which is quite far from the average, the terms $X_i - \overline{X}$ and/or $Y_i - \overline{Y}$ may be much larger, in absolute value, than the other terms in the sum. This issue is often called "King Kong" or "Big Apple" effect.[18] The following example shows the care that should be taken in presence of a peculiar observation.

Example A.19 Consider a sample consisting of ten observations of temperature and demand for an item, as displayed in table A.1. These data are depicted in figure A.19, which suggests a negative correlation; the sample correlation is indeed -0.47, and this would support the belief that temperature has a negative effect on demand for this item.

But now suppose that we include another observation, as in table A.2 and the corresponding figure A.20. Based on the sample correlation, which is now positive (0.85), conclusions could be quite different. Actually, this would be

---

[18]Statistics on towns in the USA may be affected by the inclusion of NY, which has peculiar characteristics.

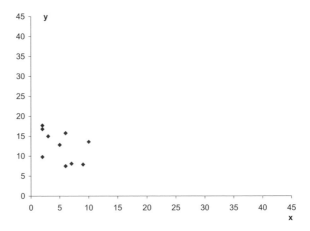

Fig. A.19 Pictorial representation of the data in table A.1; sample correlation is −0.47.

Table A.2   Example of positive correlation due to the King Kong effect

i	1	2	3	4	5	6	7	8	9	10	11
x	9	6	10	2	7	3	2	5	6	2	40
y	7.9	7.5	13.6	16.8	8.1	15	9.8	12.8	15.8	17.7	42.5

somewhat careless: What we can say is that in the one observation in which we had a very high temperature, we observed a large demand. However, we cannot really conclude that in the normal range, a larger temperature leads to an increase in demand.   ⬜

A.8.2   Confidence intervals

The sample mean is a point estimator for the expected value; since it is subject to some variance, it would be nice to have some measure of how much we can trust that single number. The same consideration applies to any estimator, and the typical additional information which is associated comes in the form of a **confidence interval**. Roughly speaking, a confidence interval is a range in which the true, unknown parameter should lie with some probability. This probability is known as **confidence level**; if the confidence level is $1 - \alpha$, where α is a relatively small value such as 0.01 or 0.05, then we can say that the confidence interval contains the "true" value with probability $1 - \alpha$. The following definition formalizes the definition of a confidence interval.

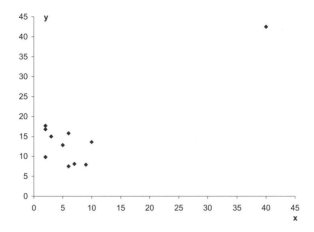

Fig. A.20 Pictorial representation of the data in table A.2; sample correlation is now 0.85.

DEFINITION A.15 *Let the confidence level be* $1 - \alpha$. *A confidence interval for a parameter* θ *is a pair of statistics* T_1 *and* T_2, *such that*[19]

$$P\{T_1 \leq \theta \leq T_2\} = 1 - \alpha.$$

This is a general definition of a confidence interval, which takes a more specific form depending on how we estimate the parameter and on possible additional assumptions on the distribution of samples.

If we are dealing with a normal sample, with unknown expected value and variance, which are estimated by sample mean and sample variance, we observed that the statistic

$$\frac{\overline{X} - \mu}{S/\sqrt{n}}$$

has t distribution with $n - 1$ degrees of freedom. Hence, if we denote by $t_{1-\alpha/2,n-1}$ the $(1 - \alpha/2)$-quantile of t distribution, we have

$$P\left\{-t_{1-\alpha/2,n-1} \leq \frac{\overline{X} - \mu}{S/\sqrt{n}} \leq t_{1-\alpha/2,n-1}\right\} = 1 - \alpha,$$

which can be rearranged as

$$P\left\{\overline{X} - t_{1-\alpha/2,n-1}\frac{S}{\sqrt{n}} \leq \mu \leq \overline{X} + t_{1-\alpha/2,n-1}\frac{S}{\sqrt{n}}\right\} = 1 - \alpha.$$

---

[19]In the statement we take for granted that $T_1 \leq T_2$. In general, an inequality involving random variables is potentially critical, and we should require that it holds with probability one.

We see that the interval

$$\left(\overline{X} - t_{1-\alpha/2,n-1}\frac{S}{\sqrt{n}}, \qquad \overline{X} + t_{1-\alpha/2,n-1}\frac{S}{\sqrt{n}}\right) \qquad (A.20)$$

is a confidence interval for the expected value μ, with confidence level $1 - \alpha$. It is essential to grasp the meaning of a confidence interval, which has a probabilistic interpretation. In fact, the confidence interval is the outcome of a random sampling experiment. All we can say is that if we repeat the random sampling a large number of times, the true value would fall within the confidence interval in a fraction $(1 - \alpha)\%$ of the experiments. There is a probability α that μ lies outside the interval, either to the left or to the right. We consider quantiles with probability $1 - \alpha/2$ because symmetry of Student's distribution implies that the probabilities associated to the right and left tails are equal.

Example A.20 Let us consider the random sample

$$\{43, \quad 79, \quad 26, \quad 137, \quad 45, \quad 55, \quad 93, \quad 52, \quad 46, \quad 17\},$$

under the assumption that it comes from a normal distribution, and let us compute a 95% confidence interval for the expected value. We have

$$n = 10, \qquad \overline{X} = 59.3, \qquad S \approx 35.2422,$$

and from statistical tables (or from a suitable piece of software) we may get

$$t_{1-\alpha/2,n-1} = t_{0.975,9} \approx 2.2622.$$

By straightforward application of (A.20), we obtain the confidence interval (34.0893, 84.5107). □

From a qualitative point of view, we can observe the following:

- The larger the confidence level $1 - \alpha$, the larger the confidence interval; in other words, a wider interval is required to be "almost sure" that it includes the true value.

- The interval is large when the underlying variability σ of the elements of the sample is large.

- The interval shrinks when we increase the number of samples.

- When the sample is very large, we may use the quantiles $z_{1-\alpha/2}$ from the standard normal distribution.

Given these observations, we could conclude that when we really need a tight interval, we must accept the cost of a large sample. Actually, this need not be always true. In stochastic simulation on a computer, we may sometimes exploit different sampling mechanisms to reduce the variance σ^2 of the elements in the sample. In other cases, there might be competing ways to estimate a parameter; it turns out that, sometimes, there is a biased estimator with lower variance. If the bias tends to disappear when the size of the sample increases, this estimator may be preferable.

Once again, we stress that all we have said for confidence intervals holds under the assumption of a normal sample, consisting of independent random variables. If the sample is not normal, then we may invoke the central limit theorem and say that the confidence interval could be a fairly accurate approximation for a large sample; otherwise, different forms of intervals should be devised, particularly if the underlying distribution is very skewed and the sample is small. The next example shows how the general theory can be adapted to cope with a specific case.

Example A.21 Say that we are interested in estimating the fraction of a population meeting a certain condition, e.g., they like a certain product. Of course, what we should do is take a sample, ask a question, and calculate the fraction of "yes" answers over the total. More formally, what we are doing is estimating the parameter p of a Bernoulli random variable (see section A.3.1). If we denote the size of the sample by n, we know that the number X of "yes" answers is a binomial variable, with expected value np and variance $np(1-p)$. For a suitably large sample, thanks to the central limit theorem, we may say that

$$\frac{X - np}{\sqrt{np(1 - p)}} \sim N(0, 1),$$

from which we may build an *approximate* confidence interval,

$$\hat{p} \pm z_{1-\alpha/2}\sqrt{\hat{p}(1 - \hat{p})/n},$$

where $\hat{p} = X/n$. In this case, we see a different way to estimate variance, since expected value and variance are related in a very specific way. However, we may still use much of what we know, at least when the sample is large.   ⬚

The following example shows that care is needed to ensure independence.

Example A.22 We use again example A.17, where a simple queuing system was considered. A typical problem in this field is determining the number of servers, i.e., clerks, in such a way to avoid long waiting times that lead to customer dissatisfaction. What we can do, among other things, is simulating the queuing system for different numbers of servers, in order to assess the tradeoff between system cost and service quality. A possible measure of service quality is the expected value of the waiting time. This can be estimated by a suitably long experiment, but how long exactly? We could simulate the process until n customers have been served, collect the waiting time W_k for each customer $k = 1, \ldots, n$, and use the formula (A.20) to check if the confidence interval is small enough. In doing so, we might make at least three mistakes at the same time.

1. To begin with, if we start our simulation with an empty system, we have a transient phase that may affect our statistics. We should wonder if we should discard the first data to avoid this issue. A similar problem must be dealt with when simulating inventory control systems, as the initial inventory may play a role, unless the simulation experiment is very long. In the shop case, this may be not an issue if the shop opens, say, at 9 a.m. and closes at 5 p.m., because in such a case reaching the steady state is not an issue.

2. A more general issue is that the waiting times are unlikely to be normally distributed, and the confidence interval will only be an approximation; as we have said, however, this is fairly good for a large sample.

3. Actually, the really serious mistake is that, as we have pointed out in example A.17, waiting time of successive customers are *not* independent random variables. What may happen is that sample variance underestimates the true variance, and the width of confidence interval is underestimated as well. The net result is that we are overconfident in our conclusions.

In practice, the way out of the last issue is the *batch* method. We simulate m samples, each one consisting of n customers, amounting to nm customers for the whole experiment, and we consider the m sample means

$$\overline{W}^j = \sum_{k=(j-1)n+1}^{jn} W_k, \qquad j = 1, \ldots, m.$$

Each sample mean $\overline{W}^j$ is, at least approximately, independent from the other ones, and we may apply the standard procedure on them. The good news is that they should have a rather small variance and, thanks to the central limit theorem, they are more or less normal, providing further justification for the approach. □

A.9 HYPOTHESIS TESTING

Parameter estimation can be used to address questions like "What is the average demand for a given item?" or "What is the average useful life of this product?" If a sample is available, we may also build a confidence interval to assess the reliability of the estimate. A different but related issue must be tackled when, for instance, a manufacturer claims that his product has an average life of $\mu = 100$ hours, and a skeptical customer does not trust his claim. One thing she could do is run a statistical experiment in order to have a check. Suppose that the estimated average life is 99 hours; can she sue the manufacturer? Not really, since the sample mean is a random variable, and it might well be the case that an "unlucky" sample leads to a result which is smaller than the true value. On the other hand, if the result were 101, we would be willing to trust the manufacturer, but we could be wrong as well. However, if the sample mean were 50 hours, we could be somewhat suspicious about the truth of the claim, because such a large discrepancy with respect to the claim is hardly attributable to randomness in the sample.

When facing such an issue, we run a twofold risk: On the one hand, we could reject a claim which is in fact true; on the other one, we could accept a claim which is false. Hence, we need a sound theoretical basis to make well-informed decisions. We illustrate here a procedure for **hypothesis testing**. In the literature, the term *significance testing* is also used, e.g., when we want to check if the sample correlation coefficient is significant. Here we mainly deal with hypotheses about the mean of a normally distributed population, but we will also outline different tests. Formally, we postulate a **null hypothesis** H_0, which is tested against an alternative hypothesis H_1. The overall idea is analyzing the properties of a certain statistic under the assumption that the null hypothesis is true. In our case, we use properties of the sample mean assuming that the random sample comes from a normal population with expected value $\mu = \mu_0$, where μ_0 comes from the null hypothesis. Then, by checking the sample mean, we see if it is consistent with H_0; if it looks severely inconsistent, we reject the null hypothesis, keeping under some control the probability of rejecting H_0 when, in fact, it is true (i.e., the probability of a type I error). We do not consider here the risk of accepting a false hypothesis (a type II error).

To formalize the problem in a more general setting, we consider a population which is distributed according to a density (or cumulative distribution) F_θ, depending on an unknown parameter θ, which in our case is the expected value, but it need not be in general. We formulate a null hypothesis, denoted by H_0, such as $H_0 : \theta = 1$, or $H_0 : \theta \leq 1$. Then, we take a random sample of size n from the population, denoted by $(X_1, X_2, \ldots, X_n)$. To ascertain if the sample is "compatible" with the null hypothesis, we build a region $\mathbf{C} \in \mathbb{R}^n$, called **rejection region**, according to a suitable criterion; then we accept H_0 if $(X_1, X_2, \ldots, X_n) \notin \mathbf{C}$, or we reject it if $(X_1, X_2, \ldots, X_n) \in \mathbf{C}$. We build the rejection region in such a way that we have a small probability α of rejecting a true hypothesis; in other words, if H_0 is indeed true, the sample may happen to fall in the rejection region, but it is unlikely. The number α is called **significance level**.

To be more concrete, let us consider testing a hypothesis about the expected value μ of a normally distributed population, assuming that the variance σ^2 is unknown as well. Say that the null hypothesis is

$$H_0 : \mu = \mu_0,$$

which we test against the alternative

$$H_1 : \mu \neq \mu_0.$$

Table A.3 Sample data for hypothesis testing

5.00	1.82	15.95	−13.74	9.28	13.96	12.31	10.78	5.40	11.77

Intuitively, we should reject H_0 if the sample mean $\overline{X}$ is "far" from μ_0 (both larger or smaller). To make "far" clearer, we build the test statistic

$$\text{TS} = \frac{\sqrt{n}(\overline{X} - \mu_0)}{S},$$

where S is the sample standard deviation. From section A.8 we know that *if the null hypothesis is true* and if the population is actually normal, the statistic TS has t distribution with $n - 1$ degrees of freedom. Hence,

$$P_{\mu_0}\left\{-t_{1-\alpha/2,n-1} \leq \frac{\sqrt{n}(\overline{X} - \mu_0)}{S} \leq t_{1-\alpha/2,n-1}\right\} = 1 - \alpha,$$

where we use the notation P_{μ_0} to emphasize that we compute this probability under the probability measure assumed in H_0. Wrapping everything up, the procedure prescribes the following, for a given significance level α:

$$\text{accept } H_0 \text{ if} \qquad \left|\frac{\sqrt{n}(\overline{X} - \mu_0)}{S}\right| \leq t_{1-\alpha/2,n-1};$$

$$\text{reject } H_0 \text{ if} \qquad \left|\frac{\sqrt{n}(\overline{X} - \mu_0)}{S}\right| > t_{1-\alpha/2,n-1}.$$

Example A.23 Consider the data listed in table A.3. Actually, these data have been obtained by running a generator of pseudorandom variates, and they are samples from a normal distribution with expected value 5 and standard deviation 10. Now, suppose we forget about what we know, and let us test the hypothesis that $\mu = 5$, with a significance level $\alpha = 0.1$. To begin with, we compute the sample statistics:

$$n = 10, \qquad \overline{X} = 7.253, \qquad S = 8.5757.$$

Note that the sample mean looks rather large with respect to $\mu_0 = 5$, but this intuitive feeling must be carefully checked, taking the large variability and the limited sample size into account. The test statistic is

$$\text{TS} = \frac{\sqrt{10}(7.253 - 5)}{8.5757} = 0.8308,$$

and this should be compared with the quantile $t_{1-\alpha/2,n-1} = t_{0.95,9} = 1.8331$. Since TS < 1.8331, we *cannot* reject the hypothesis with that significance level. ▯

We may notice some similarity with confidence intervals. Indeed, in the case above we might argue that we reject the hypothesis if μ_0 does not fall within the confidence interval. However, the underlying thinking is a bit different, at least in principle. Furthermore, we should use a different rejection region if the hypothesis is different, leading to a *one-sided*, rather than *two-sided*, test. Consider, for instance, the following null hypothesis

$$H_0 : \mu \leq \mu_0$$

against the alternative

$$H_1 : \mu > \mu_0.$$

In this case we build the test under the assumption $\mu = \mu_0$, but the rejection region is only one of the two tails:

$$\text{accept } H_0 \text{ if} \quad \frac{\sqrt{n}(\overline{X} - \mu_0)}{S} \leq t_{1-\alpha,n-1};$$

$$\text{reject } H_0 \text{ if} \quad \frac{\sqrt{n}(\overline{X} - \mu_0)}{S} > t_{1-\alpha,n-1}.$$

Intuitively, in this case we reject H_0 if the sample mean is suspiciously large. Also note the use of a quantile with probability level $(1-\alpha)$ instead of $(1-\alpha/2)$. We trust that the reader will now find the symmetric case rather easy to figure out. What may be not so easy to figure out is, Which is the appropriate hypothesis to use when tackling a real-life business problem? Sometimes the answer is obvious. For instance, if the unknown parameter is the average life of an item, we should not complain if this is larger than claimed. Hence, we may argue that in such a case we should test something like $H_0 : \mu \geq \mu_0$, complaining with the manufacturer only if the test statistic gets stranded on the left tail. In other cases, the answer might not be that obvious.

Another tricky point is finding a suitable value of α. Note that the larger the value of α, the easier it is to reject the null hypothesis. This happens because the rejection region increases with α. So, we could find a case in which we accept (or, better said, we cannot reject) the null hypothesis if $\alpha = 0.05$, but we reject it if we set $\alpha = 0.06$. This is clearly a critical situation, because the right confidence level is nowhere engraved on a rock. A useful concept from this point of view is the *p*-value. In a two-sided test, the *p*-value is given by the probability that a variable T_{n-1}, i.e., following a t distribution with $n-1$ degrees of freedom, is in absolute value larger than the value of the statistic TS:

$$p = 2 \cdot P\{T_{n-1} \geq |t|\},$$

if TS $= t$. In practice, the *p*-value is a "critical" significance level, in the sense that the hypothesis would be accepted for all significance levels smaller than p.

Example A.24 In example A.23, we could not reject the null hypothesis with a significance level $\alpha = 0.1$. With a different significance level, we could

Table A.4   Hypothesis testing about the mean of a normal population, when variance is unknown: TS is the test statistic and α is the significance level

H_0	H_1	TS	Test with Level α	p-Value if TS $= t$
$\mu = \mu_0$	$\mu \neq \mu_0$	$\dfrac{\sqrt{n}(\overline{X} - \mu_0)}{S}$	reject if $|\text{TS}| > t_{1-\alpha/2, n-1}$	$2\text{P}\{T_{n-1} \geq |t|\}$
$\mu \leq \mu_0$	$\mu > \mu_0$	$\dfrac{\sqrt{n}(\overline{X} - \mu_0)}{S}$	reject if $\text{TS} > t_{1-\alpha, n-1}$	$\text{P}\{T_{n-1} \geq t\}$
$\mu \geq \mu_0$	$\mu < \mu_0$	$\dfrac{\sqrt{n}(\overline{X} - \mu_0)}{S}$	reject if $\text{TS} < -t_{1-\alpha, n-1}$	$\text{P}\{T_{n-1} \leq t\}$

reach a different conclusion. For instance, if $\alpha = 0.5$, we must use the quantile $t_{0.75,9} = 0.7027$, resulting in a rejection. However, we have a 50% probability of rejecting a true hypothesis, because the rejection region is large. To spot the "critical" significance level, we may compute the p-value:

$$p = 2 \cdot \text{P}\{T_9 \geq 0.8308\} = 2 \cdot 0.2138 = 0.4276.$$

We see that we may reject the hypothesis only if we accept at least a 42.76% probability of a type I error. ⬜

The p-value for one-sided tests is found using a similar logic, where only one tail of the Student distribution is considered. Table A.4 summarizes what we have discussed. If the variance were known, the reasoning is again the same, but we should use the quantiles from the standard normal distribution.

A.9.1   An example of a nonparametric test: the chi-square test

In the main body of the book, we often assume that the demand for a certain item is normally distributed. However, this should not be taken for granted, and the claim should be tested in some way. When we test if experimental data fit a given probability distribution, we are not really testing a hypothesis about a parameter or two; in fact, we are running a nonparametric test. The chi-square test is one example of such a test. The idea is fairly intuitive, although the technicalities may require some care. We could divide the range of realizations in J disjoint intervals, and compute the probability that a random variable distributed according to that distribution falls in each interval. Then, we may calculate the number E_j of observations (out of n) that should fall in interval j, $j = 1, \ldots, J$, if the assumed distribution is indeed the true one. This number should be compared against the number O_j of observations which actually fall in interval j; a large discrepancy would suggest that the hypothesis about the underlying distribution should be rejected.

Like any statistical test, the chi-square test relies on a distributional property of a statistic. It can be shown that for a large number of samples, the statistic

$$\chi^2 = \sum_{j=1}^{J} \frac{(O_j - E_j)^2}{E_j}$$

has (approximately) a chi-square distribution. But what about the degrees of freedom? This is the tricky part of the procedure, as this depends on the number of parameters of the distribution that we have estimated using the data. If no parameter has been estimated, the degrees of freedom are $J - 1$. Otherwise, the distributional results are more complicated. Whatever the case, the intuitive idea of the test is that if $\chi^2 > \chi^2_{k,\alpha}$, where α is the level of significance and k is a suitable number of degrees of freedom, the hypothesis should be rejected. Indeed, if the hypothesis were true, χ^2 should be close to zero; a large value leads to rejection of the hypothesis.

A.9.2   Testing hypotheses about the difference in the mean of two populations

Sometimes, we have to run a test concerning two (or more) populations. For instance, we could wonder if two markets for a given product are really different in terms of expected demand. Alternatively, after the reengineering of our business processes, we could wonder if the performances are significantly different. In both cases, the rationalization of the problem calls for assessing the difference between two expected values, $\mu_1 - \mu_2$, where μ_1 and μ_2 are the expected values of two random variables. We consider here how we can build a confidence interval for this difference; running a test of hypothesis requires a fairly straightforward adaptation. What we should do exactly depends on a number of issues:

- Is the number of samples, from both populations, large or small?

- Are the two variances known? If they are not, can we assume that they are equal?

- Are the samples from the two populations independent?

- Are the two populations normal?

Depending on the answers to these questions, we may exploit certain distributional results about the statistic

$$\overline{X}_1 - \overline{X}_2, \tag{A.21}$$

i.e., the difference between the two sample means. Let n_1 and n_2 be the two sample sizes, respectively.

If the two samples are both large and mutually independent, the statistic (A.21) is, at least approximately, normally distributed. Furthermore, independence allows to estimate the standard deviation of the difference by

$$S_{\overline{X}_1 - \overline{X}_2} = \sqrt{\frac{S_1^2}{n_1} + \frac{S_2^2}{n_2}},$$

where S_1^2 and S_2^2 are the two sample variances. Then, the following confidence interval can be built:

$$(\overline{X}_1 - \overline{X}_2) \pm z_{1-\alpha/2} S_{\overline{X}_1 - \overline{X}_2}.$$

Based on these estimates, it is also easy to test if the two populations are significantly different (in this case, the test boils down to checking if the origin lies within the confidence interval).

With small samples (say $n_1, n_2 < 30$), the procedure is not as simple. A relatively easy case is when we may assume that the two variances in the two populations are the same. To estimate the standard deviation, we may pool the observations,

$$S_p = \sqrt{\frac{(n_1 - 1)S_1^2 + (n_2 - 1)S_2^2}{n_1 + n_2 - 2}},$$

and use

$$S_{\overline{X}_1 - \overline{X}_2} = S_p \sqrt{\frac{1}{n_1} + \frac{1}{n_2}}$$

to build a confidence interval

$$(\overline{X}_1 - \overline{X}_2) \pm t_{n_1 + n_2 - 2, 1 - \alpha/2} \cdot S_{\overline{X}_1 - \overline{X}_2}.$$

We see that we are relying on Student's t distribution; we know that, strictly speaking, this requires that the two populations are normal. Also note the number of degrees of freedom. If the two variances are different, we may also use the same distribution, but we must estimate the degrees of freedom. A (nontrivial) distributional result justifies the following estimate:

$$\hat{f} = \frac{\left(\dfrac{S_1^2}{N_1} + \dfrac{S_2^2}{N_2} \right)^2}{\dfrac{1}{N_1 - 1} \left(\dfrac{S_1^2}{N_1} \right)^2 + \dfrac{1}{N_2 - 1} \left(\dfrac{S_2^2}{N_2} \right)^2}.$$

Since $\hat{f}$ need not be an integer, we may round it down (which makes sense because with fewer degrees of freedom the confidence interval is larger and more conservative) and build the confidence interval

$$(\overline{X}_1 - \overline{X}_2) \pm t_{\hat{f}, 1 - \alpha/2} \sqrt{\frac{S_1^2}{n_1} + \frac{S_2^2}{n_2}}.$$

All of the above procedures rely on the independence between the two populations. Now assume, on the contrary, that the samples are strictly related. Such a case occurs when the observations actually *paired*. For instance, if we draw random demand scenarios and we evaluate the performance of two management policies, we have paired samples $X_k^{(1)}$ and $X_k^{(2)}$, which are the performances of policies 1 and 2, respectively, in scenario k. The case of paired samples requires working directly on the differences

$$D_k \equiv X_k^{(1)} - X_k^{(2)},$$

computing the statistics

$$\overline{D} = \frac{1}{n} \sum_{k=1}^{n} D_k,$$

$$S_D = \sqrt{\frac{1}{n-1} \left(\sum_{k=1}^{n} D_k^2 - n\overline{D}^2 \right)},$$

and building the confidence interval

$$\overline{D} \pm t_{n-1,1-\alpha/2} \frac{S_D}{\sqrt{n}}.$$

A.10   SIMPLE LINEAR REGRESSION

We considered the sample correlation coefficient in section A.8.1 as a way to assess the possible role of a variable X in explaining the dynamics governing a phenomenon measured by another variable Y. However, from a forecasting point of view, this is not enough, as we would also like to predict which value of Y we may expect corresponding to a certain level of X_0 the independent variable X. In order to do so, we must come up with an explicit link, in the form of a functional dependence, between Y (e.g., demand) and the explanatory variable X (e.g., outside temperature). The simplest tool to analyze such links is **simple linear regression**, which assumes a functional relationship such as

$$Y = a + bX. \tag{A.22}$$

We speak of "regression" because we try to identify suitable values a and b in such a way that the model is consistent with a set of empirical observations. It is "linear" because we are using a linear[20] function to model the relationship between variables; we should recall that correlation captures linear associations, but it may fail to point out nonlinear associations. It is

---

[20] Well, we should really say *affine*.

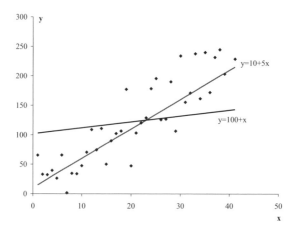

Fig. A.21   Regression lines with different levels of "sample fit."

"simple" because there is only one explanatory variable; we may (and typically should) also build multiple regression models in which more factors are
used to explain the values assumed by a variable of interest.

In the rest of this appendix we illustrate simple linear regression in some
detail. Multiple and nonlinear regression will not be considered in detail: We
just describe a simple way to use linear regression to estimate a nonlinear
relationship in section A.10.6, and we outline multiple linear regression in
web section W.A.11. We also emphasize that the mathematical tools we will
use, *per se*, do not assign a precise role to variables: The model captures
an association, and *we* interpret X as the cause and Y as the effect; but we
could switch these two roles, and the mathematics would just be the same.
Regression is a very useful but dangerous tool, as by playing with numbers
one may build models which have really no meaning at all. Hence, we will try
to point out all the pitfalls of simple-minded linear regression.

The starting point of linear regression is a set of empirical observations, i.e.,
pairs (X_i, Y_i). In general, if we have three or more points, it will be impossible
to find a pair of coefficients a and b ensuring a perfect fit by a linear function.[21]
What we can do, for instance, is to find the "best" coefficients a and b in such a
way that the theoretical model, the regression line, is as close as possible to the
empirical data. For instance, a look at figure A.21 shows a set of data which
is more consistent with line $Y = 10 + 5X$ than line $Y = 100 + X$. To make
the idea more concrete, we must explicitly specify a distance measure between

---

[21] Of course, we might consider the idea of assuming a complicated functional form, with
a lot of coefficients which ensure enough degrees of freedom to get an almost perfect fit.
Unfortunately, this "overfitting" process is sensitive to noise, among other things, and it is
rarely advisable.

the prediction we get from the theoretical model, $Y_i = a + bX_i$, and the n empirical observations of X ed Y; we will use the sum of squared deviations as a measure, leading to the least squares method. In the next section we do so without referring to statistical concepts at all, as the approach can be cast within the framework of function approximation, which is an important branch of Numerical Analysis. Then, we will introduce statistical concepts, which are needed to evaluate the reliability of the predictions we obtain from a linear regression model.

A.10.1 Best fitting by least squares

Say that we have a set of n points (x_i, y_i), $i = 1, \ldots, n$. These points can be the result of empirical measures or simulation experiments, but for now we treat them as *numbers*. Please note the use of lowercase letters; we are not (yet) considering our data points as the outcome of random sampling. We postulate a functional form, say $y = f(x)$, and we look for the function which yields the best approximation of the given data, within some class of functions and with respect to a given criterion. In the case of linear regression, we consider the class of affine functions like

$$y = f(x) = a + bx,$$

and we look for the "optimal" pair of coefficients a and b. In general, a perfect fit is impossible to obtain with a reasonably simple model, and we will have some deviation between the theoretical prediction and the empirical data. We define a **residual** e_i as

$$e_i = y_i - f(x_i) = y_i - (a + bx_i). \tag{A.23}$$

We should aggregate the n residuals in order to come up with a single number playing the role of a distance. There are different and sensible ways for doing so, but the most common one is by summing the *squared* residuals:

$$\mathrm{SS} \equiv \sum_{i=1}^{n} e_i^2 = \sum_{i=1}^{n} (y_i - a - bx_i)^2, \tag{A.24}$$

where SS stands for Sum of Squared residuals. Then, the approximation problem boils down to an optimization problem requiring the minimization of SS with respect to a and b. Of course, we square the residuals in order to avoid cancelation between positive and negative values, but we could also take absolute values. One reason to prefer squared residuals is that they lead to an analytical solution, whereas absolute deviations call for a numerical solution by linear programming. An analytical solution allows for an easier interpretation, and it paves the way for the application of statistical tools. Summing squared residuals is equivalent to taking an average, as dividing the objective function by n does not change the solution of the optimization problem. We could also consider a *worst-case* error, leading to a min–max optimization problem; this is more appropriate in other applications, but it calls again for a numerical solution.

To find the optimal values of a and b, we just need to enforce the first-order optimality conditions.[22] The first condition, with respect to a, is

$$
\begin{aligned}
\frac{\partial SS}{\partial a} &= -\sum_{i=1}^{n} 2\left(y_i - a - bx_i\right) \\
&= -2\left(\sum_{i=1}^{n} y_i - \sum_{i=1}^{n} a - \sum_{i=1}^{n} bx_i\right) \\
&= -2\left(\sum_{i=1}^{n} y_i - na - b\sum_{i=1}^{n} x_i\right) = 0,
\end{aligned}
$$

which yields

$$
a^* = \frac{1}{n}\sum_{i=1}^{n} y_i - b\frac{1}{n}\sum_{i=1}^{n} x_i = \bar{y} - b\bar{x}, \tag{A.25}
$$

where $\bar{x}$ and $\bar{y}$ are the average values of x and y; *formally*, they are similar to sample means. This condition, by the way, tells us that the barycenter $(\bar{x}, \bar{y})$ of the experimental data lies on the regression line, which does make sense. The optimality condition with respect to b reads

$$
\frac{\partial SS}{\partial b} = -2\sum_{i=1}^{n} x_i\left(y_i - a - bx_i\right) = 0.
$$

We can plug the optimal value a^* into this condition:

$$
\begin{aligned}
\frac{\partial SS}{\partial b} &= -2\sum_{i=1}^{n} x_i\left[y_i - \left(\frac{1}{n}\sum_{j=1}^{n} y_j - b\frac{1}{n}\sum_{j=1}^{n} x_j\right) - bx_i\right] \\
&= -\frac{2}{n}\left(n\sum_{i=1}^{n} x_i y_i - \sum_{i=1}^{n} x_i \cdot \sum_{i=1}^{n} y_i + b\sum_{i=1}^{n} x_i \cdot \sum_{i=1}^{n} x_i - nb\sum_{i=1}^{n} x_i^2\right) = 0.
\end{aligned}
$$

Rearranging this condition, we get

$$
b^* = \frac{n\sum_{i=1}^{n} x_i y_i - \sum_{i=1}^{n} x_i \cdot \sum_{i=1}^{n} y_i}{n\sum_{i=1}^{n} x_i^2 - \left(\sum_{i=1}^{n} x_i\right)^2} = \frac{\sum_{i=1}^{n} x_i y_i - n\bar{x}\bar{y}}{\sum_{i=1}^{n} x_i^2 - n\bar{x}^2}. \tag{A.26}
$$

---

[22]They are sufficient conditions, as the objective function is convex with respect to the decision variables; see appendix B.

The second rewriting is easily obtained by dividing both numerator and denominator by n, and it may be easier to remember. As we anticipated, we find explicit expressions for a^* and b^*, which can be interpreted intuitively to improve our understanding of the results. To ease the notation, we will drop the asterisk * and denote the optimal value of coefficients by a and b.

We begin by rewriting the formula (A.26). Using the same trick, i.e., by dividing both numerator and denominator by n, we may see that

$$b = \frac{\sum\limits_{i=1}^{n} x_i y_i - \sum\limits_{i=1}^{n} x_i \bar{y}}{\sum\limits_{i=1}^{n} x_i^2 - \sum\limits_{i=1}^{n} x_i \bar{x}} = \frac{\sum\limits_{i=1}^{n} x_i (y_i - \bar{y})}{\sum\limits_{i=1}^{n} x_i (x_i - \bar{x})}.$$

Now, we can use the rather obvious identities

$$\sum_{i=1}^{n} (x_i - \bar{x}) \equiv 0 \qquad \text{and} \qquad \sum_{i=1}^{n} (x_i - \bar{x}) \equiv 0$$

to subtract a zero term from both the numerator and denominator of the fraction and to rearrange:

$$\begin{aligned} b &= \frac{\sum\limits_{i=1}^{n} x_i (y_i - \bar{y}) - \sum\limits_{i=1}^{n} \bar{x} (y_i - \bar{y})}{\sum\limits_{i=1}^{n} x_i (x_i - \bar{x}) - \sum\limits_{i=1}^{n} \bar{x} (x_i - \bar{x})} \\[2mm] &= \frac{\sum\limits_{i=1}^{n} (x_i - \bar{x})(y_i - \bar{y})}{\sum\limits_{i=1}^{n} (x_i - \bar{x})(x_i - \bar{x})} = \frac{\frac{1}{n-1} \sum\limits_{i=1}^{n} (x_i - \bar{x})(y_i - \bar{y})}{\frac{1}{n-1} \sum\limits_{i=1}^{n} (x_i - \bar{x})^2} \\[2mm] &= \frac{S_{xy}}{S_x^2} = \frac{r_{xy} S_x S_y}{S_x^2} = \frac{r_{xy} S_y}{S_x}. \end{aligned} \qquad (A.27)$$

In the last line, we have used the notations S_x, S_y, S_{xy}, and r_{xy}. Formally, these quantities are similar to sample standard deviations, covariances, and correlation coefficients, but the interpretation is different here as we are dealing with numbers and not with random samples; still, the notation is quite handy and tells us a lot. When we cast regression within a statistical modeling framework, we will point out what should be considered as a random variable by using uppercase letters when appropriate.

This way of writing the regression coefficient b suggests an interesting geometrical interpretation. In fact, it plays the role of an incremental ratio

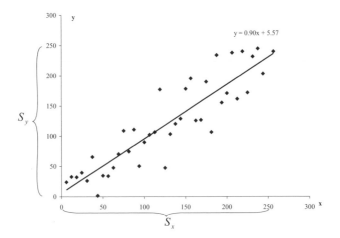

Fig. A.22   Geometrical interpretation of coefficient b (case 1).

between y and x. In elementary analytical geometry, the slope of a line passing through points $P_1(x_1; y_1)$ and $P_2(x_2; y_2)$ is

$$b = \frac{y_2 - y_1}{x_2 - x_1} = \frac{\Delta_y}{\Delta_x}. \tag{A.28}$$

In linear regression, the terms Δ_y and Δ_x are replaced by the terms S_y and S_x, which basically measure the observed variability in y and x. Moreover, the "correlation coefficient" tells us if, for increasing values of x, the values of y tend to increase (positive correlation), to decrease (negative correlation), or are not affected at all (no correlation). Clearly, we have a positive slope in the first case, a negative slope in the second case, and a horizontal line in the last case. The last case does not necessarily imply that there is no link between the two variables: We should recall here figure A.18 on page 489; correlation is just a measure of linear association, and in that case linear regression will not pick up the link between the two variables.

To get a feeling for equation (A.28), it is also useful to fix a correlation coefficient and to see how b changes when the ratio between S_y and S_y varies. In figures A.22, A.23, and A.24 we show three cases with high correlation, namely 0.9; however, in the first case, S_y is definitely larger than S_x, they are practically equal in the second one, and S_x is larger than S_y in the last case.

A.10.2   Analyzing properties of regression estimators

So far, we have dealt with linear regression as an approximation problem, without any reference to statistical concepts. However, when we plug a value x_0 in the regression line to obtain a forecast, we would like to see something like a confidence interval. Before doing so, we should also check if the re-

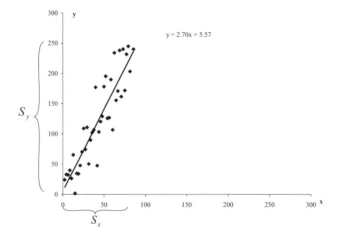

Fig. A.23 Geometrical interpretation of coefficient b (case 2).

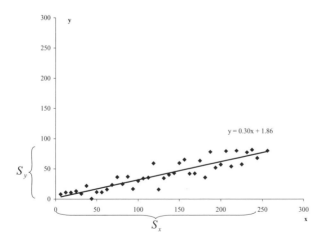

Fig. A.24 Geometrical interpretation of coefficient b (case 3).

gression has some significance and if the coefficients a and b can be trusted. In order to do so, we must make some assumptions about the way data are generated. The statistical model we will consider is the following:

$$Y_i = \alpha + \beta x_i + \epsilon_i, \qquad i = 1, \ldots, n, \tag{A.29}$$

where α and β are unknown parameters we have to estimate, and the ϵ_i are *random variables* meeting the following conditions:

- All the variables ϵ_i are mutually independent and identically distributed with expected value zero and standard deviation $\sigma_i = \sigma_\epsilon$, for all i.

- Each random variable ϵ_i is also independent from x_i, which is considered as a *number*.

The last remark deserves some comment. We consider the variable as a number, which makes sense if it represents a quantity, such as price, which is under the control of a decision maker. It is also possible to build regression models in which random variables X_i are considered. This second approach makes sense when X is not under our control, as in the case of outside temperature, or when we measure a quantity x subject to some measurement error. We will deal with the easier case for the sake of simplicity, but the basic results are the same, if the random variables ϵ_i and X_i are independent. Whatever the choice, Y_i is definitely a random variable and, given the assumptions above,

$$E[Y_i] = \alpha + \beta x_i, \qquad \text{Var}(Y_i) = \sigma_\epsilon^2$$

for some value of x_i. Basically, random variables ϵ_i play the role of an **error**. Errors, in this model, are a sort of catch-all for what is intrinsically random or what we are not able (or willing) to take into account.[23] The assumptions above on errors have the basic meaning that we are not missing some significant pattern with our very simple model.

From a notational point of view, it would be advisable to denote the errors by $\tilde{\epsilon}_i$ in order to separate the random variable from its realization. We refrain from doing so to avoid heavy notation. What is really important is that, in this context, Greek letters are associated what *what we don't know* and *what we cannot observe*. We assume that the "true" data generating process is given as in equation (A.29), but we ignore the values of the parameters α and β, which are unknown numbers. When we take a random sample, consisting of pairs of observations (x_i, Y_i), we are implicitly sampling ϵ_i, but the realizations of the errors are not directly observable. What we can do is build *estimators* a and b, which are random variables, for the unknown parameters α and β. Again, we are departing from our usual notation and use lowercase a and b

---

[23] In Physics, there has been a very heated debate between the likes of Einstein and Bohr about randomness. Is it an intrinsic feature of Nature, or is it the result of our lack of knowledge?

to denote random variables, just to avoid unnecessary burden. Given values of the estimators, we may evaluate the **residuals**

$$e_i = Y_i - (a + bx_i).$$

To summarize, we use Roman letters (a, b, e_i) to refer to what we may estimate and observe and ultimately use for forecasting; Greek letters $\alpha, \beta, \epsilon_i$ refer to unknown numbers and unobservable random variables.

Now the natural questions to ask concern the quality of a and b as estimators of α and β. We should check bias first, i.e., we verify that $E[b] = \beta$ and $E[a] = \alpha$, given the assumptions above about the data generating process. Then, we should investigate their variability and their effects on the predictive ability of our model. To this aim, we refer again to formulas (A.25) and (A.26), where we plug random variables Y_i where numbers y_i occur.

Biasedness The first issue we tackle is whether $E[b] = \beta$ or not. The first step is rewriting the formula (A.27) in our context, where a random variable Y_i replaces the number y_i:

$$
\begin{aligned}
b &= \frac{S_{xY}}{S_x^2} = \frac{\sum_{i=1}^{n} (x_i - \bar{x})(Y_i - \overline{Y})}{\sum_{i=1}^{n} (x_i - \bar{x})^2} \\[2em]
&= \frac{\sum_{i=1}^{n} (x_i - \bar{x}) \cdot [\alpha + \beta x_i + \epsilon_i - (\alpha + \beta \bar{x} + \bar{\epsilon})]}{\sum_{i=1}^{n} (x_i - \bar{x})^2} \\[2em]
&= \frac{\sum_{i=1}^{n} (x_i - \bar{x}) \cdot [\beta(x_i - \bar{x}) + (\epsilon_i - \bar{\epsilon})]}{\sum_{i=1}^{n} (x_i - \bar{x})^2} \\[2em]
&= \beta + \frac{\sum_{i=1}^{n} (x_i - \bar{x})(\epsilon_i - \bar{\epsilon})}{\sum_{i=1}^{n} (x_i - \bar{x})^2}.
\end{aligned}
\tag{A.30}
$$

Here $\bar{x}$ is the average of the x_i, and $\bar{\epsilon}$ is the sample mean of the errors, over the n observations. We see that b is given by the sum of β and a random term depending on the errors ϵ_i. We should essentially prove that the expected value of this random term is zero:

$$
\mathrm{E}\left[b\right] = \mathrm{E}\left[\beta + \frac{\sum_{i=1}^{n}(x_i - \bar{x})(\epsilon_i - \bar{\epsilon})}{\sum_{i=1}^{n}(x_i - \bar{x})^2}\right] = \beta + \mathrm{E}\left[\frac{\sum_{i=1}^{n}(x_i - \bar{x})(\epsilon_i - \bar{\epsilon})}{\sum_{i=1}^{n}(x_i - \bar{x})^2}\right]
$$

$$
= \beta + \frac{\sum_{i=1}^{n}(x_i - \bar{x}) \cdot \mathrm{E}\left[\epsilon_i - \bar{\epsilon}\right]}{\sum_{i=1}^{n}(x_i - \bar{x})^2} = \beta + 0.
$$

In the manipulations above, we have used the fact that β and x_i are numbers and can be taken outside the expectation; then we rely on the assumption that the expected value of the errors is zero, as well as the expected value of their sample mean.

The same line of reasoning can be adopted to prove the desired property of a. We rewrite (A.25) in the assumed context:

$$
a = \bar{Y} - b\bar{x} = \frac{1}{n}\sum_{i=1}^{n}Y_i - \frac{b}{n}\sum_{i=1}^{n}x_i
$$

$$
= \frac{1}{n}\sum_{i=1}^{n}(\alpha + \beta x_i + \epsilon_i) - \frac{b}{n}\sum_{i=1}^{n}x_i
$$

$$
= \alpha + \frac{1}{n}\sum_{i=1}^{n}(\beta - b)\,x_i + \frac{1}{n}\sum_{i=1}^{n}\epsilon_i.
$$

Now we must take the expected value of both the terms above. We note that a is given by the sum of three terms. The first one is a number and can be taken outside the expectation; in the second term, the x_i are numbers as well, and we may use the just proven fact that $\mathrm{E}[b] = \beta$; finally we have the sum of errors, whose expected value is zero. Wrapping everything up, we have

$$
\mathrm{E}\left[a\right] = \mathrm{E}\left[\alpha + \frac{1}{n}\sum_{i=1}^{n}(\beta - b)\,x_i + \frac{1}{n}\sum_{i=1}^{n}\epsilon_i\right]
$$

$$
= \alpha + \frac{1}{n}\sum_{i=1}^{n}\mathrm{E}\left[\beta - b\right]x_i + \frac{1}{n}\sum_{i=1}^{n}\mathrm{E}\left[\epsilon_i\right] = \alpha.
$$

Estimation errors   Now we have some guarantee about the quality of the estimators we build by the linear regression approach: Provided that the assumptions about the data generating process are true, the estimators are unbiased. This is good news, but we should also have an idea about their variability. A useful way to frame the issue is considering the **standard error of estimate**, denoted by See and defined as

$$\text{See}_b \equiv \sqrt{\text{E}\left[(b-\beta)^2\right]}, \qquad \text{See}_a \equiv \sqrt{\text{E}\left[(a-\alpha)^2\right]}.$$

Recalling that $\text{E}[Z^2] = \text{Var}(Z) + \text{E}^2[Z]$, for any random variable Z, and that our estimators are unbiased, we also see that

$$\text{See}_b = \sqrt{\text{Var}(b-\beta) + \text{E}^2[b-\beta]} = \sqrt{\text{Var}(b-\beta)}.$$

A similar relationship holds for See_a, a, and α.

From the (ideal) point of view of someone endowed with knowledge of the parameters α and β, See measures the dispersion of the estimator around its expected value. From our (real) point of view, See is a valuable tool in going beyond a point estimator; we can build confidence intervals, test hypotheses, and even try building a probability distribution for a predicted outcome. The bottom line is that if See_a and See_b are small enough, then we could start consider the possibility of trusting our model; otherwise, great care must be taken in taking business decisions.

To quantify the estimation error, we start with the parameter β. Recalling equation (A.30),

$$b = \beta + \frac{\displaystyle\sum_{i=1}^{n}(x_i - \bar{x})(\epsilon_i - \bar{\epsilon})}{\displaystyle\sum_{i=1}^{n}(x_i - \bar{x})^2} = \beta + \frac{\displaystyle\sum_{i=1}^{n}(x_i - \bar{x})\,\epsilon_i}{\displaystyle\sum_{i=1}^{n}(x_i - \bar{x})^2},$$

we may proceed as follows:

$$\text{See}_b = \sqrt{\text{Var}(b-\beta)} = \sqrt{\text{Var}\left[\frac{\displaystyle\sum_{i=1}^{n}(x_i - \bar{x})\,\epsilon_i}{\displaystyle\sum_{i=1}^{n}(x_i - \bar{x})^2}\right]}$$

$$= \sqrt{\frac{\text{Var}\left[\displaystyle\sum_{i=1}^{n}(x_i - \bar{x})\,\epsilon_i\right]}{\left(\displaystyle\sum_{i=1}^{n}(x_i - \bar{x})^2\right)^2}} = \sqrt{\frac{\displaystyle\sum_{i=1}^{n}\text{Var}\left[(x_i - \bar{x})\,\epsilon_i\right]}{\left(\displaystyle\sum_{i=1}^{n}(x_i - \bar{x})^2\right)^2}}$$

$$= \sqrt{\frac{\displaystyle\sum_{i=1}^{n}(x_i - \bar{x})^2 \cdot \text{Var}[\epsilon_i]}{\left(\displaystyle\sum_{i=1}^{n}(x_i - \bar{x})^2\right)^2}} = \sqrt{\frac{\displaystyle\sum_{i=1}^{n}(x_i - \bar{x})^2\,\sigma_\epsilon^{\,2}}{\left(\displaystyle\sum_{i=1}^{n}(x_i - \bar{x})^2\right)^2}}$$

$$= \sqrt{\frac{\sigma_\epsilon^2 \sum_{i=1}^n (x_i - \bar{x})^2}{\left(\sum_{i=1}^n (x_i - \bar{x})^2 \right)^2}} = \frac{\sigma_\epsilon}{\sqrt{\sum_{i=1}^n (x_i - \bar{x})^2}}.$$

In the manipulations above we have taken advantage of the nature of the x_i (numbers) and of the errors ϵ_i (mutually independent and with fixed standard deviation σ_ϵ).

The careful reader will notice that there is a problem with the formula above: How do we know σ_ϵ, if we cannot observe the errors but just the residuals? The answer is that we should estimate this standard deviation as well, but this does not prevent a practical use of See. Before doing so, it is useful to interpret the result we have obtained.

- As expected, the reliability of our estimate of the slope of the linear law describing our phenomenon depends on intrinsic variability of the phenomenon itself. If random variability is low, and the n observations are very close to the line $Y = \alpha + \beta x$, then estimating the slope is a fairly easy task. Indeed, we see that See_b is proportional to σ_ϵ.

- Another fairly intuitive observation is that the more observations we have, the better. In fact, for each observation we have, the denominator of the ratio above increases, reducing See_b.

- A less obvious observation is that our ability to estimate the slope depends on where the observations are placed. We note that at the denominator of the ratio there is a term similar to a variance, which is in fact the (nonrandom) variability of the observations x_i. If the points x_i are close to each other, i.e., they are close to their average $\bar{x}$, we have a small denominator. It is difficult to see the impact of small variations of x on Y, because this effect is "buried" in noise. If the observed range of x is wide enough, assessing the impact of x on Y is easier.

The last point is illustrated in the following example.

Example A.25 Say that we want to use simple linear regression to investigate the relationship between outside temperature and ice cream consumption. Simple linear regression is but the simplest approach, as other factors may play a role, leading to multiple regression; furthermore, the relationship need not be linear. Leaving these caveats aside, our task is certainly difficult if all we have is a set of data whereby temperature, measured in Celsius degrees, lies between $22°$ and $22.5°$. Even if the relationship is linear, with slope β, we should expect a rather small difference $\beta \cdot 0.5$ in the consumption for the two extreme temperatures in the range. The slope β should be positive, but in a random sample we might have enough noise to get a negative estimated slope b. The task would be probably easier if we had a sample in the range between $10°$ and $30°$. The difference $\beta(30 - 10)$ should be large enough to avoid, at least, a negative estimated slope because of randomness. □

The example suggests that we should have observations over a large range of the explanatory variable x, in order to get a good estimate of slope. However, it is worth noting that in many cases a linear relationship may hold over a limited range; if the "real" model is nonlinear, a linear one may be at best a suitable *local* approximation. Hence, by taking a wide sample we might run into a different kind of trouble, namely poor fit. Furthermore, we should be very careful when we extrapolate a prediction, i.e., when we consider something like $\hat{Y}_0 = a + bx_0$, where x_0 lies outside the range of observed values. For instance, the model considered in the example above would probably suggest that on very cold days we have negative ice cream consumption!

In order to assess the estimation error for α, we may follow the same route we took for the slope. The starting point is

$$a = \overline{Y} - b\bar{x} = \frac{1}{n} \sum_{i=1}^{n} (\alpha + \beta x_i + \epsilon_i) - b\bar{x} = \alpha + (\beta - b)\bar{x} + \frac{1}{n} \sum_{i=1}^{n} \epsilon_i.$$

To get closer to See_a we observe that

$$
\begin{aligned}
\mathrm{Var}(a - \alpha) &= \mathrm{Var}\left[(\beta - b)\bar{x} + \frac{1}{n} \sum_{i=1}^{n} \epsilon_i \right] \\
&= (\mathrm{See}_b \cdot \bar{x})^2 + \frac{1}{n^2} \sum_{i=1}^{n} \mathrm{Var}(\epsilon_i) + 2\mathrm{Cov}\left((\beta - b)\bar{x}, \frac{1}{n} \sum_{i=1}^{n} \epsilon_i \right) \\
&= \mathrm{See}_b^2 \cdot \bar{x}^2 + \frac{\sigma_\epsilon^2}{n} - 2\frac{\bar{x}}{n} \mathrm{Cov}\left(b - \beta, \sum_{i=1}^{n} \epsilon_i \right).
\end{aligned}
$$

This expression is quite complex, but actually we can show that the last term is null. Focusing on this term, and rewriting $b - \beta$ in order to see the contribution of errors, we obtain

$$
\begin{aligned}
\mathrm{Cov}\left(b - \beta, \sum_{i=1}^{n} \epsilon_i \right) &= \mathrm{Cov}\left(\frac{\sum_{i=1}^{n} (x_i - \bar{x}) \epsilon_i}{\sum_{i=1}^{n} (x_i - \bar{x})^2}, \sum_{j=1}^{n} \epsilon_j \right) \\
&= \frac{\sum_{i=1}^{n} \sum_{j=1}^{n} \mathrm{Cov}\left((x_i - \bar{x}) \epsilon_i, \epsilon_j \right)}{\sum_{i=1}^{n} (x_i - \bar{x})^2}
\end{aligned}
$$

$$= \frac{\sum\limits_{i=1}^{n} \mathrm{Cov}\left(\left(x_i - \bar{x}\right)\epsilon_i, \epsilon_i\right)}{\sum\limits_{i=1}^{n} \left(x_i - \bar{x}\right)^2} = \frac{\sum\limits_{i=1}^{n} \sigma_\epsilon^2 \left(x_i - \bar{x}\right)}{\sum\limits_{i=1}^{n} \left(x_i - \bar{x}\right)^2} = 0,$$

where we have exploited the mutual independence between the errors ϵ_i and the fact that their variance does not depend on the observations. Now, plugging the expression of See$_b$, we obtain

$$\mathrm{See}_a = \sigma_\epsilon \sqrt{\frac{\bar{x}^2}{\sum_{i=1}^{n} \left(x_i - \bar{x}\right)^2} + \frac{1}{n}}. \tag{A.31}$$

This formula too lends itself to a useful interpretation, as we see that we have two terms under the square root. The second one is basically linked to the average contribution of random error ϵ, whereas the first one is linked to the error in the estimate of β. Let us get a closer look at both terms.

- Suppose, for a moment, that we have been pretty good at estimating the slope, i.e., $b = \beta$. If in the sample we observed, the errors have been positive on average ($\sum_{i=1}^{n} \epsilon_i > 0$), we will arguably tend to estimate an intercept a which is larger than the true one, α, because we observed a set of points which are on average above the ideal line $Y = \alpha + \beta x$. On the contrary, if the errors have been negative on average ($\sum_{i=1}^{n} \epsilon_i < 0$), we have basically observed a set of points below the ideal line, and the estimate a will turn out smaller than the true intercept α; in this last case, the estimated line will lie below the ideal one. This error in the "vertical" placement of the line will be smaller when the number n of samples is large. This is basically what the term $\sigma_\epsilon/\sqrt{n}$ in See$_a$ tells us. The idea is illustrated in figure A.25, where we see that the observations of the dependent variable Y are more often than not above the ideal line; this is due to an "excess" of positive observations of the random error, leading to an estimated line above the ideal one.

- To see the second phenomenon contributing to the error in estimating α, let us assume that we had an "ideal" sample of errors, $\sum_{i=1}^{n} \epsilon_i = 0$. The first term in See$_a$ tells us that even in this case, we will have an error in the estimate of a due to errors in the estimate of the slope. As we see from the expression of the estimator a, the regression line goes through the barycenter $(\bar{x}, \overline{Y})$ of the n observations. This implies that if the estimated value b is larger than the true slope β, the estimate a will be smaller than α, for the case $\bar{x} > 0$. On the contrary, if we underestimate the slope β, i.e., the estimated line is "flatter" than the ideal one, the estimate a will be larger than α, for a positive average value $\bar{x}$. Basically, if the average error is zero, then we have a rotation of the ideal line around the barycenter of the observation, which in this

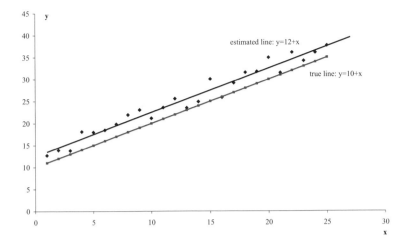

Fig. A.25   Sources of estimation errors for α: sampling the error ϵ.

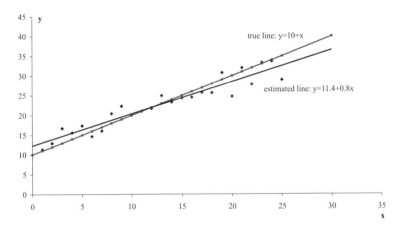

Fig. A.26   The effect of an error in the estimate of the slope β on the estimate of the intercept α (1).

case lies on both the ideal and the estimated line. We may get a better picture of this phenomenon by having a look at figure A.26. Clearly, the larger the error in the estimate of the slope, the larger the error in a, but this also depends on the barycenter $\bar{x}$ of the observations of the independent variable. If this is large, the slope error $(\beta - b)$ has a larger impact because the "lever arm," i.e., the distance between the vertical coordinate axis and the center of rotation, is large. In the opposite limit case, i.e., when $(\bar{x} = 0)$, the impact of this rotation in the intercept a is null. We can see this in figure A.27, where the barycenter of the observations, $(\bar{x}, \overline{Y})$, lies on the vertical axis.

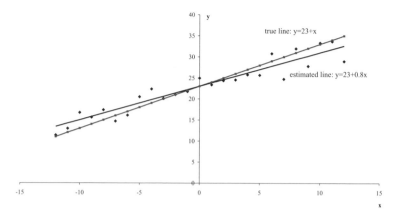

Fig. A.27   The effect of an error in the estimate of the slope β on the estimate of the intercept α (2).

The last missing piece in the puzzle is how we can estimate the standard deviation σ_ϵ of the random errors, which are not directly observable. The only viable approach we have is to rely on the residuals e_i as a proxy for the errors ϵ_i. Each residual is, for a given value x_i, the deviation between the theoretically predicted value, $\hat{Y}_i = a + bx_i$, and the observed one, Y_i. Note that if we trust the estimated model, $\hat{Y}_i$ is the expected value of Y_i; hence, in order to assess the variability of the errors, it is reasonable to consider the variability of the observations with respect to their expected value. Another way to get the picture is by noting that the assumptions behind the statistical model imply that $\sigma_\epsilon^2 = \text{Var}(Y_i)$, but the variance of the observed value Y_i is a squared deviation with respect to an expected value which depends on x_i. This reasoning leads to the following estimate:

$$\hat{\sigma}_\epsilon = \sqrt{\frac{\sum_{i=1}^n \left(Y_i - \hat{Y}_i\right)^2}{n-2}}. \tag{A.32}$$

This is basically a sample standard deviation; the only point which could raise a few eyebrows is the denominator, which is $n-2$ rather than $n-1$. The serious way to see this is by showing that with that term the estimator of σ_ϵ is unbiased. A first (very) intuitive argument runs as follows. What is the minimal number of observations to analyze the standard error? If we had just $n=2$ observations, we could say nothing, because in such a case the observed points would exactly spot one line, and there would be no deviation between Y_i and $\hat{Y}_i$. If we had n or $n-1$ at the denominator, the conclusion would be that whenever we have $n=2$, estimated random variability is zero, which makes no sense. The term $n-2$ points out that with just two observations we cannot say anything, as we have a ratio $0/0$. It is only from the third sample on that we can say something. Of course, this argument is just an interpretation and

Table A.5   First data set for example A.26

i	1	2	3	4	5	6	7	8	9	10	11
x	0	1	2	3	4	5	6	7	8	9	10
y	100	104	111	114	121	125	129	133	141	147	150

Table A.6   Second data set for example A.26

i	1	2	3	4	5	6	7	8	9	10	11
x	0	1	2	3	4	5	6	7	8	9	10
y	73	125	149	91	95	175	93	118	125	193	134

not a proof. Another intuitive way to interpret the formula runs in terms of degrees of freedom. In this case, we are estimating *two* parameters, α and β, of the model, and this consumes some available information in the observed sample; this results in the loss of two degrees of freedom.

Now, we are ready to evaluate the standard estimation errors of the regression parameters. This turns out to be essential in building confidence intervals and testing hypotheses.

Example A.26 Reliability of estimators a and b. To build some intuition as far as See is concerned, let us consider the two data sets given in tables A.5 and A.6. To distinguish the two cases, we will use subscript 1 for the estimates referred to the first data set of table A.5 and subscript 2 for the data set of table A.6. Using equations (A.25) and (A.26), we obtain the following estimators:

$$a_1 = 99.64, \qquad b_1 = 5.07, \qquad a_2 = 99.32, \qquad b_2 = 5.06.$$

We wee that the regression coefficients are essentially the same, even if the data look quite different. How different exactly can be seen in figures A.28 and A.29. In the first case, the observed points lie almost exactly on a line; in the second one, we see a lot of variability. The two situations are clearly discriminated when we compute the See for the four parameters:

$$\text{See}_{a_1} = 0.688, \qquad \text{See}_{b_1} = 0.116, \qquad \text{See}_{a_2} = 19.53, \qquad \text{See}_{b_2} = 3.30.$$

We see that large standard errors, with respect to the value of the estimators, are associated with the second data set. We could even wonder if the linear relationship we have estimated is statistically significant. In the next section we consider such an issue in a general setting.   ⬜

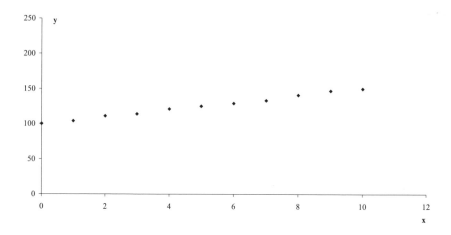

Fig. A.28   Plot of the first data set for example A.26.

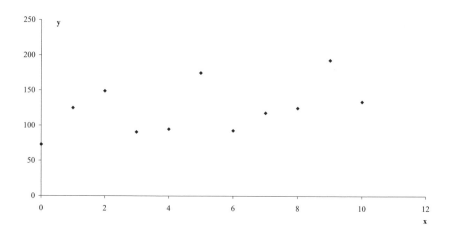

Fig. A.29   Plot of the second data set for example A.26.

A.10.3 Confidence intervals and hypothesis testing for regression estimators

To use regression in a sensible and responsible way, we should assess the probability distribution of the estimators a and b. This distribution depends on the distribution of the errors ϵ_i. If we assume that errors are normally distributed, since the estimates a and b basically involve sums of the errors, they will be normal too. We use this knowledge in order to:

- compute confidence intervals

- test hypotheses

These related activities are essential in order to assess the validity of the regression model (if the confidence interval for b includes both positive and negative values, we are unsure about the effect of the independent variable on the output) and to use regression as a forecasting tool (a one-number forecast may be extremely dangerous, and its uncertainty must be qualified).

In doing so, we must rely on estimates of the volatility σ_ϵ. As we have seen in section A.8.2, Student's t distribution is involved; taking into account the form of the estimator (A.32), we may also see that we should use the t distribution with $n - 2$ degrees of freedom.

Confidence intervals Although we will never know the exact value of the parameters α and β, we may use See to build a confidence interval including the unknown values with probability p; to avoid an ambiguous notation, here we avoid denoting the confidence level by $1 - \alpha$.

Example A.27 To illustrate the idea, we use again the data set in tables A.5 and A.6, to build confidence intervals for the estimates of α and β, with confidence level $p = 95\%$.

We have eleven data points; from statistical tables or numerical software we obtain the relevant quantile for the t distribution with nine degrees of freedom: $t_{0.975,9} \approx 2.26$. Hence, using the standard errors of estimate we obtained in example A.26, we may say that, with probability 0.95, we have in the first case

$$a - t_{0.975,9} \cdot \text{See}_a = 99.64 - 2.26 \cdot 0.688 = 98.09 \leq \alpha \leq$$
$$a + t_{0.975,9} \cdot \text{See}_a = 99.64 + 2.26 \cdot 0.688 = 101.19$$
$$b - t_{0.975,9} \cdot \text{See}_b = 5.07 - 2.26 \cdot 0.116 = 4.81 \leq \beta \leq$$
$$b + t_{0.975,9} \cdot \text{See}_b = 5.07 + 2.26 \cdot 0.116 = 5.33,$$

whereas in the second one we have

$$a - t_{0.975,9} \cdot \text{See}_a = 99.32 - 2.26 \cdot 19.53 = 52.13 \leq \alpha \leq$$
$$a + t_{0.975,9} \cdot \text{See}_a = 99.32 - 2.26 \cdot 19.53 = 146.51$$
$$b - t_{0.975,9} \cdot \text{See}_b = 5.06 - 2.26 \cdot 3.30 = -2.79 \leq \beta \leq$$
$$b + t_{0.975,9} \cdot \text{See}_b = 5.06 + 2.26 \cdot 3.30 = 12.93.$$

We see that, as expected, the confidence intervals for the second data set are much larger. What is more important is that even the sign of the slope is somewhat dubious. □

We should remark that the confidence intervals we obtain rely on several assumptions about the random errors in the underlying statistical model, and on our ability to estimate their variance. Nevertheless, they are useful approximations showing that, for a given probability p, the standard error of estimate has an important impact on our knowledge of parameters α and β.

Hypothesis testing   Armed with some knowledge about the distribution of the estimators, we may also run some hypothesis testing on parameters α and β. The conceptual background has been given in section A.9, and here we provide the reader with a few examples.

Typical questions we want to answer concern the impact of the explanatory variable on the predicted variable. For instance, if we cannot reject the hypothesis $\beta = 0$, we cannot trust the model too much. Sometimes, we want to check if the effect has some sign; for instance, if we look for support to the hypothesis that a reduction in price has a significant impact on sales, we should consider the hypothesis $\beta < 0$. Sometimes, it is not trivial to decide if the test is one- or two-sided. Similar questions can be asked as far as α is concerned, and they are typically of two types:

1. Can we say that, with some probability p, the parameter (α or β) is positive (or negative, or non-null, depending on our problem formulation)?

2. What is the maximum confidence level with which we can state some property about the parameter?

If our aim is checking whether α or β is nonzero (which amounts to saying, in the second case, that x has a statistically significant impact on Y), with some probability p, we must essentially check if, at confidence level p, the confidence interval includes 0 or not. Referring to example A.27, we may say that, in the first case, both parameters α and β are different from zero with probability 95%; in the second case we cannot.

If we are wondering with what probability we can say that a parameter is different from zero, we must reason on the maximum width of the confidence interval such that the origin is not included; in this case, this is equivalent to finding a p-value. If a and b are positive, as in our examples so far, we should set the left endpoint of the interval to zero to obtain the corresponding quantile. In the case of α we have

$$a - t_{p,n-2} \cdot \text{See}_a = 0 \qquad \Rightarrow \qquad t_{p,n-2} = \frac{a}{\text{See}_a}.$$

Given the quantile, we can read the probability level p, checking tables for the t distribution with $n - 2$ degrees of freedom.

Example A.28 The data set of table A.5 yields two values $t_{p,9}$, for α and β, equal to 144.84 and 43.71, respectively. This implies that the two parameters are nonzero with (practically) 100% probability.

The case of the second set, in table A.6, is different, as the two quantiles are 5.08 and 1.53. This means that α is positive with (practically) 100% probability, whereas the probability for β is only 92%. ⬜

A.10.4 Performance measures for linear regression

We have discussed important properties of the estimators a and b of parameters α and β, but of course the real deal is using the explanatory variable x to predict the output Y. For instance, regression analysis could support the view that the average price of soft drinks has an impact on the demand for salted snacks, because reducing the price of drinks stimulates people to offer an aperitif. Based on a suitable amount of data, it is likely that we could indeed support this, but of course it is unlikely that this is the only or the main factor explaining sales volume of salted snacks. Other factors could be the price, the weather, proximity to social or sports events, etc. So, on the one hand, the view that the β coefficient of the relationship between soft-drink price and salted snack sales is significantly negative could be supported by proper statistical analysis, but, on the other hand, this does not imply that such a model is a good one. We should try to measure how much variability of the output variable can be attributed to an explanatory variable.

To make this idea concrete, we can measure the correlation between Y_i and $\hat{Y}_i$. By considering the squared correlation coefficient, we may define the following R^2 statistic, which is based on the sample correlation coefficient (A.19):

$$R^2 = r_{Y\hat{Y}}^2 = \frac{\left[\sum_{i=1}^{n} \left(Y_i - \overline{Y}\right)\left(\hat{Y}_i - \overline{\hat{Y}}\right)\right]^2}{\sum_{i=1}^{n} \left(Y_i - \overline{Y}\right)^2 \cdot \sum_{i=1}^{n} \left(\hat{Y}_i - \overline{\hat{Y}}\right)^2}.$$

By its very definition, R^2 is a measure bounded between 0 and 1. It is also interesting to rewrite this expression in order to shed some more light on its meaning, paving the way for useful interpretations. We know that

$$\overline{\hat{Y}} = \frac{1}{n}\sum_{i=1}^{n}\hat{Y}_i = \frac{1}{n}\sum_{i=1}^{n}(a + bx_i) = a + b\overline{x} = \overline{Y};$$

hence, we may rewrite R^2 as

$$R^2 = \frac{\left[\sum\limits_{i=1}^{n} (Y_i - \overline{Y})(\hat{Y}_i - \overline{Y})\right]^2}{\sum\limits_{i=1}^{n} (Y_i - \overline{Y})^2 \cdot \sum\limits_{i=1}^{n} (\hat{Y}_i - \overline{Y})^2}$$

$$= \frac{\left[\sum\limits_{i=1}^{n} \left[(Y_i - \hat{Y}_i) + (\hat{Y}_i - \overline{Y})\right](\hat{Y}_i - \overline{Y})\right]^2}{\sum\limits_{i=1}^{n} (Y_i - \overline{Y})^2 \cdot \sum\limits_{i=1}^{n} (\hat{Y}_i - \overline{Y})^2}$$

$$= \frac{\left[\sum\limits_{i=1}^{n} (Y_i - \hat{Y}_i)(\hat{Y}_i - \overline{Y}) + \sum\limits_{i=1}^{n} (\hat{Y}_i - \overline{Y})(\hat{Y}_i - \overline{Y})\right]^2}{\sum\limits_{i=1}^{n} (Y_i - \overline{Y})^2 \cdot \sum\limits_{i=1}^{n} (\hat{Y}_i - \overline{Y})^2}.$$

Now we prove that the first term in the numerator is actually zero:

$$\sum_{i=1}^{n} \left(Y_i - \hat{Y}_i\right)\left(\hat{Y}_i - \overline{Y}\right)$$

$$= \sum_{i=1}^{n} [Y_i - (a + bx_i)](a + bx_i - a - b\overline{x})$$

$$= \sum_{i=1}^{n} \left[Y_i - (\overline{Y} - b\overline{x}) - bx_i\right](bx_i - b\overline{x})$$

$$= b\sum_{i=1}^{n} \left[(Y_i - \overline{Y}) - b(x_i - \overline{x})\right](x_i - \overline{x})$$

$$= b\left[(n-1)S_{xY} - b(n-1)S_x^2\right]$$

$$= b(n-1)\left[S_{xY} - \frac{S_{xY}}{S_x^2}S_x^2\right] = 0.$$

So we end up with the following formula for R^2:

$$R^2 = \frac{\sum\limits_{i=1}^{n} \left(\hat{Y}_i - \overline{Y}\right)^2}{\sum\limits_{i=1}^{n} (Y_i - \overline{Y})^2} = \frac{S_{\hat{Y}}^2}{S_Y^2}.$$

We see that R^2 can be interpreted as the ratio of two (sample) variances: the "variance" of the forecasts $\hat{Y}_i$ and the variance of real observed data Y_i. We

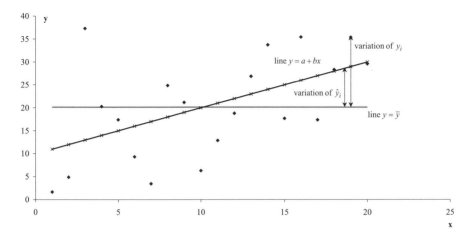

Fig. A.30   A geometrical interpretation of the R^2 coefficient: a case with $R^2 = 0.29$.

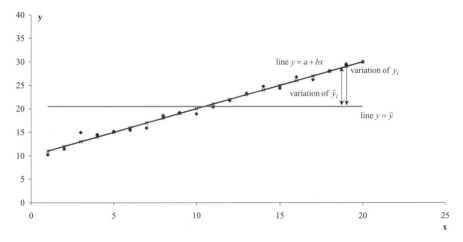

Fig. A.31   A geometrical interpretation of the R^2 coefficient: a case with $R^2 = 0.98$.

should note that, by definition, the forecasts lie on a line; when we speak of "variance," we actually mean the (nonrandom) variability with respect to the average ordinate along the line. In other words, R^2 measures the fraction of variability which is explained by the regression model $Y = a + bx$. Referring back to our snack sales example, the R^2 of a regression against the average soft-drink price is likely to explain a small fraction of variability.

To improve our feeling for R^2, we may have a geometrical look. In figure A.30, the forecasts $\hat{Y}_i$ display a lower level of variability than observed data Y_i; in fact, observations are somewhat placed around the line, but we see a lot of variability beyond the variability due to the slope of the line. Hence, there is a lot of variability which is not explained by the model. By comparing figure A.30 against figure A.31, we see that when R^2 increases we have a much larger ability of explaining the observed data, and unexplained variability is quite limited.

A.10.5   Verification of the underlying assumptions

In running a diagnostic test on the explanatory/predictive ability of a regression model, we should not forget that all we have said depends heavily on a set of precise assumptions on the random errors. It is therefore essential that these assumptions are tested, at least informally, against the observed data. We recall the assumptions:

- ϵ_i is a random variable with expected value zero.

- The random variables ϵ_i are mutually independent and identically distributed; in particular they have the same standard deviation.

- The distribution of the error ϵ_i does not depend on x_i.

- We have perfect knowledge of x, which is not a random variable but is, instead, a number and can be measured with no uncertainty.

There are specific procedures to check the validity of these assumptions. We illustrate here graphical checks, which are useful for a rough-cut analysis and reinforce the concepts.

The assumptions above say that the noise ϵ_i in the model can be regarded as (a) a stochastic process with expected value zero, (b) stationary, (c) not autocorrelated, and (d) independent of x. To really check this, we should observe the errors, which is not possible, because we do not have knowledge of the ideal line $\alpha + \beta x$. We must settle for a *proxy* of ϵ_i, i.e., the residual $e_i = Y_i - (a + bx)$.

The assumption about the expected value is automatically met, since the estimators are such that $a = \overline{Y} - b\overline{x}$. To check stationarity and lack of autocorrelation, we may plot the residuals. If the assumptions are compatible with the data, the plot of the residuals e_i should look like figure A.32, where we see that they reasonably behave like pure noise. On the contrary, the next three figures display plots of residuals which do not support the underlying assumptions.

In figure A.33 we see a pattern which is typically associated to positively correlated errors. If we draw a positive error at observation number i, the next observation $i + 1$ is likely to be affected by a positive error as well. The same holds for negative errors, and we see "waves" of positive vs. negative residuals. Such a pattern may be observed for at least a couple of reasons. The first one is that there is indeed some correlation between consecutive observations in time. In such a case subscript i really refers to time and to the order in which observations were taken; the obvious case is when time is the explanatory variable. Another possible reason has really little to do

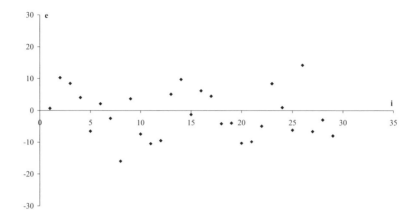

Fig. A.32   Plot of residuals coherent with the regression model assumptions.

with Statistics: We may also observe a pattern like this when there are non-linearities in the observed phenomenon. Consider for instance the nonlinear function in figure A.34, and assume we use linear regression to approximate it, using a few sample points (possibly affected by noise). The nonlinear curve is somehow cut by the regression line, and this results in a nonrandom pattern in the residuals: They have one sign in the middle range of the interval of x, and the opposite sign near the extreme points.

Of course, this second case has more to do with the appropriateness of selected functional form, and we are somewhat improperly using statistical concepts as a diagnostic tool. If so, we should clarify the meaning of the subscript i associated with an observation. If the explanatory variable is time, subscript i refers to the position in the chronological sequence of observations, and there is no ambiguity. But if we are regressing sales against price, we might wish to sort observations according to the value of the explanatory variable; in this case, subscript i should *not* refer to the order in which we took our samples.

Actually, we are talking about two different issues, and we should pay due attention to both of them. Whatever the case, a quantitative check can be run by estimating the correlation between e_i and e_{i+1}.

Another check concerns the stationarity of the error process. In section A.7 we have considered weak stationarity of the second order, which basically says that we should (at least) check stationarity of the mean and the variance of the error process. In figure A.35 we see that the mean error seems dependent on i. If i is actually the time of the observation (but it need not be), we should consider running a multiple regression in which time is an explanatory variable. A quantitative check can be run by estimating the "correlation" between e_i and i. Figure A.36 displays a case in which the errors in the first observa-

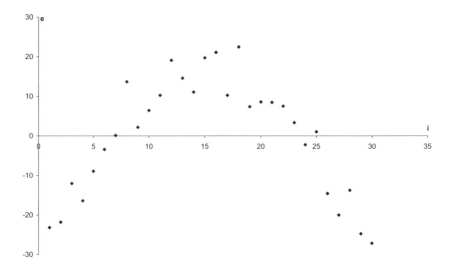

Fig. A.33   Plot of residuals suggesting autocorrelation in the errors.

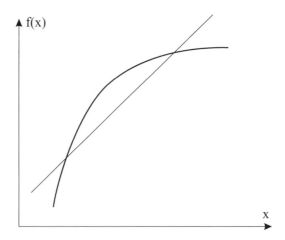

Fig. A.34   Using linear regression with a nonlinear underlying function results in "au-tocorrelated" residuals.

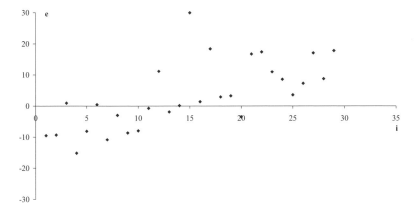

Fig. A.35 Plot of residuals suggesting that the mean of the error process is not stationary.

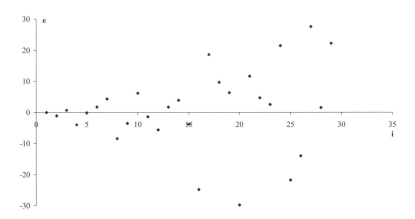

Fig. A.36 Plot of residuals suggesting that the variance of the error process is not stationary.

tions look much smaller, in absolute value, than in the last observations. This raises some doubt on the stationarity assumption for variance. A technical word used in Econometrics to refer to a situation like this is *heteroskedasticity*. We could probably circumvent the difficulty by assigning more weights to observations with a larger amount of information (i.e., less noise), but this goes beyond the scope of this book. One way to have a quantitative check is by estimating the "correlation" between ϵ_i^2 and i.

Finally, we should check whether residuals do not depend on x. To this aim, we can produce plots of residuals like those in figures A.35 and A.36, where the independent variable is x, rather than the observation index or time.

A.10.6   Using linear regression to estimate nonlinear relationships

After the long list of checks and caveats we have seen so far, the reader might have the feeling that linear regression is a rather rigid tool, whose practical use is hindered by a plethora of assumptions. There are statistical issues, which can be somehow circumvented by using more sophisticated techniques. We refer the reader to a book on statistics or econometrics to appreciate the richness of this field.

Apart from statistical issues, linearity itself is certainly a possibly strong limitation, as both Nature and Business are nonlinear. For instance, if we want to study the relationship between the amount we spend in advertisements and revenue, a linear model could lead to the following two difficulties (at least):

1. A linear relationship fails to capture thresholds effects: Below a certain minimal effort, the message is not perceived at all.

2. A linear relationship fails to capture saturation effects: Increasing the effort beyond a certain limit is useless because of intrinsic limitation in the market size and because of competition. The bottom line is that there are diminishing marginal returns from the investment, whereas a linear model used to take decisions could suggest the opportunity of increasing the effort too much (at least, if the advertisement costs are linear or marginally decreasing).

If we have to deal with a nonlinear phenomenon, one possibility is to forget about a simple linear law and to develop a theory of nonlinear regression. This has indeed been done, but nonlinearity can introduce an array of technical complications. Hence, we could try to use linear regression as a tool to estimate nonlinear relationships. This may seem to be a hopeless endeavor, but an example can prove that we can actually resort to suitable data transformations to identify the parameters of a postulated nonlinear relationship between x and Y.

Example A.29 Suppose we have a data set displaying a strong nonlinearity, as depicted in figure A.37. In such a case, enforcing a linear regression would be less than advisable, but we can try to work with a nonlinear functional form lending itself to a data transformation, such that the familiar tool can be used. Since we have just dealt with simple linear regression, whereby we estimated two parameters, we should use a nonlinear function linking Y and x, in which two parameters are used. A typical choice is an exponential functional form like

$$Y = kx^{\gamma}. \tag{A.33}$$

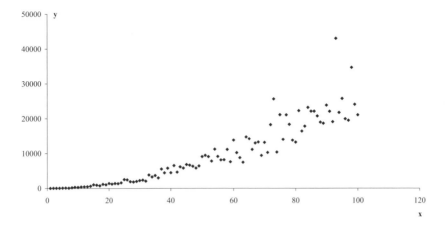

Fig. A.37 A nonlinear relationship.

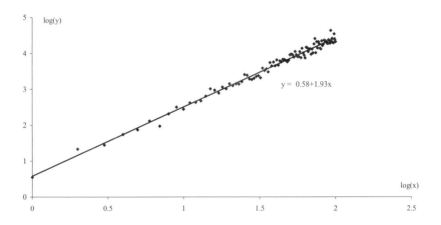

Fig. A.38 Linearizing a nonlinear relationship by a logarithmic transformation.

To linearize this relationship, we can exploit a logarithmic transformation:

$$\log Y = \log\left(kx^{\gamma}\right) = \log k + \gamma \log x.$$

We see that, on a logarithmic scale with coordinates $\log Y$ and $\log x$, we have a linear relationship. This should be checked against the actual data by plotting them according to the transformed scales. An example of this rough-cut check is displayed in figure A.38. This plot suggests that the relationship between $\log Y$ and $\log x$ can be captured by a linear model, which can be estimated

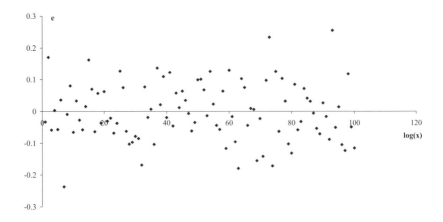

Fig. A.39   Verifying the fit of the assumed functional form.

by the following linear regression model:

$$\log Y = \alpha + \beta \log x + \epsilon, \tag{A.34}$$

where

$$\alpha = \log k, \tag{A.35}$$
$$\beta = \gamma. \tag{A.36}$$

While the nonlinear transformation is a rather simple trick from a technical point of view, the real trouble comes when we consider the statistical side of the coin. Since we are using the linear regression machinery on the transformed variables (logarithmic, in the case above), we should check the familiar assumptions within the transformed model. To check the residuals, we can analyze a plot like the one in figure A.39. We will also find confidence intervals, but they will refer to the *transformed* variables; to get confidence intervals in terms of the original variables, we must invert the transformation. For instance, the data set we are considering yields the following estimated model:[24]

$$\log Y = 0.58 + 1.93 \log x + e, \tag{A.37}$$

which can be transformed back in terms of the original variables Y and x:[25]

$$Y = \left(10^{0.58}\right) x^{1.93} \cdot 10^{\epsilon} = 3.80 \cdot x^{1.93} \cdot 10^{\epsilon}. \tag{A.38}$$

---

[24] In the estimated model we use the residual e rather than the unobservable error ϵ.
[25] From this equation we see that decimal logarithms have been used; natural logarithms can be used as well, but one advantage of decimal logarithms is that their value is more readable in terms of order of magnitude of their argument.

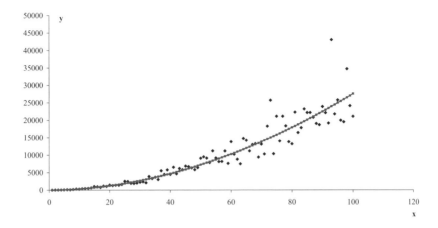

Fig. A.40   Estimated nonlinear relationship.

The fit of the resulting model can be seen in figure A.40. In this case, it turns out that the See for a is 0.04. Let us assume that, given the degrees of freedom and the required confidence level, the right quantile is $t = 2$; then, the confidence interval for a is $0.58 - 2 \cdot 0.04 = 0.5 \le a \le 0.58 + 2 \cdot 0.04 = 0.66$. Actually, this is not a confidence interval for the original demand model (A.33), but for the transformed one instead. To get confidence intervals for k, we must transform the confidence interval for a back to the original parameter k, using equation (A.35). Therefore, we can say that the parameter k of model (A.33) lies, with that confidence level, in the range from $k_{\min} = 10^{0.58-2 \cdot 0.04} \approx 3.16$ to $k_{\max} = 10^{0.58+2 \cdot 0.04} \approx 4.57$. We immediately see that this confidence interval, unlike those we are used to, is not symmetrical around the point 3.80; this happens just because of the nonlinear transformation, which more often than not results in a lack of symmetry.

Finally, it is important to understand the role of the error ϵ. In the transformed model (A.37) the error is additive with respect to the estimated line. When we use the inverse transformation to get back to the original variables x and Y, we see from equation (A.38) that the error is not additive; in fact, we have a multiplicative factor 10^ϵ. From a statistical point of view, we assume that the error ϵ is independent of x in the underlying model (A.34); however, when we switch back to the original variables, we see that the residuals in the original model tend to grow with x (see figure A.41). Note that by residuals in the original model we mean

$$e_i^o = Y_i - k(x_i)^\gamma.$$

In the figure, we see that the residuals tend to increase, in absolute value, when x increases. From a statistical point of view, if we assume that the usual assumptions hold for the linearized model, this happens because there

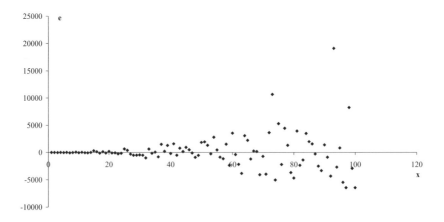

Fig. A.41   Residual plot for model (A.38).

is a random variable which is not additive in the original model, but it gets multiplied by x. From the point of view of least-squares optimization, we did not really minimize the sum of squared residuals in the original model, but in the transformed model. The residual in the transformed model can be written as:

$$e_i = \log Y_i - \log (kx^\gamma) = \log \left(\frac{Y_i}{kx_i^\gamma} \right).$$

The bottom line is that, with respect to the residuals in the original model, we assign less weight to the residuals in the transformed model when x is large, which results in the effect of figure A.41.   □

The example above shows that using suitable variable transformations is a nice way to extend the range of applicability of linear regression; however, this raises issues as far as confidence intervals and errors are concerned. This is why such transformations should be taken with care, and full-fledged nonlinear regression might be a better alternative.

We close this section by noting that if the error ϵ in the underlying statistical model is normally distributed, the multiplicative term 10^ϵ has lognormal distribution. To see this, note that

$$10^\epsilon = e^{\ln 10^\epsilon} = e^{\epsilon \cdot \ln 10}$$

and $\epsilon \cdot \ln 10$ has normal distribution as well. If you use natural rather than decimal logarithms, the result follows directly.

W.A.11 MULTIPLE LINEAR REGRESSION

In simple linear regression we use one explanatory variable, whereas in practice multiple factors could contribute to explain a phenomenon of interest. This suggests the adoption of a multiple regression model. If we want to preserve linearity, we may resort to the following statistical model:

$$Y = \alpha + \sum_{j=1}^{m} \beta_j x_j + \epsilon,$$

where m is the number of explanatory variables, each one associated with its β_j. Solving the related least squares problem is not too difficult, but this extension is characterized by a few issues and opportunities:

- It is tempting to use as many factors as possible, but this may be a bad idea if they are somehow related; this raises issues such as collinearity and ill-conditioned regression.

- Increasing the number of factors cannot decrease the R^2 statistic; however, we should come up with a more refined measure, able to capture the tradeoff between a possibly increased fit and the issues above when using too many factors. In other words, we want to understand whether a sort of "cost/benefit" ratio justifies the use of more explanatory variables.

- We may also use "categorical" variables, i.e., variables related to the presence or lack of a certain feature. For instance, we may relate sales to the presence of promotions or to peculiar events. This is accomplished by using *dummy* explanatory variables, taking values which can be either 0 or 1.

In the web section we deal with all of the points above.

A.12 FOR FURTHER READING

- Readers interested in a practical approach to probability theory may read, e.g., [5], which is rich of examples; some of them provided an inspiration for examples in this appendix.

- Example A.12 has been taken from [1], which is a readable introduction for those interested in a more rigorous introduction to the axiomatic approach to probabilities.

- As far as Statistics is concerned, there are many books, but an excellent introduction is [4], which points out the probabilistic foundation of

Statistics, offers complete and readable proofs, and is at an adequate level for a reasonably quantitative-minded reader.

- The use of linear regression for forecasting purposes is well illustrated in [3].

- Readers interested in a serious introduction to parameter estimation and regression analysis can have a look at [2], which also deals with nonlinear regression. See also [6].

REFERENCES

1. M. Capiński and T. Zastawniak. *Probability through Problems*. Springer-Verlag, Berlin, 2000.

2. W.H. Greene. *Econometric Analysis (5th Ed.)*. Prentice Hall, Upper Saddle River, NJ, 2002.

3. S.G. Makridakis, S.C. Wheelwright, and R.J. Hyndman. *Forecasting: Methods and Applications*. Wiley, New York, 1997.

4. S. Ross. *Introduction to Probability and Statistics for Engineers and Scientists (2nd Ed.)*. Academic Press, San Diego, CA, 2000.

5. S. Ross. *Introduction to Probability Models (8th Ed.)*. Academic Press, San Diego, CA, 2002.

6. G.A.F. Seber and C.J. Wild. *Nonlinear Regression*. Wiley, Hoboken, NJ, 2003.

Appendix B
An Even Quicker Tour in Mathematical Programming

The objective of this appendix is to get the reader acquainted with unconstrained and constrained optimization models, limiting the exposition to what is strictly necessary for the main body of the book. We will use the term *Mathematical Programming* to refer to that subset of Optimization Theory which deals with finite-dimensional problems. With respect to a full-fledged treatment, we will cut corners as far as mathematical rigor is concerned; furthermore, emphasis is on optimization *models*, rather than optimization *methods*. On the one hand, treating computational optimization methods would require too much background and detail; on the other hand, we are typically users of off-the-shelf optimization software, and what we normally need is just some knowledge of the underlying methods in order to choose the right one within a library and to understand diagnostics when something goes wrong.

Another good reason to present some basic optimization concepts is that, sometimes, practitioners use approaches which can be somewhat justified (or criticized) as the crude simplification of an optimization model. Given the computational speed of state-of-the-art optimization software and their integration as components in commercially available packages, such practices may have lost their original justification. Taking for granted that old practices should still be used, even without a solid reason, may lead to poor performance. A well-known example is the overstatement of fixed ordering costs, which leads to large lots with an unnecessary increase in on-hand inventory. Sometimes, *just in time* practices may make lots obsolete altogether, but in other cases the real answer should come from a suitable quantitative model. In section 4.6.2 we illustrate this point in detail for a multi-item inventory management problem. In this case, we use optimization as a means to an end, to illustrate a problem, rather than as a decision-making tool. This "conceptual" use of optimization models, which should be contrasted to the "computational" approach, is quite common in Economics. We will sometimes use optimization as a framework to clarify issues, even if the result is not really implemented. Again, to accomplish this limited aim, excessive mathematical finesse and deep algorithmic knowledge are not really needed.

In section B.1 we introduce the basic elements of optimization model building, using a toy production planning problem. A more formal treatment is offered in section B.2. Intuitively, we should expect that the larger an optimization model, the more CPU time is needed for its solution. In practice, this need not be true; a major factor in the difficulty of tackling an optimization model is its *convexity*, or lack thereof. We introduce basic convexity concepts, i.e., convex sets and convex functions, in section B.3. Another important factor is linearity vs. nonlinearity of the model. We consider nonlinear programming models in section B.4, whose main aim is introducing the *shadow price* concept; from a theoretical point of view, getting a grasp of shadow prices is essential for an economic understanding of optimality conditions, whereas from a computational point of view this is important in interpreting the solution of a model. We deal with linear programming with continuous decision variables in section B.5; if modeling requires the introduction of integrality restrictions on the decision variables, solving the model is more difficult, and we will see why in section B.6.

As we have hinted at, optimization modeling in the past had a reputation of a pretty academic subject. The situation has changed in recent years, because of several reasons. The obvious one (maybe *too* obvious) is the availability of more and more powerful hardware at decreasing costs, which paves the way for the solution of optimization models which were beyond our reach. But also *software* has improved considerably. To begin with, more efficient software has been developed. The simplex method for linear programming was invented in 1947, whereas branch and bound methods for integer programming date back to the early 1960's; still, there is an ongoing and amazing progress in new commercial software releases. Nevertheless, you may have powerful hardware and lightning-fast software, but all of this is useless if you lack data. As the old adage says, garbage in, garbage out. Nowadays, optimization libraries exploit software engineering approaches and come in the form of object-oriented libraries, which can be integrated as components of a firm-wide information system. Databases can be accessed, and visualization libraries and management tools complete the picture. Indeed, optimization libraries are part of well-established ERP packages, which provide the required data backbone for a successful application.

All of this is good, but it does not warrant the conclusion that optimization models are a panacea. We should keep in mind at least a couple of fundamental limitations. The first one is that some objectives a manager has in mind cannot be easily quantified; furthermore, an optimization model assumes that all of the (possibly conflicting) objectives are put in the same basket of a single objective function that we maximize or minimize. Sometimes, assessing tradeoffs is a thorny issue, which cannot be solved by attaching some weight factor to each single facet of the overall problem. As a typical example, consider the tradeoff between on-hand inventory and customer service level. Assuming the first objective is easy to quantify, the second is not: How much does an angry customer cost? Hard to tell, isn't it? Sometimes, late delivery penalties are explicitly written in a contract, but loss-of-goodwill is harder to assess. Hence, we may need some help in visualizing the tradeoff, and this is the reason for the inclusion of section B.7 on multiobjective optimization, whose aim is to spot some "reasonable" solutions, leaving to the decision-maker the task of selecting the most preferred one.

Another limitation of the models we consider in the appendix is that they are all deterministic, i.e., they rely on perfect knowledge of data. This is hardly the case in practice, especially in distribution logistics, where demand uncertainty is *the* problem. In the main body of the text we will hint at ways of extending deterministic optimization models to cope with uncertainty represented as a set of alternative scenarios (see sections 1.5.2 and 2.2.3).

B.1 ROLE AND LIMITATIONS OF OPTIMIZATION MODELS

The best way to get acquainted with optimization models is by a little toy example, whereby we want to optimize a production mix.[1]

Example B.1 Suppose we have been hired by a rather small firm, which manufactures and sells just two items, P_1 and P_2. Each item requires a given manufacturing cycle, which involves use of four resource types (machine A, B, C, and D). Note that A refers to a machine *type*, and not to a physical machine; what matters is the overall availability of each resource type, which may constrain the amount produced. Table B.1 displays the single operation times for each item on each machine $(T_A, \ldots, T_D)$ measured, e.g., in minutes

---

[1]The numerical values are taken by an illustrative example available on http://www.factory-physics.com, as a companion to an excellent text in manufacturing systems modeling and management, [3].

Table B.1   Data for the production mix example

Item	T_A	T_B	T_C	T_D	Cost	Price	Demand
P_1	15	15	15	25	45	90	100
P_2	10	35	5	14	40	100	50

per piece. We assume that the working calendar and the number of machines per type are such that the weekly availability for each resource type is 2400 minutes per week. Producing a single piece of any item has some cost, which may include raw materials and variable manufacturing costs, expressed in some monetary unit, say €; this cost is also shown in the table, and we should also add a fixed cost of 5000€ per week. This cost is incurred anyway, unless we shut the plant down, but this decision is not considered at our level. What we should decide is the production mix, i.e., how many pieces of each type we manufacture each week. This approach makes sense if both capacity and demand are constant over time, which rules out building inventories. The weekly demand for both items is given in the last column of the table, which also includes the price at which we may sell each item. Our aim is maximizing profit.

One simple and intuitive way of tackling such a problem is by checking which item looks most profitable. If we look at profit contribution, P_2 looks definitely better than P_1 ($100 - 40 > 90 - 45$); hence, one possible idea is maximizing the amount we produce of the more profitable item. Let us denote by x_i the amount produced for item $i = 1, 2$. What are the factors limiting x_2? One factor is available capacity. We have four resources, but one of them is the most critical (the bottleneck) as far as P_2 is concerned. A look at table B.1 shows that the largest requirements of item P_2 is on machine B. Hence, to figure out the maximum amount we can produce of P_2, we can consider the following inequality:

$$35x_2 \leq 2400 \Rightarrow x_2 \leq 68.57.$$

Actually, market limitations tell us that we cannot sell more than 50 pieces, so we set $x_2 = 50$. This decision leaves some room to produce an amount x_1 of item P_1. To find how much we can produce, we should compute the residual availability of each resource, given what we must reserve to P_2. We can write down a system of inequalities, in order to find out the binding one:

$$15x_1 + 10 \cdot 50 \leq 2400 \quad \Rightarrow \quad x_1 \leq 126.67,$$
$$15x_1 + 35 \cdot 50 \leq 2400 \quad \Rightarrow \quad x_1 \leq 43.33,$$
$$15x_1 + 5 \cdot 50 \leq 2400 \quad \Rightarrow \quad x_1 \leq 143.33,$$
$$25x_1 + 15 \cdot 50 \leq 2400 \quad \Rightarrow \quad x_1 \leq 66.$$

We see that machine B is still the critical factor; the largest amount we can produce is $x_1 = 43.33$. The reader might object that we can just make 43 pieces, and that all variables x_i should be restricted to integer values. This is certainly true for discrete items produced in small quantities; when we produce "continuous" items, such as paint, we may consider decisions modeled by real variables, and this is also true for large-volume discrete items, since accepting real-valued quantities is a negligible modeling error. We will see that, whenever possible, continuous variables make our life much easier; for now, let us neglect such issues and say that real-valued variables are acceptable. Since $x_1 = 43.33$ does not exceed the market limit for P_1, we can take this as a possible solution and calculate profit:

$$(90 - 45) \times 43.33 + (100 - 40) \times 50 - 5000 = -50.$$

The bad news is that profit is negative, so what went wrong? A possible idea is that we have not considered resource usage. True, P_2 is more profitable, but it uses a larger amount of the bottleneck machine B. Maybe, had we taken this into account, P_1 would have looked much better than P_2. We can try this conjecture immediately, by maximizing production of P_1. If we repeat a similar reasoning, we get an even worse solution; in fact, we would come up with the solution $x_1 = 96$ and $x_2 = 0$, which yields a worse profit: $45 \cdot 96 - 5000 = -680$. Now, it seems that shutting the plant down is the only option, unless we can reduce costs or ask for a higher price.

Actually, we have the option of building an optimization model, whereby we explicitly maximize profit, subject to relevant constraints:

$$
\begin{aligned}
\max \quad & 45x_1 + 60x_2 & \text{(B.1)} \\
\text{s.t.} \quad & 15x_1 + 10x_2 \leq 2400, \\
& 15x_1 + 35x_2 \leq 2400, \\
& 15x_1 + 5x_2 \leq 2400, \\
& 25x_1 + 15x_2 \leq 2400, \\
& 0 \leq x_1 \leq 100, \\
& 0 \leq x_2 \leq 50.
\end{aligned}
$$

This model includes:

- two decision variables, x_1 and x_2, which are restricted to non-negative values[2];

- four capacity constraints, one per resource type;

- two market bounds on production;

---

[2]Unless you are *really* bad with marketing, you do not sell negative quantities.

- an objective function which we wish to maximize.

The objective function, as it is written, is not really profit, which should be written as $(90 - 45)x_1 + (100 - 40)x_2 - 5000$. Nevertheless, it is easy to see that adding or subtracting a constant term to or from any objective function does not change the optimal solution: Shifting the graph of a function up or down does not change its minima and maxima.

The model above is an example of **linear programming** model. It is linear because the decision variables occur linearly in the objective function and the constraints: You do not see terms such as x_1^2, $x_1 x_2$, or $\sin x_2$, which would make the model nonlinear. There is a wide array of software packages to solve such a problem numerically. We will discuss their algorithmic foundation very briefly in section B.5, but using any of them, we get the following solution: $x_1^* = 73.85$, $x_2^* = 36.92$. This solution looks like it came out of the blue, but the good news is now profit is positive: 538.46. If you find fractional quantities way too annoying, we can add an integrality restriction on decision variables, which results in the integer solution $x_1^* = 73$ and $x_2^* = 37$, with profit 505. The reduction in profit should not be surprising: Whenever we add a restriction to a maximization problem, the value of the objective cannot increase. It is tempting to believe that whenever variables are restricted to integer values, all we have to do is finding the optimal solution in the continuous domain, and then round it in a sensible way; unfortunately, we will see in section B.6 that this is not the case; solving integer programming problems can be very hard.   ⬚

The optimal production mix is a typical linear programming model, and it can be easily solved for a large number of items and resources. Still, it looks too simplistic as a real-life production planning model, and indeed it is. The first difficulty we would face is demand variability; of course we could apply the model weekly, updating the demand data, but this would be quite myopic and it would not ensure satisfaction of demand. To begin with, if capacity is limited, we could build up inventories when demand is low, in order to meet demand when this is larger than capacity. In other words, there is a tradeoff between the cost of inventories and the cost of capacity. If we consider capacity as given, which is reasonable for short-term planning, we should generalize the model (B.1) to a multiperiod model taking demand variability into account.

When facing a nontrivial modeling problem, the starting point is thinking of the decisions we must take, and how they can be represented by decision variables. Then we try to express constraints on decision variables and to write the objective function (e.g., profit to maximize or cost to minimize). In the process, we might discover that additional variables are required to express a constraint or the objective, and the process may need to be iterated. The main ingredient in a multiperiod model is, of course, time. Say that, for our purposes, we need a production plan stating weekly production quantities for N items, over a planning horizon of T weeks. There are several reasons why we cannot stretch the planning horizon beyond some limit. To begin with, the model could be too hard to solve, but there is a more important issue: uncertainty. We deal with forecasting in chapter 3; it is clear that no demand forecast is reliable if it is too far in time. Given uncertainty in demand, we should solve the model on a rolling horizon basis, revising decisions on the basis of new information; hence stretching the forecasts beyond a certain limit is useless, if not dangerous. Also, the choice of the time bucket must be made sensibly; we are assuming that a weekly time bucket is enough, leaving to detailed execution level the timing of the single activities.

Since what we need is a time-bucketed production plan, we see that the most relevant decision variables are the manufactured quantities x_{it}, for item $i = 1, \ldots, N$ and time bucket $t = 1, \ldots, T$. We can immediately generalize the capacity constraints of example B.1. If we have M resource types, indexed by $m = 1, \ldots, M$, we should write a set of inequalities:

$$\sum_{i=1}^{N} r_{im} x_{it} \leq R_{mt}, \qquad m = 1, \ldots, M, \ t = 1, \ldots, T,$$

where r_{im} is the amount of resource m required for the production of one unit of item i and R_{mt} is the availability of resource m during time bucket t. We are considering this availability as given, and it could change over time because of planned maintenance, holidays, etc. In a different problem setting, available capacity can be a decision variable, rather than a given parameter. Typically, there is some uncertainty in execution, because of possible machine break-downs; hence, we should leave some slack when stating resource requirements and/or availability. All of this may not be relevant for distribution logistics, however. What is certainly common in distribution as well is demand uncertainty (chapter 5 deals with inventory management under uncertainty). For the sake of simplicity, we assume that we have quite reliable demand forecasts, which allows us to represent the demand for item i during time bucket t by a parameter d_{it}. Note that the demand d_{it} should not be just directly related to manufactured quantities x_{it}, since production and demand are partially decoupled by inventories; in a given week, we may sell less than demand (if we have not enough capacity and inventory), and we may produce more or less than demand. Hence, we need to introduce two more variables: the sold amount z_{it}, for each item i and time bucket t, and the inventory level I_{it}, *at the end* of time bucket t, after adding an amount x_{it} to inventory and selling an amount z_{it}. This last type of clarification is often essential in discrete-time models, where we consider the value of variables only at the beginning and at the end of time intervals, but not during the time intervals themselves.

Now we can write down the second type of constraint we met in the optimal mix model, i.e., sales cannot exceed demand:

$$z_{it} \leq d_{it}, \qquad \forall i, t.$$

Note that unlike the static model, this bound involves sales variables z_{it}, not production variables x_{it}. Now we certainly need some constraint linking the three groups of variables. This is an inventory balance constraint, stating that the inventory level at the end of a time bucket is given by what was available at the beginning of the time bucket, plus what has been produced, less what has been sold:

$$I_{it} = I_{i,t-1} + x_{it} - z_{it}, \qquad \forall i, t.$$

Strictly speaking, we made a mistake as far as time bucket $t = 1$ is concerned: for the first time bucket, the constraint involves the initial inventory level I_{i0}, which is *not* a decision variable, as it is given. This is an essential consideration when we really implement decision models using a software tool, but we will not pay attention to such issues. Of course, all the involved decision variables are restricted to non-negative values.

Now we are ready to write the last missing piece, i.e., the objective function. This must include the weekly inventory holding cost, as well as the variable cost and the selling price for each item; if we denote them by h_i, c_i, and p_i, respectively, we arrive at the following linear programming model:

$$\max \quad \sum_{i=1}^{N}\sum_{t=1}^{T} p_i z_{it} - \sum_{i=1}^{N}\sum_{t=1}^{T} c_i x_{it} - \sum_{i=1}^{N}\sum_{t=1}^{T} h_i I_{it} \qquad \text{(B.2)}$$

$$\text{s.t.} \quad I_{it} = I_{i,t-1} + x_{it} - z_{it}, \qquad i = 1, \ldots, N, \ \ t = 1, \ldots, T,$$

$$\sum_{i=1}^{N} r_{im} x_{it} \le R_{mt}, \qquad m = 1, \ldots, M, \ \ t = 1, \ldots, T,$$

$$z_{it} \le d_{it}, \qquad i = 1, \ldots, N, \ \ t = 1, \ldots, T,$$

$$x_{it}, \ z_{it}, \ I_{it} \ge 0.$$

What we have here is a simple version of a common model for production planning. In chapter 2 we show that this can be the basis of models which are also relevant in distribution logistics, even if purchasing rather then manufacturing is the core problem. In principle, the model can be extended to cope with more complex problems, involving the production of components and their assembly into end items. Furthermore, we will see that we should also consider possible economies of scale, e.g., due to fixed costs associated to the setup of each machine. In distribution logistics, we may have similar issues with fixed ordering or transportation costs, and they can be tackled by introducing binary decision variables, as we illustrate in section 2.2. Clearly, each extension increases the computational requirements of solving the model, which can spell trouble for large problem instances we meet in real life. Yet, there are some more basic issues that we must be well aware of.

- In the model above we have assumed perfect knowledge of all the involved data, most notably demand. There are a few lucky cases in which one may afford the luxury of making strictly to order. In such a case, assuming a known demand may make sense. In general, demand uncertainty may be very critical. One possibility to tackle the issue is by extending the linear programming framework to deal with uncertain scenarios. This is a very challenging and demanding approach, both for modeling and computing; we give an example of such an approach in section 1.5.2.

- Some data are not uncertain in the stochastic sense, but they may still be hard to quantify. The model above is based on a lost-sales assumption: If we are not able to serve a customer order immediately from stock, it is lost. In practice, it may be difficult to assess if what we lose is just the order or the customer; in the second case, data depend on our decisions. Sometimes, backorders are considered: Formally, negative inventory levels may be feasible, and they correspond to orders which are waiting to be fulfilled. Negative inventories should not be penalized by holding costs but by backlog costs. Unfortunately, it may be quite difficult to quantify a backlog cost; we deal with such issues in chapter 5.

- We have considered sales of different items as unrelated: What we sell or not for an item does not influence sales of other items. This is not necessarily true, especially if products are complements or substitutes. If cross-effects are well understood, we can try to extend the model; otherwise, the task may be too difficult.

- A multiperiod model may suffer from "end-of-horizon" effects. In the model above, it is easy to see that whatever the optimal solution in intermediate periods, we have $I_{iT}^* = 0$; i.e., inventories are depleted during the last time bucket. This happens because, from the model's point of view, there is no reason to keep inventories available after the end of the world. We may think of circumventing the difficulty by taking a suitably large time horizon and using the model in a rolling horizon fashion.[3] By doing so, it is reasonable to expect that the border effect will be less critical, but we pay a price in terms of increased computational burden; furthermore, if the planning horizon is too long, we lack reliable demand data.

All of the above limitations can be tackled using sophisticated approaches which are beyond the scope of an introductory book. The message we want to deliver is that the use of optimization models is always a means to an end, and it must be framed within a *decision process*. Optimization models may be a very useful support in this process, but they are not *the* decision process.

---

[3]This means that if our planning horizon consists of T time buckets, we do not wait until the end of this horizon to replan, but instead we replan immediately at the beginning of the next time period. In practice only the decisions pertaining to the first time bucket are actually implemented. The planning horizon rolls forward, in the sense that it involves initially time buckets $(1, \ldots, T)$, then $(2, \ldots, T+1)$, $(3, \ldots, T+2)$, and so on.

The danger is to find an optimal solution for a model, which is quite poor for the real problem. This may happen if we neglect implicit constraints, which may be difficult to formalize but make the "optimal" solution hard to implement. Moreover, no optimization model, however sophisticated, can overcome defects in the *organizational structure*. If the business process is flawed, mathematical modeling will hardly help. For instance, if the optimization model is used in one room by a planner, but in the next room the marketing manager decides to launch a campaign based on discounted prices without feeling the need of informing anyone, the likely spike in demand will make plans quickly obsolete and unusable. On the other hand, sale plans should be compatible with available capacity; if economic incentives lead the sales office to promise unrealistic due dates for order fulfillment, long-term customer relationships will suffer.

B.2   OPTIMIZATION MODELS

The building blocks of an optimization model are:

- A set of decision variables. In general, decision variables are collected in a vector within some multidimensional space; in the case of the optimal mix model, the size of this space is just the number of end items. In complex models, the size of the vector can be very large, but finite nonetheless. Typically, the term *mathematical programming* is reserved to optimization models in finite-dimensional spaces.[4]

- An objective function, which may be a cost to minimize or a profit to maximize.

- A feasible set, which constrains the decision variables; the feasible set depends on technological, economical, and commercial constraints.

In abstract terms, an optimization model is something like:

$$\min \quad f(\mathbf{x}) \tag{B.3}$$
$$\text{s.t.} \quad \mathbf{x} \in S \subseteq \mathbb{R}^n,$$

where f is the objective function, $\mathbf{x}$ is the vector of decision variables, and S is the feasible set, which is a subset of the $n-$dimensional space $\mathbb{R}^n$. If $S \equiv \mathbb{R}^n$, we have an **unconstrained problem**. There is no loss of generality in considering only minimization problems; a maximization problem can easily be transformed into an equivalent minimization problem by noting that

---

[4]We are ruling out optimal control problems in continuous time; in such a case, we deal with functions $u(t)$ whose domain, $t \in [0, T]$, is not even countable. Also, infinite-horizon discrete-time problems, whose feasible set is countable but infinite, are not considered here because of their limited role in supply chain management.

$\max f(\mathbf{x}) = -\min(-f(\mathbf{x}))$; by "equivalent," we mean that the transformed problem has the same set of optimal solutions (we should not take for granted that there is a unique optimal solution; actually, we should not even take for granted that there is an optimal solution).

Constrained optimization models are built by representing the feasible set in a practical way. We have seen the following constraints in the previous examples:

- Equality constraints, i.e., equations linking decision variables.

- Inequality constraints.

- Constraints stating that some variables should belong to specific sets, such as $x \in \mathbb{Z}_+ = \{0, 1, 2, 3, \ldots\}$ for integer variables.

Solving an optimization problem like (B.3) means finding a global optimizer.

DEFINITION B.1 *Given the optimization problem (B.3), a point $\mathbf{x}^* \in S$ is said to be a* **global optimizer** *if $f(\mathbf{x}^*) \leq f(\mathbf{x})$, for all $\mathbf{x} \in S$. We have a* **local optimizer** *if the condition only holds in the intersection between S and a neighborhood of $\mathbf{x}^*$.*

When one speaks of an "optimum," a little ambiguity arises, because it is not quite clear if we mean the optimizer $\mathbf{x}^*$ or the optimal value $f(\mathbf{x}^*)$; usually, the context clarifies what we really mean. We may also use the notation $\mathbf{x}^* = \arg\min_{\mathbf{x} \in S} f(\mathbf{x})$.

Given an optimization model, the following cases may arise[5]:

1. There is no optimal solution because the feasible set S is empty; this may happen, e.g., when production capacity is small with respect to demand and we do not admit lost sales.

2. There is no optimal solution because the optimum goes to infinity (typically, this is due to a modeling error).

3. There exists a unique optimizer.

4. We have multiple equivalent optimizers; in such a case, we might wish to come up with a "secondary" criterion in order to discriminate between solutions which are equally good from the main point of view.

The first case is not as unlikely as one might think. It may be the result of excessively stringent requirements on the solution. But even if it happens only every now and then, you may imagine the future of a decision support

---

[5]We do not consider rather pathological cases, such as $\min x$ subject to $x > 2$, in which there is no optimal solution because the feasible set is open; in this case $x^* = 2$ is not the optimizer, but it solves the related problem $\inf x$.

system which occasionally informs a planner that he should shut everything down because there is no solution. The least we should offer is some diagnostic about where constraints can be relaxed in order to restore feasibility. This task can be accomplished, e.g., by adopting "elastic" model formulations.

Example B.2 In section B.1 we developed a multiperiod planning model allowing for lost sales. Now assume that your boss insists that demand must be met at all costs. That model could be modified as follows in order to forbid both lost sales and backorders:

$$\min \quad \sum_{i=1}^{N} \sum_{t=1}^{T} h_i I_{it}$$

$$\text{s.t.} \quad I_{it} = I_{i,t-1} + x_{it} - d_{it}, \qquad i = 1, \ldots, N, \; t = 1, \ldots, T,$$

$$\sum_{i=1}^{N} r_{im} x_{it} \le R_{mt}, \qquad m = 1, \ldots, M, \; t = 1, \ldots, T,$$

$$x_{it}, \; I_{it} \ge 0.$$

Here we want to meet demand at minimum (inventory holding) cost; since we assume that demand must be met, revenue is fixed and profit maximization is equivalent to cost minimization (we assume pricing is an outside decision). By the way, a typical newcomer's mistake, when trying to capture the requirement "demand must be met," is writing down a constraint like $x_{it} \ge d_{it}$. But such a constraint does not make any sense, because it rules out inventory holding, and it can only be satisfied by the trivial solution $x_{it} = d_{it}$, which is probably not feasible if capacity constraints do matter. Actually, what makes sure that demand is met is the non-negativity constraint on inventory level, which together with the inventory balance constraint entails $I_{i,t-1} + x_{it} \ge d_{it}$.

It is clear that meeting demand can be a hard constraint when capacity is tight, and we may be unable to find a solution. What we should do is to help the decision maker in figuring out where the critical requirements are (which demand is too high and when, or which resource is too scarce and when). Then, it is up to the decision maker to find a way out. One way to do so is relaxing a constraint by a suitable *penalty function*. For instance, if we define a suitably high penalty coefficient β for lost sales, we may write the following elastic model formulation:

$$\min \quad \sum_{i=1}^{N} \sum_{t=1}^{T} h_i I_{it} + \beta \sum_{i=1}^{N} \sum_{t=1}^{T} z_{it}$$

$$\text{s.t.} \quad I_{it} = I_{i,t-1} + x_{it} - d_{it} + z_{it}, \qquad i = 1, \ldots, N, \; t = 1, \ldots, T,$$

$$\sum_{i=1}^{N} r_{im} x_{it} \le R_{mt}, \qquad m = 1, \ldots, M, \; t = 1, \ldots, T,$$

$$x_{it}, \; z_{it}, \; I_{it} \ge 0,$$

where z_{it} represent a lost sale for item i in time bucket t. If the penalty β is high enough, whenever the original model is feasible, the optimal solution of the elastic formulation is the same as in the previous model, i.e., $z_{it}^* \equiv 0$. If we have some $z_{it}^* > 0$, this means that there is a critical order somewhere. Having an idea of where to look is helpful in supporting a negotiation process with some customers, who may be willing to wait a little more for delivery or may accept a substitute product. All of these adjustments can actually be modeled and explicitly represented in an optimization model, but some decision makers may find themselves in trouble when required to quantify the costs of these actions.

By the same token, we may consider relaxing capacity constraints as follows:

$$\sum_{i=1}^{N} r_{im} x_{it} \leq R_{mt} + O_{mt}, \qquad m = 1, \ldots, M, \ \ t = 1, \ldots, T,$$

where $O_{mt} \geq 0$ plays the role of overtime capacity. If this new variable is penalized by a true economic cost, the resulting model actually represents a joint production and capacity planning model; if it is penalized by a large coefficient with no real monetary value, we have a true penalty function. □

Now that we have an idea of how to formulate an optimization model, let us consider how we can solve one. What we know from Calculus is that setting the first-order derivative to zero, i.e., enforcing a stationarity condition, may be a starting point.

Example B.3 Let us consider the problem

$$\min f(x) = Ax + \frac{B}{x}, \quad \text{s.t. } x \geq 0,$$

where A and B are strictly positive parameters. In the main body of the text, we show that this is the form of a well-known model to find an "optimal" quantity to order. The feasible set includes a troublesome point, $x = 0$, where the objective function is not defined. We could rewrite the feasible set as $x > 0$, but we prefer to avoid the trouble and note that when x gets smaller and smaller, the objective function grows without bound; hence the optimal solution must lie in the interior of the domain and the constraint is actually irrelevant.

Setting the first-order derivative with respect to x to zero, we get the stationarity condition:

$$f'(x) = A - \frac{B}{x^2} = 0,$$

which yields the candidate solution $x^* = \sqrt{B/A}$. We see that indeed $x^* \geq 0$, but to make sure this is a true minimum, we must check the second-order derivative too:

$$f''(x) = 2\frac{B}{x^3}.$$

Since this is positive on the domain of interest, the candidate point is actually the global minimizer.   ▯

When we have an optimization problem involving several decision variables, the stationarity condition would involve all of the first-order derivatives, yielding a possibly awkward system of nonlinear equations:

$$\frac{\partial f}{\partial x_i}(\mathbf{x}^*) = 0, \qquad i = 1, \ldots, n. \tag{B.4}$$

We recall that we may collect the first order derivatives of a function in a vector, called gradient and denoted by $\nabla f(\mathbf{x}^*)$. Hence, we may rewrite the stationarity conditions in the compact form $\nabla f(\mathbf{x}^*) = \mathbf{0}$. However, leaving aside the possible difficulty of solving this system of equations, this is not what we need to solve most optimization problems.

- To begin with, stationarity conditions assume that the objective function is differentiable, which should not be taken for granted; a practical case of a nondifferentiable objective function featuring kinks arises in purchasing decisions, when quantity discounts are offered.

- If the decision variables are discrete, e.g., because they are restricted to integer values, we cannot rely on the derivative concept.

- We know that first-order conditions are not sufficient, as they do not discriminate mimima, maxima, and saddle points, but they are actually not even necessary in constrained optimization. In a linear programming problem, stationarity cannot play any significant role; to see why, consider an objective function like $f(x_1, x_2) = 45x_1 + 60x_2$, and notice that its gradient is constant and will never vanish.

In general, it is hard to come up with simple global optimality conditions that are both sufficient and necessary. Typically, we settle for weaker conditions which are just necessary for local optimality. Despite their limitations, such conditions are the starting point for the development of numerical optimization procedures which are widely used.[6] There are, however, practically relevant cases in which some difficulties can be avoided. These cases exploit some properties linked to *convexity*, which is the subject of next section.

B.3   CONVEX SETS AND FUNCTIONS

The difficulty in solving an optimization problem does depend on the number of variables and constraints, but this need not be the main driving factor.

---

[6]We should remark that typical optimization routines offered in spreadsheets and numerical libraries aim at *local* optimization; the end result may depend on the starting point provided by the user to the search algorithm.

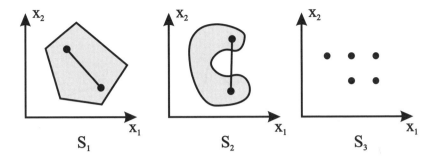

Fig. B.1   Illustrating the concept of convex set.

Nonlinearity is typically more troublesome than linearity, but the main discriminating feature is *convexity*, both in the feasible region and the objective function. We introduce convex sets first, then convex functions. The first concept is useful to investigate properties of the feasible set of an optimization problem; the second concept is useful to characterize the objective function.

Informally, a set $S \subset \mathbb{R}^n$ is convex if, taken any pair of points $\mathbf{x}$ and $\mathbf{y}$ in S, all points on the line segment joining $\mathbf{x}$ and $\mathbf{y}$ lie in S as well. For instance, the set S_1 in figure B.1 is convex, whereas S_2 and S_3 are not. It is worth noting that S_3 is an example of the kind of feasible sets we deal with when tackling an integer programming problem. Formally, the segment joining two points can be described as a linear combination of them, such that the weights are non-negative and their sum is 1; such a linear combination is called **convex combination**:

$$\lambda \mathbf{x} + (1 - \lambda)\mathbf{y}$$

for $\lambda \in [0, 1]$. We see that when λ is 0 and 1 we get the two extreme points of the segment. Now we may formally define a convex set.

DEFINITION B.2 (Convex set) *A set $S \subset \mathbb{R}^n$ is said convex if, for any* $\mathbf{x}, \mathbf{y} \in S$, *we have*

$$\lambda \mathbf{x} + (1 - \lambda)\mathbf{y} \in S$$

for $0 \leq \lambda \leq 1$.

Let us consider again set S_1 in figure B.1; this set is a **polyhedron**, i.e., the intersection of a finite number of half spaces (in two dimensions, the intersection of half planes). A **half space** is the subset of points in $\mathbb{R}^n$ lying on one side of a hyperplane, i.e., the set of points satisfying a linear inequality like

$$\sum_{i=1}^{n} \alpha_i x_i = \boldsymbol{\alpha}' \mathbf{x} \leq \beta.$$

It is easy to see that a half space is a convex set; a polyhedron is convex because intersection is an operation preserving convexity. Proving this is helpful in reinforcing the concepts above.

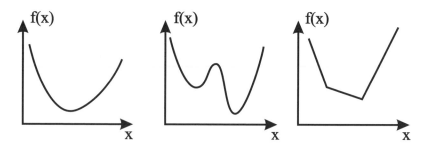

Fig. B.2   Illustrating the concept of convex function.

Example B.4 Let us prove that the intersection of convex sets is a convex set. Consider m convex sets S_j, $j = 1, \ldots, m$ and their intersection:

$$S = \bigcap_{j=1}^{m} S_j.$$

We know that, taken any pair of points $\mathbf{x}$ and $\mathbf{y}$ in set S_j, for any j, convexity implies that $\lambda \mathbf{x} + (1 - \lambda)\mathbf{y} \in S_j$, for any λ between 0 and 1. If we take two points in the intersection S, then the two points belong to *all* of the sets S_j. But then also $\lambda \mathbf{x} + (1 - \lambda)\mathbf{y}$ belong, for any $\lambda \in [0, 1]$, to all sets S_j; hence, this combination belongs to the intersection S too.   ▯

Convexity of sets is readily extended to convexity of functions. The first function in figure B.2 is convex, but the second is not. If we regard these functions as costs to be minimized, we see that the first function has one local minimizer that is global as well, whereas the second one has two local minimizers, and only one of them is the global one. We expect that local minima are a complicating factor both for optimization algorithms and for the development of optimality conditions. For instance, stationarity cannot discriminate local vs. global optimizers. However, stationarity is a concept requiring differentiability, whereas convexity does not. The third function in the figure is in fact convex, but it is not everywhere differentiable. Convexity of a function is actually linked to the convexity of its epigraph, i.e., the set of points above the function graph. If the epigraph of a function is a convex set, the function is convex too; this can be formalized as follows.

DEFINITION B.3 (Convex function) *A function $f : S \to R$ is convex on S if, for any $\mathbf{x}, \mathbf{y} \in S$, we have*

$$f\left(\lambda \mathbf{x} + (1 - \lambda)\mathbf{y}\right) \leq \lambda f(\mathbf{x}) + (1 - \lambda)f(\mathbf{y})$$

for $0 \leq \lambda \leq 1$.

From a geometrical point of view, the condition above tells that a function is convex if, given any pair of points on its graph, the line segment joining them

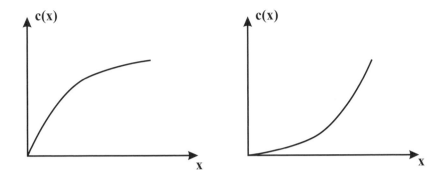

Fig. B.3   Economy and diseconomy of scale.

lies above the graph of the function. From an economical point of view, convex functions are associated to a diseconomy of scale. The functions in figure B.3 represent a total cost $c(x)$ as a function of the level x of some activity, such as a produced, purchased, or transported quantity. As expected, the cost is increasing with respect to x, but in the function on the right the marginal cost is increasing too,[7] which is exactly what diseconomies of scale are about. On the left, we see a function displaying an economy of scale (decreasing marginal cost); a typical case occurs when quantity discounts on some purchased item are offered by the supplier. This type of function is, in some sense, a convex function turned upside down. We say that a function $f(\mathbf{x})$ is **concave** when $-f(\mathbf{x})$ is convex. Indeed, concave functions are used, among other things, to model economies of scale.[8]

We have given a general definition of a convex function, which may be hard to check. It is useful to mention that, when the function $f(x)$ depends on a single variable and is twice differentiable, convexity is equivalent to non-negativity of the second-order derivative. If the function represents a cost, the condition $f''(x) \geq 0$ for $x \in S$ basically says that marginal cost is increasing. Reversing the inequality, we characterize concave functions. The related condition for functions of several variables is a bit more involved. Here we just point out that an affine function, such as $f(\mathbf{x}) = \mathbf{a}'\mathbf{x} + b$, is a very peculiar one, as it is both convex and concave.

In optimization, functions are used both to represent the objective function and to describe the feasible set. It is interesting to shed some light on the relationship between convex functions and convex sets when dealing with an inequality constraint.

---

[7]If the cost function is differentiable, the marginal cost is the first-order derivative $c'(x)$.
[8]A relevant example of a concave cost function is the total cost in the economic order quantity model, as given in equation (1.2) on page 23; there is an economy of scale with respect to demand, and this is important in distribution logistics.

Example B.5 We show that the region S described by the inequality constraint

$$g(\mathbf{x}) \leq 0$$

is a convex set if g is a convex function.

If $\mathbf{x} \in S$, it means that $g(\mathbf{x}) \leq 0$; by the same token, if $\mathbf{y} \in S$, then $g(\mathbf{y}) \leq 0$. What we want to prove is that $\lambda\mathbf{x} + (1 - \lambda)\mathbf{y} \in S$, for all $\lambda \in [0, 1]$, i.e., that $g(\lambda\mathbf{x} + (1 - \lambda)\mathbf{y}) \leq 0$. But, since g is convex, we have

$$g(\lambda\mathbf{x} + (1 - \lambda)\mathbf{y}) \leq \lambda g(\mathbf{x}) + (1 - \lambda)g(\mathbf{y}) \leq 0,$$

where the last inequality depends on the fact that we are summing non-positive terms, which are obtained by multiplying a non-positive quantity by a non-negative coefficient. This proves that $\lambda\mathbf{x} + (1 - \lambda)\mathbf{y}$ is in S. ▯

The properties we have proved in examples B.4 and B.5 are useful when we characterize a feasibility region by inequality constraints, but what about the equality constraints $h(\mathbf{x}) = 0$? We can regard an equality constraint as a pair of inequality constraints $h(\mathbf{x}) \leq 0$ and $h(\mathbf{x}) \geq 0$. This implies that an equality constraint describes a convex set only if $h(\mathbf{x})$ both convex and concave and, as we have said, this happens only for an affine function[9]:

$$h(\mathbf{x}) = \sum_{i=1}^{n} a_i x_i + b.$$

Convexity is a property that makes optimization problems relatively easy. For instance, if we want to minimize a convex differentiable function, we get a simple necessary and sufficient condition for global optimality in unconstrained optimization.

THEOREM B.4 *Consider the unconstrained optimization problem*

$$\min_{\mathbf{x}} f(\mathbf{x}).$$

If the objective function is convex and differentiable, the stationarity condition

$$\nabla f(\mathbf{x}) = \mathbf{0}$$

is necessary and sufficient for global optimality.

In fact, the function we considered in example B.3 is convex, and in that case stationarity yields the (unique) global minimizer, without the need of bothering about local minima, maxima, nor saddle points. If we deal with a maximization problem, it is easy to see that the theorem can be applied by requiring concavity of the objective.

---

[9]We typically speak of linear constraints, but the function is actually linear only if $b \equiv 0$.

When dealing with constrained optimization, stationarity of the objective is not necessarily related to optimality. However, we can exploit stationarity again by using a trick based on the so called Lagrange multipliers. We do this in the next section, where we also see that in constrained optimization we need both a convex objective and a convex feasible set to get a relatively easy problem.

B.4 NONLINEAR PROGRAMMING

The constrained optimization problem

$$\begin{aligned}
\min \quad & f(\mathbf{x}) & \text{(B.5)}\\
\text{s.t.} \quad & h_j(\mathbf{x}) = 0, \quad && j = 1, \ldots, m,\\
& g_k(\mathbf{x}) \le 0, \quad && k = 1, \ldots, l
\end{aligned}$$

is a nonlinear programming problem if even one function among f, h_j, or g_k is nonlinear. The stationarity condition (B.4) for the objective function does not help in finding an optimizer (ruling out trivial cases); to see why, a look at the following counterexample suffices:

$$\min_{2 \le x \le 3} x^2.$$

The obvious optimizer $x^* = 2$ is not a stationarity point, because it is the lower bound on x that determines the optimal solution; the function is stationary at the origin, but this point is outside the feasible region. However, assuming that all of the involved functions are well-behaved enough, in terms of differentiability, we can try to use stationarity concepts to find candidate optimal points.

For the sake of simplicity, we start considering the equality constrained case:

$$\begin{aligned}
\min \quad & f(\mathbf{x}) & \text{(B.6)}\\
\text{s.t.} \quad & h_j(\mathbf{x}) = 0, \quad && j = 1, \ldots, m,
\end{aligned}$$

which can be dealt with by the classical **Lagrange multipliers** method.

THEOREM B.5 *Assume that the functions f and h_j in problem (B.6) meet some differentiability requirements, that the point $\mathbf{x}^*$ is feasible, and that the constraints satisfy a suitable regularity property in $\mathbf{x}^*$. Then, a necessary condition for local optimality of $\mathbf{x}^*$ is that there exist numbers λ_j^*, $j = 1, \ldots, m$, called Lagrange multipliers, such that*

$$\nabla f(\mathbf{x}^*) + \sum_{j=1}^{m} \lambda_j^* \nabla h_j(\mathbf{x}^*) = \mathbf{0}.$$

The reader has certainly noticed that we have been very loose in stating the conditions of the theorem. In fact, what we need for the main body of the text is the *concept* of Lagrange multiplier, and since in this book we deal with relatively simple problems, we can do without too many technicalities. However, it is important to realize that, in general, the theorem is somewhat weak. It holds under technical conditions,[10] which we do not describe in detail; furthermore, it is only a *necessary* (hence, not sufficient) condition for *local* (hence, not global) optimality. The good news is that it can be shown that the condition of the theorem is necessary and sufficient for a **convex optimization problem**. We say that a minimization problem is convex, if its feasible set and objective function are both convex.

To interpret the condition above, we may observe that it generalizes the stationarity condition; the trick is requiring stationarity not for the objective function, but for the following **Lagrangian function**:

$$\mathcal{L}(\mathbf{x}, \boldsymbol{\lambda}) = f(\mathbf{x}) + \sum_{j=1}^{m} \lambda_j h_j(\mathbf{x}) = f(\mathbf{x}) + \boldsymbol{\lambda}' \mathbf{h}(\mathbf{x}). \tag{B.7}$$

In practice, the "recipe" requires us to augment the objective function by the constraints, which are multiplied by the Lagrange multipliers, and to enforce stationarity both with respect to the decision variables $\mathbf{x}$:

$$\nabla_{\mathbf{x}} \mathcal{L}(\mathbf{x}, \boldsymbol{\lambda}) = \nabla f(\mathbf{x}) + \sum_{j=1}^{m} \lambda_j \nabla h_j(\mathbf{x}) = \mathbf{0}, \tag{B.8}$$

and with respect to the multipliers, which actually yields the constraints again:

$$\nabla_{\boldsymbol{\lambda}} \mathcal{L}(\mathbf{x}, \boldsymbol{\lambda}) = \begin{bmatrix} h_1(\mathbf{x}) \\ h_2(\mathbf{x}) \\ \vdots \end{bmatrix} = \mathbf{0}, \tag{B.9}$$

The mechanism can be best clarified by an example, but it is important to check that the conditions above are consistent. We have n decision variables and m equality constraints ($m < n$); equations (B.8) and (B.9) yield a system of $n + m$ (possibly) nonlinear equations to find the n values x_i^* and the m multipliers λ_j^*.

---

[10] That differentiability is required is clear, otherwise we cannot take the derivatives involved in the condition. The "regularity" conditions are known in the literature as constraint qualification conditions and take many forms. One such condition is that in $\mathbf{x}^*$ the gradients of functions h_j are linearly independent. That this condition makes some sense is not too difficult to understand. The stationarity condition in theorem B.5 states that the gradient of the objective can be expressed as a linear combination of the gradient of the constraints; in pathological cases this may be impossible if gradients of the constraints are, e.g., parallel.

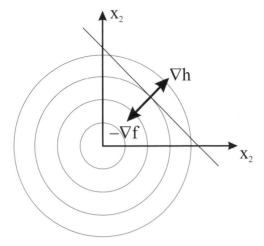

Fig. B.4 A quadratic programming example: geometrical interpretation of Lagrange conditions.

Example B.6 Consider the quadratic programming problem:

$$\min \quad x_1^2 + x_2^2$$
$$\text{s.t.} \quad x_1 + x_2 = 4.$$

This is called "quadratic" programming because the objective function is a quadratic form and the constraints are linear; assuming that the objective function is convex, this is the easiest case of nonlinear programming. Since this quadratic form is indeed convex, we may use theorem B.5 to find the global optimum. We associate the constraint with one multiplier λ, and form the Lagrangian function:

$$\mathcal{L}(x_1, x_2, \lambda) = x_1^2 + x_2^2 + \lambda(x_1 + x_2 - 4).$$

The stationarity conditions:

$$\frac{\partial \mathcal{L}}{\partial x_1} = 2x_1 + \lambda = 0,$$
$$\frac{\partial \mathcal{L}}{\partial x_2} = 2x_2 + \lambda = 0,$$
$$\frac{\partial \mathcal{L}}{\partial \lambda} = x_1 + x_2 - 4 = 0,$$

are a system of linear equations, yielding $x_1^* = x_2^* = 2$ and $\lambda^* = -4$. We may also notice that the equality constraint can also be written as $4 - x_1 - x_2 = 0$; if we do so, we just have a change in the sign of the multiplier.

We may get an intuitive feeling for the conditions by taking a look at figure B.4, where we see the level curves of the objective function (the concentric circles) and the feasible region (a line). From a geometrical point of view, the problem calls for finding the closest point to the origin on the line $x_1 + x_2 = 4$. We note that the optimizer is where this line is tangent to the level curve associated to the lowest value of the objective. From an analytical point of view, the gradient of the objective function $f(\mathbf{x}) = x_1^2 + x_2^2$ is

$$\nabla f(x_1, x_2) = \begin{bmatrix} \dfrac{\partial f}{\partial x_1} \\[2ex] \dfrac{\partial f}{\partial x_2} \end{bmatrix} = \begin{bmatrix} 2x_1 \\[1ex] 2x_2 \end{bmatrix}.$$

This gradient, changed in sign, is a vector pointing toward the origin, which is the steepest descent direction for the distance. At point $\mathbf{x}^* = (2, 2)$ the gradient is $[4, 4]'$. The gradient of the constraint $h(\mathbf{x}) = x_1 + x_2 - 4$ is

$$\nabla h(x_1, x_2) = \begin{bmatrix} \dfrac{\partial h}{\partial x_1} \\[2ex] \dfrac{\partial h}{\partial x_2} \end{bmatrix} = \begin{bmatrix} 1 \\[1ex] 1 \end{bmatrix}.$$

Note that this vector is orthogonal to the feasible region and is parallel to the gradient of the objective at the optimizer. If we multiply the gradient of the constraint by $\lambda^* = -4$ and we add the result to the gradient of the objective, we get the null vector, as required. Actually, all of this boils down to requiring that the gradient $-\nabla f^*$, i.e., the descent direction for the objective, is orthogonal to the constraints at the optimizer; this means that further improvements could only be obtained by going out of the feasible region, which is forbidden. The last condition is what characterizes the optimizer.

□

B.4.1   The case of inequality constraints

The case of inequality constraints can be tackled by an approach which is similar to the Lagrange multiplier method, even though historically it has been developed much later. The basic theorem here is known under the names of Kuhn and Tucker. Given our limited aim, we will try to justify their result intuitively. Let us consider a problem like

$$\max \quad \sum_{i=1}^{n} f_i(x_i) \tag{B.10}$$

$$\text{s.t.} \quad \sum_{i=1}^{n} g_i(x_i) \le b, \tag{B.11}$$

with a single inequality constraint. To be concrete, we interpret the decision variables x_i, $i = 1, \ldots, n$, as activities yielding a profit $f_i(x_i)$ and consuming a resource amount $g_i(x_i)$. The objective function (B.10) is total profit, and the *budget* constraint (B.11) says that total resource consumption cannot exceed its availability b. Note that the objective function is measured in monetary terms, whereas b is measured in resource units.

Now, there are two possible cases: Either the resource is fully utilized in an optimal solution, or it is not. In the case of a resource budget constraint, it is likely that the first case will apply, i.e., the constraint is *active* in the optimal solution; if we accept this hypothesis, for the moment, the inequality constraint can be treated as an equality. Then, we can think of applying the same approach we have seen for equality constraints. Introducing a multiplier μ, we write the Lagrangian function:

$$\mathcal{L}(\mathbf{x}, \mu) = \sum_{i=1}^{n} f_i(x_i) + \mu\left(b - \sum_{i=1}^{n} g_i(x_i)\right) = \sum_{i=1}^{n}[f_i(x_i) - \mu g_i(x_i)] + \mu b.$$

In the case of equality constraint, the way we include the constraints in the Lagrangian function is irrelevant. In the case of inequality constraints, this is not true. Using economic intuition, we may try to see why the way we have included the constraint makes sense. To this aim, let us interpret the Lagrangian function as a profit, for a *given* value of the multiplier μ. This function includes a term, μb, which is constant for a given value of the multiplier and can be disregarded. The Lagrangian includes the sum of profits, minus a coefficient μ times the consumption of the resource. From a dimensional point of view, we immediately see that the multiplier is a price: money per unit resource. If we interpret the multiplier this way, it is also clear that it cannot be negative. Indeed, when dealing with equality constraints, the sign of the multiplier is not restricted and the constraint can be introduced in the Lagrangian in both possible ways; in the case of an inequality constraint, the multiplier is restricted in sign and we must pay attention to the sense of the inequality.

Remember that, for the moment, we are assuming that the resource budget is fully utilized at the optimum. What is the right price μ^* associated with the optimal solution? We should find a resource price such that the overall consumption is equal to the resource availability, no more, no less. In fact, for a given price, the optimization problem could be decomposed into a set of n unrelated unconstrained problems, one for each decision variable x_i. Each problem corresponds to the optimization of a single activity in which resource availability is taken into account by a resource price which should discourage excessive resource usage. Someone should coordinate all of the individual decisions, by pricing the resource so that the n independent agents in charge of each single activity use the resource in such a way to exactly consume the available budget b. If we exceed the budget, the price should be increased; if we do not saturate the resource, then we should decrease the price (if the budget constraint is active at the optimum). We illustrate this type of interpretation in a few places in the main body of the text.

But what if the constraint is not active at the optimum? In such a case, we have

$$\sum_{i=1}^{n} g_i(x_i^*) < b,$$

for the optimal solution $\mathbf{x}^*$. We may get this solution by assigning a price $\mu^* = 0$ and solving the unconstrained problem.[11]

We will further pursue the economic interpretation in the next section, but what we have seen suggests a few intuitive conclusions:

- When dealing with an inequality constraint $g(\mathbf{x}) \leq 0$, the sign of the multiplier is restricted and the sense of the inequality is important when we add the constraint to the Lagrangian function.

- If the multiplier is strictly positive, $\mu^* > 0$, then the constraint is active, $g(\mathbf{x}^*) = 0$; if the constraint is inactive, $g(\mathbf{x}^*) < 0$, then the multiplier must be zero. In other words, at least one of them must be zero, which is summarized by the **complementary slackness** condition $\mu^* g(\mathbf{x}^*) = 0$.

All of these intuitive (and far from rigorous) arguments can be summarized in the following theorem, which we state for an inequality-constrained minimization problem like

$$\begin{aligned} \min \quad & f(\mathbf{x}) && \text{(B.12)} \\ \text{s.t.} \quad & g_k(\mathbf{x}) \leq 0, && k = 1, \dots, l. \end{aligned}$$

THEOREM B.6 *Assume that the functions f and g_j in problem (B.12) are suitably differentiable, that point $\mathbf{x}^*$ is feasible, and that the constraints satisfy a regularity condition in $\mathbf{x}^*$. Then, a necessary condition for the local optimality of $\mathbf{x}^*$ is that there exist numbers $\mu_k^* \geq 0$, $k = 1, \dots, l$, such that*

$$\nabla f(\mathbf{x}^*) + \sum_{k=1}^{l} \mu_k^* \nabla g_k(\mathbf{x}^*) = \mathbf{0}$$

and

$$\mu_k^* g_k(\mathbf{x}^*) = 0, \qquad k = 1, \dots, l.$$

These conditions are known as **Kuhn–Tucker conditions** and are a generalization of the Lagrange conditions for equality-constrained problems. The first

---

[11] It is tempting to say that if a constraint is not active at the optimal solution, the constraint can be eliminated from the model. Actually this not always true. It is possible to build counterexamples, such as *nonconvex* problems in which by eliminating an inactive constraint the optimal solution is still a local optimum, but another point becomes feasible and is the new global optimum. What we can say is that small perturbations of an inactive constraint do not change the optimal solutions; we will see more of this interpretation in the next section.

condition can be interpreted as the stationarity of the Lagrangian function

$$\mathcal{L}(\mathbf{x}, \boldsymbol{\mu}) = f(\mathbf{x}) + \sum_{k=1}^{l} \mu_k g_k(\mathbf{x}) = f(\mathbf{x}) + \boldsymbol{\mu}' \mathbf{g}(\mathbf{x}).$$

With respect to the equality-constrained case, we must also require non-negativity of the multipliers and complementary slackness.

The remarks we have made about the limitations of theorem B.5 apply here as well; in the convex and differentiable case, these conditions enable us to find the global optimum. If we have both equality and inequality constraints, we form a Lagrangian using all of them, but apply the additional Kuhn–Tucker restrictions to inequalities only.

The theorem applies to a minimization problem, but what if we have a maximization problem? The answer can be found in our previous intuitive reasoning. In this case, the Lagrangian function should be

$$\mathcal{L}(\mathbf{x}, \boldsymbol{\mu}) = f(\mathbf{x}) - \boldsymbol{\mu}' \mathbf{g}(\mathbf{x}).$$

This can be proved mathematically, but it is consistent with economic intuition. For instance, we build the Lagrangian function this way in section 5.2.1, when we tackle a multi-item newsvendor problem.

We close this section by warning readers against a common misunderstanding. We have justified Kuhn–Tucker conditions by an economic argument, whereby the Lagrangian function was interpreted as a profit. For a *given* multiplier μ, it makes sense to maximize the profit. Now, in this case, it is tempting to say that the stationarity conditions on the Lagrangian function are conditions for a maximum of the Lagrangian. By the same token, in an equality-constrained problem such as $\min f(\mathbf{x})$, subject to $\mathbf{h}(\mathbf{x}) = \mathbf{0}$, it is tempting to say that we look for the minimum of the Lagrangian function $\mathcal{L}(\mathbf{x}, \boldsymbol{\lambda}) = f(\mathbf{x}) + \boldsymbol{\lambda}' \mathbf{h}(\mathbf{x})$. But it is easy to see that this is wrong, as the minimum of this Lagrangian function is always $-\infty$. To see this, fix an arbitrary point $\mathbf{x}^o$; we may drive the value of the Lagrangian function to $-\infty$ just by setting λ_j to $+\infty$ or $-\infty$, depending on the sign of each constraint $h_j(\mathbf{x}^o)$. In fact, a deeper study of the subject, leading to duality theory, shows that we should minimize the Lagrangian function with respect to the original variables $\mathbf{x}$, but we should maximize it with respect to Lagrange multipliers (which are restricted in sign for inequality constraints). All of this is beyond the scope of this book.

B.4.2 An economic interpretation of Lagrange multipliers: shadow prices

To introduce Kuhn–Tucker conditions we have suggested an economic interpretation, i.e., that Lagrange multipliers express how much a constraint is "important" at the optimum. In this section, we would like to dig a bit deeper into this interpretation.

Example B.7 Consider the quadratic programming problem:

$$\begin{aligned} \min \quad & (x_1 - 2)^2 + (x_2 - 2)^2 \\ \text{s.t.} \quad & x_1 + x_2 = b, \end{aligned}$$

where b is a parameter, and let us investigate how the optimal value changes as a function of b. In fact, the optimal value of the objective is a function $q(b) = f(x_1^*, x_2^*; b)$, and in this very simple case we may find this function explicitly. To this aim, we may eliminate the constraint in order to get an equivalent unconstrained problem. From the constraint we get $x_2 = b - x_1$ and plug this into the objective function to obtain

$$(x_1 - 2)^2 + (b - 2 - x_1)^2.$$

Then, setting the first-order derivative with respect to x_1 to zero, we get $x_1^* = b/2$. This also implies $x_2^* = b/2$, which can be easily checked geometrically, since the problem asks for finding a point on the line $x_1 + x_2 = b$, such that the distance from point $(2, 2)$ is minimal. The optimal value as a function of b is

$$q(b) = 2 \left(\frac{b}{2} - 2 \right)^2,$$

and if we take the derivative with respect to b we obtain

$$\frac{dq}{db} = b - 4.$$

This shows that the optimal value will decrease, if we increase b when the line is below the point $(2, 2)$ (the line gets closer to the point); if the line is above that point, increasing b will increase the distance. If we neglect all of this and apply the Lagrange multiplier approach, we build the Lagrangian first:

$$\mathcal{L}(x_1, x_2, \lambda) = (x_1 - 2)^2 + (x_2 - 2)^2 + \lambda(x_1 + x_2 - b),$$

and the stationarity conditions are

$$\begin{aligned} \frac{\partial \mathcal{L}}{\partial x_1} &= 2(x_1 - 2) + \lambda = 0, \\ \frac{\partial \mathcal{L}}{\partial x_2} &= 2(x_2 - 2) + \lambda = 0, \\ \frac{\partial \mathcal{L}}{\partial \lambda} &= x_1 + x_2 - b = 0, \end{aligned}$$

which yield

$$x_1^* = x_2^* = \frac{b}{2}, \qquad \lambda^* = 4 - b.$$

We see that, apart from a change in sign, the multiplier is actually the derivative of the optimal value with respect to b. ◻

The result of the example suggests that a multiplier measures the sensitivity of the optimal value with respect to a perturbation of the right-hand side of the associated constraint. Indeed, this can be proved in a more general setting. The change in sign is not really relevant, as it depends on how we build the Lagrangian function, but differentiability of the value function $q(b)$ is an issue.

Thanks to the complementary slackness condition, we can also extend the result to the case of an inequality constraint. If the constraint is inactive, the sensitivity is zero, because small perturbations of the constraint have no effect. Otherwise, the constraint can be substituted by an equality constraint and we are back to the case above, with the additional caveat concerning the sign of the multiplier. We may also see why the non-negativity condition on the multiplier makes sense: If b is the resource availability and we are minimizing cost, an increase in the availability can only decrease the cost (remember the change in sign). We have also seen that if q is money and b is measured in resource units, the multiplier has the dimension of a price; indeed, an alternative term to indicate a Lagrange multiplier is **shadow price**.

The shadow price tells, to a first-order approximation, how much the optimal value would change if we could increase the availability of a resource. Hence, this also tells the maximum unit price we should be willing to pay for one more unit of that resource. If the resource price is larger than the shadow price, then the increase in overall cost term in the objective function would be more than offset by the increase in profit. If the resource is not fully utilized, then there is no point in getting more (the shadow price is zero).

Of course, this interpretation must be taken with care, as it holds for small perturbations, just like any first-order approximation based on a first-order derivative. It can be seen as a sort of marginal analysis, which can be very useful in interpreting optimality conditions.

Example B.8 Let us consider again problem (B.10). Given the additive form of both the objective and the constraint, the stationarity conditions on the Lagrangian function can be written separately for each activity $i = 1, \ldots, n$:

$$\frac{\partial f_i}{\partial x_i}(\mathbf{x}^*) - \mu \frac{\partial g_i}{\partial x_i}(\mathbf{x}^*) = 0,$$

which also yields

$$\frac{\dfrac{\partial f_i}{\partial x_i}(\mathbf{x}^*)}{\dfrac{\partial g_i}{\partial x_i}(\mathbf{x}^*)} = \mu \qquad \forall i. \tag{B.13}$$

The ratio of partial derivatives is the ratio between the (marginal) increment in profit from activity i, if we increase its level, and the increment in resource consumption. Equation (B.13) tells that this ratio is given by the multiplier, but above all it tells us that in the optimal solution this ratio must be *the same* for all of the activities. In fact, if this were not the case, i.e., if there were two activities j and k such that

$$\frac{\frac{\partial f_j}{\partial x_j}(\mathbf{x}^*)}{\frac{\partial g_j}{\partial x_j}(\mathbf{x}^*)} > \frac{\frac{\partial f_k}{\partial x_k}(\mathbf{x}^*)}{\frac{\partial g_k}{\partial x_k}(\mathbf{x}^*)},$$

then $\mathbf{x}^*$ could not be the optimal solution, because we could increase the overall profit by reallocating part of the resource from k to j.   ⬜

This type of marginal analysis proves very useful in the chapters dealing with inventory management.

B.5 LINEAR PROGRAMMING

The optimization model (B.5) is a linear programming problem if all of the involved functions are linear (strictly speaking, affine). In this case, the model has the following form:

$$\min \quad \sum_{i=1}^{n} c_i x_i$$

$$\text{s.t.} \quad \sum_{i=1}^{n} \alpha_{ij} x_i = \beta_j, \quad j = 1, \ldots, m,$$

$$\sum_{i=1}^{n} \gamma_{ik} x_i \leq \delta_k, \quad k = 1, \ldots, l.$$

Linear unconstrained optimization does not make sense, because the gradient is constant and the stationarity condition cannot be met; the optimum is unbounded. Linear programming problems are "easy," in the sense that quite efficient and reliable algorithms are available to solve them. This depends partly on the convexity properties of linear programming and partly on its geometrical features: The feasible set is a polyhedron, and we can find an optimal solution by looking only at the extreme points (vertices) of the polyhedron.

Example B.9 Let us consider the optimal mix problem (B.1) again and try a geometrical solution. A good starting point is observing that the problem can be simplified as follows:

$$
\begin{aligned}
\max \quad & 45x_1 + 60x_2 \\
\text{s.t.} \quad & 15x_1 + 35x_2 \leq 2400, \\
& 25x_1 + 15x_2 \leq 2400, \\
& 0 \leq x_2 \leq 50.
\end{aligned}
$$

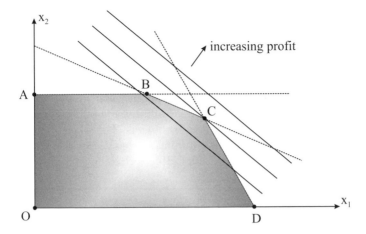

Fig. B.5   Feasible set for the optimal mix model.

In fact, it is easy to verify that any mix satisfying the second capacity constraint will also satisfy the first and the third one. To see this, observe that item P_1 requires the same resource amount on machine groups A, B, and C, while P_2 has a higher requirement on B; the availability of resources A, B, and C is the same, but only the second one is critical.[12] A similar consideration applies to the market limit for item P_1: We cannot produce 100 pieces of it anyway, because there is not enough capacity on resource D.

We can also verify these observations geometrically, by drawing the feasible set and checking for redundant constraints. In practice, good software solvers do this automatically. In figure B.5 we see the feasible region, which is the intersection of three half spaces linked to the three relevant constraints. The vertices of this polyhedron are the origin $O(0,0)$ and points $A(0,50)$, $B(43.33, 50)$, $C(73.85, 36.92)$, and $D(96,0)$. The figure also displays the level curves of the profit function, which is increasing along the shown direction. We also see that the highest level of profit is associated to point C, which is indeed the optimal solution. Note that we cannot find the optimal solution by reasoning along the lines of example B.1, by focusing on one product at a time; in fact, this yields points B and D, for which profit is negative.

We should also note that by changing selling prices, we could also change the optimal solution, but this would be a vertex anyway. If the profit level curves were parallel to a constraint, we would have an infinite number of equivalent optimal solutions on a face of the polyhedron. $\square$

---

[12]We could say that B is the bottleneck, and we should expect that the shadow prices for A and C turn out zero. There is a whole managerial approach, the *Theory of constraints*, revolving around the idea of just focusing on relevant constraints. However, spotting relevant constraints may be difficult in a very dynamic and complex setting.

The classical algorithm to tackle linear programming problems is the **simplex** method, which explores vertices of the feasible polyhedron by a clever strategy, until a locally optimal solution is found, which is also a global one, courtesy the convexity of the problem. The simplex method was born in 1947, but it is continuously improved from a computational point of view. We also mention that alternative strategies are available, which explore the *interior* of the feasible region; these **interior point** methods may be more efficient than standard simplex on some type of problems. For all the purposes of this book, we may consider linear programming models as problems which can be tackled by a reliable, mature, and affordable technology.

B.6   INTEGER LINEAR PROGRAMMING

When some or all of the decision variables in a linear programming model are restricted to integer values, we have an **integer linear programming** problem. To be more precise, we have a *pure* integer programming problem when the restriction must be enforced on all of the decision variables, and we have a **mixed-integer** programming problem when some variables are continuous. A generic integer variable takes values in the set $\mathbb{Z}_+ = \{0, 1, 2, 3, 4, \ldots\}$; the case of negative integers is quite rare in applications. Actually the most common case involves **binary** decision variables, which are restricted to the set $\{0, 1\}$; these variables are so common because they model logical decisions, such as "should we open a new distribution center in that city, or not?" Clearly, this is an all-or-nothing decision; we cannot open 75% of a distribution center.

Unlike continuous linear programming, integer programming can be tough. The main issue is that the feasible set is nonconvex. In convex programming, we have a suitable characterization of an optimal solution; this means that if we are handed an optimal solution, it is fairly easy to check that this is really the optimal one; for linear programming, this requires adapting the Kuhn–Tucker conditions. With integer programming, even if we are handed a feasible solution by someone swearing its optimality, there is no easy way to check this claim. Apparently, a trivial strategy can be applied: Disregard the integrality restrictions, solve the *continuous relaxation* of the problem, and then round the solution. Indeed, in the very simple optimal mix instance we have considered, such a strategy would work. The continuous solution is $x_1 = 73.85$ and $x_2 = 36.92$, which is not too far from the optimal integer solution $x_1 = 73$ and $x_2 = 37$. But the following counterexample proves that this is not a generally viable approach.

Example B.10 Consider the following pure integer programming problem:

$$
\begin{aligned}
\max \quad & x_1 + x_2 \\
\text{s.t.} \quad & 10x_1 - 8x_2 \leq 13, \\
& 2x_1 - 2x_2 \geq 1,
\end{aligned}
$$

$$x_1, x_2 \in \mathbb{Z}_+.$$

If we relax the integrality requirement, i.e., we just require $x_1, x_2 \geq 0$, we can apply the simplex method to find

$$x_1^* = 4.5, \qquad x_2^* = 4,$$

with an optimal objective value 8.5. By relaxing integrality constraints this way, we obtain the **continuous relaxation** of the integer program. The reader is invited to try rounding the above solution; unfortunately, the trivially rounded solutions are not feasible, and the integer optimum is

$$x_1^* = 2, \qquad x_2^* = 1$$

with optimal value 3. We see that the continuous solution is quite far from the true integer optimum; in this case it is even difficult to find a *feasible* solution by rounding, let alone the optimal one. We invite the reader to check the situation graphically to figure out where the trouble is: The feasible solution consists only of points $(1, 0)$ and $(2, 1)$, whereas the convex polyhedron in the continuous relaxation is very narrow and includes a lot of non-integer points far from these two feasible solutions. □

The bottom line is that in integer programming we must resort to some type of enumeration in order to find the optimal solution. Certainly, enumerating *all* of the feasible integer solutions is typically out of the question; even if they are finite, the number of feasible integer solutions can be staggering in a real-life problem. Luckily enough, the example above does suggest a way to avoid complete enumeration of the feasible set. We may see that, in a maximization problem, the continuous relaxation yields an optimistic estimate of the optimal value of the objective, i.e., an *upper bound*. For a minimization problem, the continuous relaxation yields a *lower bound*. This observation is the starting point of a class of methods which are collectively known under the label of *branch and bound* methods and are discussed in the following section. Making such a method work requires quite a bit of finesse, but getting a grasp of the overall idea is not too difficult. We should do so in order to appreciate the potential trouble in solving a large mixed-integer program. However, from our point of view, the ability of *building* a mixed-integer linear programming model is more important than the ability of solving it; this task can be left to state-of-the-art software packages which are commercially available and are continuously improved. Some mixed-integer programs which were way beyond the reach of very powerful computers a few years ago can now be routinely solved on a desktop PC. Mixed-integer programming models are among the main topics of chapter 2 on network design. For this class of applications, commercial packages based on branch and bound are actually able to solve real problems, at least to near-optimality. Unfortunately, other types of problems, such as the vehicle routing problems described in chapter 8, cannot be tackled

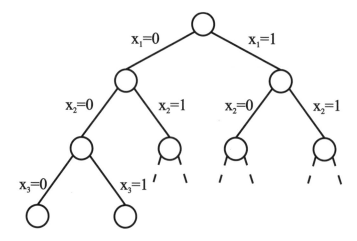

Fig. B.6 Search tree for a pure binary programming problem.

by the same approach. In principle, a mixed-integer programming model can be built, but it proves too hard to solve at optimality by branch and bound, unless very sophisticated and specialized approaches are used. This is why, in that case, we typically resort to heuristic approaches.

B.6.1   Branch and bound methods

From a certain point of view, some integer programming problems could look easier than continuous linear programming models. Consider a pure binary problem, i.e., a linear programming problem where all of the variables are restricted by $x_i \in \{0, 1\}$, $i = 1, \ldots, n$. In such a case, the number of feasible solutions is clearly finite, and we could think of enumerating all of them to spot the optimal one. We can visualize the search process as in figure B.6, where we show a search tree; at each node of the tree we branch on a decision variable, which basically amounts to partitioning the feasible set into disjoint regions. However, this idea is not really feasible but for small values of n; in fact, we might have up to 2^n candidates to test. Many of them would be ruled out by constraints, but we see that the complexity of such an algorithm is exponential. Furthermore, it is not yet clear how to branch on general integer variables (opening a branch for every possible integer value is out of the question), nor how to cope with mixed-integer problems.

   What we should aim at is exploring only a small part of the search tree, avoiding portions of the feasible set in which we cannot find the optimal solution. In other words, we should prune the tree in order to avoid wasting computational effort, but how can we be sure that we cannot find the optimal solution in the portion of tree below a certain node? We have already noted that if we relax integrality constraints, for a minimization problem, we get a lower bound on the optimal value of the objective. In a pure binary problem, this means that if we relax the restriction $x_i \in \{0, 1\}$ to $x_i \in [0, 1]$, we can apply the simplex method to get such a lower bound. Now suppose that, while wandering up and down the tree, we already came across a feasible solution whose cost is, say, 100. At present we are considering a node in the tree, in which variables x_i, $i = 1, \ldots, k$, have been fixed to a binary value, and we have solved the relaxation of the problem with respect to the remaining variables x_i, $i = k+1, \ldots, n$. The optimal value of the relaxation turned out to be 102; should we branch on that node? Not really, since the best solution we can hope to find in the subtree below that node has cost 102; maybe the optimal solution in that subset of the feasible region will turn out even worse than that. Hence, by exploring that subtree we cannot improve what we already have in hand; so, we may safely prune that branch.

This line of reasoning, suitably adapted to more general integer programming problems, is the foundation of **branch and bound** methods. Actually, there are many variations on the theme, but all of them rely on two basic ingredients:

1. A *branching* strategy, which builds a search tree by generating finer and finer partitions of the original feasible set. Each branching spawns two (or more) subtrees, corresponding to subsets of the feasible region. We should not miss any opportunity; hence, the subsets must be collectively exhaustive of the feasible region. For the stake of efficiency, it is also advisable that they are mutually exclusive. This is easy to achieve when branching on binary decision variables. A different approach is taken when dealing with a generic integer variable. Say that solving the continuous relaxation yields a fractional value $x_j^* = 3.7$. We can branch on this variable by creating two subproblems, one subject to the bound $x \leq 3$ and the other one to $x \geq 4$, respectively. Ruling out pathological cases which may happen when the original feasible region is unbounded, sooner or later we will end up with a subproblem such that the simplex method returns an integer solution. This subproblem is a "leaf" of the tree, and we may resume search somewhere else (recording the newly found integer solution if it provides us with the best solution so far).

2. A *bounding* strategy, which in commercial libraries is based on the continuous relaxation of the integer problem.[13] A bounding strategy helps in limiting the search process, during which we keep an *incumbent* solution, i.e., the best integer solution found so far; the value of the incumbent solution is, for a minimization problem, an upper bound on the optimal value. We compare this upper bound against the lower bound of a node to understand if the latter is worth digging deeper. We must

---

[13]Several *ad hoc* strategies have been devised for specific problems, but they have no place in a commercial package for general mixed-integer models.

stress the fact that we *cannot* prune a node by comparing two lower bounds. To see why, consider a subproblem P_1 with lower bound 190 and a subproblem P_2 with lower bound 180. P_2 may look more promising, and maybe we could consider branching on it first, but we cannot eliminate P_1 from further consideration; it may well be the case that the optimal solution of P_1 has cost 191, whereas P_2 yields a solution with cost 195.

An efficient branch and bound method also requires dealing with other issues: how to select the next subproblem to tackle; how to solve efficiently a linear program after adding a constraint, without starting everything from scratch; the selection of the variable we should branch on. Sometimes, we can tackle a rather large model by a branch and bound method; somewhat surprisingly, there may be smaller problems which prove a much harder nut to crack. The main factor is the quality of lower bounds: If they are tight, the pruning process can be effective; otherwise, a lot of useless branches will be explored. This is why sometimes we may have to settle for heuristics, i.e., faster algorithms to find a hopefully "good" solution with no guarantee about its optimality. We outline a few heuristics for the vehicle routing problem in chapter 8; we have to resort to heuristics in this case because straightforward mixed-integer model formulations of this problem yield weak continuous relaxations. Still, the range of real-life models that we can tackle by commercial libraries has been considerably expanded, also thanks to ways to reinforce the quality of bounds. We will not cover these sophisticated methods, but we illustrate the basic approach by a complete example.

Example B.11 Solving an integer linear program by branch and bound. A good candidate to illustrate a full run of branch and bound is the optimal mix problem (B.1), whose continuous solution is $x_1^* = 73.84$, $x_2^* = 36.92$, with profit 538.46; this is an upper bound on the profit from the optimal integer solution, since this is a maximization problem. Let P_0 the root problem in the tree, i.e., the problem with the feasible set depicted in figure B.5. It is convenient to associate each subproblem with two vectors of lower and upper bounds on variables, respectively; for P_0 we have the original market demand bounds

$$L_0 = [0 \quad 0]', \qquad U_0 = [100 \quad 50]'.$$

To partition the feasible region, we may start branching on variable x_1, generating the two subproblems P_1 and P_2, resulting from the addition of constraints $x_1 \leq 73$ and $x_1 \geq 74$, respectively, which do not eliminate any integer solution. The whole search tree is depicted in figure B.7; each subproblem corresponds to a rectangle defined by the bounds on variables; we see the additional constraints on each branch, whereas each node contains the optimal solution of the relaxed subproblem along with its profit.

Subproblem P_1 is characterized by the following bounds on variables:

$$L_1 = [0 \quad 0]', \qquad U_1 = [73 \quad 50]',$$

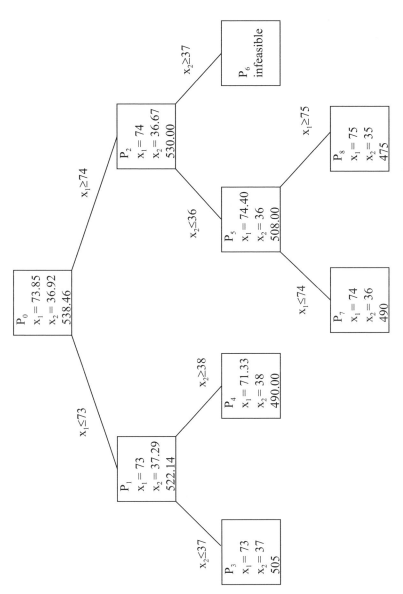

Fig. B.7 Search tree for example B.11.

whereas P_2 is associated with

$$L_2 = [74 \quad 0]', \qquad U_2 = [100 \quad 50]'.$$

Solving P_1 yields

$$x_1^* = 73, \qquad x_2^* = 37.29,$$

with profit 522.14. From P_2 we get

$$x_1^* = 74, \qquad x_2^* = 36.67$$

with profit 530.00 (rounded within two decimal digits). Both profits are smaller than the profit in P_0, which is natural since they are obtained after including a further restriction. We cannot eliminate either subproblem. If we branch from P_1, we create subproblem P_3 with condition $x_2 \leq 37$; its bounds on variables are

$$L_3 = [0 \quad 0]', \qquad U_3 = [73 \quad 37]',$$

and by applying the simplex method we get our first integer solution:

$$x_1^* = 73, \qquad x_2^* = 37$$

with profit 505. This is a feasible solution, corresponding to a leaf of the tree, and it is not necessarily optimal; it just gives a lower bound on the optimal profit against which we may compare upper bounds from relaxations.

From P_1, adding the constraint $x_2 \geq 38$, we get subproblem P_4, with bounds on variables

$$L_4 = [0 \quad 38]', \qquad U_4 = [73 \quad 50]'.$$

Its continuous solution is

$$x_1^* = 71.33, \qquad x_2^* = 38$$

with profit 490.00. This solution is not integer, but we can get rid of the subproblem, since its upper bound is smaller than the lower bound 505; going down this branch of the tree, we cannot improve the incumbent solution. It is, however, necessary to branch from subproblem P_2, which looks promising given its upper bound 530. From P_2 we generate subproblems P_5 and P_6. Imposing $x_2 \leq 36$, i.e.,

$$L_5 = [74 \quad 0]', \qquad U_5 = [100 \quad 36]',$$

yields

$$x_1^* = 74.40, \qquad x_2^* = 36$$

with profit 508, which still looks better than 505. From the other subproblem, where $x_2 \geq 37$, we do not get anything as the problem is infeasible; by adding more and more restrictions, we ended up with an empty feasible set. In fact, we may see that the rectangle

$$L_6 = [74 \quad 37]', \qquad U_6 = [100 \quad 50]'$$

has no intersection with the feasible set, as the lower bounds on production are such that capacity constraints are violated for machines B and D.

We still have to explore subproblem P_5, by branching on the fractional variable x_1. By imposing $x_1 \leq 74$, we get another integer solution,

$$x_1^* = 74, \qquad x_2^* = 36,$$

whose profit is 490 and is not better than the incumbent. The constraint $x_1 \geq 75$ yields another integer solution

$$x_1^* = 75, \qquad x_2^* = 35,$$

with profit 475, which is of no use.

Now we can conclude that the integer solution

$$x_1^* = 73, \qquad x_2^* = 37$$

is indeed the optimal one. We see that, in this case, we did a lot of work just to prove that the first integer solution we met was the optimal one. This may also happen in practical problems, even though a sequence of improving incumbents is normally visited before proving optimality. ☐

B.6.2   Model building in integer programming

As we have already pointed out, integer programming models may pop up when there is a need to restrict purchase or production decisions to integer quantities, maybe multiples of a standard batch. However, the most common reason for using such models is by far the inclusion of logical decisions. In the remainder of this section we illustrate a few examples of modeling decisions by binary variables. The ability of using binary variables is essential in modeling distribution network design problems, as we illustrate in chapter 2.

Example B.12 Lot sizing with setup times and costs. In this example we illustrate the use of binary variables to model fixed charges. In the multi-period planning model (B.2) we did not consider the need for machine setup before starting production. Suppose that in order to produce a lot of item i, we need to spend a setup time r'_{im} for each resource m. This setup time does not depend on the lot size, and it gives us an incentive to stock an item. By the same token, we may have a fixed cost f_i associated to each setup for item i; this may depend, e.g., on material which is scrapped at the beginning of a lot because of the need of adjusting machines. In purchasing, setup times play no role, but we may need to tackle similar issues, e.g., when there is a fixed component in the transportation cost. The decision of starting a lot of item i at the beginning of time bucket t is a logical decision; either we do it or not. In principle, we could introduce a step function such as

$$\delta(x) = \begin{cases} 1 & \text{if } x > 0 , \\ 0 & \text{if } x = 0 , \end{cases}$$

which allows us to express fixed charges. Unfortunately, this is a nonlinear and discontinuous function; If we want to stick to commercial linear programming packages, we must introduce a binary decision variable:

$$\delta_{it} = \begin{cases} 1 & \text{if we carry out a setup for item } i \text{ during time bucket } t, \\ 0 & \text{otherwise,} \end{cases}$$

and we must figure out a way to link x_{it} and δ_{it} using linear constraints. We can do this by a typical trick of the trade, based on the aptly named "big-M." Let M be a suitably large constant; more precisely, it should be an upper bound on the amount that we can or should produce during a time bucket; assuming we can quantify the big-M, we write the following constraint:

$$x_{it} \leq M\delta_{it}.$$

If $\delta_{it} = 0$, however big the constant M is, this boils down to $x_{it} \leq 0$; since production variables are non-negative, this means that if we do not carry out the setup, we cannot produce that item. If $\delta_{it} = 1$, we get $x_{it} \leq M$, which is a redundant constraint if M is large enough. In practice, one way to quantify the big-M is to consider that there is no economic reason to produce more than we can sell in the remaining time to the end of planning horizon:

$$x_{it} \leq \left(\sum_{\tau=t}^{T} d_{i\tau} \right) \delta_{it}. \tag{B.14}$$

An alternative approach is based on capacity constraints; if we carry out the setup for item i, the largest amount we can produce is bounded as follows:

$$x_{it} \leq \left(\frac{R_{mt} - r'_{im}}{r_{im}} \right) \delta_{it}.$$

This bound is obtained by thinking of allocating to item i all of the available capacity on resource m, minus the setup time. But which resource exactly? And how should we choose between this idea and (B.14), which is based on demand? From a computational point of view, the smaller the big-M, the better. It is not difficult to understand why. If we use a large constant, there is nothing wrong logically, but this weakens the bound we get from continuous relaxation of the binary setup variables. So, we should select the smallest big-M we can, provided this still yields a redundant constraints when $\delta_{it} = 1$.[14]

---

[14]Actually, the best strategy is to reformulate the model completely, using less intuitive decision variables, such as the amount we produce during one time bucket to meet demand

The resulting model is a fairly straightforward extension of (B.2):

$$\max \quad \sum_{i=1}^{N}\sum_{t=1}^{T} p_i z_{it} - \sum_{i=1}^{N}\sum_{t=1}^{T} c_i x_{it} - \sum_{i=1}^{N}\sum_{t=1}^{T} h_i I_{it} - \sum_{i=1}^{N}\sum_{t=1}^{T} f_i \delta_{it} \quad \text{(B.15)}$$

$$\text{s.t.} \quad I_{it} = I_{i,t-1} + x_{it} - z_{it}, \qquad i = 1, \ldots, N, \quad t = 1, \ldots, T,$$

$$\sum_{i=1}^{N} (r_{im} x_{it} + r'_{im} x_{it}) \le R_{mt}, \qquad m = 1, \ldots, M, \quad t = 1, \ldots, T,$$

$$z_{it} \le d_{it}, \qquad i = 1, \ldots, N, \quad t = 1, \ldots, T,$$

$$x_{it} \le M \delta_{it}, \qquad i = 1, \ldots, N, \quad t = 1, \ldots, T,$$

$$x_{it}, \ z_{it}, \ I_{it} \ge 0, \qquad \delta_{it} \in \{0, 1\}.$$

The careful reader could raise one possible objection: Constraint (B.14) is not really an exact translation of what we could write using the step function $\delta(x)$. In fact, the model above allows a useless setup, as we can set $\delta = 1$ and $x = 0$. However, such a solution is feasible, but it will never be optimal if there is a setup cost. $\square$

Example B.13 Multiple choices. In this example we illustrate the use of binary variables to model mutually exclusive logical decisions. Suppose we have L alternative suppliers for a raw material that we need to feed a manufacturing process. Each supplier $l = 1, \ldots, L$ asks a unit price p_l, which need not be the same for all suppliers; apart from this variable cost, we should also take into account a fixed ordering cost f_l, which may depend on the geographic distance of the supplier (think of a fixed component of the transportation cost). It may well be the case that lower unit prices are associated with a distant supplier, so that the tradeoff is not obvious. Assume further that we prefer using one supplier for the whole planning horizon; this can be justified by organizational reasons and by the need to establish a trustworthy relationship.

The decision of how much, when, and from whom to buy will be a part of a possibly large multiperiod model, but let us focus on this purchase decision. We need first a decision variable x_{lt} expressing how much we buy from supplier l in time bucket t. Just like the lot sizing model of the previous example, we need a binary decision variable to model the fixed cost component: Let δ_{lt} be 1 if we buy from supplier l in time bucket t, 0 otherwise. To link the two decision variables, we need a big-M constraint, such as

$$x_{lt} \le M \delta_{lt}.$$

in a future time bucket. These disaggregated decision variables allow for smaller big-Ms. Sophisticated model formulation is beyond the scope of this book, and we refer the interested reader, e.g., to [2].

The objective function includes a cost component like

$$\sum_{t=1}^{T} \sum_{l=1}^{L} (p_l x_{lt} + f_l \delta_{lt}).$$

Now, we must enforce the selection of (at most) one supplier over the whole time horizon. To this aim, we introduce another set of binary variables γ_l, set to 1 if supplier l is the lucky one. The selection of at most one supplier is modeled by the inequality

$$\sum_{l=1}^{L} \gamma_l \le 1,$$

where we are using an inequality to allow for the selection of no supplier; if we insist on selecting exactly one supplier, we can rewrite the constraint as an equality. However, there should be no trouble with the inequality, which is somewhat more general, as other demand satisfaction constraints will probably enforce the selection of a supplier. The last step is linking the two sets of binary variables:

$$\sum_{t=1}^{T} \delta_{lt} \le T\gamma_l, \qquad l = 1, \ldots, L. \tag{B.16}$$

Note that we must multiply γ_l by the number of time buckets, to allow for purchasing whenever we want, and not just once over the planning horizon. We should note that there is an alternative way to express this link, which is logically equivalent to (B.16):

$$\delta_{lt} \le \gamma_l, \qquad l = 1, \ldots, L, \ t = 1, \ldots, T. \tag{B.17}$$

It is easy to see that both constraints do their job. It could be argued that the first idea is better since it involves much less constraints, whereas the form (B.17) is disaggregated. In fact, good software libraries, when handed an aggregate form like (B.16), reformulate the model automatically by disaggregating that constraint into the form (B.17). To see why this is a good idea, observe that the aggregate form is obtained by summing each single inequality over t. In general, when we sum constraints, we enlarge the feasibility region.[15] In our case, the two feasible sets are the same in the discrete domain; otherwise one of the two formulations would not be correct. However, when relaxing the integrality constraint, we get a weaker continuous relaxation when using the aggregate form. Sometimes, the difference in the quality of the bounds we get is so large that branch and bound efficiency is remarkably improved by disaggregation. ▯

---

[15] For instance, all the points that satisfy both inequalities $g_1(\mathbf{x}) \le 0$ and $g_2(\mathbf{x}) \le 0$, also satisfy $g_1(\mathbf{x}) + g_2(\mathbf{x}) \le 0$, but not vice versa.

B.7   ELEMENTS OF MULTIOBJECTIVE OPTIMIZATION

All of the optimization concepts and models we have illustrated in this appendix rely on two (quite limiting) assumptions:

1. All data are known with certainty.

2. All of the requirements we have on the solution, which are typically in mutual conflict, can be aggregated into one objective function.

Both assumptions are certainly not to be taken for granted. In distribution logistics, uncertainty does typically affect demand. At a more strategical level, when long-term decisions must be taken within a globalized context, exchange rates, prices of raw materials, and selling prices for end items are also affected by uncertainty. In the main body of the text we give a few clues on how to cope with uncertainty.

Also the second assumption is open to quite some bit of criticism. To begin with, not everything can be translated to monetary terms. In an era of pollution and greenhouse effect, the environmental impact of a transportation policy has to play some role, which is not easy to trade off against the pure optimization of transportation costs; also the social impact of opening a huge commercial center in a neighborhood may be significant, and this is important to design a retail network. But even if we stay within the bounds of low-level tactical decisions, the economic impact of a lost sale or a delayed shipment may be hard to assess. If there is a contract stating precise penalties for late delivery, it is easy to bring everything to a common monetary measure. However, the damage to your image with customers and its long-term impact is hard to assess. In a risky situation, it may be difficult to assess our own degree of risk aversion.

The bottom line is that it may be hard to come up with a single objective function capturing all of the facets of a complex decision problem. This is why multicriteria decision-making and multiobjective optimization techniques were born. In this section we illustrate a couple of basic concepts in multiobjective optimization, which are essential, e.g., to appreciate different modeling approaches we use when managing inventories under uncertainty. For the sake of simplicity, assume that we have just two objective functions, f_1 and f_2, both to be minimized, which cannot be aggregated into one objective function. From a mathematical point of view, each feasible solution is characterized by a vector of objective values; hence, we could consider a "vector" optimization problem:

$$\text{``min''} \quad \begin{bmatrix} f_1(\mathbf{x}) \\ f_2(\mathbf{x}) \end{bmatrix} \tag{B.18}$$

$$\text{s.t.} \quad \mathbf{x} \in S. \tag{B.19}$$

However, stated as such, the problem has no meaning, and this is why we use "min." Indeed, vectors are not a well-ordered set, unlike the real line.

We cannot say that a vector is better than another one: The number 5 is larger than the number 2, but we cannot compare vectors $[10 \ 1]'$ and $[2 \ 5]'$ that easily. True, we can say that a vector is longer, by referring, e.g., to the Euclidean concept of vector length, but this amounts to choosing a specific norm to scalarize the problem; it is this scalarization that is difficult to specify.

As a concrete example, consider the task of selecting a household electric appliance, such as a fridge or a washing machine. Among the many factors playing a role in our choice, cost is certainly an important one. However, we may also consider esthetics of design, reliability, quality of post-sale services, capacity (for both fridge and washing machine), and washing time (in the second case only). All of these factors are hard to express as money. But even factors that have a definite economic impact are hard to assess for us. Think of resource consumption; sure, the energy class of the appliance has an impact on our energy bill, but this is difficult to measure because it is a future and uncertain cost. Last but not least, some customers are more sensitive than others to the ethics of energy consumption. So, it is difficult to choose between a cheap appliance that consumes a lot of energy and a more expensive one which allows a significant energy saving. Nevertheless, one thing should be clear, if we consider only price and energy consumption: We should not prefer an expensive appliance that consumes huge amounts of energy over an alternative that is less expensive and saves a lot. This observation alone may not help us in spotting one "optimal" solution, but at least it eliminates unreasonable alternatives from further consideration. In other words, we should just concentrate on *efficient* solutions.

DEFINITION B.7 *Given the vector optimization problem (B.18), a feasible solution* $\mathbf{x}^*$ *is said* **efficient**[16] *or* **nondominated** *solution, if there is no other solution* $\tilde{\mathbf{x}} \in S$ *such that*

$$f_1(\tilde{\mathbf{x}}) \leq f_1(\mathbf{x}^*) \qquad and \qquad f_2(\tilde{\mathbf{x}}) \leq f_2(\mathbf{x}^*)$$

with a strict inequality for at least one of the two objectives. The set of nondominated solutions is called **efficient frontier**.

The idea may be easily grasped by having a look at figure B.8. We see that in the case of the figure the is not necessarily *one* optimal solution, but rather a set of "reasonable" solutions to which we may restrict the choice, ruling out dominated alternatives. What we can do to help the decision maker is to generate a set of reasonable alternatives. To this aim, we can scalarize the problem according to some strategy, boiling the vector problem down to a family of single-objective optimization problems.

---

[16]Often we speak of Pareto efficiency, in honor of Italian economist Vilfredo Pareto, who studied the allocation of goods among economic agents in these terms. By the way, it is worth noting that although he is best remembered as an economist, he had a degree in Engineering. In the 1950s, many scholars who eventually made a big name in Economics worked on inventory management and workforce planning.

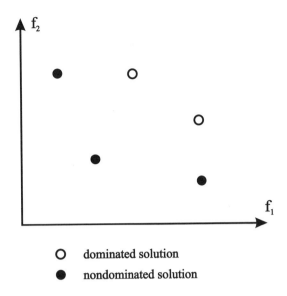

$\bigcirc$ dominated solution

$\bullet$ nondominated solution

Fig. B.8   Illustrating the concept of efficient solution.

The first, perhaps more intuitive, approach is to devise a weighted linear combination of the two objectives: We define a parameter λ, bounded by 0 and 1, which expresses the relative importance of the objectives; letting λ span its range, we define and solve a sequence of problems:

$$\text{min} \quad \lambda f_1(\mathbf{x}) + (1 - \lambda) f_2(\mathbf{x})$$
$$\text{s.t.} \quad \mathbf{x} \in S.$$

Note that the parameter λ has no precise economic meaning, as it is just a tool to span the efficient frontier. This approach is clearly intuitive and related to the idea of varying a set of weights. We have the guarantee that all of the solutions we generate this way are efficient; however, it does not guarantee in general that *all* of the efficient solutions will be generated, unless some condition related to convexity is satisfied.[17] An alternative approach is based on the idea of transforming one objective into a constraint. In other words, we can optimize f_1, subject to the constraint that f_2 cannot exceed some limit (or vice versa):

$$\text{min} \quad f_1(\mathbf{x})$$

---

[17]To see why, try the following exercise, with reference to figure B.8. Imagine drawing the level curves of the linear combination of objectives when λ varies; changing this parameter implies a rotation on these lines, and all of the three efficient solutions are optimal for some interval of λ. But what happens if the second efficient solution, i.e., the one closest to the origin, moves up along the north–east direction?

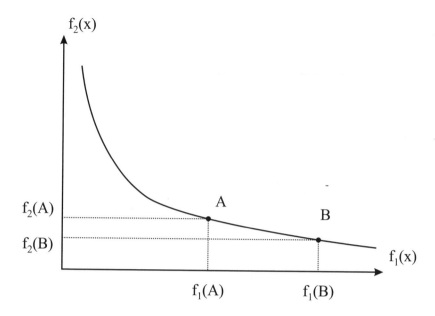

Fig. B.9 The efficient frontier.

$$\text{s.t.} \quad \mathbf{x} \in S,$$
$$f_2(\mathbf{x}) \le \bar{f}_2.$$

Solving a family of scalar problems for varying values of $\bar{f}_2$, we may trace the efficient frontier. If we have an optimization problem over continuous decision variables, the efficient frontier will look something like figure B.9. It is worth noting that this second approach does not suffer from the aforementioned difficulty with the weighted combination approach, but the most important feature, arguably, is that it is more "readable" for a decision maker.[18] While the parameter λ has no clear managerial meaning, the parameter $\bar{f}_2$ is much clearer. It is a threshold level, that might be chosen by having a look at what competitors do. For instance, if we have to trade off service level against the cost of our inventories, having an idea of what service level is offered by our competitors helps a lot in choosing a sensible threshold.

We clearly see that optimization modeling in such a context is a tool to *support* decision makers, and not to replace them. It is up to an informed manager to compare alternatives A and B in figure B.9, to assess the involved tradeoffs, and to decide if the improvement of solution B with respect to A, in terms of the second objective, is enough to compensate the loss in terms of the first one

---

[18]On the other hand, we should also mention that sometimes the model resulting from a convex combination of objectives may be easier to solve from a computational point of view.

B.8 FOR FURTHER READING

- Example B.10 is taken from [5], which is good introductory reading for solution methods.

- A more complete treatment can be found, e.g., in [1], whereas [7] is an excellent reference as far as integer programming is concerned.

- Readers interested in optimization models for manufacturing management may also have a look at [2].

- The bibliography on optimization methods is quite rich, but the same cannot be said when it comes to model *building*. A welcome exception is [6], which deals with a wide class of applications; [4] may also be useful reading, and it is more focused on supply chain management.

- From a practical point of view, optimization modeling is of no use if it is not complemented by a working knowledge of commercial optimization software. We suggest having a look at `http://www.ilog.com`, which also offers interesting material describing real-life applications. Other useful links are:

  `http://www.informs.org`

  `http://www.optimization-online.org`

REFERENCES

1. M.S. Bazaraa, C.M. Shetty, and H.D. Sherali. *Non-Linear Programming: Theory and Algorithms.* Wiley, Chichester, 1993.

2. P. Brandimarte and A. Villa. *Advanced Models for Manufacturing Systems Management.* CRC Press, Boca Raton, FL, 1995.

3. W. Hopp and M. Spearman. *Factory Physics (2nd Ed.).* McGraw-Hill, New York, 2000.

4. J.F. Shapiro. *Modeling the Supply Chain.* Duxbury/Thomson Learning, Pacific Grove, CA, 2001.

5. H.P. Williams. *Model Solving in Mathematical Programming.* Wiley, Chichester, 1993.

6. H.P. Williams. *Model Building in Mathematical Programming (4th Ed.).* Wiley, Chichester, 1999.

7. L.A. Wolsey. *Integer Programming.* Wiley, Chichester, 1998.

Index

STATISTICS IN PRACTICE

Advisory Editor

Peter Bloomfield
North Carolina State University, USA

Founding Editor

Vic Barnett
Nottingham Trent University, UK

Statistics in Practice is an important international series of texts which provide detailed coverage of statistical concepts, methods and worked case studies in specific fields of investigation and study.

With sound motivation and many worked practical examples, the books show in down-to-earth terms how to select and use an appropriate range of statistical techniques in a particular practical field within each title's special topic area.

The books provide statistical support for professionals and research workers across a range of employment fields and research environments. Subject areas covered include medicine and pharmaceutics; industry, finance and commerce; public services; the earth and environmental sciences, and so on.

The books also provide support to students studying statistical courses applied to the above areas. The demand for graduates to be equipped for the work environment has led to such courses becoming increasingly prevalent at universities and colleges.

It is our aim to present judiciously chosen and well-written workbooks to meet everyday practical needs. Feedback of views from readers will be most valuable to monitor the success of this aim.

A complete list of titles in this series appears at the end of the volume.

STATISTICS IN PRACTICE

Human and Biological Sciences

Brown and Prescott · Applied Mixed Models in Medicine
Ellenberg, Fleming and DeMets · Data Monitoring Committees in Clinical Trials:
    A Practical Perspective
Lawson, Browne and Vidal Rodeiro · Disease Mapping With WinBUGS and MLwiN
Lui · Statistical Estimation of Epidemiological Risk
*Marubini and Valsecchi · Analysing Survival Data from Clinical Trials and
    Observation Studies
Parmigiani · Modeling in Medical Decision Making: A Bayesian Approach
Senn · Cross-over Trials in Clinical Research, *Second Edition*
Senn · Statistical Issues in Drug Development
Spiegelhalter, Abrams and Myles · Bayesian Approaches to Clinical Trials and Health-
    Care Evaluation
Turner · New Drug Development: Design, Methodology, and Analysis
Whitehead · Design and Analysis of Sequential Clinical Trials, *Revised Second Edition*
Whitehead · Meta-Analysis of Controlled Clinical Trials

Earth and Environmental Sciences

Buck, Cavanagh and Litton · Bayesian Approach to Interpreting Archaeological Data
Glasbey and Horgan · Image Analysis in the Biological Sciences
Helsel · Nondetects and Data Analysis: Statistics for Censored Environmental Data
McBride · Using Statistical Methods for Water Quality Management: Issues, Problems
    and Solutions
Webster and Oliver · Geostatistics for Environmental Scientists

Industry, Commerce and Finance

Aitken and Taroni · Statistics and the Evaluation of Evidence for Forensic Scientists,
    Second Edition
Brandimarte · Numerical Methods in Finance and Economics: A MATLAB-Based
    Introduction, *Second Edition*
Brandimarte and Zotteri · Introduction to Distribution Logistics
Chan and Wong · Simulation Techniques in Financial Risk Management
Lehtonen and Pahkinen · Practical Methods for Design and Analysis of Complex Surveys,
    Second Edition
Ohser and Mücklich · Statistical Analysis of Microstructures in Materials Science

*Now available in paperback.